Fungal Biology

Series Editors

Fungal biology has an integral role to play in the development of the biotechnology and biomedical sectors. It has become a subject of increasing importance as new fungi and their associated biomolecules are identified. The interaction between fungi and their environment is central to many natural processes that occur in the biosphere. The hosts and habitats of these eukaryotic microorganisms are very diverse; fungi are present in every ecosystem on Earth. The fungal kingdom is equally diverse, consisting of seven different known phyla. Yet detailed knowledge is limited to relatively few species. The relationship between fungi and humans has been characterized by the juxtaposed viewpoints of fungi as infectious agents of much dread and their exploitation as highly versatile systems for a range of economically important biotechnological applications. Understanding the biology of different fungi in diverse ecosystems as well as their interactions with living and non-living is essential to underpin effective and innovative technological developments.

This series will provide a detailed compendium of methods and information used to investigate different aspects of mycology, including fungal biology and biochemistry, genetics, phylogenetics, genomics, proteomics, molecular enzymology, and biotechnological applications in a manner that reflects the many recent developments of relevance to researchers and scientists investigating the Kingdom Fungi. Rapid screening techniques based on screening specific regions in the DNA of fungi have been used in species comparison and identification, and are now being extended across fungal phyla. The majorities of fungi are multicellular eukaryotic systems and therefore may be excellent model systems by which to answer fundamental biological questions.

A greater understanding of the cell biology of these versatile eukaryotes will underpin efforts to engineer certain fungal species to provide novel cell factories for production of proteins for pharmaceutical applications. Renewed interest in all aspects of the biology and biotechnology of fungi may also enable the development of "one pot" microbial cell factories to meet consumer energy needs in the 21st century. To realize this potential and to truly understand the diversity and biology of these eukaryotes, continued development of scientific tools and techniques is essential. As a professional reference, this series will be very helpful to all people who work with fungi and should be useful both to academic institutions and research teams, as well as to teachers, and graduate and postgraduate students with its information on the continuous developments in fungal biology with the publication of each volume.

Arti Gupta
Editor

Prospects of Fungal Biotechnologies for Livestock Volume 2

Fungal Biotechnologies for Animal Cell Lines

Springer

Editor
Arti Gupta
Department of Zoology
Shri Awadh Raj Singh Degree College
Ram Manohar Lohia University
Ayodhya, India

ISSN 2198-7777 ISSN 2198-7785 (electronic)
Fungal Biology
ISBN 978-3-032-06477-6 ISBN 978-3-032-06478-3 (eBook)
https://doi.org/10.1007/978-3-032-06478-3

This Springer imprint is published by the registered company Springer Nature Switzerland AG
The registered company address is: Gewerbestrasse 11, 6330 Cham, Switzerland

Preface

The remarkable diversity and versatility of fungal biotechnologies have positioned fungi at the forefront of sustainable innovation across sectors—from environmental remediation to industrial protein production. This volume, *Prospects of Fungal Biotechnologies for Livestock Volume 2: Fungal Biotechnologies for Animal Cell Lines*, presents a curated collection of chapters that showcase the multifaceted roles of fungi in biotechnology, with particular emphasis on their potential contributions to livestock and cellular agriculture.

This volume brings together pioneering research on the evolving role of fungi in cellular agriculture, with a particular focus on applications in animal cell lines. By harnessing the unique metabolic capabilities and engineering potential of fungi, the chapters explore innovative strategies for sustainable protein production, bioconversion, and bioprocess optimization.

The opening chapters ("Structural and Functional Evolution of Streamlined Microsporidia" and "Biology and Evolution of Highly Reduced Microsporidia") delve into the evolutionary and biological intricacies of Microsporidia, offering insights into their structural and functional streamlining. This lays a molecular foundation for the subsequent chapters that explore how fungal systems are engineered (the chapters "Engineering Interventions in Fungal Cell Factories" and "Filamentous Fungi: Heterologous Protein Expression") and optimized for applications such as heterologous protein expression and cellular factory design (the chapter "Engineered Fungal Cell Factories: Pioneering Sustainable Protein and Peptide Production"). The chapters "Fungal Biotechnology: Contribution to the Circular Economy. Recycling of Waste. Sustainability" and "Fungal Biotechnology and the Circular Economy" examine the role of fungal biotechnology in advancing the circular economy, highlighting sustainable practices such as waste recycling and resource valorization—further expanded in the chapter "Fungi as Bioconverters: Advancing Circular Economy Through Waste Recycling and Resource Optimization" with fungal bioconversion strategies. These themes are complemented by the exploration of green bioeconomy principles in the chapter "Recent Innovations in Myco-Biotechnology: Valourisation of Fungal Resources for a Sustainable and a Circular Economy" and the review of fermented fungal food proteins in the chapter

"Fermented Food Fungal Proteins and Peptides: Past, Present and Future." As the volume progresses, the focus transitions toward industrial applications. The chapters "The Fungal Edible Proteins' Industrial Biotechnology: Where Are We?" and "An Overview About Pharmaceutical Grade Fungal Protein and Peptides" investigate the production and quality of edible and pharmaceutical-grade fungal proteins, drawing attention to both current innovations and future outlooks.

Each chapter reflects a collaborative effort among researchers from diverse regions and disciplines. The authors blend foundational science with applied research, underscoring the global momentum toward environmentally responsible and scalable fungal solutions. Through contributions from global experts, the book bridges fundamental biology with applied biotechnology, offering insights into how fungal systems can be leveraged for future-ready, eco-conscious solutions in livestock and food industries. Together, these contributions illuminate how fungal biotechnologies can support the development of novel feed systems, high-value bioactive compounds, and sustainable protein alternatives—making this volume an essential resource for researchers, practitioners, and policymakers aiming to integrate fungal innovations into livestock and cellular systems.

We extend our heartfelt gratitude to all contributors whose dedication and insights have shaped this work. It is our hope that the knowledge compiled here will stimulate further research, foster interdisciplinary collaboration, and inspire tangible advancements in fungal biotechnology.

Ayodhya, India Arti Gupta

Acknowledgments

First and foremost, I would like to thank God for his never-ending grace, mercy, and provision during what ended up being one of the toughest times of my life.

I am indebted to the many people who helped bring this book to light. I wish to thank series editor Dr. Vijai Kumar Gupta, Assistant Editor Swati Sharma, Production Editor Arun Siva Shanmugam, and Springer Nature for their generous assistance, constant support, and patience in initializing the series volume. The editor is very thankful to Vice Chancellor of Faizabad University, Ayodhaya, for their kind support and constant encouragement. Special thanks are due to well-wishers, family members, and friends.

The author owes a debt of gratitude to her wonderful husband, Saket Agrawal, and their son, Reyansh. Their unrelenting support and motivation have been pivotal in inspiring her to undertake this endeavor, overcome all obstacles, and assemble all the pieces into a coherent whole. I also give special thanks to all faculty colleagues and senior faculty members of our team.

Contents

Editor and Contributors

About the Editor

Arti Gupta obtained her Ph.D. from Mahatma Jyotiba Phule Rohilkhand University, Bareilly, India, in 2010 in Animal Science. Her current research interests are animal biotechnology, molecular plant biotechnology, molecular animal biotechnology, bioprocess technology, and microbiology. From July 2014 to till date, she has been serving at Shri Avadh Raj Singh Smarak Degree College, Gonda, Uttar Pradesh, India, and has been engaged in some editorial work with Springer Nature, CRC Press, and De Gruyter. She has 13 years of teaching and research experiences in Plant and Animal Biotechnology and Microbial Biotechnology. She has worked as an external examiner for central evaluation in 2023–24 at MJP Rohilkhand University, Bareilly.

She has received the following awards: University Topper (Gold Medal), M.Sc. (Biotechnology) by Ch.C.S. University, Meerut; Young Scientist Award (Gold Medal) by the Zoological Society of India, Lucknow; Best Poster Presenter by Asian Journal of Experimental Science, Jaipur; Best Poster Presenter by International Consortium of Contemporary Biologists (ICCB) and Madhawai-Shyam Educational Trust, Ranchi; Fellowship Award by International Consortium of Contemporary Biologist (FICCB) and Madhawi Shyam Educational Trust (FMSET); and Dr. V.P. Agarwal (Gold Medal) by D.A.V. (P.G.) College, Muzaffarnagar. She has been involved with the following publications: 2 monographs, 18 edited books with Springer Nature, Switzerland; CRC Press Taylor & Francis, USA; De Gruyter GmBH Germany, 10 book chapters, and around 50 national and international research papers. She has presented many abstracts in national and international symposia/seminars/conference/workshops. She has lifetime membership of the Indian Science Congress Association, Biotech Research Society of India, Zoological Society of India, and International Consortium of Contemporary Biologist. She has been serving as a member on several editorial boards. Previously, she served as a member of Teaching Personnel, G.B. Pantnagar University, Pantnagar, Uttarakhand, India. Visiting Academic Researcher, Biorefining and Advanced Material Research Center (BAMRC), Scotland's Rural College (SRUC), University of Edinburgh, Scotland, United Kingdom.

Contributors

June Susan Abraham Department of Life Sciences, School of Sciences, Christ University, Bengaluru, India

Rini Abraham Department of Biotechnology, Jyoti Nivas Autonomous College, Bengaluru, India

Jameel M. Al-Khayri Department of Agricultural Biotechnology, College of Agriculture and Food Sciences, King Faisal University, Al-Ahsa, Saudi Arabia

Özlem Erdal Altıntaş Faculty of Health Sciences, Department of Nutrition and Dietetics, Afyonkarahisar Health Sciences University, Afyonkarahisar, Turkey

Amro Abd Al Fattah Amara Protein Research Department, Genetic Engineering and Biotechnology Research Institute [GEBRI], City of Scientific Research and Technological Applications [SRTA-City], Universities and Research Center District, Alex, Egypt

Asha Anand Department of Life Sciences, School of Sciences, Christ University, Bengaluru, India

R. S. Arvind Bharani Institute of Obstetrics and Gynaecology, Madras Medical College, Chennai, Tamil Nadu, India

Tanmay S. Bachhav School of Life Sciences, Kavayitri Bahinabai Chaudhari, North Maharashtra University, Jalgaon, Maharashtra, India

Susheela Biranjia-Hurdoyal Department of Health Sciences, Faculty of Medicine and Health Sciences, University of Mauritius, Reduit, Mauritius

Ahmet Çabuk Department of Biology, Eskisehir Osmangazi University, Eskisehir, Turkey

Pinar Aytar Çelik Environmental Protection and Control Program, Eskisehir, Osmangazi University, Eskisehir, Turkey

Shikha Baghel Chauhan Amity Institute of Pharmacy, Amity University, Noida, Uttar Pradesh, India

Blaise Manga Enuh Great Lakes Bioenergy Research Centre, Wisconsin Energy Institute, Madison, WI, USA

Sweta Gaikwad Department of Life-Science, Shri Shankaracharya Professional University, Bhilai, Chhattisgarh, India

Arti Gupta Department of Zoology, Shri Awadh Raj Singh Degree College, Ram Manohar Lohia University, Ayodhya, India

Charu Gupta Amity Institute of Herbal Research and Studies, Amity University, Noida, Uttar Pradesh, India

J. Helan Chandra Ampersand Academy, Ashok Nagar, Chennai, Tamil Nadu, India

Tomas Hasek Department of Biotechnology, University of Chemistry and Technology, Prague, Dejvice, Czechia

Şule Inci Health Services Vocational, Istanbul Rumeli University, Istanbul, Turkey

Joshny Joseph Department of Life Sciences, School of Sciences, Christ University, Bengaluru, India

Sevda Kirbağ Science Faculty, Department of Biology, Firat University, Elazig, Turkey

Sunil H. Koli Department of Microbiology, Yashwantrao Chavan College of Science, Karad, Maharashtra, India

R. Lekhavani Centre for Materials Engineering and Regenerative Medicine, Bharath Institute of Higher Education and Research, Chennai, Tamil Nadu, India

Bhavana V. Mohite Department of Microbiology, Bajaj College of Science, Wardha, Maharashtra, India

Praveen Nagella Department of Life Sciences, School of Sciences, Christ University, Bengaluru, India

Satish V. Patil School of Life Sciences, Kavayitri Bahinabai Chaudhari, North Maharashtra University, Jalgaon, Maharashtra, India

S. Ramalakshmi Department of Bioengineering, Vels Institute of Science, Technology and Advanced Studies, Pallavaram, Chennai, India

K. Sivasubramanian Centre for Materials Engineering and Regenerative Medicine, Bharath Institute of Higher Education and Research, Selaiyur, Chennai, Tamil Nadu, India

K. Suguna Centre for Materials Engineering and Regenerative Medicine, Bharath Institute of Higher Education and Research, Chennai, Tamil Nadu, India

Belma Nural Yaman Department of Biomedical Engineering, Eskisehir, Osmangazi University, Eskisehir, Turkey

Structural and Functional Evolution of Streamlined Microsporidia

June Susan Abraham, Rini Abraham, Jameel M. Al-Khayri, and Praveen Nagella

Abstract Microsporidia represent a distinct lineage of obligatory intracellular parasitic eukaryotes, characterized by their exclusive dependence on host cells for survival and replication. Despite their significant chromosomal reduction and divergence, evolutionary studies indicate that microsporidia and cryptomycotan fungus, especially *Rozella allomyces*, have a common ancestor. Genomic reduction and the presence of infection apparatus, polar tube, reflects the evolutionary adaptation of the microbes towards a parasitic lifestyle. Microsporidia have a distinctive life cycle that enables them to survive and also to be transmitted across their host cells. K-mer-based analysis have revealed that polyploidy is a common feature in these microbes. The chapter explores the diverse strategies employed by these intracellular parasites in establishing and maintaining a niche within the host population. Although they are frequently found in insects, microsporidia parasitize a variety of animal hosts, including protists and mammals. The polar tube's structural elements and their molecular interactions facilitate microsporidia's adhesion to host cells. The lifecycle phases of several eukaryotic intracellular parasitic groups, such as Apicomplexa, Kinetoplastida, and the Holomycotan lineage, which includes Microsporidia, are also examined and compared in this chapter. These discoveries broaden our knowledge of eukaryotic parasitism and demonstrate the significance of Microsporidia as a model organism for investigating evolutionary specialization and extreme genome reduction.

J. S. Abraham · P. Nagella (✉)
Department of Life Sciences, School of Sciences, Christ University, Bengaluru, India
e-mail: susan.thomas@res.christuniversity.in; praveen.n@christuniversity.in

R. Abraham
Department of Biotechnology, Jyoti Nivas Autonomous College, Bengaluru, India

J. M. Al-Khayri (✉)
Department of Agricultural Biotechnology, College of Agriculture and Food Sciences, King Faisal University, Al-Ahsa, Saudi Arabia
e-mail: jkhayri@kfu.edu.sa

A. Gupta (ed.), *Prospects of Fungal Biotechnologies for Livestock Volume 2*,
Fungal Biology, https://doi.org/10.1007/978-3-032-06478-3_1

Keywords Genome reduction · Holomycota · Intracellular parasites · Microsporidia · Mitosome · Polar tube

1 Introduction

Microsporidia are related to fungi and are obligate intracellular parasites. They are categorized as eukaryotes due to the presence of key cellular features such as a nucleus enclosed by a nuclear envelope, an internal membrane system, chromosome segregation via mitotic spindles, a vesicle-based Golgi apparatus, and a mitosome—an organelle derived from mitochondria that lacks its own genome (Han & Weiss, 2017). More than 150 years ago, the first microsporidia which was identified, *Nosema bombycis,* is responsible for the infection, pebrine observed in the silk worms (Han et al., 2021). Highly reduced microsporidia belong to a group, namely, Holomycota, which has branched out from the eukaryotic supergroup, namely, Opisthokonta (Galindo et al., 2021). Microsporidia are parasitic organisms that can infect a wide range of hosts, including mammals and humans, as well as unicellular protists. They live in a variety of environments, such as soil, freshwater, and marine habitats. Significant economic harm can be caused by infections in vertebrates like fish and important invertebrates, namely, honeybees, shrimp, and silkworms. Furthermore, in organisms such as nematodes, *Daphnia,* locusts, mosquitoes, and honeybees, microsporidian parasites can have a major effect on host population dynamics (Han et al., 2020). Both immunocompetent and immunodeficient humans, such as organ transplant recipients, individuals with severe HIV infection, and patients undergoing immune-modulatory treatments have greater chances for being infected by microsporidia (Han et al., 2021).

The existence of a distinctive structure known as the polar tube facilitates the infection of Microsporidia. The polar tube emerges from the spore, punctures a cell membrane, and facilitates the entry of sporoplasm into the new host cell when the proper environmental stimuli are made available (Xu & Weiss, 2005). The proteins, which are the major constituents of the polar tube, are highly specific and can be utilized for diagnosis of the disease, microsporidiasis (Chen et al., 2023). Significant gene loss and the acquisition of specific genes, necessary for host exploitation occurred due to its adaptation for intracellular parasitic lifestyle (Nakjang et al., 2013). Lack of vital organelle, mitochondria, and the presence of transporter gene are few of the evidences which were reported (Burri et al., 2006; Dean et al., 2018). Due to selection pressure, most of the eukaryotic ribosomal RNA (rRNA) expansion regions were eliminated, but the functional core was conserved (Barandun et al., 2019). The evolution of parasitism in microsporidia sheds light into various aspects such as extreme reduction in its genome, coevolution of the host and the parasite as well as emergence of diseases (Han et al., 2021; Nakjang et al., 2013; Andreadis et al., 2012). By examining the biology, evolutionary background, and genomic adaptations of microsporidia, this review illuminates how their distinct position in the eukaryotic tree of life has been moulded by extreme parasitism.

2 Microsporidia as Intracellular Parasite

2.1 Mechanisms of Infection and Host Cell Invasion

Microsporidia are obligate intracellular, unicellular eukaryotes that generate spores and are categorized as protists. They have been identified as parasites in all major animal groups, including mammals, humans, and other protists. They are important from an economic and medical perspective and can be found in freshwater, marine, and terrestrial environments. Since they are intracellular parasites, they must have access to food, and they have evolved a number of strategies to get the metabolites they need, which has decreased the amount of biochemicals they must create to grow. Their physiological machinery has decreased, and distinct organelles and metabolic pathways have formed as a result of their decreased requirement to create their own metabolites. Microsporidia aggressively infect host cells by a special method. It includes piercing the plasma membrane without causing phagosome development or destruction. Within the intact spore, the polar filament, also known as the polar tube after germination, is coiled and fixed by the anterior attachment complex. It is barely 0.1 μm in external diameter and frequently exceeds 100 μm in length. During spore germination, the active polar structure is everted through the spore wall at this point. If the polar tube pierces a cell, the sporoplasm, made up of cytoplasm and nucleus surrounded by a membrane, goes through the tube and is injected into the cytoplasm of the host cell. This action confirms that the parasite is mostly located in the host cell cytoplasm, not in a plasma membrane-derived phagosome vacuole, which is typically the situation for parasites that are ingested by endocytosis. In numerous commercial, medical, and economic host groups, microsporidia are significant parasites. Even though the vast majority of species are known from insects and fish, 17 species have been found in humans which can cause serious ailments (Cali et al., 2017).

Humans are infected by about a dozen kinds of microsporidia, which can cause a range of illnesses and even death, particularly in those with weakened immune systems. The most prevalent symptom is an intestine infection, which is typically caused by either *Enterocytozoon bieneusi* or *Enterocytozoon intestinalis*. The majority of human infections are caused by *Enterocytozoon bieneusi*, which also infects the small intestine and biliary tract, resulting in persistent diarrhoea (Flores et al., 2021).

Microsporidia that infect arthropods have a major role in their significance for both public health and the economy. *Nosema bombycis*, the causative agent of a disease called "pébrine" in farmed silkworms (*Bombyx mori*, Lepidoptera), was actually the first microsporidian to be described. This disease led to massive breakdowns in the silk industry all over the world. The honeybee *Apis mellifera* is infected with the microsporidian *Nosema apis*, which is linked to worker lifetime reduction and colony failure. Numerous microsporidian species have been found in *Anopheles*, and the majority of these have been linked to reduced longevity, fecundity, and larval mortality. In recent times, the microsporidians *Vavraia culicis* and "Microsporidia MB" were shown to be related to a reduction in *Plasmodium* transmission in infected *Anopheles*, suggesting that Microsporidia may represent a potential route to malaria control (Khalaf et al., 2024a).

Microsporidia infections in vertebrates like fish and economically significant invertebrate hosts like shrimp, honeybees, and silkworm can result in large economic losses. Microsporidia can proliferate in nature in a number of ways. Vertical transmission, whereby the infection is directly transferred from the parent to offspring and horizontal transmission, by which the infection spreads among the individuals of the same generation due to transmission of spores by contaminated food and water. Most microsporidia infections in humans are believed to be zoonotic and they are spread by spores consumed through food or water. When Microsporidia infect host cells, the sporoplasm is transmitted from spores through the polar tube (PT), causing the illness to spread. The sporoplasm passes through the PT during this phase, functions as a droplet at the PT's distal end, and stays attached to the PT for a few minutes. It is likely that the sporoplasm interacts with the host cell within the safe environment of the invasion synapse during penetration. When sporoplasm enters the host cell, it begins a reproduction cycle that consists of meronts (proliferative forms), sporoblasts, sporonts, and ends with the mature spores. During spore germination and host cell penetration, the sporoplasm remains completely bound to the PT. The development of the invasion synapse is most likely essential to the sporoplasm's survival after leaving the PT since it is extremely susceptible to osmotic stress. Microsporidian spores usually adhere to host cells or the surrounding tissue as the initial stage of infection (Han et al., 2020).

Numerous proteins found in microsporidia's spore walls are thought to have a part in invasion. Several spore wall proteins (SWPs), which can interact with host cells by binding to the heparin-binding motif (HBM) and host cell surface sulphated glycosaminoglycans (GAGs), have been identified from *Nosema bombycis*, *Encephalitozoon* spp., and *Antonospora locustae*. In *N. bombycis*, six spore wall proteins (SWP7, SWP9, SWP11, SWP12, SWP16, and SWP26), many with heparin-binding domains, are required for spore adherence and host cell infection. In addition, the heparin-binding domain of the *Enterocytozoon hepatopenaei* spore wall protein EhSWP1 is necessary for binding to GAGs (Tamim El Jarkass & Reinke, 2020).

An infection with microsporidia can significantly affect human metabolism. Compared to uninfected bees, parasite-infected honeybees exhibit greater levels of hunger and sugar consumption. Moreover, certain metabolites, such as lipids, carbohydrates, and nucleotides, are less abundant in shrimp infected with *E. hepatopenaei* (Ning et al., 2019). *N. bombycis*-infected silkworms exhibit reduced ATP and protein levels, but many lipid levels remain unchanged (Li et al., 2018).

Among animals, lagomorphs such as the European rabbit (*Oryctolagus cuniculus*), European brown hare (*Lepus europaeus*), and Eastern cottontail rabbit (*Sylvilagus floridanus*) have been found to harbour zoonotic microsporidia species in some European nations. Although microsporidiosis infections in rabbits are typically asymptomatic, cases have been documented that result in neurological or renal problems and, in rare instances, death. Four recognized genotypes (I, II, III, and IV) of *Encephalitozoon cuniculi*, a species that is especially related to rabbits, infect a range of species. Genotypes I and II have been detected in rabbits from Slovakia and China. However, genotype III has only been identified in Slovakian domestic rabbits (Abdoli et al., 2025).

2.2 *Structure of Microsporidia Spores and Their Lifecycle*

In 1894, Thélohan observed the general structure of the spore (Han et al., 2021). Since those initial findings, studies have defined the internal structure of these spores, including the polar tube, anchoring disk, polaroplast, spore coat, posterior vacuole, and nucleus (Han & Weiss, 2017). The endospore serves as the link between the exospore and plasma membrane and is primarily made of chitin. Many proteins present in the spore coat possess adhesion domains that are probably involved in the interaction of microsporidian spores with either the cell surface or mucus of the gastrointestinal tract prior to germination. The anchoring disk is situated at the anterior pole of the spore, where the endospore is thinnest in the anterior region. In addition, the anchoring disc is where the polar tube attaches to the spore wall and is released during spore germination (Southern et al., 2007; Bigliardi & Sacchi, 2001).

Prior to germination, the polar tube is coiled around the nucleus in the spore (the number of coils depends on the specific microsporidian species). During germination, the polar tube is extruded from the anchoring disk of the spore and forms a hollow tube that acts as a bridge between the microsporidian spore and host cell, allowing transfer of the nucleus and sporoplasm into the host cell cytoplasm, resulting in cell infection (Xu & Weiss, 2005; Lom, 1972).

The merogony, or proliferating stage, and the sporogony, or infecting or mature stage, are the two separate stages of microsporidia's life cycle. Through inhalation or ingestion, spores can enter the host. The polar tube, a very special, specialized invasive structure, extends during germination and injects sporoplasm into the cytoplasm of the host cell during this process. In certain species, the glycocalyx also contains the injected sporoplasm. This structure invades the host cells and gives the microsporidia nourishment (Vávra & Lukeš, 2013). Later, sporoplasm transforms into schizonts, the maturation stage, when multinucleate plasmodial production and fusion facilitate replication. This cell membrane thickens and forms sporonts. After dividing and developing into sporoblasts, the sporonts eventually reach a developed stage of spore without further multiplication. The mature form is equipped with the

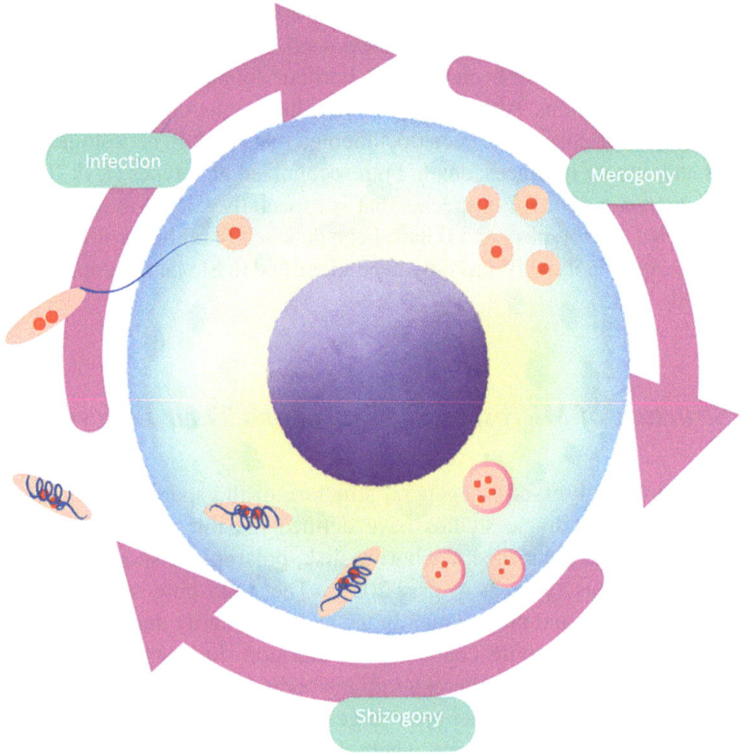

Fig. 1 Life cycle of Microsporidia

injection apparatus. The presence of the injection apparatus in the spore is a characteristic that sharply defines microsporidia as a monophyletic taxon.

Eventually, the developed spores cause the host cells to enlarge, burst, and release the spores into the outside world or adjacent body cells (Fadhilah et al., 2025). The life cycle of microsporidia is shown in Fig. 1.

2.3 *Evolution of Microsporidia*

2.3.1 Phylogenetic Relationships of Microsporidia with Other Eukaryotes

The evolution of fungi, the diverse kingdom of eukaryotic organisms, has likely begun in water bodies with ancestors of parasitic nature, possessed flagella, and gradually transitioned to terrestrial environment. During the process of adaptation, there would have been evolutionary changes such as disappearance of flagella and development of hyphae (Naranjo-Ortiz & Gabaldón, 2019). Sequencing of the genomes of microsporidia has revealed the similarity with a group of intracellular parasitic group of fungi, Cryptomycota or Rozellomycota. *Rozella allomyces*, a species of Rozellomycota, differed from microsporidia due to the presence of mitochondria and higher proportion of proteins which are highly conserved with those of eukaryotes (Wadi & Reinke, 2020). The above findings were further refined by the identification of several parasitic forms which were similar to the earlier types of microsporidia, but also had mitochondria and genes similar to that of fungi (Corsaro, 2022). Lee et al. suggest that the presence of identical genes (TPT, HMG, as well as RNA helicase) and the conserved gene order within the sex locus of both microsporidia and Zygomycetes provides strong evidence supporting the hypothesis that microsporidia are true fungi, having evolved from ancestral Zygomycetes (Lee et al., 2008).

2.3.2 Evolution of Parasitism in Microsporidia

2.3.2.1 Vertical and Horizontal Transmission

Microsporidia parasitize multiple animal hosts ranging from Protists to Mammals, although it is commonly found in insects. These parasites gain entry into the hosts either through vertical or horizontal transmission as well as a combination of both the modes. Horizontal mode of transmission is associated with multiple infective stages which can result in the death of the host, whereas the vertical mode ensures the survival of the host. Evolutionary forces may lead to the loss of genes which are required for horizontal transmission and can result in vertical transmission as well as vice versa in microsporidia (Smith, 2009). As vertical transmission is commonly observed in microsporidia, which can also distort the sex ratio in the host, it is important to consider the evolutionary interactions between these parasites and their hosts (Terry et al., 2004). Molecular interactions play a significant role in the adaptation of the parasite with a specific host which is essential in vertical transmission. Various evidences report that the microsporidia species remain associated with the same host, and host switching is rare (Smith, 2009; Solter, 2006). It has also been reported by Haag et al. that the microsporidia, *Hamiltosporidium* which are transmitted by both vertical and horizontal modes have more number of genes and longer intergenic regions compared to *Ordospora* which are solely transmitted horizontally. Non-synonymous substitution in the parasitic forms with mixed form of transmission is not exclusively due to the absence of recombination, but can be due to genetic drift. These results imply that *Hamiltosporidium* species, which have transitioned from horizontal to both the modes of transmission, probably result in genetic bottleneck, decreasing the efficacy of adaptive process and permitting nonadaptive processes, such as genetic drift to significantly shape their genomic features (Haag et al., 2020).

2.3.2.2 Horizontal Gene Transfer

Horizontal gene transfer (HGT), observed in the ancestors of microsporidia, have resulted in gaining nucleotide transporter proteins, which help in transporting ATP across cell membranes. This event led to the evolution of genome in microsporidia as it facilitated stealing of ATP from host cells. The duplication and evolution of transporter genes also enabled the parasites to transport other energy-rich molecules such as GTP and NAD^+ from the host cells. Thus, they lost numerous biosynthetic and energy production pathways, which were essential for the survival of eukaryotes. As a result, their cells and genomes were drastically decreased, which is a crucial characteristic of contemporary microsporidia (Dean et al., 2018). Microsporidia gain entry into host mainly, by using its polar tube as an injection needle or by endocytosis. Genomic sequencing of *Ordospora colligata* has revealed the presence of a protein, septin, which is analogous to septin 7, found in *Candida albicans*. *Candida albicans* gain entry into the host by endocytosis, with the help of septin 7. It was observed that *O. colligata* gained septin from the host, *Daphnia* sp. through HGT. Therefore, it can be assumed that *O. colligate* gains entry into the host by endocytic infection facilitated by septin (Pombert et al., 2015). The majority of microsporidian genes are either exclusive to microsporidia or frequently present in other eukaryotes. However, six gene families that most likely originated from species outside of their typical group (Opisthokonta) were discovered by researchers. These include phosphoribosyl transferase, an aspartate-ammonia ligase, and well-known HGTs like the NTTs. Additionally, they identified two novel potential HGTs: one involving a CAP-domain protein and another for glycosyl transferase (Nakjang et al., 2013).

2.3.2.3 Evolution of Cellular Structures: The Polar Tube

A solitary polar tube is present in microsporidia irrespective of species, location, or host. This implies that polar tube did not arise by convergent evolution, but originated prior to the diversification of microsporidia. The polar tube is thought to have developed from Endoplasmic Reticulum (ER) and vacuoles produced from Golgi apparatus. According to preliminary research, Golgi saccules may be the source of the filament's core and ER may be the source of the outer layer. A Golgi and ER origin is supported by enzymatic markers such as nucleoside diphosphatase and thiamine pyrophosphatase. Three polar tube proteins, known as ptp1, ptp2, and ptp3, have been identified through the use of molecular and immunological approaches (Xu & Weiss, 2005). The core amino acid repeat sections of all the cloned PTP1s are mostly hydrophilic. But the quantity and nature of the repeats vary. There is a chance that this area serves as an immune mask and is not crucial for polar tube construction. Similar internal sequence duplication has been observed in malaria and other protozoan genes during evolution, and the microsporidia ptp1 may be affected by this mechanism (Xu & Weiss, 2005; Rich & Ayala, 2000). The variation in the number of repeats within the central region of the *ptp1* gene among different *Encephalitozoon hellem* isolates provides supporting evidence for this hypothesis (Xu & Weiss, 2005; Peuvel et al., 2000; Weiss, 2001; Xiao et al., 2001; Haro et al., 2003). In *E. cuniculi*, ptp1 as well as ptp2 are found as a gene cluster on chromosome VI. The orientation and spacing of this gene cluster are conserved in *E. hellem* and *E. intestinalis*. Recent investigations of microsporidian genomes have reported the conservation of linear gene arrangement, which might be the result of evolutionary restrictions on these compact eukaryotic genomes (Xu & Weiss, 2005). Gaining additional knowledge about the structural properties of PTPs helps to expose the PTP assembly on the polar tube of microsporidia and sheds light on the evolution and functionality of the above proteins (Chen et al., 2023).

2.3.2.4 Evolution of Reproductive Strategies in Microsporidia

Spore formation is an integral phase during the microsporidian life cycle, during which the parasites are resistant and have the capability to survive outside the host. After infecting the host, the spores undergo intracellular replication, therefore forming multiple copies of the parasite (Nakjang et al., 2013). These resistant structures are composed of an inner plasma membrane, an electron-lucent endospore made of chitin, and an outward electron-dense exospore made of glycoprotein (Didier et al., 2004). A distant relative of microsporidia, *Rozella allomyces* differs from early diverging groups such as *Paramicrosporidium saccamoebae* and *Mitosporidium daphnia*, as the latter does not have flagellated spores, instead have a polar filament instead of flagella (Wadi & Reinke, 2020). Proteomics of *Trachipleistophora hominis* and *Encephalitozoon cuniculi* spores showcased the differences in lineage-specific gene families, indicating the adaptation of the spores for infecting specific hosts. The occurrence of the enzyme, protein mannosyl transferase, and also the mannosylation of spore wall proteins in *E. cuniculi* illustrates the evolution of microsporidia towards a parasitic lifestyle, as it is crucial for virulence, adhesion, and motility (Albuquerque-Wendt et al., 2020). Various alterations in the spore structure during the evolution of microsporidia improved its ability to infect the host (Nakjang et al., 2013).

Loss of sexual reproduction has been pivotal during the evolution of microsporidia. Both the development of gametes and the fertilization process can result in increased expenditures associated with sex. Considering that many species depend on effective vertical transmission from female hosts to their progeny, *Nosema* may be especially impacted by such expenses. The host bears the higher energy costs of sexual reproduction, which lowers its fertility and/or survival and, in turn, lowers the parasite's vertical transmission. Other lifestyle changes in microsporidia such as lack of requirement for the overwintering spores or an intermediate host may also be the reason for transformation to asexual reproduction (Ironside, 2007). Although microsporidia favour asexual reproduction, there are evidences of the presence of genes which are homologous to certain meiotic genes, which is required for recombination, crossing over, etc. in eukaryotes (Ang'ang'o et al., 2024). Recombination in apparently asexual microsporidia raises the possibility that they may reproduce sexually discreetly. This might significantly impact how their virulence, capacity to infect various hosts, and drug resistance develop (Ironside, 2013).

2.3.2.5 Loss of Non-essential Genes and Organelles

The obligate intracellular parasitic lifestyle has led to the loss of several genes and organelles in microsporidia. Research on the genomes of microsporidia reveals that they only retain roughly 800 essential proteins, having lost many protein families present in other eukaryotes. The primary roles of the proteins they preserved are in fundamental cellular processes including transcription, translation, and DNA replication. Numerous others were lost, such as regulatory proteins, metabolic enzymes, and those responsible for intracellular transport, such as clathrin and TOR (Target of Rapamycin) (Galindo et al., 2018). Additionally, compared to other fungi, microsporidia have shorter proteins. *E. cuniculi* proteins, for instance, are roughly 85% longer than yeast proteins of the same length. Leucyl-tRNA synthetase (LeuRS) is one example of an aminoacyl-tRNA synthetase that has lost many of its regulatory components (Wadi & Reinke, 2020).

Additionally, microsporidia have smaller noncoding RNAs, such as the 18s RNA, which produces the smallest eukaryotic ribosomes and is only about two-thirds as large as in other eukaryotes (Melnikov et al., 2018). The majority of conserved ribosomal proteins can still bind, despite the elimination of many binding sites, as shown by structural characterization and mass spectrometric profiling of ribosomal complexes from *Vairimorpha necatrix* (Barandun et al., 2019). In ribosomes of related microsporidia with moderately reduced genomes, such as *Paranosema locustae* and *Vairimorpha necatrix*, the loss of rRNA segments may be offset by the emergence of novel ribosomal proteins (e.g., msL13) or novel protein-protein interactions. Fewer than 2000 protein-encoding genes and even smaller genomes (~2.5 Mb) are seen in *E. cuniculi* and also *Ordospora*. In addition to eukaryote-specific rRNA expansion segments, their ribosomes are expected to lack four ribosomal proteins since their genes are not present in the *E. cuniculi* genome. *E. cuniculi* ribosomes may use novel molecular strategies to carry on, despite these losses brought on by this significant genome reduction (Nicholson et al., 2022).

Reduced organelles, such as mitosomes, which are mitochondrial remains, are also present in microsporidia. Microsporidia must import ATP from their host since these mitosomes lack the mitochondrial genome and cannot perform oxidative phosphorylation (Dean et al., 2018; Williams et al., 2002). The mitosome imports proteins from the cytosol. In contrast to fungi, which employ intricate import machinery, microsporidia retain a fraction of components. Targeting signals that resemble those in yeast are still present in some imported proteins. These proteins are processed by a simpler peptidase in *Antonospora locustae*, but in *E. cuniculi*, the processing complex has been eliminated, leaving the proteins unprocessed (Burri et al., 2006). When *E. cuniculi* genome was compared to 423, known yeast mitochondrial proteins, researchers were able to identify genes that might be related to mitochondrial genes. Six of the 22 genes they discovered shared a lot of similarities with proteins from α-proteobacteria, which are the ancestors of mitochondria. The yeast proteins NFS1, ATM1, ISU1/ISU2, PDB1, SSQ1, and YAH1 are homologs of these six genes. Iron–sulphur (Fe–S) cluster assembly is a crucial mitochondrial activity that involves five of them. These mitochondrial-type proteins appear to be functional based on the presence of essential domains and amino acids. Five of them are predicted by PSORT analysis to have amino-terminal sequences for targeting (Katinka et al., 2001).

Microsporidia have a simplified Golgi apparatus without clear structural polarity but still essential for spore formation. Early N-glycosylation steps may occur, but complex processing is absent. O-glycosylation mainly uses mannose via two protein mannosyl transferase (PMT) family enzymes, and the absence of mannose-phosphorylating enzymes suggests lysosomal sorting is absent. A limited collection of putative proteins for Golgi and post-Golgi trafficking sustains membrane fusion and recognition of certain target membrane activities. Certain distinctive Rab proteins may be involved in the constitutive secretion pathway that leads to the plasma membrane. Vacuolar protein sorting-associated proteins, Vps1-like dynamin, and β1-adaptin are a few possible partners for trans-Golgi and endosome transport, demonstrating the functional polarization of the Golgi-like apparatus (Katinka et al., 2001).

2.4 Current Advances in Microsporidia Research

A widespread genus of unicellular fungal parasites, *Nosema* infects a wide range of arthropods. Researchers observed that the genomes of seven species of *Nosema* which infects wasps, bees, moths, amphipods, and butterflies were significantly reduced. Loss of genes involved in oxidative phosphorylation and a notable decrease in genes linked to endocytosis.G-C concentration is lower than that of other microsporidia. Near start codons are distinct DNA motifs (5–12 bp) that most likely regulate gene expression. These motifs were connected to *Nosema muscidifuracis* genes, which were significantly expressed. Two main *Nosema* clades were identified by phylogenomic analysis, and there was a discrepancy between the phylogeny of the *Nosema* species and that of their hosts, suggesting possible host swapping events (Xiong et al., 2024).

Ploidy for all microsporidian data sets, using short k-mer spectrum analyses, showed that polyploid state is common in Microsporidia and that ploidy change is dynamic within the group. It has been demonstrated that polyploid microsporidian genomes are comparatively homozygous, using genome-wide heterozygosity estimates (Khalaf et al., 2024b). Researchers examined the polar tube in the human microsporidian *Encephalitozoon intestinalis* using cryoelectron tomography. They found that there are four primary layers in the PT:

Two layers of proteins. A membrane that surrounds a thick core. Protein filaments aid in preserving the structure of the PT. A 3D model that illustrates the connection between the polaroplast, an organelle close to the PT, and its outer membrane was also suggested by the study. By providing the first in-depth examination of the PT's internal structure, this study advances our knowledge of how microsporidia infiltrate host cells (Usmani et al., 2025).

2.5 Parasitism in Microsporidia and Other Intracellular Parasites

Symbiosis is a persistent connection between two or more distinct organisms that can be harmful or beneficial. Parasitism is a particularly intriguing aspect of symbiosis (Carlson et al., 2020). Three groups of intracellular parasites such as Kinetoplastida, Holomycota, and Apicomplexa can be used for comparative analysis as they are distantly related single-celled eukaryotes (Bartošová-Sojková et al., 2024). The evolution of these forms into intracellular parasites is discussed here.

The phylum Euglenozoa contains the varied and extensively researched group Kinetoplastida. Although many of its distinctive traits have subsequently been discovered in other eukaryotes, they include the paraflagellar rod, polycistronic transcription, trans-splicing, circular mitochondrial DNA, and RNA editing. The parasitic association of two non-photosynthetic protozoans are discussed here. The common ancestor of Perkinsela belonging to Kinetoplastia was an ectoparasite on a marine animal which has also been infected by amoeba. Since both parasites were closely associated, the common ancestor of Perkinsela became an intracellular parasite of amoeba (Yazaki et al., 2017). Perkinsela lost flagella and the genes required for flagella formation during its transition to intracellular lifestyle. The other changes include intertwining of metabolic pathway between the host and parasite as well as loss of ancestral kinetoplastid genes (Tanifuji et al., 2017). Trypanosomatids were also ectoparasites of arthropods, occasionally were transmitted into vertebrates during a blood meal and therefore developed a complex life cycle inhabiting two hosts. The cytostome–cytopharynx complex, which is the ancestral feeding system for phagocytic food intake, has been preserved in certain intracellular species, such as *T. cruzi*. On the other hand, Leishmania has evolved to rely on endocytosis via the flagellar pocket and has lost this structure in favour of osmotrophic feeding (Galindo et al., 2023). Trypanosomatids have been reported to utilize several molecular tools such as proteins rather than morphological structures for internalization into the hosts (de Castro Neto et al., 2021).

Holomycota is a widespread clade of the Opisthokonta that includes fungi, microsporidia, and several other lineages. This group includes organisms with a variety of lifestyles, such as parasitoid, mutualistic, free-living, and authentic parasitic forms (Bartošová-Sojková et al., 2024). Nucleariids, including Nuclearia and Fonticula, were the first holomycotan group to appear. These are free-living, non-flagellated amoebae that ingest bacteria and green algae (phagotrophy) (Galindo et al., 2021, 2023). The holomycotans are divided into several lineages. Microsporidians and Rozellids are members of the group Opisthophagea, which evolved into parasites that mostly infect hosts that contain chitin. While aphelids and rozellids continue to be phagotrophs, fungi and microsporidians have shifted to osmotrophy, fungi being external parasites and microsporidians internal parasites (Bartošová-Sojková et al., 2024). While their shortened genomes still contain some functional long introns, microsporidians have lost a large number of spliceosomal introns. It has been suggested that differences in Microsporidia‘s ways of transmission are connected to this genetic diversification (Haag et al., 2020; Žárský et al., 2023).

The apical complex, a structure that aids in host cell invasion, is the hallmark of the diverse group of protists known as Apicomplexa (Bartošová-Sojková et al., 2024). Different forms of parasitism have emerged within Apicomplexa. Sporozoa includes well-known disease-causing organisms such as those that cause malaria, toxoplasmosis, and cryptosporidiosis. All of these species are obligatory parasites of animals (Votýpka et al., 2016). Apicomonada, also known as chrompodellids, on the other hand, exhibit a variety of lifestyles. These imply that parasitism developed separately within this group. Only chromerids continue to carry out photosynthesis, despite the fact that all apicomplexans descended from photosynthetic ancestors. The apicoplast, a non-photosynthetic plastid that is essential to their metabolism and a possible target for medication, is retained by the majority of other organisms, including parasites (Bartošová-Sojková et al., 2024).

In order to adapt to their hosts, many protist species independently evolved into intracellular parasites and acquired distinctive characteristics. Loss of flagella, osmotrophy, cyst development, and specialized transporters are common characteristics despite their diversity. Their rapid growth and plasticity, which allow them to flourish in a variety of habitats, are supported by gene expansions, horizontal gene transfers, and genomic rearrangements (Bartošová-Sojková et al., 2024). Table 1 denotes parasitism in microsporidia and other intracellular parasites.

2.6 *Conclusion*

One of the most specialized and highly reduced kinds of intracellular parasites among fungi are microsporidia. Significant genome shrinkage, the loss of metabolic pathways, and the emergence of special invasion machinery like the polar tube have all been characteristics of their evolution. Microsporidia have successfully colonized a variety of hosts in a range of conditions despite their simplicity. In addition to offering insights into the mechanisms underlying extreme parasitism, knowledge

Table 1 Parasitism in microsporidia and other intracellular parasites

Group	Parasite	Infectious form	Entry into the host	Life cycle stages	Host interaction	References
Holomycota	Microsporidia	Spore with an electron-rich outermost spore coat which covers a thicker inner lucent coat, and a membrane encloses the spore contents	Sporoplasm is injected into the host cell using a special polar tube by the spore	Infection of host cell via spore, merogony and formation of mature spore	Highly dependent on host as the genome is highly reduced, imports ATP from host	Han et al. (2020)
Apicomplexa	*Plasmodium*	Sporozoite—single nucleus, lancet-shaped	Mosquito bite	Sporozoite, shizont, merozoite, gametocyte	Infects RBCs and hepatocytes. Evades host immune system—Kupffer cells	Crutcher and Hoffman (1996), Bargieri et al. (2014), and Chandley et al. (2023)
	Cryptosporidium	Oocyst	Oral route—after excystation, sporozoites penetrate mucus layer and attach to enterocyte	Oocyst, sporozoite, trophozoite, merozoite, microgamont, macrogamont, zygote	Can access nutrients from host cell cytoplasm—Feeder organelle	Leitch and He (2012)
	Toxoplasma	Tachyzoites, bradyzoites within tissue cyst, sporozoites within oocyst	Oral, blood transfusion, vertical	Tachyzoite, tissue cyst, and oocyst	Infects any nucleated cell. Can dysregulate host's cell cycle	Al-Malki (2021), Hu et al. (2006), and Blader and Saeij (2009)
	Babessia spp.	Sporozoite	Tick bite—sporozoite endocytosed into erythrocytes	Sporozoite, trophozoite, merozoite, gametocytes, zygote, ookinete, kinete	Replication within RBC	Mirzayee et al. (2024)

Kinetoplastida	*Leishmania* spp.	Metacyclic promastigotes	Sandfly bite	Promastigotes, amastigotes	Infects neutrophils, Langerhans cells, fibroblasts	Teixeira et al. (2013)
	Trypanosoma cruzi	Metacyclic trypomastigotes	Parasites in the excreta of triatomine vector enter through bite wounds	Trypomastigote, amstigote, epimastogotes	Infects phagocytic and non-phagocytic cells, cell membrane lysis	Martín-Escolano et al. (2022)
	Endotrypanum	Promastigote	Sandfly bite	Promastigote, epimastigote	Infects erythrocyte of mammals	Franco and Grimaldi (1999)

of their biology and evolutionary history may also help identify possible targets for the control of infections in economically significant species, people, and animals.

Acknowledgements The authors express appreciation for the support provided by the Deanship of Scientific Research, Vice Presidency for Graduate Studies and Scientific Research, King Faisal University, Saudi Arabia [Grant No. KFU253322].

References

Abdoli, A., Olfatifar, M., Zaki, L., et al. (2025). The global prevalence of microsporidia infection in rabbits as a neglected public health concern: A systematic review and meta-analysis. *Preventive Veterinary Medicine, 234*, 106380. https://doi.org/10.1016/j.prevetmed.2024.106380

Al-Malki, E. S. (2021). Toxoplasmosis: Stages of the protozoan life cycle and risk assessment in humans and animals for an enhanced awareness and an improved socio-economic status. *Saudi Journal of Biological Sciences, 28*, 962–969. https://doi.org/10.1016/j.sjbs.2020.11.007

Albuquerque-Wendt, A., Jacot, D., Dos Santos Pacheco, N., et al. (2020). C-Mannosylation of toxoplasma gondii proteins promotes attachment to host cells and parasite virulence. *Journal of Biological Chemistry, 295*, 1066–1076. https://doi.org/10.1016/S0021-9258(17)49916-9

Andreadis, T. G., Simakova, A. V., Vossbrinck, C. R., et al. (2012). Ultrastructural characterization and comparative phylogenetic analysis of new microsporidia from Siberian mosquitoes: Evidence for coevolution and host switching. *Journal of Invertebrate Pathology, 109*, 59–75. https://doi.org/10.1016/j.jip.2011.09.011

Ang'ang'o, L. M., Herren, J. K., & Tastan Bishop, Ö. (2024). Bioinformatics analysis of the microsporidia sp. MB genome: A malaria transmission-blocking symbiont of the anopheles arabiensis mosquito. *BMC Genomics, 25*, 1132. https://doi.org/10.1186/s12864-024-11046-y

Barandun, J., Hunziker, M., Vossbrinck, C. R., & Klinge, S. (2019). Evolutionary compaction and adaptation visualized by the structure of the dormant microsporidian ribosome. *Nature Microbiology, 4*, 1798–1804. https://doi.org/10.1038/s41564-019-0514-6

Bargieri, D., Lagal, V., Andenmatten, N., et al. (2014). Host cell invasion by apicomplexan parasites: The junction conundrum. *PLoS Pathogens, 10*, e1004273. https://doi.org/10.1371/journal.ppat.1004273

Bartošová-Sojková, P., Butenko, A., Richtová, J., et al. (2024). Inside the host: Understanding the evolutionary trajectories of intracellular parasitism. *Annual Review of Microbiology, 78*, 39–59. https://doi.org/10.1146/annurev-micro-041222-025305

Bigliardi, E., & Sacchi, L. (2001). Cell biology and invasion of the microsporidia. *Microbes and Infection, 3*, 373–379. https://doi.org/10.1016/S1286-4579(01)01393-4

Blader, I. J., & Saeij, J. P. (2009). Communication between *Toxoplasma gondii* and its host: Impact on parasite growth, development, immune evasion, and virulence. *APMIS, 117*, 458–476. https://doi.org/10.1111/j.1600-0463.2009.02453.x

Burri, L., Williams, B. A. P., Bursac, D., et al. (2006). Microsporidian mitosomes retain elements of the general mitochondrial targeting system. *Proceedings of the National Academy of Sciences, 103*, 15916–15920. https://doi.org/10.1073/pnas.0604109103

Cali, A., Becnel, J. J., & Takvorian, P. M. (2017). Microsporidia. In *Handbook of the protists* (pp. 1559–1618). Springer.

Carlson, C. J., Dallas, T. A., Alexander, L. W., et al. (2020). What would it take to describe the global diversity of parasites? *Proceedings of the Royal Society B: Biological Sciences, 287*, 20201841. https://doi.org/10.1098/rspb.2020.1841

Chandley, P., Ranjan, R., Kumar, S., & Rohatgi, S. (2023). Host-parasite interactions during plasmodium infection: Implications for immunotherapies. *Frontiers in Immunology, 13*. https://doi.org/10.3389/fimmu.2022.1091961

Chen, Y., Lv, Q., Liao, H., et al. (2023). The microsporidian polar tube: Origin, structure, composition, function, and application. *Parasites & Vectors, 16*, 305. https://doi.org/10.1186/s13071-023-05908-9

Corsaro, D. (2022). Insights into microsporidia evolution from early diverging microsporidia. In *Microsporidia* (pp. 71–90). Springer.

Crutcher, J. M., & Hoffman, S. L. (1996). Malaria. In: Baron S, editor. Medical Microbiology. 4th ed. Galveston (TX): University of Texas Medical Branch at Galveston; 1996.

de Castro Neto, A. L., da Silveira, J. F., & Mortara, R. A. (2021). Comparative analysis of virulence mechanisms of Trypanosomatids pathogenic to humans. *Frontiers in Cellular and Infection Microbiology, 11*. https://doi.org/10.3389/fcimb.2021.669079

Dean, P., Sendra, K. M., Williams, T. A., et al. (2018). Transporter gene acquisition and innovation in the evolution of Microsporidia intracellular parasites. *Nature Communications, 9*, 1709. https://doi.org/10.1038/s41467-018-03923-4

Didier, E. S., Stovall, M. E., Green, L. C., et al. (2004). Epidemiology of microsporidiosis: Sources and modes of transmission. *Veterinary Parasitology, 126*, 145–166. https://doi.org/10.1016/j.vetpar.2004.09.006

Fadhilah, A., Gabbar, A., & Bokhari, A. A. (2025). *Microsporidium*. Stat Pearls.

Flores, J., Takvorian, P. M., Weiss, L. M., et al. (2021). Human microsporidian pathogen *Encephalitozoon intestinalis* impinges on enterocyte membrane trafficking and signaling. *Journal of Cell Science, 134*. https://doi.org/10.1242/jcs.253757

Franco, A. M. R., & Grimaldi, G., Jr. (1999). Characterization of Endotrypanum (Kinetoplastida: Trypanosomatidae), a unique parasite infecting the neotropical tree sloths (Edentata). *Memórias do Instituto Oswaldo Cruz, 94*, 261–268. https://doi.org/10.1590/S0074-02761999000200026

Galindo, L. J., Torruella, G., Moreira, D., et al. (2018). Evolutionary genomics of Metchnikovella incurvata (Metchnikovellidae): An early branching microsporidium. *Genome Biology and Evolution, 10*, 2736–2748. https://doi.org/10.1093/gbe/evy205

Galindo, L. J., López-García, P., Torruella, G., et al. (2021). Phylogenomics of a new fungal phylum reveals multiple waves of reductive evolution across Holomycota. *Nature Communications, 12*, 4973. https://doi.org/10.1038/s41467-021-25308-w

Galindo, L. J., Torruella, G., López-García, P., et al. (2023). Phylogenomics supports the monophyly of aphelids and fungi and identifies new molecular synapomorphies. *Systematic Biology, 72*, 505–515. https://doi.org/10.1093/sysbio/syac054

Haag, K. L., Pombert, J.-F., Sun, Y., et al. (2020). Microsporidia with vertical transmission were likely shaped by nonadaptive processes. *Genome Biology and Evolution, 12*, 3599–3614. https://doi.org/10.1093/gbe/evz270

Han, B., & Weiss, L. M. (2017). Microsporidia: Obligate intracellular pathogens within the fungal kingdom. *Microbiology Spectrum, 5*. https://doi.org/10.1128/microbiolspec.FUNK-0018-2016

Han, B., Takvorian, P. M., & Weiss, L. M. (2020). Invasion of host cells by microsporidia. *Frontiers in Microbiology, 11*. https://doi.org/10.3389/fmicb.2020.00172

Han, B., Pan, G., & Weiss, L. M. (2021). Microsporidiosis in humans. *Clinical Microbiology Reviews, 34*. https://doi.org/10.1128/CMR.00010-20

Haro, M., del Águila, C., Fenoy, S., & Henriques-Gil, N. (2003). Intraspecies genotype variability of the microsporidian parasite *Encephalitozoon hellem*. *Journal of Clinical Microbiology, 41*, 4166–4171. https://doi.org/10.1128/JCM.41.9.4166-4171.2003

Hu, K., Johnson, J., Florens, L., et al. (2006). Cytoskeletal components of an invasion machine—The apical complex of toxoplasma gondii. *PLoS Pathogens, 2*, e13. https://doi.org/10.1371/journal.ppat.0020013

Ironside, J. E. (2007). Multiple losses of sex within a single genus of microsporidia. *BMC Evolutionary Biology, 7*, 48. https://doi.org/10.1186/1471-2148-7-48

Ironside, J. E. (2013). Diversity and recombination of dispersed ribosomal DNA and protein coding genes in microsporidia. *PLoS One, 8*, e55878. https://doi.org/10.1371/journal.pone.0055878

Katinka, M. D., Duprat, S., Cornillot, E., et al. (2001). Genome sequence and gene compaction of the eukaryote parasite Encephalitozoon cuniculi. *Nature, 414*, 450–453. https://doi.org/10.1038/35106579

Khalaf, A., Francis, O., & Blaxter, M. L. (2024a). Genome evolution in intracellular parasites: Microsporidia and Apicomplexa. *Journal of Eukaryotic Microbiology, 71*. https://doi.org/10.1111/jeu.13033

Khalaf, A., Lawniczak, M. K. N., Blaxter, M. L., & Jaron, K. S. (2024b). Polyploidy is widespread in microsporidia. *Microbiology Spectrum, 12*. https://doi.org/10.1128/spectrum.03669-23

Lee, S. C., Corradi, N., Byrnes, E. J., et al. (2008). Microsporidia evolved from ancestral sexual fungi. *Current Biology, 18*, 1675–1679. https://doi.org/10.1016/j.cub.2008.09.030

Leitch, G. J., & He, Q. (2012). Cryptosporidiosis – An overview. *Journal of Biomedical Research, 25*, 1–16. https://doi.org/10.1016/S1674-8301(11)60001-8

Li, Z., Wang, Y., Wang, L., & Zhou, Z. (2018). Molecular and biochemical responses in the midgut of the silkworm, Bombyx mori, infected with Nosema bombycis. *Parasites & Vectors, 11*, 147. https://doi.org/10.1186/s13071-018-2755-2

Lom, J. (1972). On the structure of the extruded microsporidian polar filament. *Zeitschrift Parasitenkunde, 38*, 200–213. https://doi.org/10.1007/BF00329598

Martín-Escolano, J., Marín, C., Rosales, M. J., et al. (2022). An updated view of the *Trypanosoma cruzi* life cycle: Intervention points for an effective treatment. *ACS Infectious Diseases, 8*, 1107–1115. https://doi.org/10.1021/acsinfecdis.2c00123

Melnikov, S. V., Manakongtreecheep, K., Rivera, K. D., et al. (2018). Muller's ratchet and ribosome degeneration in the obligate intracellular parasites microsporidia. *International Journal of Molecular Sciences, 19*, 4125. https://doi.org/10.3390/ijms19124125

Mirzayee, A. R., Fasehi, Z., Sapi, R. M., et al. (2024). Mechanism of life cycle and transmission of Babesia. *British Journal of Biology Studies, 4*, 19–28. https://doi.org/10.32996/bjbs.2024.4.2.3

Nakjang, S., Williams, T. A., Heinz, E., et al. (2013). Reduction and expansion in microsporidian genome evolution: New insights from comparative genomics. *Genome Biology and Evolution, 5*, 2285–2303. https://doi.org/10.1093/gbe/evt184

Naranjo-Ortiz, M. A., & Gabaldón, T. (2019). Fungal evolution: Major ecological adaptations and evolutionary transitions. *Biological Reviews, 94*, 1443–1476. https://doi.org/10.1111/brv.12510

Nicholson, D., Salamina, M., Panek, J., et al. (2022). Adaptation to genome decay in the structure of the smallest eukaryotic ribosome. *Nature Communications, 13*, 591. https://doi.org/10.1038/s41467-022-28281-0

Ning, M., Wei, P., Shen, H., et al. (2019). Proteomic and metabolomic responses in hepatopancreas of whiteleg shrimp *Litopenaeus vannamei* infected by microsporidian *Enterocytozoon hepatopenaei*. *Fish & Shellfish Immunology, 87*, 534–545. https://doi.org/10.1016/j.fsi.2019.01.051

Peuvel, I., Delbac, F., Metenier, G., et al. (2000). Polymorphism of the gene encoding a major polar tube protein PTP1 in two microsporidia of the genus encephalitozoon. *Parasitology, 121*. https://doi.org/10.1017/S0031182000006910

Pombert, J.-F., Haag, K. L., Beidas, S., et al. (2015). The Ordospora colligata genome: evolution of extreme reduction in microsporidia and host-to-parasite horizontal gene transfer. *mBio, 6*. https://doi.org/10.1128/mBio.02400-14

Rich, S. M., & Ayala, F. J. (2000). Population structure and recent evolution of *Plasmodium falciparum*. *Proceedings of the National Academy of Sciences, 97*, 6994–7001. https://doi.org/10.1073/pnas.97.13.6994

Smith, J. E. (2009). The ecology and evolution of microsporidian parasites. *Parasitology, 136*, 1901–1914. https://doi.org/10.1017/S0031182009991818

Solter, L. F. (2006). Transmission as a predictor of ecological host specificity with a focus on vertical transmission of microsporidia. *Journal of Invertebrate Pathology, 92*, 132–140. https://doi.org/10.1016/j.jip.2006.03.008

Southern, T. R., Jolly, C. E., Lester, M. E., & Hayman, J. R. (2007). EnP1, a microsporidian spore wall protein that enables spores to adhere to and infect host cells in vitro. *Eukaryotic Cell, 6*, 1354–1362. https://doi.org/10.1128/EC.00113-07

Tamim El Jarkass, H., & Reinke, A. W. (2020). The ins and outs of host-microsporidia interactions during invasion, proliferation and exit. *Cellular Microbiology, 22*. https://doi.org/10.1111/cmi.13247

Tanifuji, G., Cenci, U., Moog, D., et al. (2017). Genome sequencing reveals metabolic and cellular interdependence in an amoeba-kinetoplastid symbiosis. *Scientific Reports, 7*, 11688. https://doi.org/10.1038/s41598-017-11866-x

Teixeira, D. E., Benchimol, M., Rodrigues, J. C. F., et al. (2013). The cell biology of Leishmania: How to teach using animations. *PLoS Pathogens, 9*, e1003594. https://doi.org/10.1371/journal.ppat.1003594

Terry, R. S., Smith, J. E., Sharpe, R. G., et al. (2004). Widespread vertical transmission and associated host sex–ratio distortion within the eukaryotic phylum Microspora. *Proceedings of the Royal Society of London - Series B: Biological Sciences, 271*, 1783–1789. https://doi.org/10.1098/rspb.2004.2793

Usmani, M., Coudray, N., Riggi, M., et al. (2025). Cryo-ET reveals the in situ architecture of the polar tube invasion apparatus from microsporidian parasites. *Proceedings of the National Academy of Sciences, 122*. https://doi.org/10.1073/pnas.2415233122

Vávra, J., & Lukeš, J. (2013). Microsporidia and 'the art of living together'. *Advances in Parasitology, 82*, 253–319.

Votýpka, J., Modrý, D., Oborník, M., et al. (2016). Apicomplexa. In *Handbook of the protists* (pp. 1–58). Springer.

Wadi, L., & Reinke, A. W. (2020). Evolution of microsporidia: An extremely successful group of eukaryotic intracellular parasites. *PLoS Pathogens, 16*, e1008276. https://doi.org/10.1371/journal.ppat.1008276

Weiss, L. M. (2001). Microsporidia: Emerging pathogenic protists. *Acta Tropica, 78*, 89–102. https://doi.org/10.1016/S0001-706X(00)00178-9

Williams, B. A. P., Hirt, R. P., Lucocq, J. M., & Embley, T. M. (2002). A mitochondrial remnant in the microsporidian *Trachipleistophora hominis*. *Nature, 418*, 865–869. https://doi.org/10.1038/nature00949

Xiao, L., Li, L., Visvesvara, G. S., et al. (2001). Genotyping *Encephalitozoon cuniculi* by multilocus analyses of genes with repetitive sequences. *Journal of Clinical Microbiology, 39*, 2248–2253. https://doi.org/10.1128/JCM.39.6.2248-2253.2001

Xiong, X., Geden, C. J., Tan, Y., et al. (2024). Genome structure, evolution, and host shift of Nosema. *Biology (Basel), 13*, 952. https://doi.org/10.3390/biology13110952

Xu, Y., & Weiss, L. M. (2005). The microsporidian polar tube: A highly specialised invasion organelle. *International Journal for Parasitology, 35*, 941–953. https://doi.org/10.1016/j.ijpara.2005.04.003

Yazaki, E., Ishikawa, S. A., Kume, K., et al. (2017). Global Kinetoplastea phylogeny inferred from a large-scale multigene alignment including parasitic species for better understanding transitions from a free-living to a parasitic lifestyle. *Genes & Genetic Systems, 92*, 35–42. https://doi.org/10.1266/ggs.16-00056

Žárský, V., Karnkowska, A., Boscaro, V., et al. (2023). Contrasting outcomes of genome reduction in mikrocytids and microsporidians. *BMC Biology, 21*, 137. https://doi.org/10.1186/s12915-023-01635-w

Biology and Evolution of Highly Reduced Microsporidia

Susheela Biranjia-Hurdoyal

Abstract Microsporidia are unique, obligate intracellular parasites closely related to fungi which have evolved highly specialized mechanisms to adapt, infect, and survive in various hosts ranging from vertebrates to invertebrates. A hallmark of microsporidia is the extrusion of a polar tube in the infection process. These parasites have evolved and undergone extreme genome reduction resulting in loss of many cellular functions, such as mitochondria-based energy production and major biosynthetic pathways. However, they have retained the essential genes for successful host invasion and energy acquisition from the host cells. The compact small-sized genome and reliance on the host demonstrate a distinct evolutionary path, making microsporidia a fascinating subject in evolutionary biology. This chapter provides an overview of microsporidia: covering their biology, life cycle, and genome reduction with a focus on how these features influence their survival and pathogenicity. Furthermore, the chapter explores the emerging biotechnological potentials of microsporidia, particularly in the field of biocontrol of vectors and promising applications in medical science. Although research in the field presents exciting prospects, several challenges remain, including the poorly understood host range, lack of reliable genetic transformational system, difficulties in culturing, and biosafety concerns that must be addressed to fully harness their potential.

Keywords Microsporidia · Polar tube · Reduced genome · Mitosome · Fungi

---

S. Biranjia-Hurdoyal (✉)
Department of Health Sciences, Faculty of Medicine and Health Sciences,
University of Mauritius, Reduit, Mauritius
e-mail: s.biranjia@uom.ac.mu

A. Gupta (ed.), *Prospects of Fungal Biotechnologies for Livestock Volume 2*,
Fungal Biology, https://doi.org/10.1007/978-3-032-06478-3_2

1 Introduction

Microsporidia are obligate, intracellular, spore-forming fungi that were previously misclassified as protozoa. Now recognized as highly specialized fungal-related eukaryotes, they are characterized by their extreme genome reduction, cellular simplification, sparse biosynthetic pathways, and unique mechanisms. Microsporidia use the highly developed structure called polar tube to inject infectious materials in the host and successfully hijack the energy and nutrients from the latter (Barandun et al., 2019). Microsporidia have a diverse host range, including insects, aquatic invertebrates, terrestrial animals, and human beings (Deplazes et al., 2000; Seatamanoch et al., 2022). Among non-human vertebrates, *Encephalitozoon* cuniculi is most commonly isolated (Lallo et al., 2016).

Infections with microsporidia can lead to ecological disasters, reduced productivity in livestock, and economic losses in aquaculture and farming sectors (Stentiford et al., 2013). In humans, *Enterocytozoon bieneusi* is the most common species, followed by *Encephalitozoon hellem, Encephalitozoon intestinalis,* and *Encephalitozoon cuniculi*. The parasites cause microsporidiosis, an emerging infectious disease, primarily affecting immunosuppressed individuals, such as those with HIV/AIDS, organ recipients, and increasingly, reported in travelers (Didier et al., 2006; Nourrisson et al., 2024; Han & Weiss, 2017; Ruan et al., 2021). Clinical manifestations depend on the species, site of infection, and immune status of the host. Intestinal infections often present with diarrhea, weight loss, and abdominal discomfort, while systemic infections can lead to pneumonia, conjunctivitis, and nephritis, among others (Han et al., 2021; Ruan et al., 2021). Microsporidian spores have been isolated from fecal matter, urine, and animal carcasses, which could potentially contaminate water sources (Moss & Snyder, 2017).

Diagnosing microsporidiosis remains challenging due to non-specific symptoms, the small size of the spores, and the limited availability of reliable molecular diagnostic tools. Treatment options are also limited, with varying efficacy reported for drugs like albendazole and fumagillin (Didier & Weiss, 2011). Microsporidia provide a unique model for studying the minimal genetic requirements for eukaryotic life. Given their wide-ranging impact on human and animal health, agriculture, ecology, and evolutionary biology, this chapter provides a comprehensive overview of microsporidia. Topics covered include their taxonomy, biology, life cycle, genome reduction, mechanisms of host invasion, disease pathogenesis, and emerging applications in biotechnology, including their potential in biocontrol and therapeutic development.

1.1 Taxonomy, Phylogeny, and Life Cycle of Microsporidia

Advances in molecular phylogenetic studies have concluded that microsporidia are more closely related to fungi. The classification of microsporidia has been challenging because of their homoplastic morphology and rapid evolutionary changes

Table 1 Taxonomy of microsporidia

Kingdom	Fungi
Subkingdom	Rozellomyceta
Phylum	Rozellomycota
Class	Microsporidia

(Keeling & McFadden, 1998). Hirt et al. (1997) published one of the pivotal studies that contributed to the reclassification of microsporidia as fungi. Initially, molecular data from conserved proteins, including RNA polymerase II, supported the fungal affinity of microsporidia. Later, molecular studies using ribosomal RNA sequences have contributed to develop phylogenetic trees to better understand microsporidian diversity (Park & Poulin, 2021). Microsporidians, encompassing about 1300 species distributed over more than 220 genera, have been extensively studied to understand their phylogeneticity. A recent study in 2024 concluded that there is still a need to revise the existing classifications within the clades (South et al., 2024). The taxonomy and phylogeny of microsporidia have been revised (Table 1). Phylogenetic analyses indicated that microsporidia are highly specialized fungi which belong to the superphylum Opisthosporidia, and they were divided into four major clades 1, 3, 4, and 5 (Park & Poulin, 2021; Corradi & Keeling, 2009). Bojko et al. (2022) proposed an informal taxonomic framework of six groups: Glugeida (previously clades 5/3), Nosematida (previously clades 4a), Enterocytozoonida (previously clades 4b), Amblyosporida (previously clades 3/5), Neopereziida (previously clades 1), and Ovavesiculida (previously clades 2).

Phylogenomic studies concluded that microsporidia have evolved from a free-living fungi ancestor that has adapted to intracellular parasitism across diverse hosts by undergoing changes at genetic and phylogenetic levels (Keeling et al., 2000).

Life Cycle of Microsporidia

Microsporidia are eukaryotic, unicellular, and spore-forming parasites, which infect both vertebrates and invertebrates (Stentiford et al., 2016). All microsporidia share key life cycle features, including the release of resistant spores in the environment, attachment to new susceptible host, germination marked by the extrusion of the polar tube, and intracellular replications (Vávra & Lukeš, 2013). The highly specialized and unique life cycle of microsporidia has two main phases: the proliferative (merogony) and the spore-forming (sporogony) phases. The life cycle begins with the presence of the infective and resistant spore outside the host. The spore can persist for months or years in the environment due to its thick protective walls (Fig. 1). Once the host has ingested or inhaled the spores, the germination process starts. The spore rapidly protrudes a polar tubule that pierces through the host membrane enabling the injection of the infective sporoplasm directly into the host cytoplasm (Fadhilah et al., 2023). Production of the polar tube during infection is a hallmark feature unique to microsporidia (Fig. 2). The next phase, merogony, involves the transformation of the sporoplasm into meronts, which then proliferate via binary fission or schizogony. This is followed by sporogony, a very crucial stage

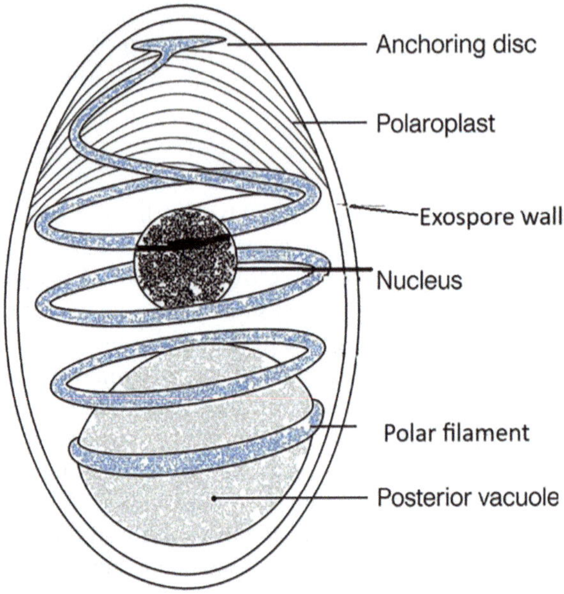

Fig. 1 Diagrammatic representation of a microsporidian spore. (Diagram was created with the help of Perplexity AI)

where the meronts differentiate into sporonts and then sporoblasts, which leads to the formation of new spores with thick chitinous wall and unique polar tubule or polar filaments (Keeling & Fast, 2002). The last stage involves rupturing of the walls of the host cell, dispersing the spores into the surrounding environment. The released mature spores will continue the cycle by infecting new susceptible host (Fig. 3).

Microsporidia exhibit diverse transmission strategies, ranging from horizontal, vertical, and mixed, which facilitate their adaptation to various hosts and environments while enhancing their virulence and evolutionary properties (Stentiford et al., 2013). In vertical transmission, the parasite is transferred from parent to offspring via transovarial or transplacental routes while in horizontal transmission, microsporidia spores are transferred between individuals of the same generation, most commonly, by ingesting contaminated food or water, or direct contact with infected tissues. Lower virulence and high host specificity have been associated with vertical transmission. In horizontal transmission, spores are released in bulk and hence, high parasite burden, increased virulence, and broad host range have been noted (Nattoh et al., 2021; Dunn & Smith, 2001).

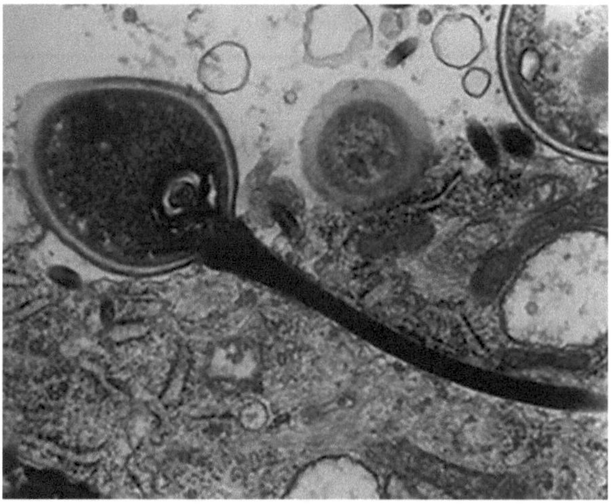

Fig. 2 Transmission electron micrograph of a microsporidian spore with an extruded polar tubule inserted into a eukaryotic cell. The spore injects the infective sporoplasms through its polar tubule. (Source: Figure courtesy of Dr. Massimo Scaglia, Laboratory of Clinical Parasitology, Institute of Infectious Diseases, University-IRCCS San Matteo, Pavia, Italy. Centre for Disease Control., n.d.-a, https://www.cdc.gov/dpdx/microsporidiosis)

1.2 Genomic and Cellular Evolution of Microsporidia

Microsporidia have drawn considerable research interest because of their distinct evolutionary traits and small compact genomes. Microsporidia have evolved to adapt to the reduced genetic material and intracellular parasitic life. The parasite has lost its mitochondrial genes, survives with mitosomes, reduces its glycolytic dependency, and hijacks the host biosynthesis mechanism to proliferate and survive intracellularly (Wiredu Boakye et al., 2017; Burri et al., 2006). The genome of microsporidia is very concise (Table 2). Non-coding DNA and gene density account for the difference between the genomic sizes. The microsporidia *Glugea atherinae* has more non-coding sequences and reduced gene density (Williams et al., 2008). Compared to *Glugea atherinae,* the free-living yeast *Saccharomyces cerevisiae* has a smaller genome size but more genes. *Saccharomyces cerevisiae* has a functional glycolytic pathway, generates its own adenosine triphosphate (ATP) and metabolizes carbohydrates, while *Glugea atherinae* features a reduced glycolytic pathway, typical of microsporidia, depending on its host for ATP and metabolites as a result of significant gene loss (Wang et al., 2024).

Genomic studies reported that microsporidia have retained the genes with greater functional significance compared to those which have been lost. Some microsporidia have very few introns while others have none, decreasing the need for a complex spliceosome (Angst et al., 2023). Microsporidian-specific protein families, which are crucial for the survival of the parasites intracellularly, have been conserved across all microsporidian genomes. The parasite has core set genes which are

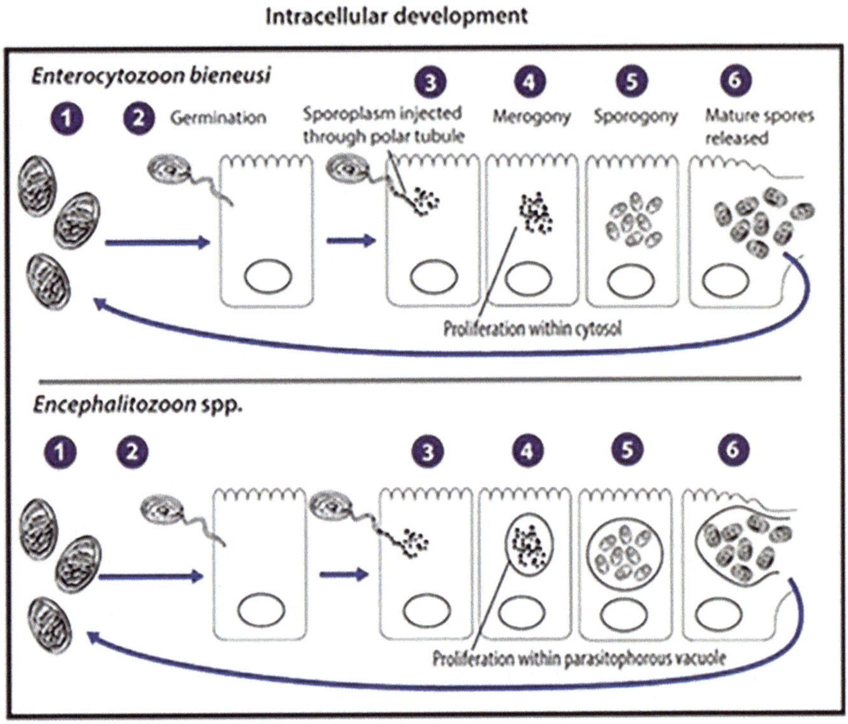

Fig. 3 Life cycle of microsporidia (Source: Centre for Disease Control, n.d.-b. Available from https://www.cdc.gov/dpdx/microsporidiosis). The infective form of microsporidia is the resistant spore, which survives in the environment for months (*1*) The spore then germinates, rapidly everting its polar tubule which contacts the host cell membrane (*2*). Through the polar tubule, the spore injects the infective sporoplasm into the host cell (*3*). Inside the cell, the sporoplasm enters the proliferative phase characterized by extensive multiplication through merogony, creating meronts (*4*). The intracellular location of this developmental stage within the host cell varies by genus; host cell cytosol (*Enterocytozoon, Nosema*), within a parasitophorous vacuole of unknown origin (Encephalitozoon), inside a parasite-secreted envelope (*Pleistophora, Trachipleistophora*), or surrounded by the host cell endoplasmic reticulum (*Endoreticulatus, Vittaforma*) (*5*). Following proliferation, meronts undergo sporogony, during which thick spore wall and invasion apparatus are formed, leading to the development of sporonts and eventually mature spores. As the number of spores increases and fills the host cytoplasm, the host cell membrane ruptures, releasing the spores into the surrounding environment. These mature spores can then infect new cells, perpetuating the cycle (*6*). These mature spores can then infect new cells, perpetuating the cycle

involved in basic cellular functions such as DNA replication, repair, protein folding, intracellular trafficking, and key enzymes such as those involved in glycolysis, trehalose metabolism, pentose phosphate pathway, and synthesis of structural components of the fungus. Microsporidia have an extremely reduced metabolism and do not synthesize fundamental molecules such as sugars, lipids, amino acids, ATP, and NADP+. The biological building blocks are acquired from the host by exploiting the wide range of cell surface transporters (Nakjang et al., 2013).

Table 2 Comparison of genome size from selected microsporidial species compared to *Saccharomyces cerevisiae*

Fungi	Genome size estimate (Mbp)	Number of genes	Gene density (kbp)	Host	References
Encephalitozoon intestinalis	2.3	1833	0.86	Mammals (human)	Peyretaillade et al. (1998) and Corradi et al. (2010)
Encephalitozoon cuniculi	2.9	1999	0.84	Mammals (human)	Biderre et al. (1995)
Nosema ceranae	8.8	2614	0.60	Honeybee mid-gut epithelial cells	Huang et al. (2021) and Cornman et al. (2009)
Glugea atherinae	19.5	2000–2500	[a]	Fish	Biderre et al. (1994)
Saccharomyces cerevisiae	12.1	~ 6000	0.5	Not a typical pathogen Used as a model eukaryotic cell	Wang et al. (2024)

[a]The exact value has not been reported. It is expected to be lower than the other microsporidians

2 Cellular Simplification

Microsporidia exhibit extreme cellular simplification, which are reflected by the reduction in genomic size, loss or modifications of key organelles, and extensive reduction in biosynthetic pathways. This simplification is driven by their obligate intracellular parasitic lifestyle, where dependence on the host cell for metabolites and energy has allowed them to discard many genes and cellular structures unnecessary for survival outside the host. One of the most notable features is the loss of typical eukaryotic organelles such as peroxisomes and the reduction of mitochondria into highly simplified forms called mitosomes (Williams & Keeling, 2011). Although a complex Golgi is unnecessary since most proteins are acquired from the host, a simplified Golgi remains present, especially during sporogony, the stage when essential structures like the spore wall and polar filament are assembled (Williams & Keeling, 2011). Furthermore, ribosomes have been detected as a major component of the cytoplasm indicating a high rate of protein synthesis necessary to sustain their rapid replication cycle within the host cells (Vávra & Lukeš, 2013). This adaptation enables them to efficiently produce the proteins required for growth and proliferation during infection.

The parasite lacks the necessary genes for biosynthetic pathways and hence depends on its host (Peyretaillade et al., 1998; Burri et al., 2006). Mitosome has some similar functions as mitochondria such as the assembly of the iron-sulfur cluster but lost the ability to produce ATP through oxidative phosphorylation or Kreb's cycle (Katinka et al., 2001). The organelle does not have genetic material, but several proteins have been identified in its cytosol, inner and outer membranes. The proteins are involved in metabolites import, iron-sulfur assembly, and export. Furthermore, enzymes for metabolic activities and superoxide dismutase for protection against oxidative stress have also been detected. Once the intracellular parasite starts replicating, the proteins are translocated from the host to ensure proper functioning of the mitosomes. The organelle has retained a simplified form of the mitochondrial protein import system which are responsible for the translocation of proteins, based on minimal recognition of specific peptide sequences (Burri et al., 2006). Despite the extreme reduction of the genome during evolution, microsporidian mitosome retained the iron-sulfur cluster, indicating its significant importance in cytosolic functions and emphasizing the need to conserve the crucial mitochondrial activity (Freibert et al., 2017). The iron-sulfur protein has a key role as a cofactor in the main processes of life, including DNA replication, DNA repair, and several enzymatic reactions (Paul & Lill, 2015).

3 Survival Mechanisms with Reduced Genomes

Genome reduction is a crucial evolutionary strategy for the survival of the intracellular parasitic microsporidia. Studies have concluded that microsporidia have shed non-vital genes and streamlined their metabolic processes to survive in diverse ecological niches. The extreme genome reduction has caused the fungi to be highly reliant on their host resources. Microsporidia save their energy outflow by having limited genes involved in complex biosynthesis such as energy production pathways, amino acids, and lipids synthesis (Akiyoshi et al., 2009). Microsporidia have lost the majority of enzymes required for nucleotide biosynthesis and hence, import the nucleotides from the host. They remain inactive in environment deficient of nutrients. *Encephalitozoon cuniculi* was found to have ATP/ADP translocases or nucleotide transporters on its surfaces, indicating the acquisition of ATP from the host rather than producing it (Tsaousis et al., 2008). RNA interference studies on ATP/ADP transporters in *Nosema ceranae* showed a decrease in parasite load in infected honeybees (Paldi et al., 2010).

The loss of redundant genes, minimization of introns, and shortening of gene length have been reported to be advantageous to the parasite. Bigger genome size would need more nutrients for its replications and more time to generate progenies. The rapid replication of the parasites contributes to the depletion of the available host resources and hijacks the system (Jespersen et al., 2022). Furthermore, immune evasion is aided by the short replication cycle and reduced genome. A shorter replication cycle would ensure the replication is complete before the immune responses get activated and a small genome will code for a limited number of antigenic determinants for the surface of the parasite (Nakjang et al., 2013).

Some microsporidia have selectively expanded their genome by gene duplication and acquired genes from their hosts by horizontal gene transfer to compensate for the reduced genome and enhanced adaptation to intracellular life. The majority of the uptake includes genes involved in DNA replication, host cell invasion, nutrient uptake, and stress defense (Nakjang et al., 2013; Pombert et al., 2015; Heinz et al., 2012). Transporters and nucleotide transporter genes have also been extensively studied as expansions of the microsporidia genome. These transporters compensate for the loss of specific functions due to genome reduction by translocating essential components such as nucleotides, ATP, sugars, and other important metabolites from the host (Nakjang et al., 2013). An increase in the diversity of transporters would lead to the uptake of a wide range of nutrients and metabolites required for the intracellular parasites (Cuomo et al., 2012). *Ordospora colligate*, a parasite of the crustacean *Daphnia* sp., has acquired the septin gene by horizontal gene transfer. The protein is similar to septin 7, which could trigger endocytosis of the parasite and hence, increasing the likelihood of infection (Pombert et al., 2015). The mechanosensitive ion channels of small conductance (MscS 2) of *Nematocida displodere* were acquired by horizontal gene transfer. The channels are integral part of the membranes and control the flow of ions across the membranes in response to changes in osmotic pressure and hence, contribute to homeostasis. The mscS2 were significantly expressed during the proliferative stages and moderately produced during sporogony and sporulation (Berg et al., 2024).

4 Host Interactions

Microsporidia have evolved and adapted to survive as obligate intracellular parasites in a wide range of hosts. In human beings, immunocompromised individuals are mostly affected while among animals, the fishery sectors, silkworm, and honeybees are known to suffer significantly (Stentiford et al., 2016; Huang et al., 2018). The fungi are economically, ecologically, and medically important as they cause diseases in animals which are crucial in the food chain. *Nosema* spp. infect honeybee colonies which are major factor in food security (Galajda et al., 2021). Among invertebrates, microsporidia could cause deformations, increase mortality rates, and decrease food intake, fertility, and lifespan. Furthermore, microsporian could switch host to adapt to evolution. Historically, the crustacean *Penaeus monodon* was infected at low prevalence by the microsporidia *Enterocytozoon hepatopenaeii*. With time, Asia produced more of the congeneric penaeid *Penaeus vannamei* and it was found that *Enterocytozoon hepatopenaeii* switched host to *Penaeus vannamei* with more prevalent and high intensity infections (Stentifort, 2013). *Pleistophora* spp. cause economic loss in aquaculture by affecting fish meat quality and mortality (Winters et al., 2016). Furthermore, with the increasing world population and potential use of insects as proteins, it is important to study the microsporidians in these mass-reared colonies to avoid the transmission to human via the food chain (Bjørnson & Oi, 2014).

4.1 Unique Mechanisms to Invade and Survive in the Host Cells

Microsporidia use unique invasion strategies to successfully invade, infect, and survive in their host cells. These distinctive and specialized mechanisms are important for exploiting the host cellular functions and evading the immune responses. The initial phase of infection typically starts when the infective spores attach to the host cells or nearby surfaces (Han et al., 2017). Microsporidia have proteins on their spore surface and use an array of secretory effectors to manipulate the host environment, facilitating their invasion and proliferation (Ran et al., 2024; Han et al., 2019). The polar tube plays a central role in microsporidian infection by delivering the parasite's contents into the host.

Spore Wall Proteins

Spore wall proteins (SWPs) are believed to initiate the initial contacts. In species such as *Nosema bombycis*, *Encephalitozoon* spp., and *Antonospora locustae*, certain SWPs have demonstrated binding affinity for host cell constituents, including heparin-binding motifs and sulfated glycosaminoglycans, facilitating attachment (Southern et al., 2007; Han et al., 2020). Several SWPs—namely SWP7, SWP9, SWP11, SWP12, SWP16, and SWP26—have been implicated in promoting spore adherence and enhancing infectivity (Tamim El Jarkass & Reinke, 2020). The protein EhSWP1 of *Enterocytozoon hepatopenaei* demonstrated specific affinity for heparin on host surface and hence, most probably contributed to the attachment process (Jaroenlak et al., 2018). Additionally, research on *N. bombycis* revealed that NbSWP26 could engage with the silkworm *Bombyx mori's* turtle-like protein (BmTLP), suggesting a role in host–parasite recognition and potential cell-to-cell signaling (Zhu et al., 2013).

Secretory Effectors

Several classes of effector molecules have been associated with the invasive strategies of microsporidia. Among them, Ricin-B-lectins are suspected to assist in early-stage adherence to host tissues and serpin-family proteins may interfere with host immune signaling, reducing the likelihood of immune detection (Ran et al., 2024). In *Nosema bombycis*, certain lectins (NbRBLs) have been shown to bind to host glycan structures, which could enhance spore attachment and facilitate entry into the host. One such protein, NbRBL28, has been localized within the host cell nucleus, where it is believed to influence the expression of genes important for cell cycle regulation, potentially creating a more favorable environment for parasite proliferation (Xu et al., 2022).

Serpins, also known as serine protease inhibitors, could inhibit the host serine proteases involved in immune signaling pathways and hence, suppress inflammatory responses and prevent the activation of immune effector mechanisms (Wang et al., 2019). Proteins with serpin domains have been identified in the secretomes of *Nosema* species, particularly *Nosema bombycis*, with the highest expression during the key phases of host invasion and intracellular development (Ran et al., 2023).

Moreover, hexokinase in microsporidia plays a dual role in infection, combining both metabolic and non-metabolic function capacities to manipulate host cells and aid in infection. Hexokinase could alter the host cell metabolism to ensure the parasite's nutritional needs are met. Hexokinase has also been identified as a constituent of the polar tube and secretome in various microsporidia, indicating its participation in invasion and host-parasite interaction. The secretory protein hexokinase of *Nosema bombycis* (NbHK) was detected in both the cytoplasm and nucleus of host cells during infection. Previous studies confirmed NbHK is secreted into host cells, with its localization varying according to the parasite's developmental stage. Silencing the NbHK suppressed the parasite proliferation and hence, highlighting its essential role in manipulating the host to ensure the survival of microsporidia (Huang et al., 2018). Hexokinase is also the first enzyme required for both glycolysis and pentose phosphate pathway. The microsporidian hexokinase could accelerate the host biosynthetic pathways for synthesis of amino acids, lipids, and nucleotides, important for its growth (Cuomo et al., 2012).

Polar Tube

Upon appropriate environmental stimulation, the spore germinates, and the coiled polar filament is projected from the spore to deliver the infective sporoplasm directly in the cytoplasm of the host (Franzen, 2008). Factors such as pH, cations, and anions have been reported to initiate the process of spore germination. The structure of the polar tube is adapted for its functions as studies have reported that it curls into a hook at its tip and consists of multilayered fibrils. The ridged surface allows an increase in diameter during the passage of the sporoplams (Takvorian et al., 2020). Seven polar tube proteins (PTPs) have been described (PTP1 to PTP7). Their positions along the propelled polar tube also vary, indicating that their functions also differ. PTP1, the major protein, has a high proline content, which may enhance the tube tensile strength and elasticity, which would be key properties required for the passage and discharge of the sporoplasm (Han et al., 2022). PTP3 acts as a scaffolding protein structure and interacts with PTP1 and PTP2 (Peuvel et al., 2002). The PTP also binds to specific receptors on the host cell surface to initiate the invasion process. PTP1 could bind to host mannose receptors (Bouzahzah & Weiss, 2010). Transferrin 1, a protein cell receptor, was found to interact with PTP4, and interference with the association led to decreased ability of *Encephalitozoon hellem* to infect susceptible hosts (Han et al., 2017). PTP4 and PTP7 have been detected at the tip of polar tubes, suggesting their possible roles in recognizing specific receptors to attach to the host cell surface (Fayet et al., 2024). The PTP6 from *Nosema bombycis* was detected on the entire polar tube and showed O-glycosylation and N-glycosylation sites, which could have cell-binding properties (Lv et al., 2024). Rodríguez-García et al. (2018) used RNA interference to suppress the expression of the ptp3 gene of *Nosema cerenae,* which is known to infect European honeybees, *Apis mellifera.* A reduction in the infection load and an improvement in the lifespan of the bees were noted. This indicates the importance of the polar tube proteins.

5 Host Manipulation and Immune Invasion

Microsporidia initiate infection through the rapid extrusion of the polar tube, a mechanism that allows them to bypass host defenses and deliver their infectious material directly into host cells (Jaroenlak et al., 2020). Once inside the host, microsporidia use several strategies to manipulate the host cell and immune defenses for their intracellular survival. Some microsporidia such as *Pleistophora hyphessobryconis*, *Nosema ceranae, Nosema bombycis*, *Loma* spp. and *Paranucleospora theridion* form xenomas, which are multinucleated host cells that serve as a dedicated place for microsporidian proliferation and development. The xenomas provide an enclosed protective environment which supplies the fungi with nutrients and contains lots of spores at various developmental stages (Franzen, 2005). Meronts and mature spores are released from the xenomas and contribute to the spreading of the parasite. Microsporidia group has the capability to manipulate the physiology of host cells to promote their own survival. They could alter histone modifications within host cells, such as interrupting H_2B monoubiquitination. This interruption impacts p53 expression levels and inhibits host ferroptosis, ultimately aiding by hindering the iron-dependent programmed cell death and hence, supporting the proliferation of the parasites (Weidner & Sibley, 1985; Franzen, 2005).

Both innate and adaptive immune responses are crucial in defending against infections caused by microsporidians. Some of the spores could inhibit phagocytosis by preventing phagosome maturation by blocking the acidification of the phagosome and merging with the secondary lysosomes (Weidner & Sibley, 1985). Some sporoplasms could evade from the developing lysosomes to the cytoplasm of the host cells (Franzen, 2005). Thus, evading the degradative environment and killing step inside the phagosomes. Furthermore, viable microsporidian spores could reduce the production of reactive oxygen species and reactive nitrogen intermediates within macrophages. Thus, reducing the ability of the macrophage to destroy ingested spores, thereby aiding the survival of the parasite (Leiro et al., 2001). Additionally, in response to the presence of the pathogen, macrophages release chemokines including CCL3 and CCL4 via the TLR2-NF-κB signaling pathway to attract more monocytes and macrophages. Microsporidia hijack the recruitment and infect the newly recruited immune cells (Leitch & Ceballos, 2008).

6 Diagnosis and Cultivation of Microsporidia

Microsporidia have been increasingly reported to cause infections in animals and human beings. The isolation of the parasite and its culture in the laboratory have proved to be challenging because of the unique intracellular life cycle. Previously, the diagnosis of microsporidia was mainly by light microscopy and the understanding of microsporidiosis was limited due to challenges with detection of the spores (Moniot et al., 2022; Garcia, 2002; Fayer & Santin-Duran, 2014. Recently, immunofluorescence, molecular methods, and cell culture have been increasingly used as the results are more sensitive and some of the tests provide species identification which is required to decide the course of treatment (Garcia, 2002; Subrungruang et al., 2004; Lallo et al., 2016).

Nosema bombycis and *Encephalitozoon cuniculi* were the first microsporidians to be cultured in vitro. Initially, the interest for in vitro culture was limited to microsporidians infecting insects and of economic importance. After the advent of HIV and AIDS, research has been intensified on the culture of microsporidia infecting HIV-positive and immunocompromised individuals (Trager, 1937; Lallo et al., 2016). Several cell lines, mammalian and non-mammalian, have been used to culture microsporidia. They include monkey kidney (E6), Madin-Darby canine kidney (MDCK), human foreskin fibroblasts (HFF), murine L929 cells, human lung fibroblasts (HLF), lung fibroblasts (MRC-5), rabbit kidney (RK-13), human cervical cancer cells (HeLa), fetal bovine lung fibroblasts (FBF), and monocytes (MDM), among others (del Aguila et al., 2001; Visvesvara, 2002; Joseph & Sharma, 2009). The effectiveness of the cell lines was found to depend on the species being cultured. Vero cell line was better for *Encephalitozoon* species while *Vittaforma corneae* grew more effectively using rabbit corneal cell line. The use of HeLa cells was hindered by rapid cell breakdown and challenges in distinguishing spores from cell debris (Joseph & Sharma, 2009). Using in vitro method, Wolk et al. (2000) studied the effects of several antimicrobial agents on microsporidia such as *E. cuniculi*, *E. hellem*, and *E. intestinalis*.

Molecular diagnosis of microsporidia from clinical samples has not always been successful because of limited reliability and the need to revise the protocols for commercial kits for molecular tests (Çetinkaya et al., 2016, 2018). The chitinous spore wall of microsporidia was reportedly hindering efficient DNA extraction, and Çetinkaya et al. (2018) recommended additional pre-treatments of the DNA extraction to reduce false negative results. There is a need for diagnostic kits with enhanced diagnostic accuracy to support disease diagnosis and epidemiological surveillance. Furthermore, direct detection of clinical samples by histological examination has low sensitivities if the spore counts were low or sparsely distributed within the tissues (Weiss & Becnel, 2014). However, once the microsporidia were successfully propagated in cell cultures, the spores could be stained and detected by light microscopy or immunofluorescence (Visvesvara, 2002).

He et al. (2020) achieved a significant breakthrough by successfully isolating the sporoplasms of *Nosema bombycis* and sustaining them in vitro, in a cell-free medium for up to 5 days. The study reported key biological processes such as ATP depletion and DNA replication in the cell-free medium. The ability to survive outside a living cell presents a crucial opportunity for in-depth exploration of microsporidian physiology, pathogenicity, and genetics. The cultivation of microsporidia is challenging but vital for the research domain. The creation of standardized cultivation protocols will be crucial for advancing both the comprehension and management of these important pathogens.

7 Biotechnological Applications of Genome Reduction

With the progress in the field of biotechnology and the growing number of microsporidia genome being sequenced, it is an opportunity to study and understand the evolutionary biology in detail and propose novel avenues in the field. The highly reduced genome and unique biology of microsporidia render them potential agents for biotechnological and synthetic biology applications. These prospects arise due to their efficient metabolic routes, remarkably condensed genomes, and unique mechanisms for interacting with hosts. Microsporidia are dependent on the host for survival and hence, they are highly host-specific and interact at lesser extent with nonspecific host. Given the genome is reduced, genomics studies focus and manipulate lesser genes making them more genetically tractable. Genetic modification could be interpreted more efficiently. Several biotechnological methods have been developed and are being used for genomic sequencing and manipulation. The most reported ones are Next-generation sequencing, CRISPR genome editing, RNA-sequencing, RNA interference (RNAi) and gene silencing, synthetic biology, transcriptomics, and the use of proteomics and secretomics (Table 3).

Potential Model for Biotechnological Applications

Microsporidia have the smallest known eukaryotic genome and, hence, can be a potential model for studying genes and their functions. It can be useful in synthetic

Table 3 Biotechnology methods and their applications

Biotechnology methods	Applications	Impact	References
Next-generation sequencing (NGS)	Rapid sequencing of large genome	Provide information of genomic size, complexity, or simplicity of the genome	Zhong et al. (2021)
Clustered regularly interspaced short palindromic repeats (CRISPR) tools	Modify the genome	Knock out or modify genes to investigate their functions	Loureiro and da Silva (2019)
Transcriptomics (RNA-seq)	Measure gene expression	Identify level of gene expression under different conditions and development stages	Du et al. (2023)
Proteomics and secretomics	Protein profile known using mass spectrometry	Identification of important proteins for different processes such as host-parasite attachment receptors.	Knecht et al. (2023)
Synthetic biology	Create engineered yeast or bacteria with incorporated new genes	Functionally characterize genes	Yan et al. (2023)
RNA interference (RNAi) and gene silencing	Knockdown of specific genes	Study the effect due to their lack of activity	Rodríguez-García et al. (2018)

biology to redesign the genome with novel genes coding for bioactive molecules. Furthermore, experiments on horizontal gene transfer could be carried out to acquire new genes for metabolic enzymes, novel transporters for drug delivery, and others.

The RNAi has proved to be effective in silencing several genes in microsporidia. Dong et al. (2025) showed that the ADP/ATP carrier 1 protein of *Nosema bombycis* could be inhibited by RNAi, which ultimately reduced the proliferation of the fungi in its host, silkworm (*Bombyx mori*). When the genome is smaller, the effects of RNAi can be studied more effectively without the interference of other cofounding variables.

Models for Host-Parasite Interactions

One of the smallest eukaryotic genome belongs to microsporidia, making them an interesting model for genomic studies. *Octosporea bayeri*, a microsporidian with an estimated size of 24 Mb and its exclusive host, the crustacean *Daphnia magna,* have been studied as excellent models for host-parasite interactions. The association between the duo helps to understand the negative frequency-dependent selection in host-parasite coevolution. Parasites infecting common host have a selective advantage as the host is more frequently infected and the host genotypes decline. The rare host genotypes are infected at lesser frequency as the parasites were not adapted to them. However, over time, the parasites evolve to target the host, and this creates a dynamic cycle of adaptation. The process is also referred as Red Queen dynamics means species have to evolve and adapt constantly in response to the changing environment and opponents (Ebert, 2008). Studies on breeding have confirmed the principle of negative frequency-dependent selection. During the life cycle of the infected female *Daphnia magna,* all asexual eggs get passed on with the parasite while sexually produced eggs get infected at lesser extent. When the offspring mate, those that undergo selfing sexual reproduction tend to have more infected offspring as compared to those that undergo outcrossing. Given that outcrossing would produce offspring with new genotypes, it brings the advantage of having a rare genotype and hence, less frequently infected (Ebert et al., 2007). The concept has potential for biotechnological applications in the fields related to agriculture, genetic engineering, and synthetic biology. New crop varieties and bacterial communities with rare genotypes can be designed to limit infections by pathogens and improve the outcome.

Use as Biological Control Agents in Vector Management

Previous studies have reported the successful use of engineer symbiotic bacteria that produced different types of effector molecules to hinder the development of *Plasmodium,* the causative agent of malaria, in the midgut of mosquitoes (Herren et al., 2020; Wang et al., 2012; Shane et al., 2018). A similar approach could be used for microsporidia, which also infect the malaria vectors. *Microsporidia MB* are known to infect the mosquito *Anopheles arabiensis*, which is a very important malaria vector (Tullu et al., 2022). During coinfection by *Microsporidia MB* and *Plasmodium species*, *Microsporidia MB* could hinder the colonization in salivary glands and development of the *Plasmodium* species in the midgut, without affecting the longevity of the mosquitoes (Herren et al., 2020). In *Anopheles gambiae,*

Nosema stegomyiae damaged the midgut which disrupted the development of *Plasmodium* binding (Fox & Weiser, 1959). In infected *Anopheles arabiensis*, midgut damage was associated with increase in serine proteases, antimicrobial peptides cecropin-2, antryp-4, and antryp-1 (Herren et al., 2020). Upregulating these genes by biotechnology would be an appropriate biocontrol method. Reduced genome tends to be more stable and robust, qualities required for engineered biocontrol applications. Less mutation and antigenic drift would ensure that the genome of *Microsporidia MB* remains stable across generations and there would not be accidental infection of other hosts, making them safe agents to be released in the ecosystem. The biotechnological solution would be a preferred one compared to chemical insecticides which are known to cause environment pollution and have associated risk of resistance among the mosquitoes. Furthermore, the biocontrol methods cause minimal ecological damage as the mosquitoes remained unharmed. *Microsporidia* MB were transmitted both vertically and horizontally, offering a self-sustaining and long-term solution.

Potential applications would be in the development of engineered *Microsporidia MB* that produced specific effector molecules to prevent *Plasmodium* from developing within the mosquito host. Furthermore, RNAi or CRISPR could be used for gene silencing or gene editing respectively. *Microsporidia MB* could be used as a delivery vehicle to silence or edit the crucial mosquito genes which contribute to *Plasmodium* attachment, development, and reproduction.

Exploring Parts of Microsporidia for Drug Targets and Delivery

Microsporidia genomes encode multiple transporter gene families to compensate for their extensive loss of biosynthetic pathways. These transporters, located on the plasma membrane, facilitate hijacking and translocating metabolites and energy substrate from the host to the parasite (Nakjang et al., 2013). Examples include nucleotide and sugar transporters that are critical for microsporidia survival. These transporters are indispensable for nutrient acquisition, and represent promising drug targets. The RNA interference technology presents a novel strategy to specifically knock down these transporter genes and disrupt the supply of energy and metabolites, which would ultimately hinder the developmental and replicative phases of the microsporidia.

During infection, the extruded polar tube ejects the sporoplasm directly in the host cytoplasm under appropriate environmental conditions. Proteomic analyses have identified various proteins as integral part of the polar tube (Han & Weiss, 2017). This unique infection mechanism opens the possibility of using the polar tube as a biotechnological tool for targeted drug delivery, whereby therapeutic agents could be transported directly into infected cells, increasing treatment specificity and efficacy.

8 Challenges and Future Directions

The main limitation in developing the efficient biotechnological solution is the non-availability of a reliable genetic transformation system for microsporidia. There is a limited number of well-established protocols available for the study of microsporidia (Didier & Weiss, 2008). Given they are obligate intracellular parasites, they can only be grown in living host which is a hurdle for genetic manipulation (Fu et al., 2024; Reinke & Troemel, 2015). Outside the *Encephalitozoon* genus, few species can be cultured in vitro, as most require propagation within live animal hosts. The tough resistant spores inhibit the uptake of DNA and require reagents for the studies (Reinke & Troemel, 2015). The presence of specialized intracellular environments such as the xenomas or parasitophorous vacuoles also hinders studies on host-parasite interactions. The compact genome and intracellular life cycle are challenges for genetic tools. Genetic engineering requires promoters and markers which are normally absent in the genome. Furthermore, the extensive horizontal gene transfer is a confounding factor in the phylogenetic analyses. Microsporidia infect both humans and animals. Biotechnological applications of microsporidia will have some biosafety risks to consider, especially if genetically modified systems are released for therapeutic needs or ecological purposes. Regulatory framework must be established to ensure safety and avoid ecological damage. Once the barriers have been overcome, synthetic biology and biotechnologies could be applied to microsporidia efficiently.

The recent progress in the field of biotechnology has created new opportunities in the study of microsporidia. Horizontal gene transfer, RNAi, genetic engineering, and synthetic biology could offer new insights into the understanding of the small genome and their possible use in therapeutics, disease control, and animal welfare.

9 Conclusion

Microsporidia are renowned for their highly reduced genome and serve as significant models in evolutionary biology. These obligate intracellular parasites employ unique and multifaceted mechanisms to successfully infect host cells. A hallmark of their adaptability is their dual strategy: direct invasion via the polar tube and evasion of host immune responses, including phagocytosis, allowing them to persist across a wide range of hosts. Progress in microsporidia cultivation methods has created novel research opportunities, especially for comprehending their biology. Recent advances in genomics and the sequencing of multiple microsporidia have underscored their value in understanding the minimal gene set required for eukaryotic life. These studies also offered new insight into various proteins, transporters, and polar tube proteins, which play critical role in interaction and pathogenesis. Beyond their relevance to infectious disease, microsporidia are increasingly recognized for their biotechnological potential. The discoveries at cellular levels have opened new avenues for the development of novel drug targets and delivery systems. Despite the existing challenges in culturing and limitations in genetic manipulation, there is an urgent need to establish standardized protocols to facilitate research. Nonetheless, microsporidia hold considerable promise in the field of vector control, targeted gene therapy, and vaccine delivery platforms.

References

Akiyoshi, D. E., Morrison, H. G., Lei, S., Feng, X., Zhang, Q., Corradi, N., Mayanja, H., Tumwine, J. K., Keeling, P. J., Weiss, L. M., & Tzipori, S. (2009). Genomic survey of the non-cultivatable opportunistic human pathogen, *Enterocytozoon bieneusi*. *PLoS Pathogens, 5*(1), e1000261. https://doi.org/10.1371/journal.ppat.1000261

Angst, P., Pombert, J.-F., Ebert, D., & Fields, P. D. (2023). Near chromosome–level genome assembly of the microsporidium *Hamiltosporidium tvaerminnensis*. *G3: Genes|Genomes|Genetics, 13*(10), jkad185. https://doi.org/10.1093/g3journal/jkad185

Barandun, J., Hunziker, M., Vossbrinck, C. R., & Klinge, S. (2019). Evolutionary compaction and adaptation visualized by the structure of the dormant microsporidian ribosome. *Nature Microbiology, 4*(11), 1798–1804. https://doi.org/10.1038/s41564-019-0514-6

Berg, A., Berntsson, R. P., & Barandun, J. (2024). *Nematocida displodere* mechanosensitive ion channel of small conductance 2 assembles into a unique 6-channel super-structure *in vitro*. *PLoS One, 19*(7), e0301951. https://doi.org/10.1371/journal.pone.0301951

Biderre, C., Metenier, G., David, D., Bata, J., Prensier, G., & Vivares, C. P. (1994). On small genomes in eukaryotic organisms: Molecular karyotypes of two microsporidian species (protozoa) parasites of vertebrates. *Comptes Rendus de l'Académie des Sciences. Série III, Sciences de la Vie, 317*(5), 399–404.

Biderre, C., Pagès, M., Méténier, G., Canning, E. U., & Vivares, C. P. (1995). Evidence for the smallest nuclear genome (2.9 Mb) in the microsporidium *Encephalitozoon cuniculi*. *Molecular and Biochemical Parasitology, 74*(2), 229–231.

Bjørnson, S., & Oi, D. (2014). Microsporidia biological control agents and pathogens of beneficial insects. In L. M. Weiss & J. J. Becnel (Eds.), *Microsporidia*. https://doi.org/10.1002/9781118395264.ch25

Bojko, J., Reinke, A. W., Stentiford, G. D., Williams, B., Rogers, M. S. J., & Bass, D. (2022). Microsporidia: A new taxonomic, evolutionary, and ecological synthesis. *Trends in Parasitology, 38*(8), 642–659. https://doi.org/10.1016/j.pt.2022.05.007

Bouzahzah, B., & Weiss, L. M. (2010). Glycosylation of the major polar tube protein of *Encephalitozoon cuniculi*. *Parasitology Research, 107*(4), 761–764. https://doi.org/10.1007/s00436-010-1950-7

Burri, L., Williams, B. A., Bursac, D., Lithgow, T., & Keeling, P. J. (2006). Microsporidian mitosomes retain elements of the general mitochondrial targeting system. *Proceedings of the National Academy of Sciences, 103*(43), 15916–15920. https://doi.org/10.1073/pnas.0604109103

Centre for Disease Control. (n.d.-a). https://www.cdc.gov/dpdx/microsporidiosis. Figure courtesy of Dr. Massimo Scaglia, Laboratory of Clinical Parasitology, Institute of Infectious Diseases, University-IRCCS San Matteo, Pavia, Italy.

Centre for Disease Control. (n.d.-b). *Life cycle of Microsporidia*. Available from https://www.cdc.gov/dpdx/microsporidiosis

Çetinkaya, Ü., Kuk, S., Sivcan, E., Kaynar, L., Arslan, D., & Şahin, İ. (2016). The high prevalence of *Encephalitozoon intestinalis* in patients receiving chemotherapy and children with growth retardation and the validity of real-time PCR in its diagnosis. *Turkish Journal of Medical Sciences, 46*(4), 1050–1058.

Çetinkaya, Ü., Charyyeva, A., Sivcan, E., Sönmez, T., Duran, M., & Şahin, İ. (2018). Evaluation of four commercial DNA extraction kits for the detection of Microsporidia and the importance of pretreatments in DNA isolation. *Acta Parasitologica, 63*(2), 386–392. https://doi.org/10.1515/ap-2018-0044

Cornman, R. S., Chen, Y. P., Schatz, M. C., Street, C., Zhao, Y., Desany, B., Egholm, M., Hutchison, S., Pettis, J. S., Lipkin, W. I., & Evans, J. D. (2009). Genomic analyses of the microsporidian *Nosema ceranae*, an emergent pathogen of honey bees. *PLoS Pathogens, 5*(6), e1000466. https://doi.org/10.1371/journal.ppat.1000466

Corradi, N., & Keeling, P. J. (2009). Microsporidia: A journey through radical taxonomical revisions. *Fungal Biology Reviews, 23*(1), 1–8. https://doi.org/10.1016/j.fbr.2009.01.001

Corradi, N., Pombert, J. F., Farinelli, L., Didier, E. S., & Keeling, P. J. (2010). The complete sequence of the smallest known nuclear genome from the microsporidian *Encephalitozoon intestinalis*. *Nature Communications, 1*, 77. https://doi.org/10.1038/ncomms1082

Cuomo, C. A., Desjardins, C. A., Bakowski, M. A., Goldberg, J., Ma, A. T., Becnel, J. J., Didier, E. S., Fan, L., Heiman, D. I., Levin, J. Z., Young, S., Zeng, Q., & Troemel, E. R. (2012). Microsporidian genome analysis reveals evolutionary strategies for obligate intracellular growth. *Genome Research, 22*(12), 2478–2488. https://doi.org/10.1101/gr.142802.112

del Aguila, C., Moura, H., Fenoy, S., Navajas, R., Lopez-Velez, R., Li, L., Xiao, L., Leitch, G. J., da Silva, A., Pieniazek, N. J., Lal, A. A., & Visvesvara, G. S. (2001). *In vitro* culture, ultrastructure, antigenic, and molecular characterization of *Encephalitozoon cuniculi* isolated from urine and sputum samples from a Spanish patient with AIDS. *Journal of Clinical Microbiology, 39*(3), 1105–1108. https://doi.org/10.1128/JCM.39.3.1105-1108.2001

Deplazes, P., Mathis, A., & Weber, R. (2000). Epidemiology and zoonotic aspects of microsporidia of mammals and birds. *Contributions to Microbiology, 6*, 236–260. https://doi.org/10.1159/000060363

Didier, E. S., & Weiss, L. M. (2008). Overview of microsporidia and microsporidiosis. *Protistology, 5*, 243–255.

Didier, E. S., & Weiss, L. M. (2011). Microsporidiosis: Not just in AIDS patients. *Current Opinion in Infectious Diseases, 24*(5), 490–495. https://doi.org/10.1097/QCO.0b013e32834aa152

Didier, P. J., Phillips, J. N., Kuebler, D. J., Nasr, M., Brindley, P. J., Stovall, M. E., Bowers, L. C., & Didier, E. S. (2006). Antimicrosporidial activities of fumagillin, TNP-470, ovalicin, and ovalicin derivatives *in vitro* and in vivo. *Antimicrobial Agents and Chemotherapy, 50*(6), 2146–2155. https://doi.org/10.1128/AAC.00020-06

Dong, Z., Wu, Q., Zhang, P., Fang, W., Lei, X., Deng, B., Hu, N., Chen, P., Huang, X., Lu, C., & Pan, M. (2025). Development of a novel anti-microsporidia strategy by inhibiting parasite and host glucose metabolism. *Pesticide Biochemistry and Physiology, 208*, 106276. https://doi.org/10.1016/j.pestbp.2024.106276

Du, J., Yang, Y. C., An, Z. J., Zhang, M. H., Fu, X. H., Huang, Z. F., Yuan, Y., & Hou, J. (2023). Advances in spatial transcriptomics and related data analysis strategies. *Journal of Translational Medicine, 21*(1), 330. https://doi.org/10.1186/s12967-023-04150-2

Dunn, A. M., & Smith, J. E. (2001). Microsporidian life cycles and diversity: The relationship between virulence and transmission. *Microbes and Infection, 3*(5), 381–388. https://doi.org/10.1016/S1286-4579(01)01394-6

Ebert, D. (2008). Host–parasite coevolution: Insights from the *Daphnia*–parasite model system. *Current Opinion in Microbiology, 11*(3), 290–301. https://doi.org/10.1016/j.mib.2008.05.012

Ebert, D., Altermatt, F., & Lass, S. (2007). A short term benefit for outcrossing in a *Daphnia* metapopulation in relation to parasitism. *Journal of the Royal Society Interface, 4*(16), 777–785. https://doi.org/10.1098/rsif.2007.0232

Fadhilah, A., Gabbar, A., & Bokhari, A. A. (2023). Microsporidium. In *StatPearls* [Internet]. Treasure Island: StatPearls Publishing. https://www.ncbi.nlm.nih.gov/books/NBK537166/

Fayer, R., & Santin-Duran, M. (2014). Epidemiology of microsporidia in human infections. In L. M. Weiss & J. J. Becnel (Eds.), *Microsporidia: Pathogens of opportunity* (pp. 135–164). Wiley Blackwell. https://doi.org/10.1002/9781118395264.ch3

Fayet, M., Long, M., Han, B., Belkorchia, A., Delbac, F., & Polonais, V. (2024). New insights into Microsporidia polar tube function and invasion mechanism. *The Journal of Eukaryotic Microbiology, 71*(5), e13043. https://doi.org/10.1111/jeu.13043

Fox, R. M., & Weiser, J. (1959). A microsporidian parasite of *Anopheles gambiae* in Liberia. *Journal of Parasitology, 45*(1), 21.

Franzen, C. (2005). How do microsporidia invade cells? *Folia Parasitologica, 52*(1), 36–40. https://doi.org/10.14411/fp.2005.005

Franzen, C. (2008). Microsporidia: A review of 150 years of research. *The Open Parasitology Journal, 2*, 1–34. https://doi.org/10.2174/1874421400802010001

Freibert, S. A., Goldberg, A. V., Hacker, C., Molik, S., Dean, P., Williams, T. A., Nakjang, S., Long, S., Sendra, K., Bill, E., Heinz, E., Hirt, R. P., Lucocq, J. M., Embley, T. M., & Lill, R. (2017). Evolutionary conservation and *in vitro* reconstitution of microsporidian iron-sulfur cluster biosynthesis. *Nature Communications, 8*, 13932. https://doi.org/10.1038/ncomms13932

Fu, M., Qu, H., Wang, Y., Guan, J., Xia, T., Zheng, K., Tang, L., Zhou, C., Zhou, H., Cong, W., Zhang, J., & Han, B. (2024). Overcoming research challenges: *In vitro* cultivation of *Ameson portunus* (phylum microsporidia). *Journal of Invertebrate Pathology, 204*, 108091. https://doi.org/10.1016/j.jip.2024.108091

Galajda, R., Valenčáková, A., Sučik, M., & Kandráčová, P. (2021). Nosema disease of European honey bees. *Journal of Fungi, 7*(9), 714. https://doi.org/10.3390/jof7090714

Garcia, L. S. (2002). Laboratory identification of the Microsporidia. *Journal of Clinical Microbiology, 40*, 1892–1901.

Han, B., & Weiss, L. M. (2017). Microsporidia: Obligate intracellular pathogens within the fungal kingdom. *Microbiology Spectrum, 5*(2). https://doi.org/10.1128/microbiolspec.FUNK-0018-201

Han, B., Takvorian, P. M., & Weiss, L. M. (2022). The Function and Structure of the Microsporidia Polar Tube. *Experientia Supplementum (2012), 114*, 179–213. https://doi.org/10.1007/978-3-030-93306-7_8

Han, B., Polonais, V., Sugi, T., Yakubu, R., Takvorian, P. M., Cali, A., et al. (2017). The role of microsporidian polar tube protein 4 (PTP4) in host cell infection. *PLoS Pathogens, 13*(6), e1006341. https://doi.org/10.1371/journal.ppat.1006341

Han, B., Ma, Y., Tu, V., Tomita, T., Mayoral, J., Williams, T., et al. (2019). Microsporidia interact with host cell mitochondria via voltage-dependent anion channels using sporoplasm surface protein 1. *MBio, 10*(4). https://doi.org/10.1128/mbio.01944-19

Han, B., Takvorian, P. M., & Weiss, L. M. (2020). Invasion of host cells by microsporidia. *Frontiers in Microbiology, 11*, 172. https://doi.org/10.3389/fmicb.2020.00172

Han, B., Pan, G., & Weiss, L. M. (2021). Microsporidiosis in humans. *Clinical Microbiology Reviews, 34*(4), e0001020. https://doi.org/10.1128/CMR.00010-20

Han, B., Takvorian, P. M., & Weiss, L. M. (2022). The function and structure of the microsporidia polar tube. In T. Tomita (Ed.), *Experimental suppl* (Vol. 114, pp. 179–213). https://doi.org/10.1007/978-3-030-93306-7_8

He, Q., Luo, J., Xu, J.-Z., Meng, X.-Z., Pan, G.-Q., Li, T., & Zhou, Z.-Y. (2020). In-vitro cultivation of *Nosema bombycis* sporoplasms: A method for potential genetic engineering of microsporidia. *Journal of Invertebrate Pathology, 174*, 107420. https://doi.org/10.1016/j.jip.2020.107420

Heinz, E., Williams, T. A., Nakjang, S., Noël, C. J., Swan, D. C., Goldberg, A. V., et al. (2012). The genome of the obligate intracellular parasite *Trachipleistophora hominis*: New insights into microsporidian genome dynamics and reductive evolution. *PLoS Pathogens, 8*(10), e1002979. https://doi.org/10.1371/journal.ppat.1002979

Herren, J. K., Mbaisi, L., Mararo, E., Machani, M. G., Kerio, L., Gitau, C. W., Mwirigi, N. W., Mutunga, J. M., Matoke-Muhia, D., & Christophides, G. K. (2020). A microsporidian impairs *Plasmodium falciparum* transmission in *Anopheles arabiensis* mosquitoes. *Nature Communications, 11*, 2187. https://doi.org/10.1038/s41467-020-16121-y

Hirt, R. P., Logsdon, J. M., Healy, B., Dorey, M. W., Doolittle, W. F., & Embley, T. M. (1997). Microsporidia are related to fungi: Evidence from the largest subunit of RNA polymerase II and other proteins. *Proceedings of the National Academy of Sciences, 94*(3), 986–991. https://doi.org/10.1073/pnas.94.3.986

Huang, Y., Zheng, S., Mei, X., Yu, B., Sun, B., Li, B., Wei, J., Chen, J., Li, T., Pan, G., Zhou, Z., & Li, C. (2018). A secretory hexokinase plays an active role in the proliferation of *Nosema bombycis*. *PeerJ, 6*, e5658. https://doi.org/10.7717/peerj.5658

Huang, Q., Wu, Z. H., Li, W. F., Guo, R., Xu, J. S., Dang, X. Q., Ma, Z. G., Chen, Y. P., & Evans, J. D. (2021). Genome and evolutionary analysis of *Nosema ceranae*: A microsporidian parasite of honey bees. *Frontiers in Microbiology, 12*, 645353. https://doi.org/10.3389/fmicb.2021.645353

Jaroenlak, P., Boakye, D. W., Vanichviriyakit, R., Williams, B. A. P., Sritunyalucksana, K., & Itsathitphaisarn, O. (2018). Identification, characterization and heparin binding capacity of a spore-wall, virulence protein from the shrimp microsporidian *Enterocytozoon hepatopenaei* (EHP). *Parasites & Vectors, 11*, 1–15.

Jaroenlak, P., Cammer, M., Davydov, A., Sall, J., Usmani, M., Liang, F. X., et al. (2020). 3-dimensional organization and dynamics of the microsporidian polar tube invasion machinery. *PLoS Pathogens, 16*(4), e1008738. https://doi.org/10.1371/journal.ppat.1008738

Jespersen, N., Monrroy, L., & Barandun, J. (2022). Impact of genome reduction in microsporidia. In L. M. Weiss & A. W. Reinke (Eds.), *Microsporidia* (Experientia Supplementum) (Vol. 114). Springer. https://doi.org/10.1007/978-3-030-93306-7_1

Joseph, J., & Sharma, S. (2009). *In vitro* culture of various species of microsporidia causing keratitis: Evaluation of three immortalized cell lines. *Indian Journal of Medical Microbiology, 27*(1), 35–39.

Katinka, M. D., Duprat, S., Cornillot, E., Méténier, G., Thomarat, F., Prensier, G., Barbe, V., Peyretaillade, E., Brottier, P., Wincker, P., Delbac, F., El Alaoui, H., Peyret, P., Saurin, W., Gouy, M., Weissenbach, J., & Vivarès, C. P. (2001). Genome sequence and gene compaction of the eukaryote parasite *Encephalitozoon cuniculi*. *Nature, 414*(6862), 450–453. https://doi.org/10.1038/35106579

Keeling, P. J., & Fast, N. M. (2002). Microsporidia: Biology and evolution of highly reduced intracellular parasites. *Annual Review of Microbiology, 56*, 93–116. https://doi.org/10.1146/annurev.micro.56.012302.160854

Keeling, P. J., & McFadden, G. I. (1998). Origins of microsporidia. *Trends in Microbiology, 6*(1), 19–23. https://doi.org/10.1016/S0966-842X(97)01185-2

Keeling, P. J., Luker, M. A., & Palmer, J. D. (2000). Evidence from beta-tubulin phylogeny that microsporidia evolved from within the fungi. *Molecular Biology and Evolution, 17*(1), 23–31. https://doi.org/10.1093/oxfordjournals.molbev.a026235

Knecht, S., Eberl, H. C., Kreisz, N., Ugwu, U. J., Starikova, T., Kuster, B., & Wilhelm, S. (2023). An introduction to analytical challenges, approaches, and applications in mass spectrometry-based secretomics. *Molecular & Cellular Proteomics, 22*(9), 100636. https://doi.org/10.1016/j.mcpro.2023.100636

Lallo, M. A., Vidoto da Costa, L. F., Alvares-Saraiva, A. M., Rocha, P. R. D., Spadacci-Morena, D. D., Konno, F. T. C., & Suffredini, I. B. (2016). Culture and propagation of microsporidia of veterinary interest. *Journal of Veterinary Medical Science, 78*(2), 171–176. https://doi.org/10.1292/jvms.15-0401

Leiro, J., Iglesias, R., Paramá, A., Sanmartín, M. L., & Ubeira, F. M. (2001). Effect of *Tetramicra brevifilum* (Microspora) infection on respiratory-burst responses of turbot (*Scophthalmus maximus* L.) phagocytes. *Fish & Shellfish Immunology, 11*(7), 639–652. https://doi.org/10.1006/fsim.2001.0340

Leitch, G. J., & Ceballos, C. (2008). Interaction of *Encephalitozoon intestinalis* with intestinal epithelial cells: Role of inflammatory cytokines and chemokines. *Journal of Eukaryotic Microbiology, 55*(1), 113–117. https://doi.org/10.1111/j.1550-7408.2008.00309.x

Loureiro, A., & da Silva, G. J. (2019). CRISPR-Cas: Converting a bacterial defence mechanism into a state-of-the-art genetic manipulation tool. *Antibiotics, 8*(1), 18. https://doi.org/10.3390/antibiotics8010018

Lv, Q., Hong, L., Qi, L., Chen, Y., Xie, Z., Liao, H., Li, C., Li, T., Meng, X., Chen, J., Bao, J., Wei, J., Han, B., Shen, Q., Weiss, L. M., Zhou, Z., Long, M., & Pan, G. (2024). Microsporidia dressing up: the spore polaroplast transport through the polar tube and transformation into the sporoplasm membrane. *mBio, 15*(2), e0274923. https://doi.org/10.1128/mbio.02749-23.

Moniot, M., Nourrisson, C., Bonnin, V., Damiani, C., Argy, N., Bonhomme, J., Fréalle, E., Angebault, C., Debourgogne, A., Sitterlé, E., Flori, P., Brunet, J., Dalle, F., Favennec, L., Poirier, P., & French Microsporidiosis Network. (2022). Evaluation of the bio-evolution microsporidia generic and typing real-time PCR assays for the diagnosis of intestinal microsporidiosis. *Parasite, 29*, 55. https://doi.org/10.1051/parasite/2022055

Moss, J. A., & Snyder, R. A. (2017). Biofilms for monitoring presence of microsporidia in environmental water. *Journal of Eukaryotic Microbiology, 64*(4), 533–538. https://doi.org/10.1111/jeu.12390

Nakjang, S., Williams, T. A., Heinz, E., Watson, A. K., Foster, P. G., Sendra, K. M., Heaps, S. E., Hirt, R. P., & Martin Embley, T. (2013). Reduction and expansion in microsporidian genome evolution: New insights from comparative genomics. *Genome Biology and Evolution, 5*(12), 2285–2303. https://doi.org/10.1093/gbe/evt184

Nattoh, G., Maina, T., Makhulu, E. E., Mbaisi, L., Mararo, E., Otieno, F. G., Bukhari, T., Onchuru, T. O., Teal, E., Paredes, J., & Bargul, J. L. (2021). Horizontal transmission of the symbiont microsporidia MB in *Anopheles arabiensis*. *Frontiers in Microbiology, 12*, 647183. https://doi.org/10.3389/fmicb.2021.647183

Nourrisson, C., Lavergne, R. A., Moniot, M., Morio, F., & Poirier, P. (2024). Enterocytozoon bieneusi, a human pathogen. *Emerging Microbes and Infections, 13*(1). https://doi.org/10.1080/22221751.2024.2406276

Paldi, N., Glick, E., Oliva, M., Zilberberg, Y., Aubin, L., Pettis, J., Chen, Y., & Evans, J. D. (2010). Effective gene silencing in a microsporidian parasite associated with honeybee (*Apis mellifera*) colony declines. *Applied and Environmental Microbiology, 76*(18), 5960–5964. https://doi.org/10.1128/AEM.00631-10

Park, E., & Poulin, R. (2021). Revisiting the phylogeny of microsporidia. *International Journal for Parasitology, 51*(10), 855–864. https://doi.org/10.1016/j.ijpara.2021.02.005

Paul, V. D., & Lill, R. (2015). Biogenesis of cytosolic and nuclear iron-sulfur proteins and their role in genome stability. *Biochimica et Biophysica Acta, 1853*(6), 1528–1539. https://doi.org/10.1016/j.bbamcr.2014.12.018

Peuvel, I., Peyret, P., Metenier, G., Vivares, C. P., & Delbac, F. (2002). The microsporidian polar tube: Evidence for a third polar tube protein (PTP3) in *Encephalitozoon cuniculi*. *Molecular and Biochemical Parasitology, 122*, 69–80. https://doi.org/10.1016/S0166-6851(02)00073-7

Peyretaillade, E., Broussolle, V., Peyret, P., Méténier, G., Gouy, M., & Vivarès, C. P. (1998). Microsporidia, a mitochondrial protists, possess a 70-kDa heat shock protein gene of mitochondrial evolutionary origin. *Molecular Biology and Evolution, 15*(6), 683–689. https://doi.org/10.1093/oxfordjournals.molbev.a025971

Pombert, J. F., Haag, K. L., Beidas, S., Ebert, D., & Keeling, P. J. (2015). The *Ordospora colligata* genome: Evolution of extreme reduction in microsporidia and host-to-parasite horizontal gene transfer. *mBio, 6*, e02400-14. https://doi.org/10.1128/mBio.02400-14

Ran, M., Shi, Y., Li, B., Xiang, H., Tao, M., Meng, X., Li, T., Li, C., Bao, J., Pan, G., et al. (2023). Genome-wide characterization and comparative genomic analysis of the serpin gene family in microsporidian *Nosema bombycis*. *International Journal of Molecular Sciences, 24*(1), 550. https://doi.org/10.3390/ijms24010550

Ran, M., Yang, W., Faryad Khan, M. U., Li, T., & Pan, G. (2024). Microsporidia secretory effectors and their roles in pathogenesis. *Journal of Eukaryotic Microbiology, 71*(5), e13046. https://doi.org/10.1111/jeu.13046

Reinke, A. W., & Troemel, E. R. (2015). The development of genetic modification techniques in intracellular parasites and potential applications to microsporidia. *PLoS Pathogens, 11*(12), e1005283. https://doi.org/10.1371/journal.ppat.1005283

Rodríguez-García, C., Evans, J. D., Li, W., Branchiccela, B., Li, J. H., Heerman, M. C., Banmeke, O., Zhao, Y., Hamilton, M., Higes, M., Martín-Hernández, R., & Chen, Y. P. (2018). Nosemosis control in European honey bees, *Apis mellifera*, by silencing the gene encoding Nosema ceranae polar tube protein 3. *Journal of Experimental Biology, 221*(19), jeb184606. https://doi.org/10.1242/jeb.184606

Ruan, Y., Xu, X., He, Q., Li, L., Guo, J., Bao, J., Pan, G., Li, T., & Zhou, Z. (2021). The largest meta-analysis on the global prevalence of microsporidia in mammals, avian and water provides insights into the epidemic features of these ubiquitous pathogens. *Parasites & Vectors, 14*(1), 186. https://doi.org/10.1186/s13071-021-04700-x

Seatamanoch, N., Kongdachalert, S., Sunantaraporn, S., Siriyasatien, P., & Brownell, N. (2022). Microsporidia, a highly adaptive organism and its host expansion to humans. *Frontiers in Cellular and Infection Microbiology, 12*, 924007. https://doi.org/10.3389/fcimb.2022.924007

Shane, J. L., Grogan, C. L., Cwalina, C., & Lazzaro, B. P. (2018). Blood meal-induced inhibition of vector-borne disease by transgenic microbiota. *Nature Communications, 9*, 4127. https://doi.org/10.1038/s41467-018-06580-9

South, L. R., Hurdeal, V. G., & Fast, N. M. (2024) Genomics and phylogenetic relationships of microsporidia and their relatives. *Journal of Eukaryotic Microbiology, 71*, e13051. Available from: https://doi.org/10.1111/jeu.13051

Southern, T. R., Jolly, C. E., Lester, M. E., & Hayman, J. R. (2007). EnP1, a microsporidian spore wall protein that enables spores to adhere to and infect host cells *in vitro*. *Eukaryotic Cell, 6*(8), 1354–1362. https://doi.org/10.1128/ec.00113-07

Stentiford, G. D., Feist, S. W., Stone, D. M., Bateman, K. S., & Dunn, A. M. (2013). Microsporidia: Diverse, dynamic, and emergent pathogens in aquatic systems. *Trends in Parasitology, 29*(11), 567–578. https://doi.org/10.1016/j.pt.2013.08.005

Stentiford, G. D., Becnel, J. J., Weiss, L. M., Keeling, P. J., Didier, E. S., Williams, B. P., Bjornson, S., Kent, M. L., Freeman, M. A., Brown, M. J. F., Troemel, E. R., Roesel, K., Sokolova, Y., Snowden, K. F., & Solter, L. (2016). Microsporidia—Emergent pathogens in the global food chain. *Trends in Parasitology, 32*(4), 336–348. https://doi.org/10.1016/j.pt.2015.12.004

Stentiford, G. D., Feist, S. W., Stone, D. M., Bateman, K. S., and Dunn, A. M. (2013). Microsporidia: diverse, dynamic, and emergent pathogens in aquatic systems. Trends in Parasitology, 29(11), 567–578. https://doi.org/10.1016/j.pt.2013.08.005

Subrungruang, I., Mungthin, M., Chavalitshewinkoon-Petmitr, P., Rangsin, R., Naaglor, T., & Leelayoova, S. (2004). Evaluation of DNA extraction and PCR methods for detection of *Enterocytozoon bieneusi* in stool specimens. *Journal of Clinical Microbiology, 42*, 3490–3494.

Takvorian, P. M., Han, B., Cali, A., Rice, W. J., Gunther, L., Macaluso, F., & Weiss, L. M. (2020). An ultrastructural study of the extruded polar tube of *Anncaliia algerae* (Microsporidia). *Journal of Eukaryotic Microbiology, 67*(1), 28–44. https://doi.org/10.1111/jeu.12751

Tamim El Jarkass, H., & Reinke, A. W. (2020). The ins and outs of host–microsporidia interactions during invasion, proliferation and exit. *Cellular Microbiology, 22*(11), e13247. https://doi.org/10.1111/cmi.13247

Trager, W. (1937). The hatching of spores of *Nosema bombycis* Nägeli and the partial development of the organism in tissue cultures. *Journal of Parasitology, 23*, 226–227. https://doi.org/10.2307/3272075

Tsaousis, A. D., Kunji, E. R., Goldberg, A. V., Lucocq, J. M., Hirt, R. P., & Embley, T. M. (2008). A novel route for ATP acquisition by the remnant mitochondria of *Encephalitozoon cuniculi*. *Nature, 453*(7194), 553–556. https://doi.org/10.1038/nature06903

Tullu, B., Pevsner, R., & Herren, J. K. (2022). Microsporidia: A promising vector control tool for residual malaria transmission. *Frontiers in Tropical Diseases, 3*, Article 957109. https://doi.org/10.3389/fitd.2022.957109

Vávra, J., & Lukeš, J. (2013). Microsporidia and "the art of living together". *Advances in Parasitology, 82*, 253–319. https://doi.org/10.1016/B978-0-12-407706-5.00004-6
Visvesvara, G. S. (2002). *In vitro* cultivation of microsporidia of clinical importance. *Clinical Microbiology Reviews, 15*(3), 401–413. https://doi.org/10.1128/CMR.15.3.401-413.2002
Wang, S., Ghosh, A. K., Bongio, N., Stebbings, K. A., Lampe, D. J., & Jacobs-Lorena, M. (2012). Fighting malaria with engineered symbiotic bacteria from vector mosquitoes. *Proceedings of the National Academy of Sciences of the United States of America, 109*(31), 12734–12739. https://doi.org/10.1073/pnas.1204158109
Wang, D., Gou, M., Hou, J., Pang, Y., & Li, Q. (2019). The role of serpin protein on the natural immune defense against pathogen infection in *Lampetra japonica. Fish & Shellfish Immunology, 92*, 196–208. https://doi.org/10.1016/j.fsi.2019.05.062
Wang, M., Li, X., Liu, X., Hou, X., He, Y., Yu, J.-H., Hu, S., Yin, H., & Xie, B.-B. (2024). Annotation of 2,507 *Saccharomyces cerevisiae* genomes. *Microbiology Spectrum, 12*(4), e0358223. https://doi.org/10.1128/spectrum.03582-23
Weidner, E., & Sibley, L. D. (1985). Phagocytized intracellular microsporidian blocks phagosome acidification and phagosome-lysosome fusion. *The Journal of Protozoology, 32*(2), 311–317. https://doi.org/10.1111/j.1550-7408.1985.tb03056.x
Weiss, L. M., & Becnel, J. J. (2014). *Microsporidia: Pathogens of opportunity*. Wiley.
Williams, B. A. P., & Keeling, P. J. (2011). Microsporidia—Highly reduced and derived relatives of fungi. In S. Pöggeler & J. Wöstemeyer (Eds.), *Evolution of fungi and fungal-like organisms* (Vol. 14, pp. 129–146). Springer. https://doi.org/10.1007/978-3-642-19974-5_2
Williams, B. A., Lee, R. C., Becnel, J. J., Weiss, L. M., Fast, N. M., & Keeling, P. J. (2008). Genome sequence surveys of *Brachiola algerae* and *Edhazardia aedis* reveal microsporidia with low gene densities. *BMC Genomics, 9*, 200. https://doi.org/10.1186/1471-2164-9-200
Winters, A. D., Langohr, I. M., Souza, M. D. A., Colodel, E. M., Soares, M. P., & Faisal, M. (2016). Ultrastructure and molecular phylogeny of *Pleistophora hyphessobryconis* (Microsporidia) infecting hybrid jundiara (*Leiarius marmoratus* × *Pseudoplatystoma reticulatum*) in a Brazilian aquaculture facility. *Parasitology, 143*(1), 41–49. https://doi.org/10.1017/S0031182015001420
Wiredu Boakye, D., Jaroenlak, P., Prachumwat, A., Williams, T. A., Bateman, K. S., Itsathitphaisarn, O., Sritunyalucksana, K., Paszkiewicz, K. H., Moore, K. A., Stentiford, G. D., & Williams, B. A. P. (2017). Decay of the glycolytic pathway and adaptation to intranuclear parasitism within Enterocytozoonidae microsporidia. *Environmental Microbiology, 19*(5), 2077–2089. https://doi.org/10.1111/1462-2920.13734
Wolk, D. M., Johnson, C. H., Rice, E. W., Marshall, M. M., Grahn, K. F., Plummer, C. B., & Sterling, C. R. (2000). A spore counting method and cell culture model for chlorine disinfection studies of *Encephalitozoon syn.* Septata intestinalis. *Applied and Environmental Microbiology, 66*, 1266–1273.
Xu, J., Luo, J., Chen, J., Vossbrinck, C. R., Li, T., & Zhou, Z. (2022). Characterization of the largest secretory protein family, Ricin B lectin-like protein, in *Nosema bombycis*: Insights into microsporidian adaptation to host. *Journal of Fungi, 8*(6), 551. https://doi.org/10.3390/jof8060551
Yan, X., Liu, X., Zhao, C., & Chen, G. Q. (2023). Applications of synthetic biology in medical and pharmaceutical fields. *Signal Transduction and Targeted Therapy, 8*(1), 199. https://doi.org/10.1038/s41392-023-01440-5
Zhong, Y., Xu, F., Wu, J., Schubert, J., & Li, M. M. (2021). Application of next generation sequencing in laboratory medicine. *Annals of Laboratory Medicine, 41*(1), 25–43. https://doi.org/10.3343/alm.2021.41.1.25
Zhu, F., Shen, Z., Hou, J., Zhang, J., Geng, T., & Tang, X. (2013). Identification of a protein interacting with the spore wall protein SWP26 of *Nosema bombycis* in a cultured BmN cell line of silkworm. *Infection, Genetics and Evolution, 17*, 38–45. https://doi.org/10.1016/j.meegid.2013.03.029

Engineering Interventions in Fungal Cell Factories

Blaise Manga Enuh, Özlem Erdal Altıntaş, Belma Nural Yaman, Pinar Aytar Çelik, and Ahmet Çabuk

Abstract Fungal cell factories are more diverse and sustainable, capable of producing pharmaceuticals, biofuels, food additives, and industrial enzymes. In the context of industrial biotechnology, this chapter gives an overview of the engineering approaches that are used to improve the functionality of fungal hosts. This progress relies heavily on genetic engineering techniques like homologous recombination and CRISPR-Cas9, which help make precise changes to the genome and metabolic pathways. Synthetic biology expands on these capabilities by allowing the development of complex metabolic circuits and heterologous pathways using modular DNA assembly techniques. Metabolic engineering projects aim to increase yields and facilitate the production of novel compounds by deregulating pathways, overexpressing essential enzymes, and expanding substrate utilization. Fermentation processes optimization, fungal morphology regulation, and unique bioreactor designs consist of advanced bioprocess engineering, which is adapted to fungal growth features and improves these genetic and metabolic improvements. High-throughput omics, including machine learning-driven strain design and the activation of latent biosynthetic gene clusters, have expedited the identification of novel metabolites. Fungi are increasingly utilized in the realms of circular bioeconomy, environmental remediation, and precision fermentation and are anticipated to significantly contrib-

B. M. Enuh
Great Lakes Bioenergy Research Centre, Wisconsin Energy Institute, Madison, WI, USA

Ö. Erdal Altıntaş
Faculty of Health Sciences, Department of Nutrition and Dietetics, Afyonkarahisar Health Sciences University, Afyonkarahisar, Turkey

B. Nural Yaman
Department of Biomedical Engineering, Eskisehir Osmangazi University, Eskisehir, Turkey

P. Aytar Çelik (✉)
Environmental Protection and Control Program, Eskisehir Osmangazi University, Eskisehir, Turkey

A. Çabuk
Department of Biology, Eskisehir Osmangazi University, Eskisehir, Turkey

A. Gupta (ed.), *Prospects of Fungal Biotechnologies for Livestock Volume 2*,
Fungal Biology, https://doi.org/10.1007/978-3-032-06478-3_3

ute to future biotechnological progress through the integration of enhanced technologies and sustainable biomanufacturing methods.

Keywords Bioprocess optimization · CRISPR-Cas9 · Fungal cell factories · Heterologous expression · Metabolic engineering · Overexpression · Synthetic biology

1 Introduction

Fungi, a diverse kingdom encompassing both yeasts and filamentous species, represent highly adaptable and industrially significant cell factories for the production of a broad spectrum of biochemicals. Their inherent advantages, such as rapid growth kinetics on inexpensive and abundant substrates, coupled with a remarkable capacity for high-level secretion of enzymes and secondary metabolites, position them as cornerstone platforms in modern industrial biotechnology (Cooray et al., 2017; Obidi & Moyosore, 2018). The escalating demand for sustainable and cost-effective production of pharmaceuticals, biofuels, food ingredients, and industrial enzymes necessitates continuous innovation in optimizing fungal bioprocesses (Lübeck & Lübeck, 2022). Achieving maximum efficiency and economic viability in fungal cell factories requires a sophisticated and integrated approach, synergistically combining advancements in genetic engineering, metabolic pathway manipulation, and bioprocess design. This chapter delves into the critical engineering interventions that underpin the development of high-performance fungal cell factories. We will explore advanced genetic manipulation techniques, strategies for rational and systems-driven metabolic engineering, innovations in bioreactor design and operation, and recent technological breakthroughs that are shaping the future of fungal biotechnology and its diverse industrial applications. The overarching goal is to provide a detailed, expert-level perspective on how engineering principles are applied to unlock the full potential of fungal systems for industrial-scale bioproduction.

2 Genetic Engineering Techniques

Genetic engineering serves as a foundational pillar in the development of fungal cell factories, enabling precise modification of existing traits and the introduction of novel functionalities. The field has progressed significantly from classical mutagenesis and screening to sophisticated, targeted genome editing and synthetic biology approaches. These modern techniques provide unprecedented control over fungal genomes, facilitating the optimization of metabolic pathways and the production of valuable biomolecules.

2.1 CRISPR-Cas9 Genome Editing

The revolutionary clustered regularly interspaced short palindromic repeats (CRISPR)-Cas9 system has transformed fungal genetics by providing a versatile and efficient platform for targeted genome editing. This RNA-guided nuclease technology allows for site-specific modifications, including gene knockouts, precise insertions, and single-nucleotide changes, across a wide range of fungal species that were previously recalcitrant to genetic manipulation (Leitão et al., 2024; Nødvig et al., 2015). The system's modularity, relying on easily designed guide RNAs, makes it significantly more accessible and rapid compared to earlier custom-enzyme-based methods like zinc-finger nucleases (ZFNs) or transcription activator-like effector nucleases (TALENs).

Implementation of CRISPR-Cas9 in fungi involves delivering the Cas9 nuclease and a specific guide RNA (gRNA) to the fungal cell. Various delivery strategies have been developed, including plasmid-based expression systems, in vitro assembled Cas9 ribonucleoprotein (RNP) complexes, and viral vectors, each optimized for different fungal species and applications (Al Abdallah et al., 2018; Chen et al., 2022a). For instance, RNP delivery has proven highly effective in achieving precise gene targeting in species like *Aspergillus fumigatus* (Al Abdallah et al., 2018). Transient expression systems and plasmid-based approaches facilitate the integration of CRISPR components into fungal genomes, enabling stable modifications (Chen et al., 2022a, b). The precision and ease of multiplexing (simultaneously targeting multiple genes) offered by CRISPR-Cas9 dramatically accelerate strain development workflows, enabling the rapid creation of complex mutant libraries and combinatorial genetic modifications. This technology is now routinely applied in diverse industrial fungal hosts, including *Fusarium*, *Trichoderma*, *Aspergillus*, and *Saccharomyces*, to optimize traits relevant to industrial production, such as improved enzyme secretion, enhanced stress tolerance, and increased metabolite yield (Wang & Coleman, 2019).

2.2 Homologous Recombination (HR)

Homologous recombination is a natural DNA repair mechanism that allows for the precise integration of exogenous DNA into a genome at a locus homologous to flanking sequences on the introduced DNA. In *Saccharomyces cerevisiae*, HR is naturally highly efficient, making it a preferred method for targeted gene knockouts and insertions (Featherstone & Jackson, 1999). However, in many filamentous fungi, the non-homologous end joining (NHEJ) pathway, which ligates broken

DNA ends without requiring homology, is the predominant DNA repair mechanism. This often leads to random integration of introduced DNA, complicating targeted genetic modifications.

A key strategy to enhance HR efficiency in filamentous fungi involves disrupting the NHEJ pathway. Deleting core components of the NHEJ machinery, such as the *ku70* and *ku80* genes, significantly reduces random integration and dramatically increases the frequency of HR-mediated gene targeting. In some engineered strains, this approach can lead to over 90% of transformations resulting in targeted integration at the desired genomic locus (Arnau et al., 2019; Fierro et al., 2022). This enables researchers to precisely knock out genes, replace native genes with modified versions, or insert entire gene cassettes at specific sites by providing a DNA repair template with homologous flanking sequences. This strategy has been successfully applied in various filamentous fungi, including *Aspergillus* species and *Monascus ruber*, to achieve high-efficiency gene targeting and streamline strain development (He et al., 2014; Nødvig et al., 2018; Tsukuda et al., 2005). By leveraging or enhancing HR, researchers can achieve precise and predictable genetic modifications essential for rational strain design (Nødvig et al., 2018).

2.3 Synthetic Biology and DNA Assembly

Synthetic biology extends genetic engineering beyond single-gene modifications to the design and construction of complex biological systems with novel functions. This field provides standardized genetic parts (e.g., promoters, coding sequences, terminators) and modular assembly strategies that enable the rapid and reliable construction of multi-gene pathways, regulatory circuits, and even artificial chromosomes in fungal hosts. Recent advancements have led to the development of sophisticated synthetic biology toolkits specifically tailored for fungal systems. For instance, toolkits like FungalBraid provide curated libraries of characterized genetic parts for filamentous fungi, allowing researchers to mix and match elements using standardized cloning methods such as Golden Gate cloning (Moreno-Giménez et al., 2023). These toolkits include libraries of promoters with varying strengths and inducibility, secretion signal peptides for directing protein export, and a range of selection markers, facilitating the construction of complex genetic constructs. FungalBraid 2.0 further expanded this resource with validated parts for applications like luciferase reporters and CRISPR activation (Moreno-Giménez et al., 2023).

Modular cloning systems, such as Golden Gate and MoClo (Modular Cloning), allow for the hierarchical assembly of genetic elements in a syntax-driven manner, significantly streamlining the construction of multi-gene pathways (Mózsik et al., 2021). Systems like MIDAS exemplify the use of standardized parts for assembling multiple genes into a single vector with control over gene order and orientation, successfully applied in fungal systems for metabolic pathway reconstruction (van Dolleweerd et al., 2018).

Synthetic biology approaches enable the design and implementation of synthetic promoters for fine-tuned gene expression, pathway redesign for enhanced metabolite flux, and the construction of synthetic gene clusters encoding entire biosynthetic pathways. These "plug-and-play" toolkits and standardized assembly methods dramatically reduce the time and effort required for constructing complex genetic circuits and pathways, accelerating the development of novel fungal cell factories with enhanced capabilities. The ability to rapidly build and test synthetic genetic constructs is equipping the fungal research community with powerful tools for innovation in fungal biotechnology.

3 Metabolic Engineering

Metabolic engineering in fungal cell factories involves the targeted modification of cellular biochemical pathways to enhance the production of desired compounds, introduce the synthesis of novel molecules, or improve the utilization of specific substrates. Fungal metabolism is inherently complex, characterized by intricate regulatory networks, including carbon catabolite repression and pathway-specific regulation, as well as the presence of often silent or cryptic secondary metabolite gene clusters (García-Estrada et al., 2018; Keller, 2019; Lai et al., 2022). Metabolic engineering strategies aim to rationally redesign these pathways to optimize flux toward the target product while maintaining cellular viability and robustness.

3.1 Pathway Deregulation and Derepression

Fungi often tightly regulate metabolic pathways, particularly those involved in secondary metabolism, to conserve resources and respond to environmental cues. Deregulating or derepressing these pathways is a key strategy to increase product yield and activate silent gene clusters. A prominent example is overcoming carbon catabolite repression (CCR), where the presence of preferred carbon sources like glucose represses the expression of genes involved in utilizing alternative carbon sources or producing secondary metabolites. Modifying or deleting key repressors of CCR, such as CreA in *Aspergillus* species, can lead to constitutive expression of target genes even in the presence of glucose, significantly boosting enzyme or metabolite production (Chen et al., 2022a, b; de Assis et al., 2021). The industrial hyperproducing *Trichoderma reesei* strain RUT-C30, for instance, owes its high cellulase production, in part, to a mutation in the Cre1 repressor (Li et al., 2016).

Beyond CCR, fungal secondary metabolism is controlled by a multilayered regulatory network involving transcriptional, epigenetic, and post-translational mechanisms. Epigenetic modifications, particularly histone methylation and acetylation, play a crucial role in controlling chromatin accessibility and the expression of biosynthetic gene clusters (BGCs) (Lai et al., 2022; Shwab et al., 2007). Manipulating these modifications, for example, by targeting histone deacetylases or methyltransferases, can lead to the derepression of silent BGCs and the discovery of novel metabolites. The velvet complex (VeA-VelB-LaeA) is a key global regulator that influences both development and secondary metabolism by modulating chromatin architecture and activating BGCs (García-Estrada et al., 2018).

Furthermore, manipulating transcription factors that respond to nutrient availability and stress conditions can also lead to pathway derepression. Disrupting or modifying global regulators like AreA (nitrogen metabolism) and PacC (pH regulation) can unlock cryptic metabolic pathways (Tudzynski, 2014; Luo et al., 2017). While derepressing pathways can significantly enhance product output, it is crucial to balance these modifications to avoid imposing excessive metabolic burden or stress on the cell (Lübeck & Lübeck, 2022). Fine-tuning regulatory gene expression using inducible promoters or partial knockdowns can help optimize the trade-off between production and cellular health. Overexpression of key enzymes or inhibition of negative regulators, such as protein kinase C (Pkc) in penicillin biosynthesis, can also enhance production (Scheckhuber et al., 2013; Ziemons et al., 2017).

3.2 Overexpressing Pathway Enzymes and Gene Clusters

A fundamental strategy in metabolic engineering is to increase the catalytic capacity of rate-limiting steps in a biosynthetic pathway by overexpressing the genes encoding the relevant enzymes. This can be achieved by increasing the copy number of the target gene, using strong constitutive or inducible promoters to drive high-level expression, or optimizing codon usage for efficient translation. In the context of secondary metabolite production, where genes are often organized in BGCs, strategies include amplifying entire gene clusters or overexpressing key regulatory or enzymatic genes within the cluster. The historical development of high-producing *Penicillium chrysogenum* strains for penicillin production, for instance, involved the accumulation of amplified copies of the penicillin BGC (Liu et al., 2001). Modern approaches utilize targeted gene amplification or strong promoters to achieve similar effects for various products, including statins and enzymes (Hasan et al., 2018).

Beyond simply increasing expression levels, metabolic engineers employ strategies to improve the efficiency of multi-enzyme pathways. Direct fusion of sequential enzymes can enhance metabolic flux by channeling intermediates between active sites, as demonstrated by the significant increase in bikaverin production through the fusion of Bik2 and Bik3 enzymes (Zhao et al., 2020). Using libraries of promoters with varying strengths allows for fine-tuning the expression of each enzyme in a pathway, optimizing flux and preventing the accumulation of toxic intermediates (Xu et al., 2021).

The ability to synthesize and assemble entire synthetic gene clusters, sometimes tens of kilobases in length, enables the introduction of complete heterologous pathways or the redesign of native pathways with optimized regulatory elements and coding sequences (Mózsik et al., 2021). Platforms like HEx facilitate the high-throughput expression of fungal BGCs in heterologous hosts like *S. cerevisiae*, accelerating the discovery and characterization of novel natural products (Harvey et al., 2018). Furthermore, addressing bottlenecks in product export is crucial. Overexpression of efflux pumps or self-resistance genes can enhance metabolite secretion and alleviate intracellular toxicity, as shown with the *mlcE* efflux pump improving statin resistance in yeast expressing compact biosynthetic genes (Ley et al., 2015).

3.3 Extending Substrate Range and Pathways

Expanding the range of substrates that a fungal cell factory can utilize is critical for reducing production costs and enabling the valorization of low-cost, renewable, or waste-derived feedstocks. Fungi exhibit remarkable metabolic versatility, and metabolic engineering can further enhance their ability to utilize complex carbon and nitrogen sources.

A classic example is the engineering of *Saccharomyces cerevisiae* to ferment xylose, a major pentose sugar present in lignocellulosic biomass but not naturally metabolized by this yeast. Introducing heterologous genes encoding xylose isomerase or the xylose reductase/xylitol dehydrogenase pathway enables efficient xylose utilization, making engineered yeast strains valuable for second-generation biofuel production from agricultural residues (Jeong et al., 2020). Oleaginous yeasts like *Yarrowia lipolytica* have been engineered to metabolize a wider range of substrates, including hemicellulose-derived sugars, crude glycerol, and hydrophobic compounds from industrial waste streams, enhancing their utility for sustainable biofuel and chemical production (Ledesma-Amaro & Nicaud, 2016).

Metabolic engineering also enables the production of non-native compounds by introducing heterologous metabolic pathways. This includes expressing plant terpene synthases in yeast for producing flavors, fragrances, or pharmaceuticals. A landmark achievement was the engineering of *S. cerevisiae* to produce artemisinic acid, a precursor to the antimalarial drug artemisinin, by introducing a dozen genes from plants and other organisms, effectively creating a yeast-based factory for a plant metabolite (Ro et al., 2006). Transferring entire BGCs from fungi that are difficult to cultivate or genetically manipulate to more tractable hosts like *Aspergillus oryzae* allows for the efficient production of valuable secondary metabolites.

Pathway engineering often involves eliminating competing branch pathways that divert precursors away from the desired product. For instance, in biofuel production, reducing glycerol formation in yeast redirects more carbon flux toward ethanol (Rokas et al., 2020). By strategically deleting genes inside routes and enhancing flux in the desired pathway, metabolic engineers can optimize the channeling of cellular resources. These interventions, guided by metabolic models and flux analysis, have enabled fungi to produce a wide array of products beyond their native capabilities, including organic acids, therapeutic proteins, and various chemicals (Oftadeh et al., 2021). Recent research continues to uncover and engineer fungal enzymes with expanded substrate scope and catalytic promiscuity, such as the dioxygenase AsqJ, which can convert diverse substrates into novel heterocyclic products, further enriching the metabolic toolkit for synthetic biology applications (Einsiedler & Gulder, 2023; Sokolova et al., 2023). Stable isotope probing provides insights into substrate-specific metabolic pathway engagement, informing targeted engineering approaches (Jabinski et al., 2024).

4 Bioprocess Engineering

Optimizing the bioprocess conditions and bioreactor design is as crucial as strain engineering for maximizing the yield, productivity, and economic viability of fungal cell factories at industrial scale. Fungal cultivation presents unique challenges compared to bacterial systems, largely due to their filamentous morphology and complex physiology.

4.1 Fermentation Mode (SmF vs SSF)

The choice of fermentation mode significantly impacts fungal growth and product formation. Submerged fermentation (SmF), where fungi are cultivated in liquid media, is widely used industrially due to its ease of control over parameters like pH, temperature, dissolved oxygen, and nutrient availability, as well as its scalability and amenability to automation (Musoni et al., 2015). However, filamentous fungi in SmF can grow as dispersed hyphae, leading to high broth viscosity and mixing challenges, or form dense pellets, which can suffer from mass transfer limitations in their core.

Solid-state fermentation (SSF), where fungi grow on moist solid substrates in the absence or near absence of free water, often mimics the natural habitat of many fungi and can result in higher product titers, particularly for enzymes and certain secondary metabolites (Mandal, 2015; Musoni et al., 2015; Rengifo et al., 2023). SSF typically requires less water and energy compared to SmF and can sometimes enhance enzyme stability and secretion. However, maintaining uniform conditions (moisture content, aeration, temperature) throughout the solid matrix is challenging, making scaling up SSF processes more complex than SmF (Musoni et al., 2015).

In practice, SmF remains more common for large-scale industrial production due to better process control and reproducibility (Shivanna & Venkateswaran, 2014). However, SSF is preferred for specific applications where it offers significant advantages in terms of yield or product characteristics. Hybrid approaches, such as semi-solid or biofilm reactors, aim to combine the benefits of both modes. Process engineers carefully consider the specific fungus, product, and desired scale when selecting and optimizing the fermentation mode.

4.2 Morphology and Culture Conditions

The morphology of filamentous fungi in liquid culture, whether as dispersed mycelia or compact pellets, has a profound influence on fermentation performance. Dispersed hyphae can lead to highly viscous broths, making mixing and oxygen transfer difficult and potentially causing issues with downstream processing. Conversely, large, dense pellets can develop nutrient and oxygen gradients within their core, leading to reduced growth and product formation due to mass transfer limitations.

Controlling fungal morphology is a critical aspect of bioprocess engineering. This can be achieved by manipulating various culture conditions, including inoculum size and morphology, pH, agitation rate (shear stress), and the addition of specific chemicals or polymers (Gopalakrishnan et al., 2024; Ibrahim et al., 2015; Musoni et al., 2015; Waldherr et al., 2024). Higher agitation rates can promote mycelial fragmentation and reduce pellet size, while adjusting spore concentration or adding divalent cations like calcium can encourage pellet formation.

Adequate aeration is also paramount in fungal fermentations due to the high oxygen demand of aerobic fungi. Insufficient dissolved oxygen can lead to reduced product yield, metabolic shifts toward less desirable byproducts (e.g., ethanol in citric acid fermentation), and even cell death. Process optimization often involves maintaining dissolved oxygen levels above a critical threshold, sometimes utilizing pure oxygen or enriched air in large-scale bioreactors, particularly for highly productive *Aspergillus* or *Trichoderma* cultures. Beyond morphology and aeration, other culture conditions such as temperature, nutrient composition, and pH profiles are carefully controlled to optimize fungal growth kinetics and product biosynthesis. The optimal morphology and culture conditions are highly specific to the fungal species and the target product.

4.3 Bioreactor Design Innovations

The unique rheological properties of fungal cultures and their tendency for wall growth and fouling necessitate specialized bioreactor designs and operational strategies. Traditional stirred-tank bioreactors, while versatile, can face challenges with highly viscous fungal broths, leading to poor mixing, uneven shear distribution, and potential damage to shear-sensitive hyphae.

Alternative bioreactor designs have been developed to address these issues. Airlift bioreactors, which rely on sparging gas to circulate the liquid, provide lower shear stress and can be advantageous for cultivating shear-sensitive fungal strains or maintaining desirable pellet morphology (Cerrone et al., 2024). Biofilm reactors, where fungi grow immobilized on surfaces or carriers, can offer advantages in terms of high cell density, improved mass transfer to the immobilized cells, and easier separation of biomass from the broth.

Innovations in bioreactor internals and operational strategies are also crucial. Designing impellers and baffles to minimize fungal attachment to surfaces helps prevent wall growth and improve mixing. Periodic reversal or pulsing of agitation can help dislodge attached biomass. For spore-forming fungi, novel bioreactor internals designed to support sporulation and retain biomass can facilitate repeated-batch processes, reducing downtime and maintaining favorable morphology over multiple cycles (Soerjawinata et al., 2022). This technique has been shown to improve oxygen transfer and maintain pellet form. Overall, bioprocess engineers continuously adapt and innovate bioreactor configurations and operating procedures to meet the specific demands of fungal cultures, ensuring consistent and high-yield production over extended fermentation runs.

5 Recent Advancements in Fungal Biotechnology

The field of fungal biotechnology is experiencing rapid progress, driven by the development of novel tools, increased understanding of fungal biology, and the integration of high-throughput technologies. These advancements are enabling more efficient strain engineering, expanding the range of accessible products, and paving the way for sustainable new applications.

5.1 High-throughput and Precision Strain Engineering

The confluence of CRISPR-Cas9 technology, advanced automation, and high-throughput omics approaches has dramatically accelerated the pace of fungal strain development. Laboratories are increasingly adopting robotic platforms to automate the construction and screening of large libraries of fungal mutants, enabling parallel testing of numerous genetic modifications (Otto et al., 2021). The ease of retargeting offered by CRISPR-Cas9, combined with readily available synthetic DNA, facilitates the rapid generation of combinatorial genetic variations.

The plummeting cost of DNA sequencing and advancements in multi-omics technologies (genomics, transcriptomics, proteomics, metabolomics) allow for rapid and comprehensive characterization of engineered strains. This provides deep insights into the molecular consequences of genetic modifications and informs subsequent rounds of engineering in a data-driven, iterative design-build-test-learn (DBTL) cycle. Machine learning algorithms are increasingly being applied to analyze large omics datasets and predict the impact of specific genetic interventions on desired phenotypes, guiding rational strain design (Fong & Wong, 2023). This shift toward high-throughput, data-driven, and precise strain engineering is bringing fungal bioprocess development in line with the rapid advancements seen in other microbial systems.

5.2 Synthetic Biology and Novel Genetic Circuits

Building upon the foundational synthetic biology toolkits, researchers are now constructing sophisticated genetic circuits in fungi to achieve dynamic and conditional control over metabolic pathways and cellular behavior. A key development is the application of CRISPR-based gene regulation, utilizing a catalytically inactive Cas9 (dCas9) fused to transcriptional activator or repressor domains (CRISPRa/CRISPRi). This allows for programmable activation or repression of endogenous fungal genes without altering the genomic sequence, enabling dynamic control over gene expression in response to specific signals or growth phases (Moreno-Giménez et al., 2023). Orthogonal CRISPR activation systems have been developed for filamentous fungi, allowing for inducible activation of target genes (Moreno-Giménez et al., 2023).

The development of synthetic promoters and transcription factors responsive to novel inducers or environmental cues provides finer control over fungal metabolism. Furthermore, researchers are designing and implementing self-regulating metabolic circuits that incorporate sensor-feedback loops. These circuits can sense the concentration of an intermediate metabolite or product and automatically adjust the expression of pathway enzymes to maintain optimal flux and prevent the accumulation of toxic intermediates. This dynamic control within the cell can lead to improved pathway efficiency and higher product yields, as demonstrated in yeast engineered with a malonyl-CoA sensor that regulates pathway gene expression (Martins-Santana et al., 2018). The ability to design, assemble, and implement such complex synthetic circuits is a significant advancement, enabling programmable control and novel phenotypes in fungal hosts that were not achievable through static genetic modifications alone.

5.3 Systems Biology and Metabolic Modeling

Systems biology plays an increasingly vital role in fungal engineering by providing a holistic framework for understanding and predicting the behavior of complex fungal cell factories. Genome-scale metabolic models (GEMs) integrate genomic, transcriptomic, proteomic, and metabolomic data to create computational representations of fungal metabolism (Oftadeh et al., 2021). These models allow researchers to simulate metabolic fluxes, identify potential bottlenecks in biosynthetic pathways, and predict the impact of genetic modifications or changes in cultivation conditions on growth and product formation.

Computational platforms and tools, such as COBRApy, OptFlux, and KBase, enable the construction, simulation, and analysis of GEMs, guiding rational strain design by identifying promising targets for genetic intervention (Oftadeh et al., 2021). Coupling these in silico predictions with experimental validation using high-throughput genetic engineering and omics analysis accelerates the DBTL cycle in metabolic engineering workflows.

Omics-driven discovery approaches, such as comparative transcriptomics, proteomics, and metabolomics of high-producing versus low-producing strains, provide valuable insights into the molecular mechanisms underlying high productivity and reveal potential targets for engineering (Arshadi et al., 2023; Fierro et al., 2022). Understanding fungal regulatory and signaling networks, including stress responses and protein secretion pathways, is also crucial for developing more robust and high-secreting strains (Chi et al., 2022; Ibe & Munro, 2021). By integrating data from various omics layers and utilizing computational modeling, systems biology approaches enable a more predictive and rational approach to fungal metabolic engineering.

5.4 Unlocking New Metabolites and Pathways

Fungal genomes harbor a vast untapped potential for producing novel secondary metabolites, many of which are encoded by "silent" or cryptic gene clusters that are not expressed under standard laboratory conditions. A significant area of recent advancement focuses on strategies to activate these silent pathways and discover new bioactive compounds with potential applications such as pharmaceuticals, agrochemicals, or industrial chemicals.

One successful approach is "chemical epigenetics," which involves treating fungi with small-molecule inhibitors of epigenetic modifiers, such as histone deacetylases or DNA methyltransferases. These inhibitors can alter chromatin structure, making previously silenced BGCs accessible to the transcriptional machinery and leading to the production of novel metabolites (Xue et al., 2023). This method has led to the discovery of new compounds from fungi that were previously unknown producers.

Another strategy is the heterologous expression of fungal BGCs in a more amenable host organism, often under the control of a strong constitutive promoter, to force their expression and facilitate the identification and characterization of the encoded metabolites (de Mattos-Shipley et al., 2022). Large-scale fungal genome sequencing projects, such as the 1000 Fungal Genomes project, provide a rich source of data for identifying putative BGCs, and bioinformatics tools are used to predict the structures of potential metabolites (Valiante, 2023). Mining these genomic datasets and employing synthetic biology approaches to activate predicted pathways is greatly expanding the chemical repertoire of fungal factories. The discovery of new antibiotic compounds by activating a silent cluster in *Aspergillus nidulans* through the introduction of a global regulator gene exemplifies this approach (Lin et al., 2018).

Adaptive laboratory evolution (ALE) is another powerful approach that complements rational engineering. By subjecting fungal populations to specific selective pressures, such as gradually increasing concentrations of a product or substrate, strains with improved tolerance or production capabilities can be evolved. Subsequent genome sequencing of these evolved strains can reveal novel genetic mutations that contribute to the desired phenotype, providing insights for rational engineering efforts (Gassler et al., 2022). These combined strategies are transforming fungal cell factories into sources of novel molecules and capabilities.

5.5 Sustainability and New Applications

Fungal biotechnology is increasingly aligned with the principles of sustainability and is being applied to address global challenges in various sectors. Precision fermentation, the use of microbes like fungi to produce specific complex organic molecules, is a key component of the move toward more sustainable and ethical production methods. Fungi, particularly *Trichoderma reesei* and engineered yeasts, are at the forefront of producing animal-free proteins for the food industry, such as dairy and egg proteins, offering cruelty-free and lower-environmental-impact alternatives to traditional animal agriculture (Boukid et al., 2023; Knychala et al., 2024).

Engineered fungi are also being explored for their potential in environmental remediation. Fungal enzymes and metabolic capabilities can be harnessed for the degradation of persistent pollutants, including plastics. Research is ongoing to enhance the plastic-degrading abilities of fungi and their enzymes for use in recycling processes. Fungi play a crucial role in the concept of a "circular bioeconomy"by their ability to convert low-value agro-industrial residues, food waste, and even gaseous emissions into valuable products like biofuels, organic acids, and enzymes, thereby closing material loops and reducing waste (Arruda et al., 2021).

In agriculture, there is growing interest in engineering endophytic fungi, which live within plant tissues, to confer beneficial traits to crops, such as enhanced drought tolerance or pest resistance, offering environmentally friendly alternatives to chemical treatments (Watts et al., 2023). While historically industrial focus has been on ascomycete fungi, basidiomycete fungi (mushroom-forming fungi) are gaining attention for their unique enzymatic capabilities, particularly in lignin degradation, and their potential to produce novel secondary metabolites. Although genetic tools for basidiomycetes have lagged, recent applications of CRISPR in species like *Ustilago maydis* are expanding their potential as industrial cell factories (Wege et al., 2021). These trends highlight the dynamic and expanding nature of fungal biotechnology, continually innovating to develop greener processes, produce novel products, and enhance production efficiency across diverse applications.

6 Industrial Applications

Fungi have been utilized as industrial cell factories for over a century, contributing significantly to the production of a wide array of valuable compounds across numerous sectors (Ding & Ye, 2023). Their metabolic diversity, ability to grow on simple media, and capacity for high-level secretion make them ideal platforms for large-scale bioproduction.

6.1 Pharmaceuticals

Fungi are indispensable in the pharmaceutical industry, serving as primary producers of numerous life-saving drugs and therapeutic compounds. A prime example is the production of β-lactam antibiotics, including penicillin from *Penicillium chrysogenum* and cephalosporins from *Acremonium* spp. (Brakhage et al., 2024; Niego et al., 2023). Other commercially important fungal antibiotics include fusidic acid (*Ramularia coccinea*) and pleuromutilin (*Clitopilus* spp.) (Niego et al., 2023). Fungal species are also the source of statins, widely used to manage cardiovascular diseases, with notable producers including *Aspergillus terreus*, *Monascus ruber*, and *Penicillium citrinum* (Niego et al., 2023). Immunosuppressive agents like cyclosporin are produced by fungi such as *Tolypocladium inflatum* (Niego et al., 2023). Echinocandins, a class of antifungal drugs, are derived from species within the *Leotiomycetes* and *Eurotiomycetes* classes (Niego et al., 2023).

Beyond traditional antibiotics and statins, fungi, particularly endophytic fungi residing within plant tissues, are a rich source of novel bioactive compounds with potential anticancer properties. Compounds like wortmannin (*Penicillium funiculosum*), which targets PI3Ks and inhibits cancer cell proliferation, and paclitaxel, originally found in the Pacific yew but also produced by endophytic fungi like *Taxomyces andreanae*, are examples of fungal contributions to cancer therapy (Pokhriyal et al., 2024; Stierle et al., 1995; Yun et al., 2012). Other anticancer compounds isolated from endophytic fungi include actinomycin, bleomycin, and rapamycin (Pokhriyal et al., 2024). The exploration of fungal biodiversity, particularly endophytic species, continues to be a promising avenue for discovering new pharmaceutical leads.

6.2 Biofuels and Chemicals

Fungi are increasingly recognized as key players in the transition toward a sustainable bio-based economy, particularly in the production of biofuels and chemicals from renewable resources. They are central to the concept of third-generation biofuels, focusing on innovative approaches for producing biofuels from non-food biomass. Lignocellulosic biomass, an abundant and cost-effective feedstock, is a primary target. Anaerobic fungi, with their potent lignocellulose-hydrolyzing enzymes, show significant potential in converting this biomass into valuable products like hydrogen and ethanol (Saye et al., 2021). Fungal mycelium and extracellular enzymes are utilized in fermentation processes to produce bioethanol from lignocellulosic substrates (Antunes et al., 2020; Pérez-Chávez et al., 2019).

Fungal cell factories are also being engineered to produce advanced biodiesel (second-generation biodiesel) from various organic wastes. Oleaginous fungi from genera such as *Mortierella*, *Rhodosporidium*, and *Yarrowia* accumulate high levels of lipids, primarily triglycerides, which can be converted into biodiesel (Tabatabaei et al., 2020). Fungal lipases are also employed as biocatalysts for the enzymatic transesterification of vegetable oils to produce biodiesel, offering a more environmentally friendly alternative to chemical catalysis (Tabatabaei et al., 2020).

Mucoromycota fungi are emerging as powerful cell factories for biorefinery applications, capable of converting inexpensive waste materials into a variety of valuable bioproducts, including lipids, pigments, chitin/chitosan, organic acids, and enzymes (Dzurendova et al., 2022). Their metabolic flexibility allows for the co-production of multiple products in a single fermentation process, enhancing the economic viability of biorefineries.

6.3 Food and Agriculture

Fungi have a long and intimate relationship with human food systems, contributing to food production, preservation, and nutrient enhancement. Fermented foods and beverages produced using fungi, such as bread (with *Saccharomyces cerevisiae*), cheese (with *Penicillium*, *Aspergillus*, *Mucor*, etc.), and traditional Asian fermented foods, are fundamental components of diets worldwide (Lübeck & Lübeck, 2022; Niego et al., 2023).

Edible mushrooms are valued for their nutritional content, including protein, dietary fiber, vitamins, and minerals, and are increasingly being used to produce mycoprotein-enriched foods. Mycoprotein, primarily derived from *Fusarium venenatum*, is a commercially important protein source (Lübeck & Lübeck, 2022; Niego et al., 2023). Cultivated mushrooms like *Agaricus bisporus*, *Pleurotus ostreatus*, and *Lentinula edodes* are grown on agricultural waste, contributing to a circular food system (Lübeck & Lübeck, 2022).

Fungi are also major producers of organic acids used extensively in the food industry as preservatives, flavor enhancers, and acidulants. *Aspergillus niger* is a primary industrial producer of citric acid and gluconic acid, while other fungi produce fumaric, itaconic, malic, and other organic acids (Książek, 2023; Niego et al., 2023). Fungal enzymes are essential biocatalysts in food processing, improving quality, shelf life, texture, and flavor. Fungal amylases are used in baking, proteases in tenderizing meat and cheese production, and pectinases in fruit juice clarification (Haile & Ayele, 2022; Naeem et al., 2022). Fungal pigments, such as *Monascus* pigments and carotenoids, serve as natural food colorants with added health benefits (Agboyibor et al., 2018; Dzurendova et al., 2022).

In agriculture, fungi play a vital role in soil health and nutrient cycling. Certain fungal species are effective in the bioremediation of heavy metals and hydrocarbon pollutants in soil through biodegradation, biotransformation, and biosorption mechanisms (Al Farraj et al., 2020; Chugh et al., 2022; Prenafeta-Boldú et al., 2019). Fungi can break down resistant organic pollutants and accumulate heavy metals in their biomass. Furthermore, fungal enzymes are used to treat agricultural waste, converting lignocellulosic biomass into valuable products or animal feed (Sun et al., 2024).

6.4 Other Industrial Sectors

Fungi contribute to various other industrial sectors. In the cosmetics industry, fungal-derived bioactive compounds like polysaccharides, glycoproteins, ergothioneine, and kojic acid are used in anti-aging, antioxidant, and skin-revitalizing products (Kalač, 2016; Phasha et al., 2022; Zi et al., 2020). Species like *Lentinula edodes* and *Ganoderma* are popular sources for these compounds.

Fungi, particularly white rot fungi, are highly effective in the bioremediation of synthetic organic pollutants in the environment due to their production of ligninolytic enzymes like lignin peroxidase and laccases (Bulkan et al., 2020; Zhen et al., 2022). These enzymes can degrade a wide range of recalcitrant compounds. Filamentous fungi also exhibit biosorption capabilities, binding toxic substances, including heavy metals and pharmaceutical compounds, to their cell walls (Ghosh et al., 2023).

7 Conclusion

The field of fungal cell factories is a dynamic and rapidly advancing area of biotechnology. The synergistic integration of cutting-edge genetic engineering techniques, sophisticated metabolic engineering strategies, and optimized bioprocess design is continuously expanding the capabilities and applications of fungal systems. Modern tools like CRISPR-Cas9, synthetic biology toolkits, and systems biology approaches enable unprecedented precision and efficiency in strain development and metabolic pathway manipulation. Innovations in bioreactor design and operation address the unique challenges of fungal cultivation at scale, leading to improved yields and process economics.

Recent advancements in unlocking cryptic metabolic pathways, coupled with the growing emphasis on sustainability and circular economy principles, are positioning fungal cell factories as key players in the production of novel pharmaceuticals, sustainable biofuels and chemicals, nutritious food ingredients, and valuable compounds for other industrial sectors. The ability of fungi to grow on diverse, low-cost substrates and their inherent metabolic versatility make them highly attractive platforms for green and cost-effective biomanufacturing. Continued interdisciplinary research and technological innovation will undoubtedly unlock even greater potential in these fascinating eukaryotic workhorses, driving the development of sustainable and efficient bioprocesses for the future.

In the future, the combination of high-throughput omics technologies, artificial intelligence, and systems biology will make it possible to engineer strains in a manner that is more rational and predictive. The activation and mining of silent biosynthetic gene clusters hold a great deal of promise for the discovery of novel compounds that have relevance in the pharmaceutical and industrial sectors. In addition, the expansion of the genetic toolbox to include less-explored fungal species, such as basidiomycetes, will enable the discovery of new enzymatic capabilities and product profiles.

In the context of the challenges that are being faced by the global community regarding sustainability, fungal biotechnology will play a significant part in the transition toward circular bioeconomies. It will provide environmentally friendly solutions to produce food, fuel, pharmaceuticals, and environmental remediation. To fully harness the potential of fungal cell factories in the construction of a resilient, bio-based future, it will be essential to continue research and innovation that spans multiple disciplines.

References

Agboyibor, C., Kong, W. B., Chen, D., Zhang, A. M., & Niu, S. Q. (2018). Monascus pigments production, composition, bioactivity and its application: A review. *Biocatalysis and Agricultural Biotechnology, 16*, 433–447. https://doi.org/10.1016/j.bcab.2018.09.012

Al Abdallah, Q., Souza, A. C. O., Martin-Vicente, A., Ge, W., & Fortwendel, J. R. (2018). Whole-genome sequencing reveals highly specific gene targeting by in vitro assembled Cas9-ribonucleoprotein complexes in *Aspergillus fumigatus*. *Fungal Biology and Biotechnology, 5*(1), 11. https://doi.org/10.1186/s40694-018-0057-2

Al Farraj, D. A., Hadibarata, T., Elshikh, M. S., Al Khulaifi, M. M., & Kristanti, R. A. (2020). Biotransformation and degradation pathway of pyrene by filamentous soil fungus *Trichoderma* sp. F03. *Water Air and Soil Pollution, 231*(4), 168. https://doi.org/10.1007/s11270-020-04514-0

Antunes, F., Marçal, S., Taofiq, O., Morais, A. M. M. B., Freitas, A. C., Ferreira, I. C. F. R., & Pintado, M. (2020). Valorization of mushroom by-products as a source of value-added compounds and potential applications. *Molecules, 25*(11), 2672. https://doi.org/10.3390/molecules25112672

Arnau, J., Yaver, D., & Hjort, C. M. (2019). Strategies and challenges for the development of industrial enzymes using fungal cell factories. In *Global challenges in fungal biotechnology* (pp. 179–210). Springer. https://doi.org/10.1007/978-3-030-29541-7_7

Arruda, A. G., Evangelista, I. V., de Menezes, L. S., Fischer, J., Cardoso, V. L., Santos, L. D., & Guidini, C. Z. (2021). Production of enzymatic complex from agro-industrial biomass and its application in combustible ethanol. *Research, Society and Development, 10*(6), e40410613705. https://doi.org/10.33448/rsd-v10i6.13705

Arshadi, N., Nouri, H., & Moghimi, H. (2023). Increasing the production of the bioactive compounds in medicinal mushrooms: An omics perspective. *Microbial Cell Factories, 22*(1), 11. https://doi.org/10.1186/s12934-022-02013-x

Boukid, F., Hassoun, A., Zouari, A., Tülbek, M. Ç., Mefleh, M., Aït-Kaddour, A., & Castellari, M. (2023). Fermentation for designing innovative plant-based meat and dairy alternatives. *Foods, 12*(5), 1005. https://doi.org/10.3390/foods12051005

Brakhage, A. A., Spröte, P., Al-Abdallah, Q., Gehrke, A., Plattner, H., & Tüncher, A. (2024). Regulation of penicillin biosynthesis in filamentous fungi. In A. A. Brakhage (Ed.), *Molecular biotechnology of fungal beta-lactam antibiotics and related peptide synthetases* (pp. 1–40). Springer. https://doi.org/10.1007/b99257

Bulkan, G., Ferreira, J. A., & Taherzadeh, M. J. (2020). Removal of organic micro-pollutants using filamentous fungi. In S. Varjani, A. Pandey, R. D. Tyagi, H. H. Nho, & C. Larroche (Eds.), *Current developments in biotechnology and bioengineering* (pp. 363–395). Elsevier.

Cerrone, F., Lochlainn, C. Ó., Callaghan, T., McDonald, P., & O'Connor, K. E. (2024). Airlift bioreactor–based strategies for prolonged semi-continuous cultivation of edible Agaricomycetes. *Applied Microbiology and Biotechnology, 108*(1), 377. https://doi.org/10.1007/s00253-024-13220-4

Chen, Y., Cai, C., Yang, J., Shi, J., Song, Y., Hu, D., Ma, J., & Ju, J. (2022a). Development of the CRISPR-Cas9 system for the marine-derived fungi *Spiromastix* sp. SCSIO F190 and *Aspergillus* sp. SCSIO SX7S7. *Journal of Fungi, 8*(7), 715. https://doi.org/10.3390/jof8070715

Chen, Y., Dong, L., Alam, M. A., Pardeshi, L., Miao, Z., Wang, F., Tan, K., Hynes, M. J., Kelly, J. M., & Wong, K. H. (2022b). Carbon catabolite repression governs diverse physiological processes and development in *Aspergillus nidulans*. *MBio, 13*(2), e03734-21. https://doi.org/10.1128/MBIO.03734-21

Chi, Z., Kong, C.-C., Wang, Z.-Z., Wang, Z., Liu, G.-L., Hu, Z., & Chi, Z.-M. (2022). The signaling pathways involved in metabolic regulation and stress responses of the yeast-like fungi *Aureobasidium* spp. *Biotechnology Advances, 55*, 107898. https://doi.org/10.1016/j.biotechadv.2021.107898

Chugh, M., Kumar, L., & Bharadvaja, N. (2022). Chapter 4 – Fungal diversity in the bioremediation of toxic effluents. In S. Rodriguez-Couto & M. Shah (Eds.), *Development in wastewater treatment research and processes* (pp. 61–88). Elsevier.

Cooray, S. T., Lee, J. J. L., Chen, W. N., (2017). Evaluation of brewers' spent grain as a novel media for yeast growth. AMB Express 7, 117. https://doi.org/10.1186/s13568-017-0414-1

de Assis, L. J., Silva, L. P., Bayram, O., Dowling, P., Kniemeyer, O., Krüger, T., Brakhage, A. A., Chen, Y., Dong, L., Tan, K., Wong, K. H., Ries, L. N. A., & Goldman, G. H. (2021). Carbon catabolite repression in filamentous fungi is regulated by phosphorylation of the transcription factor CreA. *MBio, 12*(1), e03146-20. https://doi.org/10.1128/mbio.03146-20

de Mattos-Shipley, K. M. J., Lazarus, C. M., & Williams, K. (2022). Investigating fungal biosynthetic pathways using heterologous gene expression: *Aspergillus oryzae* as a heterologous host. In E. Skellam (Ed.), *Engineering natural product biosynthesis: Methods and protocols* (pp. 23–39). Springer. https://doi.org/10.1007/978-1-0716-2273-5_2

Ding, Q., & Ye, C. (2023). Microbial cell factories based on filamentous bacteria, yeasts, and fungi. *Microbial Cell Factories, 22*(1), 20. https://doi.org/10.1186/s12934-023-02025-1

Dzurendova, S., Losada, C. B., Dupuy-Galet, B. X., et al. (2022). Mucoromycota fungi as powerful cell factories for modern biorefinery. *Applied Microbiology and Biotechnology, 106*(1), 101–115. https://doi.org/10.1007/s00253-021-11720-1

Einsiedler, M., & Gulder, T. A. M. (2023). Discovery of extended product structural space of the fungal dioxygenase AsqJ. *Nature Communications, 14*(1), 3472. https://doi.org/10.1038/s41467-023-39111-2

Featherstone, C., & Jackson, S. P. (1999). DNA double-strand break repair. *Current Biology 9*, R759–R761.https://doi.org/10.1016/S0960-9822(00)80005-6

Fierro, F., Vaca, I., Castillo, N. I., García-Rico, R. O., & Chávez, R. (2022). *Penicillium chrysogenum*, a vintage model with a cutting-edge profile in biotechnology. *Microorganisms, 10*(3), 573. https://doi.org/10.3390/microorganisms10030573

Fong, J. H. C., & Wong, A. S. L. (2023). Advancing CRISPR/Cas gene editing with machine learning. *Current Opinion in Biomedical Engineering, 28*, 100477. https://doi.org/10.1016/j.cobme.2023.100477

García-Estrada, C., Domínguez-Santos, R., Kosalková, K., & Martín, J. F. (2018). Transcription factors controlling primary and secondary metabolism in filamentous fungi: The β-lactam paradigm. *Fermentation, 4*(4), 105. https://doi.org/10.3390/fermentation4040105

Gassler, T., Baumschabl, M., Sallaberger, J., Egermeier, M., & Mattanovich, D. (2022). Adaptive laboratory evolution and reverse engineering enhances autotrophic growth in *Pichia pastoris*. *Metabolic Engineering, 69*, 112–121. https://doi.org/10.1016/j.ymben.2021.11.007

Ghosh, S., Rusyn, I., Dmytruk, O. V., Dmytruk, K. V., Onyeaka, H., Gryzenhout, M., & Gafforov, Y. (2023). Filamentous fungi for sustainable remediation of pharmaceutical compounds, heavy metal and oil hydrocarbons. *Frontiers in Bioengineering and Biotechnology, 11*, 1106973. https://doi.org/10.3389/fbioe.2023.1106973

Gopalakrishnan, A. V., Suresh, H., Mariappan, G., Kanagaraja, A., & Raman, P. (2024). Industrial microbial bioprocess development: A comprehensive overview. In P. Verma (Ed.), *Industrial microbiology and biotechnology: A new horizon of the microbial world* (pp. 89–154). Springer Nature. https://doi.org/10.1007/978-981-97-6270-5_4

Haile, S., & Ayele, A. (2022). Pectinase from microorganisms and its industrial applications. *Scientific World Journal, 2022*, 1–15. https://doi.org/10.1155/2022/1881305

Harvey, C. J. B., Tang, M., Schlecht, U., Horecka, J., Fischer, C. R., Lin, H. C., Li, J., Naughton, B., Cherry, J., Miranda, M., Li, Y. F., Chu, A. M., Hennessy, J. R., Vandova, G. A., Inglis, D., Aiyar, R. S., Steinmetz, L. M., Davis, R. W., Medema, M. H., Sattely, E., Khosla, C., Onge, R. P. S., Tang, Y., & Hillenmeyer, M. E. (2018). HEx: A heterologous expression platform for the discovery of fungal natural products. *Science Advances, 4*(4), eaar5459. https://doi.org/10.1126/sciadv.aar5459

Hasan, H., Abd Rahim, M. H., Campbell, L., Carter, D., Abbas, A., & Montoya, A. (2018). Overexpression of acetyl-CoA carboxylase in *Aspergillus terreus* to increase lovastatin production. *New Biotechnology, 44*, 55–61. https://doi.org/10.1016/j.nbt.2018.04.008

He, Y., Shao, Y., & Chen, F. (2014). Efficient gene targeting in *ligase IV*-deficient *Monascus ruber* M7 by perturbing the non-homologous end joining pathway. *Fungal Biology, 118*(10), 846–854. https://doi.org/10.1016/j.funbio.2014.07.003

Ibe, C., & Munro, C. A. (2021). Fungal cell wall proteins and signaling pathways form a cytoprotective network to combat stresses. *Journal of Fungi, 7*(9), 739. https://doi.org/10.3390/jof7090739

Ibrahim, D., Weloosamy, H., & Lim, S.-H. (2015). Effect of agitation speed on the morphology of *Aspergillus niger* HFD5A-1 hyphae and its pectinase production in submerged fermentation. *World Journal of Biological Chemistry, 6*(3), 265–271. https://doi.org/10.4331/wjbc.v6.i3.265

Jabinski, S., de Matos Rangel, W., Kopáček, M., Jílková, V., Jansa, J., & Meador, T. B. (2024). Constraining activity and growth substrate of fungal decomposers via assimilation patterns of inorganic carbon and water into lipid biomarkers. *Applied and Environmental Microbiology, 90*(4), e02065–e02023. https://doi.org/10.1128/aem.02065-23

Jeong, D., Oh, E. J., Ko, J. K., Nam, J.-O., Park, H.-S., Jin, Y.-S., Lee, E. J., & Kim, S. R. (2020). Metabolic engineering considerations for the heterologous expression of xylose-catabolic pathways in *Saccharomyces cerevisiae*. *PLoS One, 15*(7), e0236294. https://doi.org/10.1371/journal.pone.0236294

Kalač, P. (2016). Health-stimulating compounds and effects. In P. Kalač (Ed.), *Edible mushrooms* (pp. 137–153). Elsevier.

Keller, N. P. (2019). Fungal secondary metabolism: Regulation, function and drug discovery. *Nature Reviews. Microbiology, 17*(3), 167–180. https://doi.org/10.1038/s41579-018-0128-f

Knychala, M. M., Boing, L. A., Ienczak, J. L., Trichez, D., & Stambuk, B. U. (2024). Precision fermentation as an alternative to animal protein, a review. *Fermentation, 10*(6), 315. https://doi.org/10.3390/fermentation10060315

Książek, E. (2023). Citric acid: Properties, microbial production, and applications in industries. *Molecules, 29*(1), 22. https://doi.org/10.3390/molecules29010022

Lai, Y., Wang, L., Zheng, W., & Wang, S. (2022). Regulatory roles of histone modifications in filamentous fungal pathogens. *Journal of Fungi, 8*(11), 1157. https://doi.org/10.3390/jof8111157

Ledesma-Amaro, R., & Nicaud, J. M. (2016). Metabolic engineering for expanding the substrate range of *Yarrowia lipolytica*. *Trends in Biotechnology, 34*(10), 793–803. https://doi.org/10.1016/j.tibtech.2016.04.001

Leitão, M. de C., Vieira, H. L. de A., Torres, F. A. G., Perez, A. L. A., Piva, L. C., Reis, V. C. B., Coelho, C. M. (2024). Chapter 14 - Biotechnological applications of CRISPR-Cas systems in fungi. In: Joshi, S., Verma, D., Gupta, R. Kr. (Eds.), *CRISPR-Cas System in Translational Biotechnology*. Academic Press, pp. 237–263..https://doi.org/10.1016/B978-0-323-91808-4.00011-0

Ley, A., Coumou, H. C., & Frandsen, R. J. N. (2015). Heterologous expression of MlcE in *Saccharomyces cerevisiae* provides resistance to natural and semi-synthetic statins. *Metabolic Engineering Communications, 2*, 117–124. https://doi.org/10.1016/j.meteno.2015.09.003

Li, C., Lin, F., Li, Y., Wei, W., Wang, H., Qin, L., Zhou, Z., Li, B., Wu, F., & Chen, Z. (2016). A β-glucosidase hyperproduction Trichoderma reesei mutant reveals a potential role of cel3D in cellulase production. *Microbial Cell Factories, 15*, 151. https://doi.org/10.1186/s12934-016-0550-3

Lin, H., Lyu, H., Zhou, S., Yu, J., Keller, N. P., & Chen, L. (2018). Deletion of a global regulator LaeB leads to the discovery of novel polyketides in *Aspergillus nidulans*. *Organic & Biomolecular Chemistry, 16*(26), 4973–4976. https://doi.org/10.1039/C8OB01326H

Liu, G., Casqueiro, J., Gutierrez, S., Kosalkova, K., Castillo, N. I., & Martin, J. F. (2001). Elicitation of penicillin biosynthesis by alginate in *Penicillium chrysogenum*, exerted on *pcbAB*, *pcbC*, and *penDE* genes at the transcriptional level. *Applied and Environmental Microbiology, 67*(2), 812–818. https://doi.org/10.1128/AEM.67.2.812-818.2001

Lübeck, M., & Lübeck, P. S. (2022). Fungal cell factories for efficient and sustainable production of proteins and peptides. *Microorganisms, 10*(4), 753. https://doi.org/10.3390/microorganisms10040753

Luo, Z., Ren, H., Mousa, J. J., Rangel, D. E. N., Zhang, Y., Bruner, S. D., & Keyhani, N. O. (2017). The PacC transcription factor regulates secondary metabolite production and stress response but has only minor effects on virulence in the insect pathogenic fungus *Beauveria bassiana*. *Environmental Microbiology, 19*(3), 1188–1204. https://doi.org/10.1111/1462-2920.13648

Mandal, M. (2015). Production of multi-enzyme system by isolated *Aspergillus* sp under solid state fermentation and submerged fermentation. *International Journal of Latest Trends in Engineering and Technology, 5*(3), 390–398.

Martins-Santana, L., Nora, L. C., Sanches-Medeiros, A., Lovate, G. L., Cassiano, M. H. A., & Silva-Rocha, R. (2018). Systems and synthetic biology approaches to engineer fungi for fine chemical production. *Frontiers in Bioengineering and Biotechnology, 6*, 117. https://doi.org/10.3389/fbioe.2018.00117

Moreno-Giménez, E., Gandía, M., Sáez, Z., Manzanares, P., Yenush, L., Orzáez, D., Marcos, J. F., & Garrigues, S. (2023). FungalBraid 2.0: Expanding the synthetic biology toolbox for the biotechnological exploitation of filamentous fungi. *Frontiers in Bioengineering and Biotechnology, 11*. https://doi.org/10.3389/fbioe.2023.1222812

Matheus de Castro, Leitão Hugo Lins de Albuquerque, Vieira Fernando Araripe Gonçalves, Torres Ana Laura Alfonso, Perez Luíza Cesca, Piva Viviane Castelo Branco, Reis Cintia Marques, Coelho CRISPR-Cas System in Translational Biotechnology Biotechnological applications of CRISPR-Cas systems in fungi Elsevier 237-263

Mózsik, L., Pohl, C., Meyer, V., Bovenberg, R. A. L., Nygård, Y., & Driessen, A. J. M. (2021). Modular synthetic biology toolkit for filamentous fungi. *ACS Synthetic Biology, 10*(11), 2850–2861. https://doi.org/10.1021/acssynbio.1c00260

Musoni, M., Destain, J., Thonart, P., Bahama, J.-B., & Delvigne, F. (2015). Bioreactor design and implementation strategies for the cultivation of filamentous fungi and the production of fungal metabolites: From traditional methods to engineered systems. *Biotechnology, Agronomy and Society and Environment, 19*(3), 346–361.

Naeem, M., Manzoor, S., Abid, M. U. H., Tareen, M. B. K., Asad, M., Mushtaq, S., Ehsan, N., Amna, D., Xu, B., & Hazafa, A. (2022). Fungal proteases as emerging biocatalysts to meet the current challenges and recent developments in biomedical therapies: An updated review. *Journal of Fungi, 8*(2), 109. https://doi.org/10.3390/jof8020109

Niego, A. G. T., Lambert, C., Mortimer, P., Thongklang, N., Rapior, S., Grosse, M., Schrey, H., Charria-Girón, E., Walker, A., Hyde, K. D., & Stadler, M. (2023). The contribution of fungi to the global economy. *Fungal Diversity, 121*(1), 95–137. https://doi.org/10.1007/s13225-023-00520-9

Nødvig, C. S., Nielsen, J. B., Kogle, M. E., & Mortensen, U. H. (2015). A CRISPR-Cas9 system for genetic engineering of filamentous fungi. *PLoS One, 10*(7), e0133085. https://doi.org/10.1371/journal.pone.0133085

Nødvig, C. S., Hoof, J. B., Kogle, M. E., Jarczynska, Z. D., Lehmbeck, J., Klitgaard, D. K., & Mortensen, U. H. (2018). Efficient oligo nucleotide mediated CRISPR-Cas9 gene editing in Aspergilli. *Fungal Genetics and Biology, 115*, 78–89. https://doi.org/10.1016/j.fgb.2018.01.004

Obidi, O. F., & Moyosore, W. A. (2018). Growth kinetics and cheap substitute for industrial cultivation of the yeast *Saccharomyces cerevisiae*. *Biokemistri, 30*(2), 73–88.

Oftadeh, O., Salvy, P., Masid, M., Curvat, M., Miskovic, L., & Hatzimanikatis, V. (2021). A genome-scale metabolic model of *Saccharomyces cerevisiae* that integrates expression constraints and reaction thermodynamics. *Nature Communications, 12*(1), 4790. https://doi.org/10.1038/s41467-021-25158-6

Otto, M., Skrekas, C., Gossing, M., Gustafsson, J., Siewers, V., & David, F. (2021). Expansion of the yeast modular cloning toolkit for CRISPR-based applications, genomic integrations and combinatorial libraries. *ACS Synthetic Biology, 10*(12), 3461–3474. https://doi.org/10.1021/acssynbio.1c00408

Pérez-Chávez, A. M., Mayer, L., & Albertó, E. (2019). Mushroom cultivation and biogas production: A sustainable reuse of organic resources. *Energy for Sustainable Development, 50*, 50–60. https://doi.org/10.1016/j.esd.2019.02.008

Phasha, V., Senabe, J., Ndzotoyi, P., Okole, B., Fouche, G., & Chuturgoon, A. (2022). Review on the use of kojic acid—A skin-lightening ingredient. *Cosmetics, 9*(3), 64. https://doi.org/10.3390/cosmetics9030064

Pokhriyal, A., Kapoor, N., Negi, S., Sharma, G., Chandra, S., Gambhir, L., & Coutinho, H. D. M. (2024). Endophytic fungi: Cellular factories of novel medicinal chemistries. *Bioorganic Chemistry, 150*, 107576. https://doi.org/10.1016/j.bioorg.2024.107576

Prenafeta-Boldú, F. X., de Hoog, G. S., & Summerbell, R. C. (2019). Fungal communities in hydrocarbon degradation. In T. J. McGenity (Ed.), *Microbial communities utilizing hydrocarbons and lipids: Members, metagenomics and ecophysiology* (pp. 307–342). Springer.

Rengifo, L. R., Rosas, P., Méndez, N., Ludeña, Y., Sirvas, S., Samolski, I., & Villena, G. K. (2023). Comparison of pigment production by filamentous fungal strains under submerged (SmF) and surface adhesion fermentation (SAF). *Journal of Fungi, 9*(1), 48. https://doi.org/10.3390/jof9010048

Ro, D.-K., Paradise, E. M., Ouellet, M., Fisher, K. J., Newman, K. L., Ndungu, J. M., Ho, K. A., Eachus, R. A., Ham, T. S., Kirby, J., Chang, M. C. Y., Withers, S. T., Shiba, Y., Sarpong, R., & Keasling, J. D. (2006). Production of the antimalarial drug precursor artemisinic acid in engineered yeast. *Nature, 440*(7086), 940–943. https://doi.org/10.1038/nature04640

Rokas, A., Mead, M. E., Steenwyk, J. L., Raja, H. A., & Oberlies, N. H. (2020). Biosynthetic gene clusters and the evolution of fungal chemodiversity. *Natural Product Reports, 37*(6), 868–878. https://doi.org/10.1039/c9np00045c

Sachindra T., Cooray Jaslyn J. L., Lee Wei Ning, Chen (2017) Evaluation of brewers' spent grain as a novel media for yeast growth AMB Express 7(1) https://doi.org/10.1186/s13568-017-0414-1

Saye, L. M. G., Navaratna, T. A., Chong, J. P. J., O'Malley, M. A., Theodorou, M. K., & Reilly, M. (2021). The anaerobic fungi: Challenges and opportunities for industrial lignocellulosic biofuel production. *Microorganisms, 9*(4), 694. https://doi.org/10.3390/microorganisms9040694

Scheckhuber, C. Q., Veenhuis, M., & van der Klei, I. J. (2013). Improving penicillin biosynthesis in *Penicillium chrysogenum* by glyoxalase overproduction. *Metabolic Engineering, 18*, 36–43. https://doi.org/10.1016/j.ymben.2013.04.003

Shivanna, G. B., & Venkateswaran, G. (2014). Phytase production by *Aspergillus niger* CFR 335 and *Aspergillus ficuum* SGA 01 through submerged and solid-state fermentation. *Scientific World Journal, 2014*, 1–8. https://doi.org/10.1155/2014/392615

Shwab, E. K., Jin, W. B., Tribus, M., Galehr, J., Graessle, S., & Keller, N. P. (2007). Histone deacetylase activity regulates chemical diversity in *Aspergillus*. *Eukaryotic Cell, 6*(10), 1656–1665. https://doi.org/10.1128/EC.00186-07

Soerjawinata, W., Kockler, I., Wommer, L., Frank, R., Schüffler, A., Schirmeister, T., Ulber, R., & Kampeis, P. (2022). Novel bioreactor internals for the cultivation of spore-forming fungi in pellet form. *Engineering in Life Sciences, 22*(8), 474–483. https://doi.org/10.1002/elsc.202100094

Sokolova, N., Zhang, L., Deravi, S., Oerlemans, R., Groves, M. R., & Haslinger, K. (2023). Structural characterization and extended substrate scope analysis of two Mg^{2+} dependent O-methyltransferases from bacteria. *Chembiochem, 24*(16), e202300076. https://doi.org/10.1002/cbic.202300076

Stierle, A., Strobel, G., Stierle, D., Grothaus, P., & Bignami, G. (1995). The search for a taxol-producing microorganism among the endophytic fungi of the pacific yew, *Taxus brevifolia*. *Journal of Natural Products, 58*(9), 1315–1324. https://doi.org/10.1021/np50123a002

Sun, Z., Wu, Y., Long, S., Feng, S., Jia, X., Hu, Y., Ma, M., Liu, J., & Zeng, B. (2024). *Aspergillus oryzae* as a cell factory: Research and applications in industrial production. *Journal of Fungi, 10*(4), 248. https://doi.org/10.3390/jof10040248

Tabatabaei, M., Alidadi, A., Dehhaghi, M., Panahi, H. K. S., Lam, S. S., Nizami, A.-S., Aghbashlo, M., & Jouzani, G. S. (2020). Fungi as bioreactors for biodiesel production. In G. Salehi Jouzani, M. Tabatabaei, & M. Aghbashlo (Eds.), *Fungi in Fuel biotechnology* (pp. 69–101). Springer. https://doi.org/10.1007/978-3-030-44488-4_3

Tsukuda, T., Fleming, A. B., Nickoloff, J. A., & Osley, M. A. (2005). Chromatin remodelling at a DNA double-strand break site in *Saccharomyces cerevisiae*. *Nature, 438*(7066), 379–383. https://doi.org/10.1038/nature04148

Tudzynski, B. (2014). Nitrogen regulation of fungal secondary metabolism in fungi. *Frontiers in Microbiology, 5*. https://doi.org/10.3389/fmicb.2014.00656

Valiante, V. (2023). Advances in synthetic biology of fungi and contributions to the discovery of new molecules. *Chembiochem, 24*(4), e202300008. https://doi.org/10.1002/cbic.202300008

van Dolleweerd, C. J., Kessans, S. A., Van de Bittner, K. C., Bustamante, L. Y., Bundela, R., Scott, B., Nicholson, M. J., & Parker, E. J. (2018). MIDAS: A modular DNA assembly system for synthetic biology. *ACS Synthetic Biology, 7*(4), 1018–1029. https://doi.org/10.1021/acssynbio.7b00363

Waldherr, P., Bliatsiou, C., Böhm, L., & Kraume, M. (2024). Comparative study of fluid dynamic stress on and model systems. *Chemie Ingenieur Technik, 96*(4), 471–485. https://doi.org/10.1002/cite.202300193

Wang, Q., & Coleman, J. J. (2019). Progress and challenges: Development and implementation of CRISPR/Cas9 technology in filamentous fungi. *Computational and Structural Biotechnology Journal, 17*, 761–769. https://doi.org/10.1016/j.csbj.2019.06.007

Watts, D., Palombo, E. A., Jaimes Castillo, A., & Zaferanloo, B. (2023). Endophytes in agriculture: Potential to improve yields and tolerances of agricultural crops. *Microorganisms, 11*(5), 1276. https://doi.org/10.3390/microorganisms11051276

Wege, S.-M., Gejer, K., Becker, F., Bölker, M., Freitag, J., & Sandrock, B. (2021). Versatile CRISPR/Cas9 systems for genome editing in *Ustilago maydis*. *Journal of Fungi, 7*(2), 149. https://doi.org/10.3390/jof7020149

Xu, L., Liu, P., Dai, Z., Fan, F., & Zhang, X. (2021). Fine-tuning the expression of pathway gene in yeast using a regulatory library formed by fusing a synthetic minimal promoter with different Kozak variants. *Microbial Cell Factories, 20*(1), 148. https://doi.org/10.1186/s12934-021-01641-z

Xue, M., Hou, X., Fu, J., Zhang, J., Wang, J., Zhao, Z., Xu, D., Lai, D., & Zhou, L. (2023). Recent advances in search of bioactive secondary metabolites from fungi triggered by chemical epigenetic modifiers. *Journal of Fungi, 9*(2), 172. https://doi.org/10.3390/jof9020172

Yun, J., Lv, Y. G., Yao, Q., Wang, L., Li, Y. P., & Yi, J. (2012). Wortmannin inhibits proliferation and induces apoptosis of MCF-7 breast cancer cells. *European Journal of Gynaecological Oncology, 33*(4), 367–369.

Zhao, M., Zhao, Y., Yao, M., Iqbal, H., Hu, Q., Liu, H., Qiao, B., Li, C., Skovbjerg, C. A. S., Nielsen, J. C., Nielsen, J., Frandsen, R. J. N., Yuan, Y., & Boeke, J. D. (2020). Pathway engineering in yeast for synthesizing the complex polyketide bikaverin. *Nature Communications, 11*. https://doi.org/10.1038/s41467-020-19984-3

Zhen, J., Wang, J., Li, L., Du, Z., Li, G., & Yang, W. (2022). Biodegradation of PAHs by *Trametes hirsuta* zlh237 and effect of bioaugmentation on PAHs-contaminated soil. *Polish Journal of Environmental Studies, 31*(3), 2473–2484. https://doi.org/10.15244/pjoes/143010

Zi, Y., Jiang, B., He, C., & Liu, L. (2020). Lentinan inhibits oxidative stress and inflammatory cytokine production induced by benzo(a)pyrene in human keratinocytes. *Journal of Cosmetic Dermatology, 19*(2), 502–507. https://doi.org/10.1111/jocd.13005

Ziemons, S., Koutsantas, K., Becker, K., Dahlmann, T., & Kück, U. (2017). Penicillin production in industrial strain *Penicillium chrysogenum* P2niaD18 is not dependent on the copy number of biosynthesis genes. *BMC Biotechnology, 17*(1), 16. https://doi.org/10.1186/s12896-017-0335-8

Fungal Biotechnology: Contribution to the Circular Economy. Recycling of Waste. Sustainability

Şule Inci and Sevda Kirbağ

Abstract Production is increased gradually with the rapidly increasing population in the world. Reducing, recycling, or researching energy opportunities for waste generated with increased production are some of the noteworthy issues. Especially in recent years, environmental awareness has been increasing with the increase in waste. Consumers now pay attention to whether the products they use are sustainable after their expiration date and whether they do not harm the environment during and after production. For this purpose, innovative and sustainable surfaces based on fungi have been produced in recent years. Fungi biotechnology is a very important issue in terms of ensuring the sustainability of waste generated as a result of production and contributing to the economy. In particular, mushrooms have properties such as converting organic waste into valuable products, closing material cycles in the construction sector, producing biofuels, creating sustainable textile and construction materials, and bioremediation, which put them among the sought-after biomaterials. Because mushrooms are easily degradable in nature and do not harm the environment, they are also an important resource in the fight against climate change. Especially in recent years, the use of mycelium-based composites in construction materials, the use of mushroom bioreactors in cleaning polluted waters, and the use of used mushroom composts in agriculture and gardening to contribute to recycling have made them a sustainable material option in the future. Therefore, mushroom biotechnology aims to make, use, and recycle in order to achieve higher environmental gains with minimum costs, unlike the make-use-dispose application. For this purpose, in this study, the contribution of mushrooms, which have unique aromatic tastes, nutritional values, and medical effects, to waste recycling and economy will be discussed.

Keywords Fungal biotechnology · Mushroom · Mycelium · Waste recycling

Ş. Inci
Health Services Vocational, Istanbul Rumeli University, Istanbul, Turkey

S. Kirbağ (✉)
Firat University, Science Faculty, Department of Biology, Elazig, Turkey
e-mail: skirbag@firat.edu.tr

A. Gupta (ed.), *Prospects of Fungal Biotechnologies for Livestock Volume 2*,
Fungal Biology, https://doi.org/10.1007/978-3-032-06478-3_4

1 Introduction

The rapidly increasing population in the world results in increased waste production. Therefore, increasing anthropogenic activities to meet people's living standards and accommodate the ever-growing population greatly affect the ecological balance and economic development (Singh & Thakur, 2023; Kržišnik & Gonçalves, 2023). For this reason, the concept of sustainability has attracted considerable attention in developed and developing countries in recent years. The United Nations, in particular, has determined certain Sustainable Development Goals that it aims to achieve by 2030, taking into account the holistic increase in human living standards worldwide (Singh & Thakur, 2023).

Bioeconomy, which is among the sustainable development goals, reveals the full potential of all kinds of sustainably sourced biomass and organic municipal waste, including residual biomass such as crop residues, industrial side wastes, food wastes, by converting them into value-added products (FAO, 2019; Lange et al., 2021). Therefore, bioeconomy promotes economic growth through the transition from fossil-based products to bio-based products and energy. Bio-resource and bio-ecology visions are vital together with the biotechnological approach to integrate the bioeconomy (Singh & Thakur, 2023). Fungal biotechnology in particular offers great opportunities for recycling natural wastes to meet the growing demand for food, feed, fertilizer, and energy (Singh et al., 2010).

Mushrooms make significant contributions in the conversion of organic matter into various "high-value" products to address socio-economic concerns. The use of mushroom bio-based products offers sustainable approaches to secure and improve food supply for a growing population and limit greenhouse gas emissions (Tiwari & Park, 2024). In particular, fungal bioproducts enable the sustainable production of fuels, chemicals, textiles, vegan leathers, food and feed, cheese, mycoprotein (meat alternatives), food additives (polyunsaturated fatty acids), therapeutics, vitamins (β-carotene), organic acids, proteins, enzymes, transportation-based industrial materials, and composite materials (packaging and construction materials) (Meyer et al., 2020). In addition, mushrooms are not only consumed as food but also used for therapeutic purposes (Kulshreshtha et al., 2014).

For this purpose, the contribution of mushrooms to waste recycling and the economy will be emphasized in the chapter.

1.1 Use of Fungi in Bioremediation

Today, serious problems are occurring in the world with the increase in pollution. Especially, the climate change and the damage to biodiversity are just a few of the universal problems (Yadav et al., 2021). The level of pollution is increasing, especially with agricultural and industrial wastes that cause heavy metals, pharmaceuticals, xenobiotics, and various organic pollutants to accumulate in the environment. Lack of adequate waste management also accelerates the process (Barh et al., 2019).

Researchers are trying to take the necessary measures to eliminate these problems with a more suitable technology. Especially, being cost-effective and environmentally friendly technologies are one of the issues that attracts attention. Micromediation is an economical and environmentally friendly process that purifies polluted water and soil from harmful pollutants through fungal biomass (Kulshreshtha et al., 2014). The fungi used in bioremediation and the enzymes they secrete are called mycoremediation because they have the ability to reduce many stubborn environmental pollutants with a broad spectrum and to decompose industrial and agricultural industrial wastes and convert them into usable products (Gadd, 2001; Goltapeh et al., 2013; Kulshreshtha et al., 2014). Fungi use three different methods for this: biodegradation, biosorption, and biotransformation (Fig. 1).

Although fungi produce many bioactive compounds, fungal enzymes are the most prominent in bioremediation applications (Mougin et al., 2009). Most fungi do not have substrate specificity and are secreted into the substrate as exoenzymes (Vural et al., 2018). Since fungi generally provide extracellular degradation of

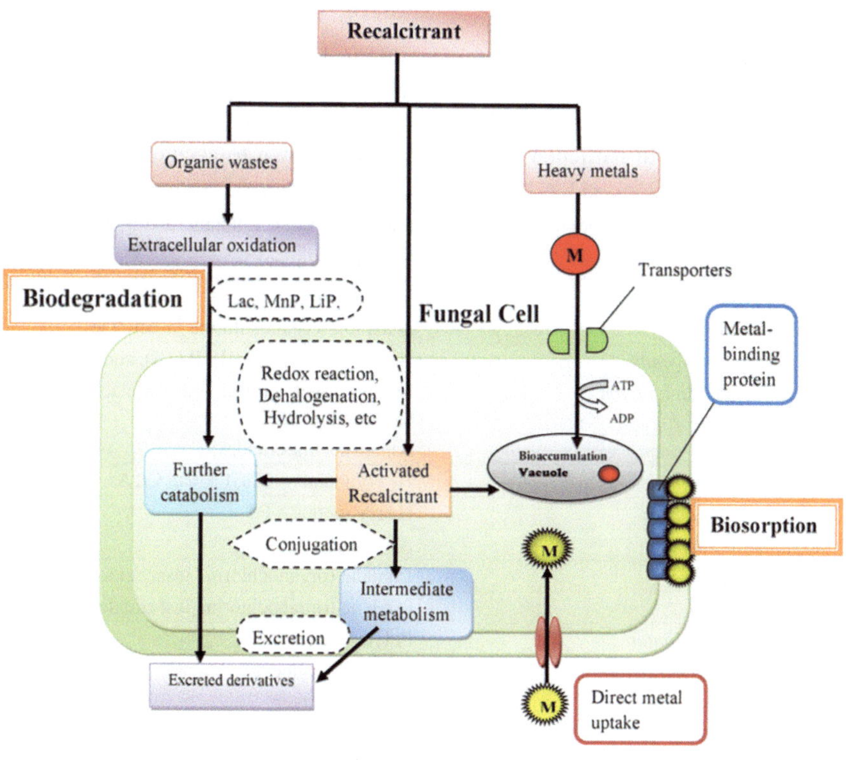

Fig. 1 Use of fungi in bioremediation. (Yadav et al., 2021)

macromolecules, they produce and secrete different enzymes to their external environment at higher rates (Kahraman & Yeşilada, 2001; Harms et al., 2011). They can absorb various substrates in their cells and metabolize them into small molecules. The fact that fungi have a well-equipped enzymatic system provides them with the ability to grow better on a wide range of synthetic substrates in nature. Due to their habitat diversity and ability to secrete a large number of enzymes, fungi are very important potential candidates for bioremediation in various areas (Deshmukh et al., 2016). In addition to their enzyme secretion ability, hyphal development also gives them the ability to overcome the problem. With their mycelial structure, they destroy and neutralize toxins in the soil before they enter the food chain (Stamets, 2005). Fungi, like other microorganisms, constitute the living biomass of the soil (Hoshino & Morimoto, 2008). However, with their high reproduction rate, complex enzymatic and secretion systems, they can sometimes replace other physical and chemical remediation agents. In addition, due to the biomass, hyphal structure, and longer life cycle of fungi, they are more resistant to bioremediation of contaminated soils and high toxin concentrations than bacteria that play a role in bioremediation. This also gives them an advantage (Evans & Hedge, 2001; Singh et al., 2015). In addition, fungi can be more easily transported, genetically modified, and propagated than other microorganisms used in bioremediation (Vural et al., 2018). Some fungal species associated with mycoremediation are given in Table 1 (Kumar et al., 2021).

The cell wall of fungi consists of mineral ions, lipids, nitrogen-containing polysaccharides, polyphosphates, proteins, and chitin. The outer part of the cell wall has the ability to act as a ligand to bind metal ions. Thus, they ensure the elimination of inorganic metals. In particular, several fungi are known to accumulate metals in

Table 1 Specific fungal species used in bioremediation of different pollutants and application areas

Fungal species	Various pollutants
Aspergillus carbonarius	Methyl blue
Aspergillus niger	Diazinon (organochlorine insecticide)
Penicilium chrysogenum	Heavy metals—Cd (II), Pd (II), and Cu (II)
Penicillium simplicissimum	Methyl violet, Crystal violet, Malachite green, Cotton blue
Fusarium proliferatum	Allethrin (Insecticides)
Ascomycota consortium	Heavy metals—Mn, Fe, Cu, Cr, As
Talaromyces islandica	Heavy metals—Pd
Bjerkandera adusta	Afrazine
Phlebia brevispora	Lindane (organochlorine insecticide)
Phlebia lindtneri	Lindane (organochlorine insecticide)
Trichoderma harzianum	Polyethylene
Cylindrocephalu maurelium	Mordant orange-1
Lasiodiploida sp.	Malachite green
Tinea versicolor	Ketoprofen

their mycelium and spores (Jha et al., 2011; Selvam et al., 2013; Gupta et al., 2015; Igir et al., 2018).

The most commonly used white-rot fungi in bioremediation are *Phanerochaete chrysosporium*, *Coriolus versicolor*, and *Trametes versicolor* (Demir et al., 2006). *Bjerkandera adjusta* and *Pleurotus* sp. fungi have been proven to be bioremediation agents in studies (dos Santos Bazanella, 2016).

It is known from studies that *Lentinula edodes*, *Agaricus bisporus*, *Irpex lacteus*, and *Pleurotus tuber-regium* white-rot fungi can decompose xenobiotic compounds (Singh, 2006; Adenipekun & Lawal, 2012).

White-rot fungi have phenoloxidase enzyme systems such as laccases and peroxidases (Reddy & Mathew, 2001; Anastasi et al., 2013). Laccases are produced to varying degrees by almost all species of white-rot fungi (Baldrian, 2006). In studies, laccases have attracted attention due to their ability to degrade pollutants such as pyrene (Anastasi et al., 2009), benzo[a]pyrene (Li et al., 2011), and chrysene (Nikiforova et al., 2010). Peroxidases, on the other hand, participate in the biodegradation of lignocelluloses by separating into enzyme systems such as lignin (LiP), manganese (MnP), and versatile peroxidases (VP) and participate in the biological transformation of more resistant compounds (Vural et al., 2018).

Some enzymes produced by fungi have been found to degrade not only natural compounds but also persistent, non-polymeric pollutants such as nitrotoluenes, synthetic and organic dyes, and pentachlorophenol (Van Acken et al., 1999; Ollikka et al., 1993; Heinfling et al., 1998; Lin et al., 1990). It has also been reported in the literature that various fungal species can degrade synthetic polymers such as plastics (Luz et al., 2013). Phylogenetic analyses suggest that fungi may have acquired this degradation capacity evolutionarily by utilizing natural carbon sources such as algae and lignocellulosic compounds before the emergence of synthetic polymers (Daly ct al., 2021).

In this context, fungi stand out as important organisms in the biodegradation of complex polymers and are considered as effective bioremediation agents to combat plastic pollution. Recent studies analyzed 18 fungal strains examining the degradation potential of common types of plastics such as polyurethane, polyethylene, and tire rubber, and revealed that *Botryotinia cinerea, Fusarium, Penicillium,* and *Trichoderma* species in particular have a high capacity to degrade these materials (Ibrahim et al., 2024).

Research shows that fungi can at least partially metabolize plastic polymers and in some cases use these materials directly as a food source. Strategies to promote the biodegradation of plastics are mostly based on enabling fungi to thrive on these materials and involving them in natural degradation processes. Despite all their chemical resistance, plastics are known to be structurally weakened over time by the action of both fungi and bacteria (Magnin et al., 2019).

Fungi are predicted to use the limited sugars available in the environment to colonize the polymer surface and produce degradation enzymes (Li et al., 2023). In order to improve the depolymerization and degradation efficiency of plastic compounds, it has been emphasized that fungi with enzymatic activities specifically developed and proven to degrade cellulose, lignin, and chitin should be selected.

Lipases, proteases, carboxylesterases, esterases, and lignocellulolytic enzymes play important roles in the degradation of plastic compounds (Srikanth et al., 2022).

In a study by Nagy et al. (2014), the effect of biocomposites based on *Agaricus bisporus* and *Lactarius piperatus* on the biosorption of cadmium ions from aqueous solution was investigated. The results showed that the fungi were effective in biosorption, but the *A. bisporus* species was more effective.

Hyde and Wood (1997) and Purnomo et al. (2008) reported the Fenton reaction as a possible degradation mechanism of pollutants by brown-rot fungi. The Fenton reaction involves the oxidation of iron (Fe) ions by hydrogen peroxide. The Fenton reaction is carried out by brown-rot fungi containing the enzyme cellobiose dehydrogenase. This enzyme has been purified by double oxidation of cellodextrins, which converts Fe3+ to Fe2+.

Brown-rot fungi *Laetiporus sulphureus* have been found to degrade phenanthurein (Sack et al., 1997), while *Gloeophyllum trabeum, Daedalea dickinsii,* and *Fomitopsis pinicola* have been found to degrade dichlorodiphenyltrichloroethane (Purnomo et al., 2011). Brown-rot fungi have been used to remove metals from wood waste treated with chromium copper arsenate, due to their copper tolerance and high production of oxalic acid (Kim et al., 2009).

Marine fungi can also produce extracellular enzymes that have lignin degradation or modification properties and are used in the degradation of many aromatic and persistent environmental pollutants (Garzoli et al., 2015).

Some studies have been conducted to enhance bioremediation of toxic and persistent organic pollutants by marine fungi. Divya et al. (2015) used *Trichoderma viride*, which they isolated from a phenolic-tolerant river mouth, in the marine ecosystem to promote laccase production tolerant to high salinity and phenolics, and obtained successful results. Similarly, it was determined that persistent organic pollutants such as PCB 118 can be degraded by two marine fungi belonging to the *Penicillium* genus (Verma et al., 2013; Gao et al., 2013).

In another study, it was observed that persistent organic pollutants and pentachlorophenol were significantly degraded by the marine fungus *Trichoderma harzianum* (Vacondio et al., 2015). Marine fungi such as *Mucor, Penicillium, Aspergillus,* and some yeasts have been found to have bioremediation potential for water-soluble crude oil distillates at concentrations of 0.01 to 0.25 mg/mL, while higher concentrations have been found to be toxic to the organisms (Hickey, 2013).

It has been determined that ectomycorrhizal fungi, *Rhizopogon roseolus* and *Suillus bovinus*, which live symbiotically in pine plant roots, are effective in removing cadmium from the environment (Sousa et al., 2014). Meharg and Cairney (2000) determined that ectomycorrhizal fungi can break down some organic pollutants (POPs) that have a persistent effect.

When literature studies are examined, the importance of bioremediation effects of fungi is quite remarkable. Considering that many industrial wastes mix with water in nature and the ecological cycle is affected by this, it is important to eliminate these pollutants. Absorbing these pollutants through natural means through fungi and cultivating mushrooms for this purpose is just one of the remarkable issues.

1.2 Use of Mushrooms as Biomaterials in the Construction Industry

Unconscious consumption is increasing day by day in the world. The consumption of materials used especially in the construction sector is among these. For this reason, the construction sector is also moving toward a circular model in order to address the environmental challenges caused by the usage mechanism of produce-use-dispose. As a result, a circular economy in the construction sector means cycles in which the end of a building's life is considered more carefully and waste is used as a resource (Bitting et al., 2022).

The use of microorganisms in the production of biomaterials, especially in the construction sector, is one of the technologies used to increase environmental sustainability in the near future (Zou & Gao, 2020). Among these microorganisms, fungi are one of the most notable. Materials obtained from fungal mycelium have potential applications in different areas including biomedical, paper, packaging, cosmetics, textile, and construction industries.

Mycelium-based composites have significant potential in the development of thermal and acoustic insulation panels due to their low cost of production, high level of sound absorption, low thermal conductivity, and fire resistance. With these properties, they can be used in various applications, especially in the construction industry (Manan et al., 2021; Sarıa et al., 2023).

Mycelia are unique in the production of bioproducts because they can easily come together and form a block-like structure during their colonization (Fig. 2). Myceliums can be easily grown in organic waste, are recyclable, and cost-free. This allows them to be used in various applications (Elkhateeb & Daba, 2019; Alemu et al., 2022).

The use of mycelium-based materials was first started by Eben Bayer and Gavin McIntyre in 2007. This company produces high-quality packaging products that are 100% recyclable and non-toxic (Dougoud, 2018; Krivanek, 2020; Sydor et al., 2021). The option to buy environmentally friendly packaging materials is especially preferred by younger generations, which encourages multinational companies and local businesses. It is known that buyers are more interested in reducing waste production and purchasing consumable products. The field of packaging has been positively impacted by the use of mycelium instead of toxic and persistent oil-based materials. Some of the mushroom-based foam products are shown in Fig. 3.

Several companies in the world such as MOGU, Evocative Design, MycoWorks (https://www.mycoworks.com, 2021), and NEFFA have started to design and commercialize mycelium-bonded composites (Neffa, 2020; Ecovative Design, 2020; Mogu, 2020).

Studies have reported that fungi such as *Agrocybe aegerita, Aspergillus nidulans, Trametler versicolor, Ganoderma sessile, Pleurotus ostreatus, Rhizopus oryzae, A. oryzae, A. terreus, Phanerochaete chrysosporium*, and *Saccharomyces cerevisiae* are used in the production of construction materials such as bioconcrete, blocks, and bioblocks (Elsacker et al., 2020; Xing et al., 2018a, b; Attias et al., 2017; Menon

Fig. 2 Mycelium-based material and concrete formwork. (Elsacker et al., 2021)

Fig. 3 Mushroom mycelium-based foam materials used in packaging. (Mojumar et al., 2021)

et al., 2019; Heisel et al., 2017; Attias et al., 2020; Joshi et al., 2020; Konwarh et al., 2020).

Ziegler et al. (2016) evaluated the potential of biocomposite material produced by Ecovative Design LLC in commercial products and packaging. They produced the test samples used in the study with different fiber-cork combinations using cotton waste, hemp pulp, and surface fiber fabric. They found that the evaluated composites had specific gravity, surface hardness, water absorption, linear thermal expansion coefficient, resistance to tensile-compression stress, and the material was extremely light, buoyant, flexible, and highly hydrophilic.

López Nava et al. (2016) aimed to develop a biobased material similar to expanded polystyrene (EPS) using wheat straw, fungi of the genus Pleurotus sp. and natural film-forming substances (carrageenan, chitosan, xanthan gum). The physical and mechanical properties of the composite material obtained in the study were evaluated; compressive strength was measured between 20 and 60 kPa and flexural strength between 4.6 and 17.9 kPa. In addition, the dimensional stability ranged between 5.1% and 7.3% and the relative density between 178.7 and 198.9 kg/m^3. In line with these results, it was concluded that the developed biomaterial is not a complete alternative to EPS in its current form, despite some of its positive properties.

Pelletier et al. (2017) used various agricultural by-products such as cottonseed meal, rice straw, corn stalks, kenaf, and sorghum to grow mushrooms on these substrates and produce a lightweight biocomposite board with natural ingredients. The research also included three reference materials such as acoustic ceiling tiles, cork flooring underlayment, and birch plywood. The obtained mycelium-based sheets were evaluated in terms of acoustic performance such as sound absorption, reflection and transmission and it was concluded that such materials can be an environmentally friendly alternative for acoustic insulation. While both studies demonstrate the potential of mycelium-based composites for developing sustainable building materials, they also show that further development is needed to achieve the desired level of performance.

Xing et al. (2018a, b) obtained insulation material using *Oxyporus latermarginatus, Megasporoporia minor, Ganoderma resinaceum* species, and wheat straw as substrate. According to the test data, they found that mycelium bricks exhibited good thermal performance.

In the study conducted by Lakk et al. (2018), the feasibility and production process of fungus-based biocomposite material for habitat structures on the Moon and Mars were investigated using construction technology. In the study, they used *Schizophyllum commune*, *Pleurotus ostreatus*, *Trametes versicolor* species, and *Azolla filiculoides* plant. In the current case, although it has much weaker mechanical properties than traditional building materials, they proved that the material can be used in load-bearing structures only when it is optimized for compressive strengths. *Schizophyllum commune* (SC) grew better in AF compared to *Pleurotus ostreatus* and *Trametes multicolor*. Tests showed that growth is possible in microgravity conditions and SC growth can be stimulated even by microgravity. Growth at high radiation levels is possible and SC has survived gamma radiation at dose levels of 200 Gy for 3 days. They have also produced small-scale AF/SC biocomposite panels for mechanical testing at ESTEC (European Space Agency).

Bruscato et al. (2019) produced biofoam using mycelia of *Pycnoporus sanguineus*, *Pleurotus albidus,* and *Lentinus velutinus* fungi on lignocellulosic wastes such as sawdust and wheat bran. The obtained samples were structurally characterized by scanning electron microscopy (SEM), thermal analysis, and Fourier Transform Infrared Spectroscopy (FTIR) and mechanical properties such as density and compressive strength were also investigated. According to the results, the compressive strength of bio-foams obtained from *P. sanguineus* and *Lentinus velutinus* species was determined as 1.3 MPa, while the compressive strength of the sample obtained from *P. albidus* species was 0.4 MPa. These values were higher than those of conventional expanded polystyrene (EPS) (0.4 MPa), especially for *P. sanguineus* and *L. velutinus* samples. However, the thermal stability of the biofoams produced was found to be lower compared to EPS. In terms of density, the biofoams are denser than EPS and have values ranging from approximately 0.3 g/cm^3 to 0.03 g/cm^3. These findings indicate that these bio-based foam materials can be a sustainable and environmentally friendly alternative to EPS in some applications.

Biopex, a composite material obtained by using the mycelium of *Ganoderma lucidum* was produced. Biopex is used instead of traditional insulation materials in air conditioning systems and also in construction sectors (Bayramoğlu et al., 2025).

Blocks were obtained from pure mycelium of *Trametes versicolor* and *Ganoderma resinaceum* and it was investigated how they could be effective in large-scale robotic production (Elsacker et al., 2021).

1.3 Potential of Mushrooms to Transform Organic Waste into Valuable Products

With the rapid population growth, it is seen that food resources are decreasing day by day in many parts of the world. Many solid by-products are formed with agriculture-industry. These include agricultural wastes such as stalk, straw, bark, cob, sawdust, bran, molasses, etc. It is seen that the majority of these by-products, which are various lignocellulose-based plant and woody, are burned, some are used as fertilizer or compost material, and a very small portion is evaluated in animal feeds (Montañez-Valdez et al., 2015).

The use of lignocellulosic agricultural wastes in mushroom cultivation increases the importance of this issue. Evaluating these wastes, which have a negative impact on the ecological environment when disposed of by stubble burning, in the production of a nutrient will support the nutritional needs of the increasing population in the world, prevent environmental pollution caused by the resulting wastes as a result of stubble burning, and contribute to the economy due to the increase in the labor force required during production. Cultivated mushroom production is considered to be only one of the ways to solve these problems (Fig. 4).

Fig. 4 Culture of *Pleurotus citrinopileatus*

Mushroom cultivation is a simple, low-cost, environmentally friendly technology that can be easily used in the culture of rural and industrial agricultural wastes. In mushroom cultivation, a large amount of protein is produced on substrates containing lignocellulose. The most critical stage in mushroom cultivation is the preparation stages of these substrates. Mycelium development, formation of fructification organs, and product quality vary from substrate to substrate (Atila 2017). Basidiomycetes class macrofungi are the best known cellulose, hemicellulose, and lignin degraders. Therefore, many agricultural and industrial wastes can be used as substrates in mushroom production (Owaid et al., 2017).

Various lignocellulosic wastes (bean straw, oak sawdust, safflower straw, and sunflower head residue) were used to cultivate *P. eryngii, P. djamor*, and *P. citrinopileatus* were cultured in suitable compost environments (Atila, 2017). In a different study, *P. djamor* was cultivated using wheat and quinoa wastes (İnci et al., 2024). In order to ensure sustainability, Yang et al. (2025) cultivated *Pleurotus* in corn straw. *Agaricus bisporus* species were cultured using wheat straw, barley and oat straw, beech sawdust, rice husks, and coffee residue (Dedousi et al., 2024). There are many studies in the literature similar to this one. It is known that especially the composts used affect the nutritional values of mushrooms. This means easy, cheap, and more accessible high-value foods. Therefore, the importance of mushrooms in the recycling of agricultural wastes attracts considerable attention.

1.4 Use of Mushrooms as Biofuel

Biofuel production from lignocellulosic waste is one of the alternative ways to ensure sustainability. For this reason, the importance of the substrate used after mushroom production has increased in recent years. Mushroom substrate is quite useful for biofuel production compared to other bio-based raw materials. Nutrients found in lignocellulosic waste can be easily degraded and converted into energy by enzyme systems found in mushrooms (Kousar et al., 2024). It is clear that biofuels, especially biodiesel and bioethanol, will be an excellent solution to the energy crisis (Tan et al., 2024). The use of compost formed after mushroom culture in bioethanol and biodiesel production can be one of the alternative ways to this solution.

A study was conducted by Lee et al. to determine the use of normal wood and mushroom compost as an alternative biomass source for energy production. The results showed that after 24 h of fermentation, 12 g/L ethanol was produced with mushroom compost, while normal wood produced 8 g/L ethanol (Lee et al., 2008).

In another study, spent mushroom substrate (SMS) from mushroom cultivation was hydrolyzed with concentrated sulfuric acid and the efficiency of the resulting hydrolysate in fermentative hydrogen production was evaluated. It was found that biohydrogen production was successfully achieved using the sugar hydrolysate obtained from this lignocellulosic waste material. Furthermore, sulfate ions present in the reduced sugar solution were removed with the help of an anion exchange resin (Li et al., 2011).

Waste mushroom composts of *Pleurotus ostreatus* and *Lentinula edodes* species have been studied as alternative fuel and biogas sources (Kapu et al., 2012; Tumwasorn et al., 1980).

Studies have been conducted on waste mushroom composts such as *Agaricus bisporus*, *Lentinula edodes,* and *Auricularia auricula*, which provide a suitable environment for biogas production. According to these studies, it has been determined that 5 kg of waste mushroom compost can meet the daily biogas needs of a family of 3–5 people (Songlin, 2002; Najafi et al., 2019).

1.5 Use of Fungi in Sustainable Textiles

Applying this to the products and services provided by consumers to ensure sustainability will lead to increased efficiency in resource use. In particular, the application of recycled products to used textile products is quite suitable for make, use, and recycle strategies (Amobonye et al., 2023). In particular, the use of fungi in biotextiles is just one of these strategies, and the use of fungi in biotextiles is increasingly being investigated as potential alternatives to traditional polymer-based materials (Jeong et al., 2022; National Institute of Horticultural and Herbal Science, 2025).

In textile production, naturally occurring pigments found in bio-based materials can affect the aesthetic and functional qualities of the final product. If these pigments are not adequately managed, they can compromise product consistency, especially in color uniformity. Therefore, in order to maintain product quality, physicochemical treatments are required to remove these pigments or to suppress their activity using decolorizers. Decolorizers are generally divided into two groups: oxidative and reductive types (da Silva et al., 2021; Im et al., 2025). However, among the reducing decolorants, sulfur compounds are disadvantaged due to environmental concerns regarding their use (Im et al., 2025). In many studies conducted for this purpose, fungal-based biotextile products have begun to be developed.

Williams et al. (2022) reported that large-scale production of mycelium-based cultured leather to replace cowhide is expected to reduce the carbon footprint of the total production process by 2.76 kg CO_2 equivalent.

Mushroom pulps belonging to the species *Agaricus bisporus, Trametes orientalis, Flammulina velutipes,* and *Pleurotus ostreatus* have been found to be promising as an environmentally friendly alternative to cotton fiber (Im et al., 2025).

In addition, the US-based Ecovative company, founded in 2007 by Eben Bayer and Gavin McIntyre, has produced and traded many products, from mushroom-based composite materials to textile materials that replace animal skin. Designer Danielle Trofe has designed 100% biodegradable, sustainable, and stylish lamps and pots using mushroom mycelium. Ecovative was the first company to use mushroom mycelium tissue as a material. Danielle Trofe shared her ideas about the lighting industry with the company by taking samples of Ecovative's unusual materials. Over the years, Danielle Trofe has managed to develop a lamp design technique by growing and molding the mycelium that forms the mushroom mycelium tissue. She has created her collection called "MushLume" from these designs (Koparan & Alkan, 2024) (Fig. 5).

MycoWorks was founded in 2013 by Philip Ross, Sofia Wang, and Eddie Pavlu. MycoWorks has succeeded in creating a new type of leather that is rapidly grown in a carbon-negative process from mycelium and agricultural by-products (Karana et al., 2018; Koparan & Alkan, 2024). MycoWorks has also taken its place in the production of clothing items such as shoes, bags, hats, and coats by producing animal leather-like products from Reishi *Ganoderma lucidum* (Sesli et al., 2020) mushroom mycelium (Koparan & Alkan, 2024).

The Dutch company NEFFA, together with many artists and designers including the founder of the company Aniela Hoitink, designed clothing using mycelium. Using a 100% biodegradable mycelium-based material called MycoTEX, they created biodegradable fabric using a three-dimensional modeling technique as part of the "Mycelium Design" project (Camere & Karana, 2017). Pure mycelium was used to make textile-like materials for clothing, and this new textile was produced by directly forming it on the body shape of the model without the need for cutting and sewing.

A similar study to the MycoTEX study conducted by Neffa was conducted by an Australian artist named Donna Franklin for the "Fiber Reactive" project, which is fungi living on fabric. In this project, the fungus known by its current name *Trametes coccinea* (Fr.) Hai J. Li & S.H. He, but known as *Pycnoporus coccineus* (Fr) Bondartsev & Singer in the study, was used in the study. Microbiology was also

Fig. 5 Examples of products produced from mushroom mycelium by the company Ecovative. (URL1, Access Date: 25.02.2024) and examples from Danielle Trofe's mushroom collection called "mushlume". (URL2, Access Date: 25.02.2024; Koparan & Alkan, 2024)

used when using this method. The appropriate environment and nutrients for the fungi were provided, and the fungus was placed on the mannequin and then transferred to this fabric. After the appropriate environment was created, the fungus developed on this fabric and covered the mannequin like an evening dress (Koparan & Alkan, 2024).

2 Conclusion

Fungal biotechnology is a strategic scientific and technological field that offers sustainable and innovative solutions to today's most critical global challenges such as environmental pollution, resource waste, food security, and climate change. Applications developed by utilizing the natural biological functions of fungi enable the establishment of environmentally friendly production systems and more efficient use of economic resources. In particular, the utilization of organic wastes as substrates in mushroom cultivation allows these wastes to be recycled into the economy without harming the environment. Re-utilization of organic wastes from sectors such as agriculture, food, and forestry in mushroom cultivation not only reduces waste, but also provides value-added products. This process not only prevents waste, but also optimizes the use of natural resources, making production systems more sustainable. This improves waste management and contributes to the zero-waste approach. Another important aspect of fungal biotechnology is biodegradable packaging, construction, and textile materials produced from fungal mycelia. These materials offer biodegradable, recyclable, and low-carbon footprint alternatives to traditional plastic and chemical-based products. The low energy and raw material requirements of these production processes reduce production costs and minimize environmental impacts. Thus, not only environmental health but also economic efficiency is directly supported.

In agriculture, the use of mycorrhizae and entomopathogenic fungal species improves crop quality and contributes to the long-term fertility of soils by reducing the use of chemical fertilizers and pesticides while maintaining plant health. This approach enables the development of sustainable agricultural practices, especially for small-scale producers, and supports rural development.

In addition, bioremediation, in which some fungal species naturally break down toxic pollutants and clean the environment, reduces the economic and ecological cost of environmental damage caused by industry and agriculture. In addition, enzymes, drugs, and biologically active compounds obtained from fungi provide high-value-added outputs in the field of biotechnological product development and make a direct economic contribution to the industry.

With all these features, fungal biotechnology contributes not only to a sustainable environment, but also to a sustainable production-consumption balance and green economic growth in line with circular economy principles. Therefore, investments in fungal biotechnology will not only provide ecological benefits, but also create new jobs, strengthen the domestic biotechnology sector, and increase economic diversity. In the future, the role of this field will become even more critical in tackling crises such as climate change, food security, and natural resource scarcity, and it will be crucial to develop research, education, and application policies with a focus on fungal biotechnology.

References

Adenipekun, C. O., & Lawal, R. (2012). Uses of mushrooms in bioremediation: A review. *Biotechnology and Molecular Biology Reviews, 7*(3), 62–68.

Alemu, D., Tafesse, M., & Mondal, A. K. (2022). Mycelium-based composite: The future sustainable biomaterial. *International Journal of Biomaterials, 2022*(1), 8401528.

Amobonye, A., Lalung, J., Awasthi, M. K., & Pillai, S. (2023). Fungal mycelium as leather alternative: A sustainable biogenic material for the fashion industry. *Sustainable Materials and Technologies, 38*, e00724.

Anastasi, A., Coppola, T., Prigione, V., & Varese, G. C. (2009). Pyrene degradation and detoxification in soil by a consortium of basidiomycetes isolated from compost: Role of laccases and peroxidases. *Journal of Hazardous Materials, 165*, 1229–1233.

Anastasi, A., Tigini, V., & Varese, G. C. (2013). The bioremediation potential of different eco-physiological groups of fungi. In *Fungi as bioremediators* (pp. 29–49). Springer.

Atila, F. (2017). Evaluation of suitability of various agro-wastes for productivity of *Pleurotus djamor, Pleurotus citrinopileatus* and *Pleurotus eryngii* mushrooms. *Journal of Experimental Agriculture International, 17*(5), 1–11.

Attias, N., Danai, O., Ezov, N., Tarazi, E., & Grobman, Y. J. (2017). *Developing novel applications of mycelium based bio-composite materials for design and architecture.* In Proceedings of building with biobased materials: Best practice and performance specification (Vol. 1, Issue 10).

Attias, N., Danai, O., Abitbol, T., Tarazi, E., Ezov, N., Pereman, I., & Grobman, Y. J. (2020). Mycelium bio-composites in industrial design and architecture: Comparative review and experimental analysis. *Journal of Cleaner Production, 246*, 119037.

Baldrian, P. (2006). Fungal laccases–occurrence and properties. *FEMS Microbiology Reviews, 30*(2), 215–242.

Barh, A., Kumari, B., Sharma, S., Annepu, S. K., Kumar, A., Kamal, S., & Sharma, V. P. (2019). Mushroom mycoremediation: Kinetics and mechanism. In *Smart bioremediation technologies* (pp. 1–22). Academic Press.

Bayramoğlu, S., Ağırtmış, Z., & Erol, S. S. (2025). Mantar Miselinden Kompozit Malzeme Üretimi: Potansiyel Uygulamalar. *Mantar Dergisi, 16*(1), 26–33.

Bitting, S., Derme, T., Lee, J., Van Mele, T., Dillenburger, B., & Block, P. (2022). Challenges and opportunities in scaling up architectural applications of mycelium-based materials with digital fabrication. *Biomimetics, 7*(2), 44.

Bruscato, C., Malvessi, E., Brandalise, R. N., & Camassola, M. (2019). High performance of macrofungi in the production of mycelium-based biofoams using sawdust – Sustainable technology for waste reduction. *Journal of Cleaner Production, 234*, 225–232.

Camere, S., & Karana, E. (2017). *Growing materials for product design.* In Alive. Active. Adaptive: Proceedings of international conference on experiential knowledge and emerging materials (EKSIG 2017) (pp. 101–115).

Chrysene bioconversion by the white rot fungus Pleurotus ostreatus D1. *Microbiology, 79*(4), 456-460

Da Silva, C. J. G., de Medeiros, A. D. L. M., de Amorim, J. D. P., do Nascimento, H. A., Converti, A., Costa, A. F. S., & Sarubbo, L. A. (2021). Bacterial cellulose biotextiles for the future of sustainable fashion: A review. *Environmental Chemistry Letters, 19*, 2967–2980.

Daly, P., Cai, F., Kubicek, C. P., Jiang, S., Grujic, M., Rahimi, M. J., Sheteiwy, M. S., Giles, R., Riaz, A., de Vries, R. P., Akcapinar, G. B., Wei, L., & Druzhinina, I. S. (2021). From lignocellulose to plastics: Knowledge transfer on the degradation approaches 10 S.S. 173188 by fungi. *Biotechnology Advances, 50*, 107770. https://doi.org/10.1016/j.biotechadv.2021.107770

Dedousi, M., Melanouri, E. M., Karayannis, D., Kaminaride, E. I., & Diamantopoulou, P. (2024). Utilization of spent substrates and waste products of mushroom cultivation to produce new crops of Pleurotus ostreatus, Pleurotus eryngii and Agaricus bisporus. *Carbon Resources Conversion, 7*(2), 100196.

Demir, G., Özcan, H. K., Elmaslar, E., & Borat, M. (2006). Decolorization of azo dyes by the white rot fungus phanerochaete chrysosporium. *Sigma, 3.*

Deshmukh, R., Khardenavis, A. A., & Purohit, H. J. (2016). Diverse metabolic capacities of fungi for bioremediation. *Indian Journal of Microbiology, 56*(3), 247–264.

Divya, L. M., Prasanth, G. K., & Sadasivan, C. (2015). Assessing the tolerance of immobilized laccase from a salttolerant strain of Trichoderma viride Pers NFCCI-2745 to heavy metal ions, detergents and copper chelating agents. *International Journal of Environmental Science and Technology, 12*(10), 3225–3234.

Dos Santos Bazanella, G. C., Araújo, C. A. V., Castoldi, R., Maria, G., Maciel, F. D. I., Marques, C. G., & Peralta, R. M. (2016). Ligninolytic enzymes from white-rot fungi and application in the removal of synthetic dyes. *Fungal Enzymes, 258.*

Dougoud, M. (2018). *Mycelium infrastructures for impermanent futures.* Research Works Archive.

Ecovative Design. (2020). https://ecovativedesign.com/

Elkhateeb, W. A., & Daba, G. M. (2019). The amazing potential of fungi in human life. *ARC Journal of Pharmaceutical Sciences, 5*(3), 12–16.

Elsacker, E., Vandelook, S., Van Wylick, A., Ruytinx, J., De Laet, L., & Peeters, E. (2020). A comprehensive framework for the production of mycelium-based lignocellulosic composites. *Science of the Total Environment, 725*, 138431.

Elsacker, E., Søndergaard, A., Van Wylick, A., Peeters, E., & De Laet, L. (2021). Growing living and multifunctional mycelium composites for large-scale formwork applications using robotic abrasive wire-cutting. *Construction and Building Materials, 283*, 122732.

Evans, C. S., & Hedge, J. N. (2001). Degradation of plant cell wall polymers. *Fungi in Bioremediation, 23*, 1–26.

FAO. (2019). *Global food waste statistics*. http://www.fao.org/platform-food-loss-waste/en/

Gadd, G. M. (2001). *Fungi in bioremediation* (481 pp.). Cambridge University Press.

Garzoli, L., Gnavi, G., Tamma, F., Tosi, S., Varese, G. C., & Picco, A. M. (2015). Sink or swim: Updated knowledge on marine fungi associated with wood substrates in the Mediterranean Sea and hints about their potential to remediate hydrocarbons. *Progress in Oceanography, 137*, 140–148.

Gao, L., Gao, F., Zhang, D., Zhang, C., Wu, G., & Chen, S. (2013). Purification and characterization of a new β-glucosidase from Penicillium piceum and its application in enzymatic degradation of delignified corn stover. *Bioresource Technology, 147*, 658–661.

Goltapeh, E. M., Danesh, Y. R., & Varma, A. (2013). Fungi as bioremediators. In *Soil biology* (32–489 pp.). Springer.

Gupta, V. K., Nayak, A., & Agarwal, S. (2015). Bioadsorbents for remediation of heavy metals: Current status and their future prospects. *Environmental Engineering Research, 20*(1), 1–18.

Harms, H., Schlosser, D., & Wick, L. Y. (2011). Untapped potential: Exploiting fungi in bioremediation of hazardous chemicals. *Nature Reviews Microbiology, 9*(3), 177.

Heinfling, M. J., Martínez, A. T., Martínez, M., & Bergbauer Szewzyk, U. (1998). Transformation of industrial dyes by manganese peroxidases from Bjerkandera adusta and Pleurotus eryngii in a manganese-independent reaction. *Applied and Environmental Microbiology, 64*, 2788–2793.

Heisel, F., Schlesier, K., Lee, J., Rippmann, M., Saeidi, N., Javadian, A., Hebel, D. E., Block, P. (2017). *Design of a load-bearing mycelium structure through informed structural engineering*. In Proceedings of the world conference on sustainable technologies (pp. 1–5).

Hickey, P. (2013). Toxicity of water soluble fractions of crude oil on some bacteria and fungi Isolated from marine water. *American Journal of Animal Research, 3*, 24–29.

Hoshino, Y. T., & Morimoto, S. (2008). Comparison of 18S rDNA primers for estimating fungal diversity in agricultural soils using polymerase chain reaction-denaturing gradient gel electrophoresis. *Soil Science & Plant Nutrition, 54*(5), 701–710.

Hyde, S. M., & Wood, P. M. (1997). A mechanism for production of hydroxyl radicals by the brown-rot fungus Coniophora puteana: Fe (III) reduction by cellobiose dehydrogenase and Fe (II) oxidation at a distance from the hyphae. *Microbiology, 143*(1), 259–266.

Ibrahim, S. S., Lonescu, D., & Grossart, H. P. (2024). Harnessing the potential of fungi: Biodegradation of plastic and rubber by powerful fungi. *Science of the Total Environment, 934*, 173188.

Igir, B. E., Okoduwa, S. I., Idoko, G. O., Akabuogu, E. P., Adey, A. O., & Ejiogu, I. K. (2018). Toxicity and bioremediation of heavy metals contaminated ecosystem from tannery wastewater: A review. *Journal of Toxicology, 2018*(1), 2568038.

Im, H. S., Tagawa, S., Jeong, J. S., & Shin, H. J. (2025). Production of decolorized mushroom pulp for nonwoven cotton composite. *Fibers, 13*(3), 30.

İnci, Ş., Kirbağ, S., & Akyü, M. (2024). Valorization of local agro-residues for the cultivation of *Pleurotus djamor* (Rumph. Ex Fr.) Boedijn and their effects on nutritional value. *Biomass Conversion and Biorefinery*, 1–10.

Jeong, Y. H., Kim, D. S., Kim, H. S., & Oh, D. S. (2022). Shin HJ (2022) Effect of culture and method and medium components on *Trametes orientalis* mycelium mat formation (Yasuda) Imazeki. *Journal of Mushrooms, 20*, 69–77.

Jha, S., Dikshit, S., & Pandy, G. (2011). Comparative study of agitation rate and stationary phase for the removal of cu2+ by *A. lentulus*. *International Journal of Pharmacy and Biological Science, 2*, 208–211.

Joshi, K., Meher, M. K., & Poluri, K. M. (2020). Fabrication and characterization of bioblocks from agricultural waste using fungal mycelium for renewable and sustainable applications. *ACS Applied Bio Materials, 3*(4), 1884–1892. https://doi.org/10.1021/acsabm.9b01047

Kahraman, S., & Yeşilada, O. (2001). Industrial and agricultural wastes as substrates for laccase production by white-rot fungi. *Folia Microbiologica, 46*(2), 133.

Kapu, N. U. S., Manning, M., Hurley, T. B., Voigt Cosgrove, D. J., & Romaine, C. P. (2012). Surfactant-assisted pretreatment and enzymatic hydrolysis of spent mushroom compost for the production of sugars. *Bioresource Technology, 114*, 399–405.

Karana, E., Blauwhoff, D., Hultink, E. J., & Camere, S. (2018). When the material grows: a case study on designing (with) mycelium-based materials. *International Journal of Design, 12*(2).

Kim, G. H., Choi, Y. S., & Kim, J. J. (2009). Improving the efficiency of metal removal from CCA-treated wood using brown rot fungi. *Environmental Technology, 30*(7), 673–679.

Konwarh, R., Palanisamy, S. B., & Jogi, P. K. (2020). A mini review on prospects and challenges of harnessing fungi for concrete-crack healing. *Material Science Research India, 17*(2), 117–128. https://doi.org/10.13005/msri/170204

Koparan, Y., & Alkan, S. (2024). Gelecekte Mantarların Kullanım Alanları: Sürdürülebilir Tasarım Ürünleri Olarak Mantarlar. *Mantar Dergisi, 15*(1), 43–49.

Kousar, A., Khan, H. A., Farid, S., Zhao, Q., & Zeb, I. (2024). Recent advances on environmentally sustainable valorization of spent mushroom substrate: A review. *Biofuels, Bioproducts and Biorefining, 18*(2), 639–651.

Krivanek, S. (2020). *Fungal mycelium; the key to a sustainable future*. University of Nebraska.

Kržišnik, D., & Gonçalves, J. (2023). Environmentally conscious technologies using fungi in a climate-changing world. *Earth, 4*(1), 69–77.

Kulshreshtha, S., Mathur, N., & Bhatnagar, P. (2014). Mushroom as a product and their role in mycoremediation. *AMB Express, 4*, 1–7.

Kumar, A., Yadav, N., Mondal, R., Kour, D., Subrahmanyam, G., Shabnam, A. A., Khan, S. A., Yadav, K. K., Sharma, G. K., & Cabral- Pinto, M. (2021). Myco-remediation: A mechanistic understanding of contaminants alleviation from natural environment and future prospect. *Chemosphere, 284*, 131325.

Lakk, H., Krijgsheld, P., Montalti, M., & Woesten, H. (2018). *Fungal based biocomposite for habitat structures on the Moon and Mars*. In 69 th International astronautical Congress.

Lange, L., Connor, K. O., Arason, S., Bundgård-Jørgensen, U., Canalis, A., Carrez, D., & Vieira, H. (2021). Developing a sustainable and circular bio-based economy in EU: By partnering across sectors, upscaling and using new knowledge faster, and for the benefit of climate, environment & biodiversity, and people & business. *Frontiers in Bioengineering and Biotechnology, 8*, 619066.

Lee, J. W., Koo, B. W., Choi, J. W., Choi, D. H., & Choi, I. G. (2008). Evaluation of waste mushroom logs as a potential biomass resource for the production of bioethanol. *Bioresource Technology, 99*, 2736–2741.

Li, Y. C., Wu, S. Y., Chu, C. Y., & Huang, H. C. (2011). Hydrogen production from mushroom farm waste with a two-step acid hydrolysis process. *International Journal of Hydrogen Energy, 36*, 14245–14251.

Li, S., Yang, Y., Yang, S., Zheng, H., Zheng, Y., Nagarajan, D., Varjani, S., & Chan, J. S. (2023). Recent advances in biodegradation of emerging contaminants-microplastics (MPs): Feasibility, mechanism, and future prospects. *Chemosphere, 331*, 138776.

Lin, J. E., Wang, H. Y., & Hickey, R. F. (1990). Degradation kinetics of pentachloro-phenol by Phanerochaete chrysosporium. *Biotechnology and Bioengineering, 35*, 1125–1134. https://doi.org/10.1002/bit.260351108

López Nava, J., Méndez González, J., Ruelas Chacón, X., & Nájera Lun, J. (2016). Assessment of edible fungi and films bio-based material simulating expanded polystyrene. *Materials and Manufacturing Processes, 31*(8), 1085–1090.

Luz, J. M. R., Paes, S. A., Nunes, M. D., da Silva, M. D. C. S., & Kasuya, M. C. M. (2013). Degradation of oxobiodegradableplastic by Pleurotus ostreatus. *Plos one, 8*(8), e69386.

Magnin, A., Hoornaert, L., Pollet, E., Laurichesse, S., Phalip, V., & Averous, L. (2019). Isolation and characterization of different promising fungi for biological waste management of polyurethanes. *Microbial Biotechnology, 12*(3), 544–555.

Manan, S., Ullah, M. W., Ul-Islam, M., Atta, O. M., & Yang, G. (2021). Synthesis and applications of fungal mycelium-based advanced functional materials. *Journal of Bioresources and Bioproducts, 6*(1), 1–10.

Meharg, A. A., & Cairney, J. W. (2000). Ectomycorrhizas—extending the capabilities of rhizosphere remediation?. *Soil Biology and Biochemistry, 32*(11–12), 1475–1484.

Menon, R. R., Luo, J., Chen, X., Zhou, H., Liu, Z., Zhou, G., Zhang, N., & Jin, C. (2019). Screening of fungi for potential application of self-healing concrete. *Scientific Reports, 9*(1), 1–12. https://doi.org/10.1038/s41598-019-39156-8. 2-s2.0-85061612274.

Meyer, V., Basenko, E. Y., Benz, J. P., Braus, G. H., Caddick, M., Csukai, M., & Wösten, H. A. (2020). Growing a circular economy with fungal biotechnology: A white paper. *Fungal Biology and Biotechnology, 7*(1), 5.

Mogu. (2020). *Radical by nature*. https://mogu.bio

Mojumdar, A., Behera, H. T., & Ray, L. (2021). Mushroom mycelia-based material: An environmental friendly alternative to synthetic packaging. In A. Vaishnav & D. K. Choudhary (Eds.), *Microbial polymers*. Springer. https://doi.org/10.1007/978-981-16-0045-6_6

Montañez-Valdez, O. D., Avellaneda-Cevallos, J. H., Guerra-Medina, C. E., Reyes-Gutiérrez, J. A., Peña-Galeas, M. M., Casanova-Ferrín, L. M., Rocío del, C., & Herrera-Herrera, R. C. (2015). Chemical composition and ruminal disappearance of maize stover treated with *Pleurotus djamor*. *Life Science Journal, 12*(2).

Mougin, C., Boukcim, H., & Jolivalt, C. (2009). Soil bioremediation strategies based on the use of fungal enzymes. In *Advances in applied bioremediation* (pp. 123–149). Springer.

Nagy, B., Măicăneanu, A., Indolean, C., Mânzatu, C., Silaghi-Dumitrescu, L., & Majdik, C. (2014). Comparative study of Cd (II) biosorption on cultivated *Agaricus bisporus* and wild *Lactarius piperatus* based biocomposites. Linear and nonlinear equilibrium modelling and kinetics. *Journal of the Taiwan Institute of Chemical Engineers, 45*(3), 921–929.

Najafi, B., Ardbili, S. F., Shamshirband, S., & Chau, K. W. (2019). Spent mushroom compost (SMC) as a source for biogas production in Iran. *Engineering Applications of Computational Fluid Mechanics, 13*(1), 967–982.

National Institute of Horticultural and Herbal Science. (2025). https://www.nihhs.go.kr/farmer/statistics/statistics.do?t_cd=0206. Accessed 10 Jan 2025.

Neffa. (2020). *Growing the future of fashion*. https://neffa.nl/.

Nikiforova, S. V., Pozdnyakova, N. N., Makarov, O. E., Chernyshova, M. P., & Turkovskaya, O. V. (2010). Chrysene bioconversion by the white rot fungus Pleurotus ostreatus D1. *Microbiology, 79*(4), 456–460.

Ollikka, P., Alhonmäki, K., Leppänen, V. M., Glumoff, T., Raijola, T., & Suominen, I. (1993). Decolorization of azo, triphenylmethane, heterocyclic, and polymeric dyes by lignin peroxidase isoenzymes from Phanerochaete chrysosporium. *Applied and Environmental Microbiology, 59*, 4010–4016.

Owaid, M. N., Al-Saeedi, S., & Abed, I. A. (2017). Antifungal activity of cultivated oyster mushrooms on various agro-wastes. *Summa Phytopathologica, 43*(1), 9–13.

Pelletier, M., Holt, G., Wanjura, J., Lara, A., Tapia-Carillo, A., McIntyr, G., & Baye, E. (2017). An evaluation study of pressure-compressed acoustic absorbers grown on agricultural by-products. *Industrial Crops and Products, 95*, 342–347.

Purnomo, A. S., Kamei, I., & Kondo, R. (2008). Degradation of 1, 1, 1-trichloro-2, 2-bis (4-chlorophenyl) ethane (DDT) by brown-rot fungi. *Journal of Bioscience and Bioengineering, 105*(6), 614–621.

Purnomo, A. S., Mori, T., Takagi, K., & Kondo, R. (2011). Bioremediation of DDT contaminated soil using brown-rot fungi. *International Biodeterioration & Biodegradation, 65*(5), 691–695.

Reddy, C. A., & Mathew, Z. (2001). Bioremediation potential of white rot fungi. *British Mycological Society Symposium Series, 23*, 52–78.

Sack, U., Heinze, T. M., Deck, J., Cerniglia, C. E., Martens, R., Zadrazil, F., & Fritsche, W. (1997). Comparison of phenanthrene and pyrene degradation by different wood-decaying fungi. *Applied and Environmental Microbiology, 63*(10), 3919–3925.

Saria, E., Cörüt, A., & Büyükakıncı, B. Y. (2023). Miselyum Kompozitlerinin Sürdürülebilir Yapı Malzemesi Olarak Kullanımı. *Mehmet Akif Ersoy Üniversitesi Fen Bilimleri Enstitüsü Dergisi, 14*(1), 196–207.

Selvam, K., Arungandhi, B., Vishnupriya, M., Shanmugapriya, T., & Yamuna, M. (2013). Biosorption of chromium (vi) from industrial effluent by wild and mutant type strain of *Saccharomyces cerevisiae* and its immobilized form. *Bioscience Discovery, 4*(1), 72–77.

Sesli, E., Asan, A., Selçuk, F., Abacı Günyar, Ö., Akata, I., Akgül, H., Aktaş, S., Alkan, S., Allı, H., Aydoğdu, H., Berikten, D., Demirel, K., Demirel, R., Doğan, H. H., Erdoğdu, M., Ergül, C. C., Eroğlu, G., Giray, G., Halikî Uztan, A., Kabaktepe, Ş., Kadaifçiler, D., Kalyoncu, Karaltı, İ., Kaşık, G., Kaya, A., Keleş, A., Kırbağ, Kıvanç, M., Ocak, İ., Ökten, S., Özkale, E., Öztürk, C., Sevindik, M., Şen, B., Şen, İ., Türkekul, İ., Ulukapı, M., Uzun, Y., Uzun, Y., & Yoltaş, A. (2020). *Türkiye Mantarları Listesi*. Ali Nihat Gökyiğit Vakfı Yayını.

Singh, H. (2006). *Mycoremediation: Fungal bioremediation*. Wiley.

Singh, I., & Thakur, P. (2023). Impact of fungi on the world economy and its sustainability: Current status and potentials. In *Fungal resources for sustainable economy: current status and future perspectives* (pp. 3–37). Springer.

Singh, M. P., Srivastva, A. K., Vishwakarma, S. K., Singh, V. K., & Pandey, V. K. (2010). Mushroom biotechnology. *Recent Trends in Biotechnology, 1*, 77–85.

Singh, M., Srivastava, P. K., Verma, P. C., Kharwar, R. N., Singh, N., & Tripathi, R. D. (2015). Soil fungi for mycoremediation of arsenic pollution in agriculture soils. *Journal of Applied Microbiology, 119*(5), 1278–1290.

Songlin, Z. (2002). Production of biogas with SMC. *Edible Fungi, 1*, 6.

Sousa, N. R., Ramos, M. A., Marques, A. P., & Castro, P. M. (2014). A genotype dependent-response to cadmium contamination in soil is displayed by Pinus pinaster in symbiosis with different mycorrhizal fungi. *Applied Soil Ecology, 76*, 7–13.

Srikanth, M., Sandeep, T. S. R. S., Sucharitha, K., & Godi, S. (2022). Biodegradation of plastic polymers by fungi: A brief review. *Bioresources and Bioprocessing, 9*(1), 42.

Stamets, P. (2005). *Mycelium running: How mushrooms can help save the world.* Ten Speed Press.

Sydor, M., Bonenberg, A., Doczekalska, B., & Cofta, G. (2021). Mycelium-based composites in art, architecture, and interior design: a review. *Polymers, 14*(1), 145.

Tan, R. Y., Ilham, Z., Wan, W. A. A., Halim-Lim, S. A., Usuldin, S. R. A., Ahmad, R., & Adlim, M. (2024). Mushroom oils: A review of their production, composition, and potential applications. *Heliyon, 10*(11).

Tiwari, P., & Park, K. I. (2024). Advanced Fungal biotechnologies in accomplishing sustainable development goals (sdgs): What do we know and what comes next? *Journal of Fungi, 10*(7), 506.

Tumwasorn, S., Chinoros, C., Easpiakdumrong, P., & Pattanasettakul, W. (1980). Effects of different rice straws and dilution rates of manures on biogas production. *Mushroom Newsletter for the Tropics, 1*(2), 6–10.

Url 1. https://www.ecovative.com/ (Erişim Tarihi: 25.02.2024).

Url 2. https://bigumigu.com/haber/mantar-kokunden-dogal-lambalar/ (Erişim Tarihi: 25.02.2024).

Vacondio, B., Birolli, W. G., Ferreira, I. M., Seleghim, M. H., Gonçalves, S., Vasconcellos, S. P., & Porto, A. L.(2015). Biodegradation of pentachlorophenol by marine-derived fungus Trichoderma harzianum CBMAI 1677isolated from ascidian Didemnun ligulum. *Biocatalysis and Agricultural Biotechnology, 4*(2), 266–275.

VanAcken, L. M., Godefroid, C. M., Peres, H., & Naveau, A. S. N. (1999). Mineralization of 14C-U ring labeled 4-hydroxylamino-2,6-dinitrotoluene by manganesedependent peroxidase of the white-rot basidiomycete Phlebia radiate. *Journal of Biotechnology, 68*, 159–169.

Verma, S., Abirami, S., & Mahalakshmi, V. (2013). Anticancer and antibacterial activity of silver nanoparticles biosynthesized by Penicillium spp. and its synergistic effect with antibiotics. *Journal of Microbiology Biotechnology Research, 3*(3), 54–71.

Vural, A., Demir, S., & Boyno, G. (2018). Biyoremediasyon ve fungusların biyoremediasyonda kullanılması. *Yuzuncu Yıl University Journal of Agricultural Sciences, 28*(4), 490–501.

Williams, E., Cenian, K., Golsteijn, L., Morris, B., & Scullin, M. L. (2022). Life Cycle Assessment of MycoWorks' Reishi™: The first low-carbon and biodegradable alternative leather. *Environmental Sciences Europe, 34*, 120.

Xing, Y., Brewer, M., El-Gharabawy, H., Griffith, G., & Jones, P. (2018a). Growing and testing mycelium bricks as building insulation materials. *IOP Conference Series: Earth and Environmental Science, 121*, 022032.

Xing, Y., Brewer, M., El-Gharabawy, H., Griffith, G., & Jones, P. (2018b). Growing and testing mycelium bricks as building insulation materials. *IOP Conference Series: Earth and Environmental Science, 022032.*

Yadav, P., Rai, S. N., Mishra, V., & Sing, M. P. (2021). Mycoremediation of environmental pollutants: A review with special emphasis on mushrooms. *Environmental Sustainability*, 1–14.

Yang, H., Lin, G., Wang, S., Wu, T., Dan, Z., Yang, J., & Zhao, Y. (2025). Integrating agricultural waste recycling with sustainable feed production: Microbial and enzymatic dynamics during *Pleurotus* Cultivation on Maize Straw. *Agronomy, 15*(5), 1171.

Ziegler, A., Bajw, S. G., Holt, G. A., McIntyre, G., & Bajwa, D. S. (2016). Evaluation of physico-mechanical properties of mycelium reinforced green biocomposites made from cellulosic fibers. *Applied Engineering in Agriculture, 32*(6), 931–938.

Zou, D., & Gao, L. (2020). Preparation and properties of fungal mycelium based on garden waste. *Science Discovery, 8*(2), 43.

Fungal Biotechnology and the Circular Economy

Satish V. Patil, Bhavana V. Mohite, Sunil H. Koli, Tomas Hasek, and Tanmay S. Bachhav

Abstract To satisfy the needs of an expanding worldwide population, industrialization will continue to play a crucial role. However, the rapid depletion of natural resources necessitates the exploration of sustainable alternatives. Adopting a circular economy model offers a promising solution to address resource exhaustion. Microorganism-based biotechnologies are undeniably pivotal in this transition, with fungi standing out due to their diverse metabolic capabilities and broad substrate utilization. Fungi are well known for producing a wide range of metabolites and for their ability to convert agro-industrial waste into valuable products, making them ideal candidates for circular economy applications. Numerous fungal species can transform agricultural residues into biofuels, offering a renewable energy source. Additionally, fungi contribute to alleviating global hunger by producing high-nutrition products such as mushrooms and single-cell proteins. In agriculture, fungal-based biopesticides and biofertilizers help reduce reliance on chemical inputs. Overall, fungi play a significant role in recycling waste into value-added products and thus contribute meaningfully to revitalizing and sustaining the global economy.

Keyword Economy · Fungi · Natural resources · Renewable energy · Recycle

S. V. Patil (✉) · T. S. Bachhav
School of Life Sciences, Kavayitri Bahinabai Chaudhari, North Maharashtra University, Jalgaon, Maharashtra, India

B. V. Mohite
Department of Microbiology, Bajaj College of Science, Wardha, Maharashtra, India

S. H. Koli
Department of Microbiology, Yashwantrao Chavan College of Science, Karad, Maharashtra, India

T. Hasek
Department of Biotechnology, University of Chemistry and Technology, Prague, Dejvice, Czechia

A. Gupta (ed.), *Prospects of Fungal Biotechnologies for Livestock Volume 2*, Fungal Biology, https://doi.org/10.1007/978-3-032-06478-3_5

1 Introduction of Economically Important Fungi

The world population is currently around 8.2 billion, and projections estimate it will reach approximately 9.7 billion by 2025. India, with a population of about 1.4 billion and continuing to grow, faces significant resource constraints due to this rapid increase. To fulfill the demands of growing population, industrialization will play an important role. However, many resources are depleting rapidly, and making it essential to find alternative solutions. Recently, the United States and Europe have embraced a circular economy model to address the environmental challenges. Similarly, India and other developing countries also need to focus on adopting the circular economy model. It is undisputed that many microorganisms-based biotechnology is of pivotal role to our daily lives. Among that fungi-based biotechnology has been used long back to produce various types of products such as organic acids, antibiotics, etc.

Fungi are a category of eukaryotic microorganisms found across the globe in a visible form as their fruiting bodies, mushrooms, or invisible, meaning they are hidden organisms in the environment (Hyde et al., 2020). Fungi grow differently depending on the species, some of them use unicellular growth like yeasts, others are filamentous (Awasthi et al., 2023). The number of known fungi species has been growing for decades and is based on expectations. It is estimated that the current number of all species could be around 13 million and most fungi are still undiscovered (Wu et al., 2019). New species are announced on a regular basis, especially as new techniques and instruments make their determination possible (Zhang et al., 2023). A large number of fungi are, among other things, endophytes, a common term for organisms that live inside plant tissues. It is derived from the Greek words endon = to be inside and phyton = plant (Vijay & Kavitha, 2022). Endophytes have the potential to be used to produce desired compounds in biotechnology to replace chemical processes (Sharma et al., 2023). Sometimes it is necessary to employ more than one specific microorganism in a system and proceed step by step to the final product due to their ability to synthesize only intermediates but not the desired product directly from the substrate (Troiano et al., 2020). Cultivation of fungi can be carried out in essentially two different ways. The first option is cultivation in submerged liquid medium with the necessary substrate, or the fungi can be cultivated on a solid state substrate (Gmoser et al., 2019). It should be borne in mind that it is not possible to cultivate all microorganisms that would eventually be used in biotechnology, as some species cannot be cultivated in artificial environments under laboratory/industrial conditions (Fadiji & Babalola, 2020).

The use of fungi and microorganisms in general to produce a myriad of compounds is most likely used by industry to make a profit. If biotechnology companies process waste as a substrate for the microorganism, they can reduce the cost of producing their preferred commodity. This is an example of the so-called circular economy, where waste is recovered into high-value products (Gmoser et al., 2019). It is not necessarily only waste, products should be made from renewable resources that are reused or recycled/biodegraded at the end of the product's life or usability

(Schilling & Weiss, 2021). However, most of the products are still obtained from cultivation using the new raw substrate. The rise of biotechnology itself is also a consequence of changes in industry around the world, as society moves from fossil sources to carbon neutrality without greenhouse gases, or at least to reducing them through renewable sources using fungi (Bhat et al., 2024). It is not just about the transition from fossil-derived feedstock, but also about feeding the population and animals. More than 1.2 billion tons of protein food is expected to be needed by 2050 and fungal biotechnology has great potential to address this problem (Boland et al., 2013). Fungi are used to synthesize a large number of compounds in many areas of industry on a daily basis and have myriad uses as listed below.

Yeasts, particularly those of the genus *Saccharomyces*, are unicellular fungi that are well known for their ability to ferment a variety of sugars, especially those used in the food industry from brewing, baking to cheese making. More than one million tons of yeasts are produced in Europe every year and this number is increasing (Parapouli et al., 2020). It is not only about beer and bread, yeast is also used in the industry to produce wine, ciders, or stronger alcohols such as whisky, vodka, and other spirits (Hittinger et al., 2018). In addition to food, yeast is used to produce bioethanol, a fuel that is not produced from fossil sources but uses plant materials, lignocellulosic sources, which are simply hydrolyzed by acid hydrolysis, fermented by yeast to produce the desired fuel, which is obtained by distillation (Guo et al., 2015). Annual bioethanol production is growing at approximately one billion gallons per year, and by 2024, more than 31 billion gallons of bioethanol had been produced, with the United States accounting for half of the total, followed by Brazil and India (RFA, 2025). *Saccharomyces cerevisiae* is one of the foremost widely used fungi in the industry, generating very high profits worldwide.

Monascus filamentous fungi are mainly applied for synthesis of secondary metabolites that are processed as food dyes in the Asian food industry and are known for red fermented rice, which is used as a food, but with overlap into medicine due to the production of monacolin K, which is a well-known statin (Chen & Hu, 2005). Fermented rice is a suitable feed for chickens that reduces the cholesterol content of eggs (Wang & Pan, 2003). The pigments produced have anti-cancer, antioxidant, and many other effects that are being extensively studied for their use in medicine. The production of pigments takes place in large industrial fermenters that allow easy scale-up from the laboratory (Dufossé et al., 2014). The application of pigments does not have to be only in food. They are used in cosmetics, for example, in sunscreens (Koli et al., 2019), anti-aging lotions (Feng et al., 2022). They are also used as non-toxic dyes in the textile industry (Yang et al., 2018).

Another filamentous genus of fungi used across the industry is *Aspergillus*. In the food industry, particularly in Japan, *A. oryzae* is used to ferment various substrates, resulting in the production of sake, miso, and soy sauce, which are indispensable ingredients in Japanese cuisine (Machida et al., 2008). The fungus *A. oryzae*, but also *A. niger*, produces many types of enzymes such as pectinases, cellulases, and others, which are essential not only for food production (Bhat, 2000). They serve as probiotics that support the intestinal microflora of food animals (Lee et al., 2006). *A. niger* also has an indispensable role in the production of citric acid in the chemical industry (Haq et al., 2003). There has been identified the potential of using *A. oryzae* as a glutamic acid-producing organism. This acid can be converted by several reactions to 2,5-furandicarboxylic acid, which is an alternative to terephthalic acid, the main component of polyethylene terephthalate (Kuivanen et al., 2016). Genetically engineered *A. nidulans* was able to upcycle polystyrene into natural secondary metabolites that are used in medicine, such as ergothioneine, pleuromutilin, and mutilin (Rabot et al., 2023).

In Asian cuisine, tempeh is undoubtedly one of the most popular foods. This soybean product, which is nowadays prepared from any legume, could not be made without the fungus *Rhizopus oryzae* (Kim et al., 2025) or *Rhizopus oligosporus*, which carries out the necessary fermentation. This type of product also helps the society to eat less meat, but while maintaining a high protein intake (Erkan et al., 2020). Other types of protein-rich food alternatives to meat are produced by the fungus *Fusarium strain flavolapis*, which are nowadays on the market as products in combination with legumes and vegetables (Furey et al., 2022). *Fusarium venenatum* is another example of a fungus that is used to produce the equivalent of the meat dishes found in shops under the brand name Quorn (Finnigan et al., 2019).

Fungi have played an important role in the treatment of diseases for centuries, even in ancient Egypt. There are records of fungi causing spoilage in foods such as jam, bread, and others, which were used to treat bacterial infections of surface wounds (Wainwright, 1989). Another example is the fungus *Gamoderma lucidum*, commercially known as Reishi mushroom, which is one of the popular natural products used in complementary cancer therapy due to its presence of anti-cancer compounds (Li et al., 2024a). Fungi are also used as biological building materials in the construction industry in an effort to mitigate environmental impacts, where the fungal mycelium serves as a binding material (Babenko et al., 2025). In the Czech Republic, the first house was built where fungal material replaced polystyrene with similar properties. In the following years, it will be possible to observe the transformation of classical construction materials into bio-based ones (Pidosky, 2025). This material can be used as an alternative packaging material in transport, and the clothing industry is also interested in using this fungus to produce a leather-like material (MycoWorks, 2025).

The most famous fungus in medicine is undoubtedly *Penicillium* (Wainwright, 2008), applied in cheese production too (Moore & Chiu, 2001), used to synthesize the antibiotic penicillin against bacterial diseases. There are many fungi that produce antimicrobial compounds, e.g., fungi of the genus *Cephalosporium, Streptomyces,* and others (Wainwright, 2008). These genera are known, but the first antibiotic described was mycophenolic acid, which was used against *Bacillus anthracis* in the late nineteenth century (Nicolaou & Rigol, 2018). Currently, scientists are trying to find new types of antibiotics because of the increasing resistance of microorganisms to these compounds. The solution is the use of specific endophytic fungi that are still effective against bacterial infection (Nguyen et al., 2025). There are also fungi, such as the parasitic fungus *Claviceps purpureus*, which are not used as antibiotic producers, but for their ability to synthesize compounds such as ergot alkaloids to treat certain diseases, migraines or to prevent, for example, postpartum hemorrhage (Králová et al., 2021).

According to the brief introduction, fungi have an irreplaceable position in the entire industrial world. Their biotechnological application can create new products that cannot be obtained from other sources. In a dynamic world, biotechnology will become more visible through the transition from fossil fuel-based industry to a green, CO_2_neutral industry with zero or reduced environmental impact.

2 Important Fungal Product in the Market

Various fungal species have been reported and traditionally utilized for the making of a variety of value-added compounds. These include antibiotics, enzymes, biocolors, organic acids, and other medicinally important metabolites. Table 1 presents an overview of fungal-originated compounds, their producing species, and associated applications. Compared to other chemical and biological sources, fungi offer a superior alternative due their ability to utilize complex agro-industrial waste as a substrate. Additionally, fungi are suitable for both submerged and solid state fermentation and are compatible with scale-up processes. Considering these advantages, fungi are often regarded as "min-factories" for the cost-effective, efficient, and eco-friendly production of valuable bioproducts.

Recent advancements in fungal biotechnology highlighted the potential use of the fungus *Neurospora intermedia* for fermentation of foods from readily available by-products into delicious and nutritious foods (Maini Rekdal et al., 2024). This process not only reduces food waste but also enhances sustainability in food production. Similarly, in the realm of enzyme production, the studies have demonstrated that substrates such as rice husk, groundnut fodder, and sugarcane bagasse can be effectively utilized to produce various enzymes. For instance, a co culture system involving *Aspergillus protuberus* and *Aspergillus unguis* achieved significant amylase yields under solid state fermentation using agro-industrial waste (Maddimadugu et al., 2024). Additionally, the valorization of lignocellulosic waste through fungal bioprocessing aligns with the principles of a circular bioeconomy. Fungi possess the unique ability to breakdown complex lignocellulosic material and convert them into valuable products such as biofuels and bioplastics.

3 How Fungal Biotechnology Plays an Important Role in Different Sectors

Fungal biotechnology is a promising tool for achieving the goals of a circular economy. The diversity and potential applications of fungi across various sectors, including the environment, agriculture, nutrition, and chemicals will enhance the adoption of fungi in sustainable practices. The following section of this chapter will focus on the various applications of fungi biotechnology.

Table 1 Commercially produced fungal products

Sr. No.	Product	Fungal species	Applications	References
1	Antibiotics			
	Penicillin	*Penicillium chrysogenum*	β-lactam class of antibiotic, FDA-approved indications for several bacterial infections	Ziemons et al. (2017), Yip and Gerriets (2025)
	Cephalosporin C	*Acremonium chrysogenum*	Use as a starting substance for the production of pharmaceutically active cephalosporin antibiotics	Zhgun (2023)
	Fusidic acid	*Fusidium coccineum*	Fusidanes class of antibiotic, antibacterial agents against gram-positive bacteria	Huang et al. (2021)
	Helvolic acid	*Aspergillus fumigatus*	Fusidanes class of antibiotic, antibacterial agents against gram-positive bacteria	Chain et al. (1943)
	Cephalosporin P1	*Acremonium chrysogenum*	Fusidanes class of antibiotic, antibacterial agents against gram-positive bacteria	Chou et al. (1967)
	Griseofulvin	*Penicillium griseofulvum*	Antifungal	Udagawa and Abe (1961), Venkata Dasu et al. (2003)
	Echinocandin B	*Aspergillus nudulans*	Precursor for semisynthetic antibiotic anidulafungin	Benz et al. (1974), Pfaller et al. (2005), Hu et al. (2020)
	Pneumocandin B_0	*Glarealozoyensis*	Precursor for semisynthetic antibiotic caspofungin	Bouffard et al. (1994), Song et al. (2018a, b)
2	Lovastatin	*Penicillium species, Monascusruber,* and *Aspergillus terreus*	Hypercholesterolemic drug	Aravindan et al. (2008)
3	Chitosan	*Rhizopus oryzae and Aspergillus niger*	Pharmaceutical areas and biotechnological applications	Pochanavanich and Suntornsuk (2002), Erdoğmuş and Altintas (2023)
4	Enzymes			
	Amylase α-amylase	*Aspergillus* spp.	α-amylase used for synthesis of high fructose corn syrup, food, ethanol, textile, detergent, and paper industries	Elyasi Far et al. (2020)
	β-amylase	*Aspergillus niger*	Applications in food and cosmetics	Ray (2004), Adeniran et al. (2010)

(continued)

Table 1 (continued)

Sr. No.	Product	Fungal species	Applications	References
	Glucoamylases	*Aspergillus* sp. *and Rhizopus* sp.	Biocatalysts used for the production of glucose and fructose on an industrial scale	Lee and Paetzel (2011), Nyamful (2013)
	Alkaline protease	*Aspergillus niger*	Tanneries, food production, and detergents.	Nouri et al. (2024)
	Acid protease	*Aspergillus oryzae* HG76	Food and beverage industries	Li et al. (2014), Song et al. (2023)
	Glucose oxidase	*Aspergillus niger*	Food industry	Hatzinikolaou et al. (1996), Dubey et al. (2017)
	Cellulase	*Humicolainsolens*, *Penicillium funiculosum*, *Trichoderma viride*	Food, textile, paper, detergent production, and waste management	Hayashida et al. (1988), Maeda et al. (2013), Lan et al. (2013)
	Invertase	*Saccharomyces cerevisiae*, *Aspergillus niger*	Food and beverage industries	Baseer and Shall (1971), Ohara et al. (2015)
	Xylanase	*Aspergillus* spp., *Trichoderma* spp.	Paper, food, pharmaceutical, and animal feed	Ghanem et al. (2000), Taneja et al. (2002), Polizeli et al. (2005)
	Pectinase	*Aspergillus niger*, *Aspergillus* spp.	Food, textiles, paper, and animal feed	Reginatto et al. (2017)
	Laccase	White rot fungus species	Textile, paper production, and biopharmaceuticals	Wang ct al. (2019)
5	Organic acids			
	Citric acid	*Aspergillus niger*	Food additives and industrial applications	Show et al. (2015)
	Gibberellic Acid	*Fusarium fujikuroi*	Plant growth regulator, phytohormone, agriculture application	Hernández Rodríguez et al. (2024)
	Itaconic acid	*Aspergillus terreus*	Plastics, textile, paint, pharmaceutical, and agro-based products	Gopaliya et al. (2021), Devi et al. (2022)
6	Pigments			
	Food color	*Monascus* spp.	Food colorants and additives	Egea et al. (2023)
7	Nutritional food			
	Mushroom	*Agaricus bisporus*, *Pleurotus* species *Lentinus edodes*, and *Volvariella* species	High-nutrient food	Assemie and Abaya (2022)

3.1 Fungal Biotechnological Approaches for Sustainable Environment

The industrial revolution is one of humanity's greatest achievements, significantly improving the quality life. This journey began with the use of natural resources, and today, many products are synthesized in laboratories. However, the economy still relies heavily on exhaustible resources, such as petroleum products. Additionally, industries contributed in global warming and environmental pollution through the discharge of pollutants into water resources, landfill waste, the release of toxic gases, and harmful byproducts. These activities impose significant burden on the economy to manage, while also negatively impacting the environment and human health. To restore environmental sustainability, fungal biotechnology, with its diverse applications, from bioremediation and waste bioconversion to biofuel production, the degradation of toxic industrial effluents and pesticides, carbon sequestration, and the removal of toxic metals from water bodies offers a promising tool for ecosystem restoration and solving environmental issues.

3.1.1 Bioremediation

The bioremediation is a process in which microorganisms employed to remove toxic pollutants from the environment. The various microbes have been studied for the remediation of toxic compounds from the different sites and it has an advantage over the various physical and chemical strategies, such as it is in situ application and environmental acceptability (Pletsch et al., 1999). The in situ applications have reduced the additional economic burden of handling, storage, and transportation of contaminated materials to the treatment plant. The methodology employed for fungal species for detoxification, treatment of contaminated water, soil, and air is known as mycoremediation. The range of fungal species, including both mycelium and higher fleshy fungi, is used for remediation of toxic heavy metals, oil hydrocarbons, and pharmaceutical compounds (Mohamadhasani & Rahimi, 2022; Ghosh et al., 2023). The mycoremediation is based on several cellular events such as adsorption, mineralization, precipitation, surfactant-based, and through the intra- or extracellular enzymes-catalyzed reactions.

3.1.2 Fungal Biotechnology for Removal of Heavy Metals

The industries and miningprocesses have discharged heavy metal-contaminated waste into water and soil which is responsible for causing the serious environmental and human health concern, and it is increased over the years. The soil and water bodies contaminated with toxic metals have negatively influenced the ecosystem present there. Although some metals are essential for the growth of microbes in environments (magnesium, manganese, copper, zinc, etc.) but only at certain concentration, some metals such as lead, mercury, cadmium, and arsenic are non-essential and highly toxic even in minute concentrations (Langer et al., 2012).

Mycoremediation can be considered an efficient and cost-effective strategy for removal of the toxic heavy metal and treatment owing to the potential of fungal adsorption, accumulation, and recovery of metals. The various filamentous fungi have been reported for their heavy metal-tolerating capability, which can decontaminate the site and recover metals from water, soil, electronic waste, and any solid waste polluted with metals (Ghosh et al., 2020, 2023; Dusengemungu et al., 2021). Among the filamentous fungi, the genus *Aspergillus* has been thoroughly explored for its capacity to tolerate a variety of metals, leaching, and recovery (Dusengemungu et al., 2021; Naveen Kumar & Prakash, 2021).

Recent studies highlighted the potential of *Aspergillus* to tolerate multi-metals such as *A. niger* and *A. fumigatus* removed Cd from polluted soil with 79 and 98% efficacy, respectively (Khan et al., 2019a). Das et al., (2019) reported the Al removal potential of *A. foetidus* from spent catalyst with highest efficacy of 88% (Das et al., 2019). Similarly, Shah et al. (2020) conducted a study to remove Al from waste of Bauxite using *Penicillium simplicissium* and *A. niger* and reported >98% removal (Shah et al., 2020). *A. nidulans* mutant effectively decontaminated (70% removal efficacy) to treat the industrial effluent contaminated with Cu (Rissoni Toledo et al., 2021). Tang et al. (2021) reported that the treatment of polluted soil with *A. tubingensis* effectively removes Pb (90%), Zn (68%), and Cr (64%) (Tang et al., 2021). The other filamentous fungi used for removal of Mg and Ag metals in mine tailings included *Cladosporium* sp. and *Penicillium chrysogenum* (Huerta-Rosas et al., 2020), Cd from polluted soil and sewage using *Penicillium rubens* and *Rhizopus oligosporium* (Ghyadh et al., 2019; Khan et al., 2019a). Perhaps, the multi-metal toxicity of heavy metals is the one serious problem and major bottleneck for remediation. However, the fungal species of *Mucor hiernalis* had potential to simultaneously tolerate multiple metals present at the site, and effectively removal of them. They have potential to simultaneously remove >81–99% of various metals present in water within 48 h. The metals removed by *Mucor* species include Ag, Al, Au, Cd, Co, Cr, Cu, Hg, Ni, Pb, Ti, U, and Zn (Hoque & Fritscher, 2019; Ghosh et al., 2023).

The higher fungi or mushrooms have also been studied for their adsorption and removal of heavy metals. Among various mushroom species, *Pleurotus* spp. was reported for high tolerance to Co, followed by Cu. The heavy exposure significantly increases the catalase activity in superoxide dismutase and also increased metal accumulation in mycelium over increasing the concentration of metals (Mohamadhasani & Rahimi, 2022). The spent mushroom substrate of *Agaricus bisporus* effectively bioremediated heavy metals such as Cr, Pb, Fe, Co, Ni, Mn, Zn, Cu, and Al present in industrial waters. Fe, Ni, and Co were effectively removed from contaminated water at acidic pH of 2.5 by spent Mushroom substrate (Corral-Bobadilla et al., 2019). *Galerinavitti formis* mushroom reported for 80–90% removal of Cd, Cr, Cu, Pb, and Zn metals from contaminated soil (Damodaran et al., 2014). Moreover, *Lactarius piperatus, Fomes fasciatus, Pleurotus platypus, Agaricus bisporus, Calocybe indica, Flammulinavelutipes, Pleurotus tube-regium, Pleurotus ostreatus,* and *Pleurotus sajor-caju* are studied for biosorption and removal of different heavy metals. In the case of mushrooms, the dead biomass has a greater capacity or similar to living cells and the major mechanism is biosorption. The mushroom-based biosorption is mediated through the different polar groups present in mushroom such as amino acids, proteins, lipids, and polysaccharides (Kulshreshtha et al., 2014).

The fungal biotechnological approach for removal of multiple metals can overcome the multi-metal toxicity issue and significantly recover the metals from environment. Additionally, mycoremediation is an environment-friendly and cost-effective approach that can significantly reduce the economic burden and exceptional impact on industrial bioremediation and help restore ecosystem health. The treated water from industries and soils can be reused for various applications. However, further research is needed for optimization of process. The mechanism of metal recovery would be helpful for commercialization of mycoremediation (Ghosh et al., 2023).

3.1.3 Fungi for Bioconversion of Agro-Industrial Waste

In recent years, the handling of industrial, agro-industrial, and household waste has become a greater concern. Among these, agriculture practices and industries have contributed to around 1.3 billion tons of waste generated annually (Kour et al., 2023) and much of this waste is either leftover or disposed of by open-field burning without proper treatment. This leads to challenges in both the environmental and economic sectors (Prado-Acebo et al., 2024). Therefore, research on the conversion of these wastes into useful products has grabbed significant attention.

The fungal species have been potentially utilized for the bioconversion of waste due to their vast diversity of enzymes that hydrolyze lignocellulosic waste. For a long time, fungi have contributed to the circular economy through the utilization of industrial and agriculture waste for the production of nutrient-rich products such as mushroom. A variety of mushrooms have been successfully cultivated on different wastes generated by industries or agriculture practices (Kulshreshtha et al., 2010, 2014; Dulay, 2012). Akter et al. (2022) successfully utilized agro-lignocellulosic waste such as corn cobs, rice husk, rice straw, wheat straw, saw dust, and sugarcane bagasse for cultivation of *Pleurotus ostreatus*. Among these, the highest cultivation was recorded in saw dust and rice husk (Akter et al., 2022).

Mushroom cultivation has the potential to utilize waste as a substrate, which reduces the burden of waste management. Additionally, the produced mushrooms are a high protein, vitamin, and minerals-rich product that generates revenue. This dual functionality helps to reduce environmental pollution caused by waste while also generating healthy food that may contribute to alleviating the global hunger problem. Similarly, different yeast and filamentous fungal species have potential for bioconversion of agro-industrial waste into valuable products such as proteins, biofuel, metabolites, pigments, enzymes, biopolymers, etc.

The production of bioethanol (a liquid biofuel) by fungal species from lignocellulosic agri-waste residues has an ideal solution to decrease both waste as well as pollution of CO_2. Recently, several studies highlighted the potential of fungi to efficiently produce bioethanol, for example, Boonchuary et al. (2021) have conducted a study in which corncob residues were utilized by thermotolerant *Saccharomyces cerevisiae* TC-5 for making ethanol. They reported both saccharification and solid state fermentation for ethanol production by *Saccharomyces cerevisiae* TC-5 that produces 31.96 g/L ethanol at elevated temperature (Boonchuay et al., 2021). The full cycle bioconversion of lignocellulosic waste into second generation ethanol has been reported through the saccharification of lignocellulosic waste using cellulolytic enzymes from *Talaromyces funiculosus* UCM F-16795 and subsequent production of 11.30 g/L of ethanol by using *Scheffersomyces stipitis* UCM Y-2810 species (Fomina et al., 2024). Besides, several fungal species including *Trichoderma*, *Aspergillus* and *Phanerochaete* having potential to produce lignocellulosic enzymes that are hydrolytic enzymes capable to breakdown the agriculture waste (Lima et al., 2024; Adav et al., 2010; Manavalan et al., 2011; Jain et al., 2017).

3.1.4 Fungi Used for Treatment of Industrial Effluents, Solid Waste, and Bioremediation

Besides the agriculture waste, the various industrial effluents have significant impact on the environment. The solid waste generated from industries contains several pollutants that harm or imbalance the soil and water ecosystem. Industrial wastes need to detoxify or undergo treatment to reduce its negative impact on the environment. The solid waste comprises various types and hospital waste is one of the dangerous solid wastes being generated by hospitals, clinical research labs, and industries working on health sector. This type of waste contains biological fluids, and various consumables were used. This waste is the most infectious, due to contamination of various microbial pathogens and therefore, it is necessary to dispose of it in a proper scientific way. However, routinely the hospital waste is efficiently disposed of by using the incineration method, in which it is destroyed in a furnace at high temperature. But, at the same time, the burned hospital waste ash contained various toxic pollutants such as toxic metals and hydrocarbons (polycyclic aromatic hydrocarbons), and disposing this ash in open landfills causes harm to the ecosystem present in soil and also contaminates the groundwater sources. To overcome this issue, several studies are attempted. Pandey and Gundevia (2008) demonstrated the potential of *Periconiella* sp. to degrade hospital waste efficiently (Pandey & Gundevia, 2008).

3.2 Fungal Biotechnology for Agriculture

The agriculture sector is the backbone of the economy in developing countries, contributing not only in meeting food demands but also in reducing poverty by providing employment opportunities. Within agriculture, the food system, in particular, creates millions of jobs at the farm level and has the potential to generate many more across the entire supply chain. Access to proper nutrition results in healthier, more productive individuals, supports entrepreneurship, and represents a crucial long-term investment in economic development. According to FAO (2023), it has been established that the share of the agricultural population in the total workforce is 27%, accounting around 4% of the global GDP, and the agricultural sector contributed USD 3.7 trillion in value in 2021.

Various inputs are used to increase agriculture yield, such as the application of fertilizers, pesticides, and manures, which include both chemical and biological sources. Among the biological sources, fungi have natural capabilities that contribute to yield improvement by enhancing nutrient uptake, acting as natural biocontrol agents, and participating in nutrient recycling. Fungi associated with plants help them withstand harsh environmental conditions, for example, *Mycorrhizae* improve soil health and crop productivity by mineralizing insoluble nutrients, while *Trichoderma* species protect crops from plant pathogens. Additionally, various fungal species can convert agricultural waste into valuable products such as compost, biofertilizers, and other useful materials, effectively utilizing organic matter. Integrating fungal biotechnological processes into farming can support resource efficiency, reduce waste, and help align agricultural practices with the principles of a circular economy.

Among various fungi, mycorrhizal fungi have been used in agriculture for a long time. They form tripartite mutualistic associations between soil, fungi, and plants, playing a crucial role in plant productivity, nutrient uptake, and protection from water scarcity and establishment of ecosystem. Recent fungal research highlighted the significant role of bioinoculants comprised of mycorrhizal and endophytic filamentous fungal consortia in improving crop yield and quality under both normal and stress conditions. These consortia help alleviate various environmental stresses effecting crop plants (Díaz-Urbano et al., 2023). Similarly, endophytic fungi form beneficial relationships with plants, enhancing their resistance to pathogens and improving overall plant quality. For instance, the endophytes *Penicillium citrinum* LWL4 and *Aspergillus teruus* LWL5 have been reported to provide resistance in sunflower against *Sclerotium rolfsii* (Waqas et al., 2015). Several entophytic fungal species are known to produce enzymes such as chitinases, lipase, cellulases, proteases, amylases, and pectinases, which play a key role in protecting plants from pathogens (Varma et al., 2017). Additionally, some fungi, such as *Trichoderma* and *Phanerochaete* species, produce lignocellulolytic enzymes that are responsible for the degradation of agricultural waste (Tiquia et al., 2002).

Fungal biotechnological approaches have the potential to reduce the dependence on chemical pesticides and fertilizers by replacing them with fungal-based products. The adaptation of fungal biopesticides and biofertilizers has been increasingly promoted in agriculture due to their targeted pathogen control activity and their ability to enhance crop yield in an eco-friendly manner. In recent years, several mycofungicides and biofertilizers have become commercially available for applications such as seed treatment and plant protection. For instance, commercial formulations such as Ketomium®, which contains *Chaetomium globosum* and *Chaetomium cupreum*, and Promote® and Trichodex® developed from *Trichoderma harzianum* (with *Trichoderma viride* included in the latter), are widely used. Similarly, SoilGard® is based on *Gliocladium virens* (Kaewchai et al., 2009). Furthermore, fungal biofertilizers include various plant growth-promoting fungal, such as *Trichoderma*, mycorrhizae and phosphate and potassium solubilizing fungi. These fungal based mycofungicides and biofertilizers offer a sustainable alternative to chemically synthesized products, eventually minimizing the adverse effects on crops and the ecosystem (Kaewchai et al., 2009).

3.3 Nutrition

Fungal biotechnology plays a crucial role in nutrition by contributing to food production, processing, and nutrient enhancement. The increasing global population, expected to reach 9 billion by 2025, necessitates the increased supply of proteins. Dietary proteins are essential as the body cannot prepare essential amino acids. The external protein supply includes animal protein as the major source along with plant proteins (Li et al., 2023). The shortage of animal proteins and scarcity of cultivable land and water cannot meet the existing global protein demand to fulfil the pressure of an increasing population and improvement of living standard. In this situation, there has been significant focus on fungal protein due to its sustainable qualities. Fungi are a sustainable food source as they require less land and resources than traditional livestock farming. Fungi are abundant source of amino acids and protein; good source of fiber, vitamins, and micronutrients, little of fat, and are nearly devoid of almost cholesterol. Fungi are also a valuable source of fiber and antioxidants. Incorporating fungi into the diet may lead to lower overall cholesterol levels, decreased insulin levels after meals, and reduced energy consumption (Amara & El-Baky, 2023). These amazing features position it as a beneficial supplement for animal and plant source proteins. Nevertheless, consumer acceptance of fungal proteins remains limited due to concerns regarding their taste and safety.

The major nutrition supplement source of fungal protein is edible mushrooms and single-cell protein. Edible mushrooms are a major source of protein particularly for vegetarians, and are free of cholesterol. Mushrooms are also rich in vitamins, minerals, and antioxidants, contributing to a balanced diet. The market for mushrooms has seen significant growth in recent years. It is projected to increase from $66 billion in 2024 to $71.62 billion in 2025, and then rise even further to $105.54 billion by 2029. The market size growth is due to boosted demand for vegetarian diet, health benefits of mushroom, and economic growth in emerging markets (Mushroom Market Report, 2025).

Although various microorganisms are used as source of single-cell protein, filamentous fungi are easy to harvest and yeast-like fungi also give the benefit of addition of vitamin B complex with fiber-rich content (Ritala et al., 2017). Fungi can reach to high protein contents from 30% to 50% by fermentation optimization for protein increase (Nasseri et al., 2011). Fungal protein market size grows from $4.06

billion in 2024 to \$4.34 billion in 2025 and will grow further \$5.71 billion in 2029 (Fungal Protein Global Market Report, 2025).

Fungi and yeast are considered excellent sources for single-cell protein (SCP) production, making them adaptable for a range of uses, including food and animal feed, as well as additional applications in other sectors like the paper industry (Li et al., 2024b). Food safety regulation concerns about the fungal-based foodstuff and supplements, due to toxic metabolites production from producing fungal strain if any, elevated nucleic acid levels accompanied by reduced cell wall digestibility, allergic reactions, and harmful substance produced from raw material. The extraction of proteins may change the nutritional content of protein, so it is considered a novel food (Amara & El-Baky, 2023). The fungal strains widely recognized for SCP production include *Saccharomyces cerevisiae,* microfungus *Fusarium venenatum* producing mycoprotein Quorn, which is the largest brand of meat alternative in the world and yeast *Yarrowia lipolytica,* which is widely used as a food supplement (Wiebe, 2004). The crude protein product obtained from fungi and yeast majorly involves *Saccharomyces* spp., *Yarrowia lipolytica*, and *Aspergillus niger* for animal feed application. Single-cell proteins (SCP) have become a viable option due to their cost-effective production and reduced resource requirements. The main sources of SCP include sugars and starch, as well as a variety of secondary and waste materials such as fruit scraps, molasses, and other combustible waste and residues like methane, methanol, and ethanol. The production of SCP through gas fermentation is particularly noteworthy, as syngas (CO/H_2) and biogas (CH_4)—or methane derived from CO_2 through methanation—can be produced from a wide array of waste biomass sources. Additionally, H_2/CO_2 facilitates the recycling of carbon dioxide from both industrial emission points and direct air capture processes (Li et al., 2024b).

3.3.1 Why Fungi as Future Food?

Apart from traditional sources of food, fungi particularly, filamentous fungi established as alternate source of protein, now an era is of technological advancement in fungi-based foods for size, shape, color, and texture (Barzee et al., 2021).

The major parameter controlling the flavor of the fungi based food is the type of fungus applied. The strongest favorable fungi can impart strong flavor while normal fungi can become more favorable with some supplements. The choice of medium used for the cultivation of fungi along with the use of spices and herbs can also improve the flavor.

The texture of fungal food product can be controlled by growth mode of fungi, addition of chemicals like albumin for production of mycoprotein (Finnigan, 2011), fungal morphology like size and porosity of fungal pellet, and after cultivation mechanical processes. The texture analysis profiling includes parameters like hardness, chewiness, springiness, gumminess, cohesiveness, and fracturability.

The color of the fungal-based food can be altered during and after cultivation, depending on the chosen dish. The ability to absorb can modify the color of fungi after cultivation and can be during cultivation, resulting from its own color and its metabolic products like enzymes.

The very important factor is the nutritional profile, which is of its own and can be modified as well. The nutritional profile could be assisted with the innate ability to harvest other microorganisms from cultivation media which may lead to nutritional as well biological effect by microorganism delivery.

Cutting-edge methods such as 3D food printing allow for precise control over the shape and ingredients of the final product, enabling the creation of functional foods including bioactive compounds. 3D food printing represents a print-and-eat approach that can produce nutritious and customized meals through additive manufacturing techniques. Examples of this technology include 3D printed items like chocolate, pizza, peanut butter, and snacks enriched with fiber derived from mushrooms (Keerthana et al., 2020), fruit-based snacks with improved energy, calcium, vitamin D. and iron (Derossi et al., 2018). There are numerous areas of knowledge still left to investigate regarding the use of 3D printing of fungal foods.

3.4 Chemical and Fermentation Industry

3.4.1 Fungi in Fermentation Industry

Fungi play a significant role in the food processing industry by providing components such as acidulants, enzymes, flavors, vitamins, colorants, and polyunsaturated fatty acids. Organic acids are extensively utilized as ingredients in contemporary food processing, primarily serving as acidulants, but also functioning as flavoring agents, buffers, preservatives, and technological enhancers. The types of organic acids include citric acid, gluconic acid, lactic acid, fumaric acid, malic acid, succinic acid, oxalic acid, and so on (Magnuson & Lasure, 2004). Filamentous fungi are key providers of enzymes utilized in food processing, improving both the efficiency and convenience of various processing stages, while also elevating the quality of the end product (Zhang et al., 2018). The use of enzymes in processing affects the flavor, aroma, visual appeal, texture, color, stability, and nutritional value of food substances. The enzyme includes amylase, protease, pectinases, invertases, galactosidases, and lactases (Copetti, 2019). Filamentous fungi can also generate pigments that encompass various chemical classes, including carotenoids, melanins, flavins, phenazines, quinones, and more specifically, monascins, violacein, or indigo (Pan & Hsu, 2014; Dufossé, 2018). Vitamin B2, riboflavin, and yellow -colored vitamin are almost exclusively produced by fungal fermentation (Stahmann, 2011).

Various fungi can store over 20% of their biomass in the form of lipids (Passoth, 2017) in the form of lipids like polyunsaturated fatty acids (PUFAs) while wild yeasts have the ability to store a higher amount of oil in their cells, reaching as much as 85% of their dry weight. One of the most remarkable and intriguing characteristics of fungi is their ability to produce volatile flavors (Fraatz & Zorn, 2011). Notable examples consist of the generation of blue cheese flavor by *Penicillium roqueforti*, which is abundant in volatile fatty acids and 2-alkanones. Additionally, bitter almond flavor is produced by white-rot fungi, including *Ischnoderma* spp. and *Saccharomyces* spp. Yeast is capable of producing 2-phenylethanol, which has a scent reminiscent of roses, when provided with this specific amino acid. Basidiomycetes fungi generate volatile flavor compounds such as cinnamic acid esters (exotic fruit), cinnamaldehyde (cinnamon), eugenol (clove), methyl benzoate (dry fruit), and benzyl acetate (jasmine).

3.4.2 Fungi in Fermented Foods and Brewing Industry

Some fungal species are safe to use in production of traditional fermented foods like *Tou-shi* (fermented black soybean paste as condiment), *Angkak* red rice (fermented steamed whole-rice kernels as bicolored food with health benefits), *Oncom* (fermented pressed peanut cake as side dish), *Tapé Ketan* (fermented glutinous rice with ragi tape as snack), and *Kochujang* (fermented soybean paste as side dish) (Nout, 2014). The use of mold in the food gives characteristic properties to the food like texture, color, flavor, and produce enzymes for desirable food modification and nutritional and health benefits. Fungi are also used in the production of other fermented foods like cheese, bread, tempeh, and so on.

The use of novel fermented food and fungal products is of great significance for future exploration (Ferrara et al. 2023). The shortcomings of crops and livestock products could be mitigated by the use of fermented foods without adverse environmental impact (Zurek et al., 2022; Pouris et al., 2024). The modification of fungal mycelium through genetic tools has been shown to improve the nutritional value, sensory attractiveness, and scalability of fungal foods for bioengineered food production (Maini Rekdal et al., 2024).

The fermented food and beverages market is projected to reach $1001.58 billion by 2032 (Fermented Foods and Beverages Market Report, 2025). There is increasing awareness of the benefits of fermented food and beverages to fuel market. Fermented food sector is expected to have larger share among the fermented food and beverages market.

Beverages and fermentation technology is one of the most popular and well-established areas for food technology for long. Fungi, particularly yeast species like *Saccharomyces cerevisiae*, are essential in the manufacturing of alcoholic drinks such as beer and wine (Koul & Farooq, 2020). Species of Saccharomyces are most tolerant to high concentrations of ethanol and osmotic pressure. In Brazil, sugarcane is used for large-scale alcohol production for automobiles. Alcoholic beverages are made from a large variety of starchy or sugary products produced using different agricultural products. Yeast metabolites play a significant role as flavor components in beverages, impacting the ultimate taste and aroma of alcoholic drinks like beer and wine, as well as other distilled products including brandy, rum, and whiskey (Niego et al., 2023).

4 Role of Fungal Biotech in Solid State Fermentations

4.1 Solid State Fermentation and Fungi

According to the popular definition of solid state fermentation (SSF), it is a fermentation method where microorganisms use water-insoluble substrates in conditions that are primarily free of water. The water activity of the substrates used in SSF typically agro-waste is below 0.6 to 0.8. This condition makes SSF particularly suitable for the growth of fungi, as the majority of filamentous fungi can grow well at 0.93 to 0.98 water activity levels. Therefore, products generated through solid-state fermentation are important when discussing the circular economy of fungi. Agricultural waste from crop processing consists mainly of solid materials, which are essentially concentrated forms of nutrients. There are various benefits of SSF-based fungi-derived products, as they are economically affordable, produced in concentrated forms, and easy to recover. A single solid surface in the process provides all the necessary nutrients for microbial growth. The major industrially important products produced through SSF fall into three main categories: products of filamentous fungi, products of yeast, and products of mushroom-forming fungi.

Compared to bacteria and yeast, fungi are considered the most suitable and effective organisms for the production of metabolites through SSF. Only a limited number of yeast species are used in metabolites production via SSF. However, some industrial processes, such as ensiling, report yeast involvement during the early stages. The application of yeast is primarily seen in traditional and commercial food fermentations, particularly in Malaysian and Thai foods (Surya, 2024). The main yeasts used in fermented foods include *Endomycopsis bartonii* and*Amylomyces rouxii*. In addition, *Saccharomyces cerevisiae* has been employed in saccharification reactions due to its amylase activity (Cripwell et al., 2019, 2020).

4.1.1 Filamentous Fungi

Filamentous fungi are the key organisms in industrial solid-state fermentation. Traditionally, well-known processes such as *koji* and *tempeh* fermentation utilize *Aspergillus oryzae*, with various substrates including wheat bran and soyabeans (Mäkelä et al., 2014; Zhang et al., 2022b). These fungi possess the ability to produce unique enzymes such as amylases, proteases, pectinases, and invertase which makes them ideal for processes that use solid agro-waste as substrate (Steudler et al., 2019).

Primarily, filamentous fungi have been responsible for value addition in starchy and cellulosic agro wastes such as cassava waste, banana peels, wheat straw, and corn stover. Their applications have expanded further; for example, *Aspergillus niger* has been used to enhance phenolic content, antioxidant activity, and lysine levels in rice, wheat, and sorghum meal (Singhania et al., 2009; Sawangwan & Saman, 2016; Yin et al., 2019). White-rot and other lignolytic fungi are commonly

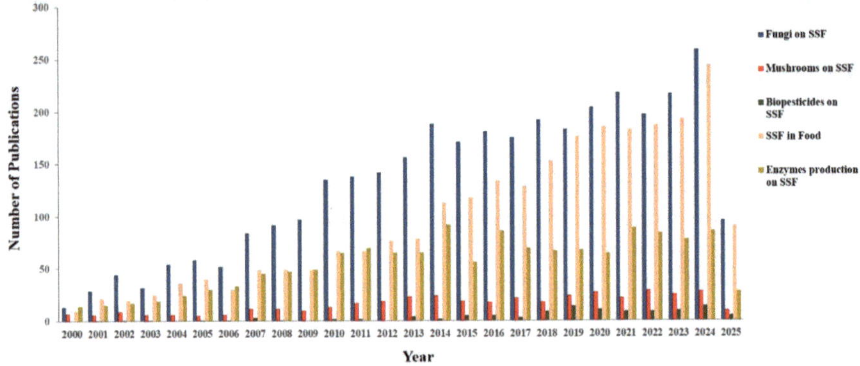

Fig. 1 Year wise Publication /research on SSF and fungi and different product produce by SSF

used to improve the digestibility of substances. While the traditional use of fungi in cheese production and coffee or tea compost fermentation is well established, filamentous fungi are now widely applied in industrial processes for the production of food, mushrooms, enzymes, biopesticides, biomass, and various primary and secondary metabolites (Fig. 1). SSF using filamentous fungi is particularly relevant to food production, value addition, and the detoxification of animal feed. Some key applications of fungi on solid substrates are discussed below.

4.1.2 Enzyme Production by SSF

Koji is a traditional method in which fungi and molds are grown on cereal grains to produce value-added food products. Enzymes produced by SSF offer various advantages, such as high production efficiency and broad stability across a range of pH and temperature (Lonsane et al., 1985). This process primarily involves enzymatic activities such as amylase, phytase, cellulase, and deaminase, produced by a variety of microorganisms's fungi being the predominant group. These organisms have been extensively studied for enzyme production both at the laboratory and commercial scales. Numerous potent fungal strains have been identified and reported for enzyme production using agro-industrial solid waste materials. For example, glucoamylase has been produced using agro-waste such as rice, soybean, corn, cassava waste, with *Aspergillus* species commonly employed. Similarly, wheat bran has been used for lipase production by *Rhizopus delemar* and *Aspergillus* spp. (Mäkelä et al., 2014; De Vries & Mäkelä, 2020). In addition, several commercially significant enzymes such as asparaginase, catalase, β-glucosidase, oxidase, and xylanase have been reported to be produced by fungi using various solid substrates. These examples are reviewed in Table 2.

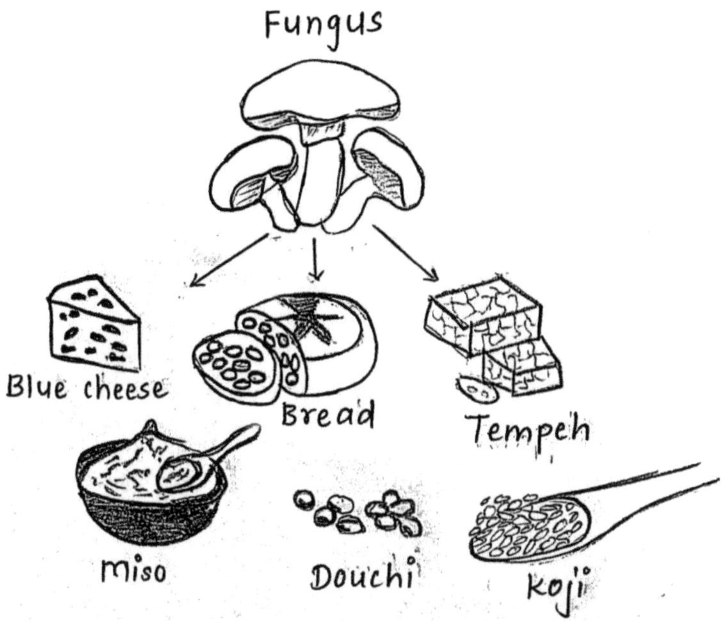

Fig. 2 Representative fungal food products produced by Solid State Fermentation (Gemini pro2.5)

4.1.3 Fungi in Food via SSF

The application of fungi in food processing and as food itself is well documented. Traditionally, fermented foods involving both bacteria and fungi are well known (Singh et al., 2020; Corbu et al., 2023). In practice, some of these have been commercialized. Often, these foods are the result of the combined activity of both fungi and bacteria. Due to their ability to grow on solid substrates, fungi play a vital role in the making of many solid-state fermented foods, either as the primary fermenting agents or by biotransforming solid materials into healthy, value-added products (Nout, 2000; Lübeck & Lübeck, 2022; Vivek & Venkitasamy, 2023). Traditionally, organisms such as *Aspergillus*, *Rhizopus*, *Penicillium,* and various yeasts have been identified for their significant roles in this process (Chen et al., 2015; Son et al., 2018; Venturini Copetti, 2019; Strong et al., 2022). A wide diversity of fungi-mediated food products exist, particularly in tropical and subtropical regions (Fig. 2). Some important examples of fungal biotransformed solid food include:

Tempeh
Mold-ripened Cheese
Ang-kak
Tempeh

Table 2 Examples of some fungal origin enzymes and their producers using solid substrates

Enzyme	Fungi	Substrate	References
Phytase	*Rhizopus oligosporus*	Rapeseed meal	Howson and Davis (1983)
Catalase	*R. niveus*	Wheat bran	Sreekantiah et al. (1968)
Ribonuclease	*A. candidus*	Wheat bran	Kumar et al. (2011)
Naringinase	*Aspergillus niger*	Citrus peel	De Oliveira et al. (2022)
Laccase	*P. oystriatus*	Sorghum stalk	Pinheiro et al. (2020)
β-galactosidase	*Rhizomucor pusillus*	Wheat bran	Raol et al. (2015), Volford et al. (2021)
Lipase	*A. niger*	*Agro waste*	Suyanto et al. (2019), Ali et al. (2023)
Chitinase	*Metarhizium*	Sugarcane bagasse	Dos Reis et al. (2018)
Pectinase	*A. niger*	Sugarcane, orange, Citrus waste	Ahmed et al. (2016), Esawy et al. (2022)
Xylanase	*A. terreus*	Wheat straw	De Carvalho et al. (2023a, b)
Chitinase	*Metarhizium anisopliae*	Wheat bran, rice bran, rice as substrate	K. Paul et al. (2022)
β-1,3-glucanase	*Metarhizium anisopliae*	Sugarcane bagasse, malt bagasse, and white rice	Aita et al. (2019)
Urease	*A. niger*	Wheat straw	Khan et al. (2019b)
Cellulase	*Trichoderma stromaticum* AM	Sugarcane bagasse, peach palm waste	Bezerra et al. (2021)
Laccase	*Trametes versicolor*	Lignocellulosic waste	Pinheiro et al. (2020), Ahmed et al. (2024)
Pectinase	Aspergillus sp.	Fruit peels banana, Grape pomace	Abreu et al. (2024), Ahsan et al. (2024)
Xylanase	*A. awamori*	Tomato pomace	Umsza-Guez et al. (2011)
Protease	*A. niger*	Lignocellulosic waste	Nouri et al. (2024)
Lipase	*Yarrowia lipolytica*	Mango agro-waste	Da S. Pereira et al. (2019), Carvalho et al. (2023)
Polyphenol oxidase	*Pleurotus pulmonarius*	Palm peach waste	De Carvalho et al. (2023a, b)
Amylase, protease	*Aspergillus oryzae*	Food waste,	Escaramboni et al. (2022)
Laccase	*T. multicolor*	Wheat bran	Unuofin (2021)

Tempeh is one of the most popular fungal fermented products in South Asia and is believed to have originated from Java–Sumatra region, although significant development and commercialization occurred in the Netherlands. *Rhizopus oligosporus* and other *Rhizopus* species are the primary microorganisms used in tempeh fermentation (Kovac & Raspor, 1997; Bavia et al., 2012). Traditionally, various leguminous beans such as soyabeans are boiled, cooled, and then inoculated with *Rhizopus* spores. The spores are often introduced using natural carriers such as hibiscus leaves, rice, or cassava flour. The mixture is incubated at around 37 °C, during which *Rhizopus* grows rapidly on the surface and penetrates the beans, producing a compact, white, mycelium-bound product known as tempeh.

Rhizopus oligosporus is a versatile fungus able to grow on a wide range of grains and pulses due to its ability to produce a variety of enzymes, including lipases, proteases, amylases, and phosphatases. These enzymes breakdown macromolecules in the beans, enriching them with amino acids, peptides, vitamins, and minerals. As a result, the final product contains nutrients in more absorbable and digestible forms (Feng et al., 2022; Zhang et al., 2022b). Research has shown that tempeh is enriched with bioactive compounds such as genistein, daidzein 6,7,4′-trihydroxy-isoflavoglycitein-7-O-β-D-(6-O-acetyl)-glucopyranoside, 2-hydroxyanthranilic (HAA), a novel antioxidant compound(Esaki et al., 1996; Kuligowski et al., 2016). These transformations highlight tempeh as a functional food with significant health benefits. *Rhizopus* species such as *R. oryaze* and *R. stolonifera* have also been used in tempeh production(Roelofsen & Talens, 1964). These fungi grow on soyabeans to form a compact chunk covered in white mycelium, resulting in nutritious and flavorful product. Tempeh exemplifies the traditional definition of fermentation, where original nutrients are transformed into more bioavailable, value-added forms.

Beyond its nutritional value, tempeh has been shown to offer several health benefits. Regular consumption has been linked to reduced cholesterol levels, improved gut function, mitigation of diarrheal symptoms, and the promotion of beneficial gut microbiota such as *Lactobacillus* and *Bifidobacterium* species. Studies also indicate that tempeh consumption increases short-chain fatty acid production, suppresses tumor development, and improves conditions such as iron-deficiency anemia (Yang et al., 2018; Ahnan-Winarno et al., 2021). Additionally, tempeh has demonstrated antimicrobial and anti-metastasis properties by alleviating oxidative stress and contributing to overall health improvement.

Mold-Ripened Cheese

Fungal-ripened cheeses have been known since ancient times and include varieties such as blue, gray, and white cheeses. Mold-ripened cheeses are generally classified into two main categories: surface mold-ripened cheeses and blue-veined cheeses. Surface-ripened cheeses typically appear coated with a whitish layer of *Penicillium* mycelia, most commonly *Penicillium camemberti* (Chávez et al., 2023). In contrast, *Penicillium roqueforti* grows within the cheese, forming blue-green veins throughout. Camembert is a well-known surface-ripened cheese originating from France, in which *Penicillium camemberti* plays a central role. Edible *Penicillium* species can produce a variety of pigments, ranging from gray to blue. *Penicillium camemberti* is particularly noted for its lipolytic and proteolytic activity, which contributes to the cheese's texture and flavor development. In addition to *P. roquefortii,* other fungi and yeasts such as *Debaryomyces hansenii* have been reported to enhance flavor and aroma (Lessard et al., 2012). These molds are responsible for producing a variety of aromatic compounds, contributing to the characteristic aroma and taste of mold-ripened cheeses (Kumura et al., 2019).

Traditionally, these fungi were not thoroughly screened for toxin production. However, it has been reported that *Penicillium* may produce small amount of cyclo-plazonic acid, a mycotoxin. Today, mutated or selected strains that do not produce mycotoxins are preferred for food production to ensure safety. During the ripening process, the molds also facilitate the growth of beneficial fermentative organisms such as *Lactococcus lactis,* and *L. cremoris,* which utilize lactose and aid curd formation by reducing the pH to around 4.0–4.5. Subsequently, yeasts such as *Kluyveromycces lactis*, *Saccharomyces cerevisiae,* and *Debaryomyces hansenii* often proliferate and contribute to further ripening alongside the molds, enriching the cheese both nutritionally and organoleptically.

Blue-veined cheeses, typically prepared from raw milk, often contain complex microbiota. However, *Penicillium camemberti* usually dominate and help control the cheese's microbial ecosystem. More recently, Monascus-fermented cheeses have gained attention for their enhanced nutritional value and unique aromatic profiles (Wang et al., 2025).

Ang-kak (Red Yeast Rice)
Red rice is similarly recognized as *Anko*, a traditional fermented product made by inoculating rice with fungi from the *Monascus* genus. Various strains of *Monascus,* including *Monascus purpureus*, *M. pilosus*, etc., have been reported for their effectiveness in this fermentation process. These fungi are known for producing a wide range of polyketide pigments, along with other health-promoting metabolites (Lin et al., 2008; Yan et al., 2023). Although *Monascus* species can produce the mycotoxin citrinin, several methods have been developed to minimize or eliminate its production during fermentation. In the traditional process, *Monascus*-inoculated rice is incubated for 2–3 weeks with occasional manual shaking. Once fermentation is complete, the pigmented rice is sterilized, dried, and often ground into a powder. This red rice powder is widely used as a natural food pigment in various applications, including soft drinks, wines, and other culinary preparations. It stands out as one of the most commercially successful fungal solid-state fermentations. One of the most important bioactive compounds produced by *Monascus* is Monacolin K, a metabolite structurally matching to lovastatin. Monacolin K has been shown to lower cholesterol by inhibiting the enzyme 3-hydroxy-3-methylglutaryl coenzyme A (HMG-CoA) reductase, which is essential for the process of cholesterol production. Additionally, *Monascus* pigments have been reported to reduce blood sugar and triglyceride levels while elevated levels of high-density lipoprotein (HDL) (Wang & Lin, 2007).

In this context of rising global diabetes prevalence, resistant starch—a kind of starch that bypasses digestion in the small intestine and is fermented by gut bacteria into short-chain fatty acids—has garnered interest for its potential in regulating blood sugar levels. However, high resistant starch rice varieties often suffer from poor palatability(Yan et al., 2023). Fermentation with edible fungi such as *Monascus* offers a promising solution by enhancing the nutritional quality and sensory appeal of rice, potentially making it a superior alternative to conventional high resistant starch rice. Fermented rice products such as *Hanggue* or *Koji* steamed rice fermented with *Monascus* are nutrient fortified preparations rich in amino acids, organic acids and decatin derivatives. Red mold rice (RMR), known for centuries in China, is recognized for its broad range of health benefits, including support for cancer prevention, osteoporosis, and diabetes management. RMR not only maintains a balanced nutritional profile but also has lower digestibility, which can be beneficial in dietary interventions. As per the latest market research, the worldwide red mold rice market was estimated to be worth around USD 3.20 billion in 2023 and is expected to increase to USD 3.93 billion by the year 2030 (Market survey report).

4.1.4 SCP (Single-Cell Protein)

Single-cell protein (SCP) is denoted as the microbial biomass that serves as a protein-rich, nutritionally valuable, and easily digestible source of protein (Ugalde & Castrillo, 2002). The fungal proteins, particularly from yeasts, have been produced for human consumption since the early 1900s. During World War I, Germany manufactured baker's yeast in large quantities for use as food and animal feed. Following this, a variety of yeast species were employed as SCP sources, including *Kluyveromyces lactis*, *Candida utilis*, *Yarrowia* spp., and *Candida tropicalis*. The first large-scale production of fungal biomass for human and animal nutrition was established by companies such as ICI and Rank Hovis McDougalL. One notable example of SCP development is the use of *Fusarium venenatum,* which is cultivated for producing Quorn, a well-known mycoprotein based meat alternative(Vivek & Venkitasamy, 2023; Lee et al., 2024).

Mushroom

Mushrooms are among the most widely accepted and commercially produced fungal products obtained through solid-state cultivation. It has been reported that approximately two million tons of mushrooms are produced annually. The primary substrates used for mushroom cultivation include agro-industrial wastes such as wheat bran and other lignocellulosic materials (Table 3). Mushroom cultivation has emerged as a successful and economically viable technology, particularly in rural areas, due to its low input requirements and high profitability (Jebapriya et al., 2013; Pandey et al., 2014).

Mushroom production is highly dependent on the selection of an appropriate substrate, as it must provide the necessary nutrients and minerals to support optimal fungal growth and yield. An extensive variety of agro wastes and related products such as straw, corn cobs, sawdust, and other lignocellulosic materials are commonly used in mushroom cultivation. The choice of substrate depends on factors such as local availability, compatibility with the fungal species, and cost-effectiveness.

Prior to cultivation, these solid substrates are often fortified with mineral supplements and nutrients to enhance productivity. The selection of the substrate must also align with the specific requirements of the fungal strain being cultivated. For instance, *Pleurotus* sp. grow efficiently on lignocellulosic materials such as straw and sawdust, while *Agaricus* generally prefer composted manure-based substrates. From an economic perspective, mushroom cultivation has significant global

Table 3 Commercially successful mushrooms cultivated on different agro-solid waste substrates

Types of mushroom	Substrate used	References
Oyster	Rice/wheat straw	Yang et al. (2013)
Lentinula edodes	Sawdust	Royse (2014)
Oyster	Corn cob	Jonathan and Fasidi (2001)
Ganoderma lucidum	Rice bran	Jo et al. (2013)
Grifolafrondosa - Agaricus bisporus	Wheat bran	Dikeman et al. (2005)
Pleurotus ostreatus	Maize stalk	Zhou et al. (2023)
Pink oyster mushroom	Rice husk	Indratmi et al. (2024)
Pleurotus ostreatus	Paddy straw	Subedi et al. (2023)
Pleurotus spp. – Lentinula Agaricales	Coffee pulp	Martínez-Carrera et al. (2000)
Pleurotus eryngii	Wheat, rice bran	Yildirim and Yildiz (2009)
Pycnoporus sp.	Sugarcane vinasse	Ahmad Zakil et al. (2020)

importance. According to the Polaris Market Survey (2023), the global mushroom market is currently valued at around USD 64.96 billion and is expected to reach USD 136.90 billion by the year 2032. A detailed summary of various fungal strains and their preferred solid substrates is provided in Table 3.

4.1.5 Fungi Produced by SSF as Biopesticides and for Soil Improvement

Fungi are among the most widely used organisms in the development of biopesticides, composting agents, and soil bio-inputs due to their remarkable abilities in substrate hydrolysis, invagination, and spore formation (El-Gendi et al., 2021). Several fungal species have been utilized as biopesticides also known as mycopesticides for the control of agricultural pests such as bollworms, thrips, and beetles (Bhuvaneswari Manivel & Subashini Rajkumar, 2018). Fungi capable of producing cellulolytic enzymes, chitinases, pectinases, xylanases, and phosphatases are especially valuable, as they promote competitiveness in soil environments. These functional traits make fungi ideal candidates not only for pest control but also for applications as soil bio-inputs and biofertilizers (Yélamos et al., 2025). Due to their ability to thrive in diverse environmental conditions, fungi are essential for breakdown of organic materials and recycling nutrients. Mycorrhizal fungi, including *Glomus* spp. *Scelrocystis, Scutellospora* spp., contribute to drought tolerance, enhanced nutrient absorption, and increased resistance to pathogens in plants (Berruti et al., 2016). Additionally, species of *Aspergillus,* and *Penicillium* are widely used for their ability to solubilize essential minerals such as phosphate, silicate, potassium (K), and zinc (Zn), thereby improving soil fertility. Other fungal species, such as *Trichoderma* spp., *Nomurea* spp., and *Beauveria* spp., are common mycopesticides known for their antagonistic effects against plant pathogens. SSF is the preferred method for producing many fungal biopesticides because the formulations are typically spore-based and require high spore density and viability. These

Table 4 Examples of reported fungal biopesticides with substrate and their applications

Fungi	Applications/target against	Solid substrate	References
Beauveria bassiana Trichoderma harzianum	*Biopesticide production*	Potato peel or orange pomace	Sala et al. (2021)
Beauveria bassiana	*Biopesticide production*	Rice fiber, rice husk, wheat straw, apple pomace, soy fiber, whisky draff, beer draff, potato peel, orange peel	Sala et al. (2020)
Trichoderma *viride*	Rhizoctonia solani-infected potato plants	Corn cobs.	Narwade et al. (2023)
Trichoderma *harzianum*	Against *F. oxysporium* R. Solani	Paddy Straw (M.)	Subash et al. (2014)
Trichoderma *brev T069*	Biopesticide	Cassava peels	Zhang et al. (2022a)
Trichoderma viridae	Rhizoctonia, Phytophthora Fusarium	Bagasse Wheat Straw	Sinha et al. (2022)
Trichoderma asperellum	Rhizoctonia disease and composting	Agrowaste compost	Trillas et al. (2006)
Metarhizium anisopliae	Biocontrolling agent Pest	Grains	Agale et al. (2018)
Metarhizium anisopliae	Biocontrolling agent	Sugarcane/Rice bagasses	Da Cunha et al. (2020)
Lenginidium giganteum	Mosquito Larvae control	Cotton seed meal Agricultural waste	Deshpande (1999)
Verticillium lecanii	Agro pest	Rice bran husk	Feng et al. (2000)
Beauveria bassiana	Musk domestica	Husk of rice, pigeon pea, wheat	Mishra et al. (2016)
Verticilium lecanii, Beauveria bassiana *Paecilomyces fumosoroseus*	Biopesticide	Wheat grain	Kumar et al. (2023)
Trichoderma reesei	Biopesticide	Food waste	Lima et al. (2024)
Trichoderma atroviride	Biopesticide	Agrowaste	Bulgari et al. (2023)

fungi are typically cultivated on agro-waste substrates, which are processed into small particles and treated through simple cooking and fermentation techniques to promote sporulation and growth. A summary of important fungal biopesticides and their agro-solid substrates is provided in Table 4.

Acknowledgments Dr. SV Patil, KBC NMU, Jalgaon kindly acknowledges Prof. Rekha Singhal, ICT, Mumbai for her academic and research suggestions as well as her work which inspired to do hard and ethical work in the biotechnology area.

References

Abreu, T., Sousa, P., Gonçalves, J., et al. (2024). Grape pomace as a renewable natural biosource of value-added compounds with potential food industrial applications. *Beverages, 10*, 45. https://doi.org/10.3390/beverages10020045

Adav, S. S., Li, A. A., Manavalan, A., Punt, P., & Sze, S. K. (2010). Quantitative iTRAQ secretome analysis of Aspergillus niger reveals novel hydrolytic enzymes. *Journal of Proteome Research, 9*(8), 3932–3940. https://doi.org/10.1021/pr100148j

Adeniran, H. A., Abiose, S. H., & Ogunsua, A. O. (2010). Production of fungal β-amylase and amyloglucosidase on some nigerian agricultural residues. *Food and Bioprocess Technology, 3*, 693–698. https://doi.org/10.1007/s11947-008-0141-3

Agale, S. V., Gopalakrishnan, S., Ambhure, K. G., et al. (2018). Mass production of entomopathogenic fungi (*Metarhizium anisopliae*) using different grains as a substrate. *International Journal of Current Microbiology and Applied Sciences, 7*, 2227–2232. https://doi.org/10.20546/ijcmas.2018.701.268

Ahmad Zakil, F., Muhammad Hassan, K. H., Mohd Sueb, M. S., & Isha, R. (2020). Growth and yield of Pleurotus ostreatus using sugarcane bagasse as an alternative substrate in Malaysia. *IOP Conference Series Materials Science and Engineering, 736*, 022021. https://doi.org/10.1088/1757-899X/736/2/022021

Ahmed, I., Zia, M. A., Hussain, M. A., et al. (2016). Bioprocessing of citrus waste peel for induced pectinase production by *Aspergillus niger*; its purification and characterization. *Journal of Radiation Research and Applied Sciences, 9*, 148–154. https://doi.org/10.1016/j.jrras.2015.11.003

Ahmed, P. M., Álvarez, A., De Figueroa, L. I. C., & Pajot, H. F. (2024). Boldly going green: Utilizing *Pycnoporus* sp. for laccase production and sustainable vinasse treatment. *International Journal of Environmental Science and Technology, 21*, 3927–3942. https://doi.org/10.1007/s13762-023-05257-1

Ahnan-Winarno, A. D., Cordeiro, L., Winarno, F. G., et al. (2021). Tempeh: A semicentennial review on its health benefits, fermentation, safety, processing, sustainability, and affordability. *Comprehensive Reviews in Food Science and Food Safety, 20*, 1717–1767. https://doi.org/10.1111/1541-4337.12710

Ahsan, M., Ashraf, H., Iahtisham-Ul-Haq, et al. (2024). Exploring pectin from ripe and unripe Banana Peel: A novel functional fat replacers in muffins. *Food Chemistry: X, 23*, 101539. https://doi.org/10.1016/j.fochx.2024.101539

Aita, B. C., Spannemberg, S. S., Schmaltz, S., et al. (2019). Production of cell-wall degrading enzymes by solid-state fermentation using agroindustrial residues as substrates. *Journal of Environmental Chemical Engineering, 7*, 103193. https://doi.org/10.1016/j.jece.2019.103193

Akter, M., Halawani, R. F., Aloufi, F. A., et al. (2022). Utilization of agro-industrial wastes for the production of quality oyster mushrooms. *Sustainability, 14*, 994. https://doi.org/10.3390/su14020994

Ali, U., Anwar, Z., Hasan, S., et al. (2023). Bioprocessing and screening of indigenous wastes for hyper production of fungal lipase. *Catalysts, 13*, 853. https://doi.org/10.3390/catal13050853

Amara, A. A., & El-Baky, N. A. (2023). Fungi as a source of edible proteins and animal feed. *Journal of Fungi, 9*, 73. https://doi.org/10.3390/jof9010073

Aravindan, R., Viruthagiri, T., Seenivasan, A., & Subhagar, S. (2008). Microbial production and biomedical applications of lovastatin. *Indian Journal of Pharmaceutical Sciences, 70*, 701. https://doi.org/10.4103/0250-474X.49087

Assemie, A., & Abaya, G. (2022). The effect of edible mushroom on health and their biochemistry. *International Journal of Microbiology, 2022*, 1–7. https://doi.org/10.1155/2022/8744788

Awasthi, M. K., Kumar, V., Hellwig, C., et al. (2023). Filamentous fungi for sustainable vegan food production systems within a circular economy: Present status and future prospects. *Food Research International, 164*, 112318. https://doi.org/10.1016/j.foodres.2022.112318

Babenko, M., Klitou, T., Klumbyte, E., & Fokaides, P. A. (2025). Environmental assessment of mycelium based straw insulation composite: A sustainability analysis at building material level. *Case Studies in Construction Materials, 22*, e04572. https://doi.org/10.1016/j.cscm.2025.e04572

Barzee, T. J., Cao, L., Pan, Z., & Zhang, R. (2021). Fungi for future foods. *Journal of Future Foods, 1*, 25–37. https://doi.org/10.1016/j.jfutfo.2021.09.002

Baseer, A., & Shall, S. (1971). Properties of the internal invertase of yeast, *Saccharomyces cerevisiae*. *Biochimica et Biophysica Acta (BBA) – Enzymology, 250*, 192–202. https://doi.org/10.1016/0005-2744(71)90133-1

Bavia, A. C. F., Silva, C. E. D., Ferreira, M. P., et al. (2012). Chemical composition of tempeh from soybean cultivars specially developed for human consumption. *Food Science and Technology, 32*, 613–620. https://doi.org/10.1590/S0101-20612012005000085

Benz, F., Knüsel, F., Nüesch, J., et al. (1974). Stoffwechselprodukte von Mikroorganismen 143. Mitteilung. Echinocandin B, ein neuartiges Polypeptid-Antibioticum aus *Aspergillus nidulans var. echinulatus* : Isolierung und Bausteine. *Helvetica Chimica Acta, 57*, 2459–2477. https://doi.org/10.1002/hlca.19740570818

Berruti, A., Lumini, E., Balestrini, R., & Bianciotto, V. (2016). Arbuscular mycorrhizal fungi as natural biofertilizers: Let's benefit from past successes. *Frontiers in Microbiology, 6*. https://doi.org/10.3389/fmicb.2015.01559

Bezerra, C. O., Carneiro, L. L., Carvalho, E. A., et al. (2021). Artificial Intelligence as a combinatorial optimization strategy for cellulase production by *Trichoderma stromaticum* AM7 using peach-palm waste under solid-state fermentation. *Bioenergy Research, 14*, 1161–1170. https://doi.org/10.1007/s12155-020-10234-4

Bhat, M. K. (2000). Cellulases and related enzymes in biotechnology. *Biotechnology Advances, 18*, 355–383. https://doi.org/10.1016/S0734-9750(00)00041-0

Bhat, M. A., Bhat, M. A., Jan, S., et al. (2024). Lignocellulosic biomass in circular economy: A techno-transition in carbon neutrality towards sustainable energy production. *Biomass and Bioenergy, 189*, 107349. https://doi.org/10.1016/j.biombioe.2024.107349

Bhuvaneswari Manivel, S., & Subashini Rajkumar, G. (2018). Mycopesticides: fungal based pesticides for sustainable agriculture. In P. Gehlot & J. Singh (Eds.), *Fungi and their role in sustainable development: Current perspectives* (pp. 183–203). Springer.

Boland, M. J., Rae, A. N., Vereijken, J. M., et al. (2013). The future supply of animal-derived protein for human consumption. *Trends in Food Science & Technology, 29*, 62–73. https://doi.org/10.1016/j.tifs.2012.07.002

Boonchuay, P., Techapun, C., Leksawasdi, N., et al. (2021). Bioethanol Production from cellulose-rich corncob residue by the Thermotolerant *Saccharomyces cerevisiae* TC-5. *Journal of Fungi, 7*, 547. https://doi.org/10.3390/jof7070547

Bouffard, F. A., Zambias, R. A., Dropinski, J. F., et al. (1994). Synthesis and antifungal activity of novel cationic pneumocandin bo derivatives. *Journal of Medicinal Chemistry, 37*, 222–225. https://doi.org/10.1021/jm00028a003

Bulgari, D., Alias, C., Peron, G., et al. (2023). Solid-state fermentation of *Trichoderma* spp.: A new way to valorize the agricultural digestate and produce value-added bioproducts. *Journal of Agricultural and Food Chemistry, 71*, 3994–4004. https://doi.org/10.1021/acs.jafc.2c07388

Carvalho, A. S. S., Sales, J. C. S., Nascimento, F. V. D., et al. (2023). Lipase production by Yarrowia lipolytica in solid-state fermentation using Amazon fruit by-products and soybean meal as substrate. *Catalysts, 13*, 289. https://doi.org/10.3390/catal13020289

Chain, E., Florey, H., Jennings, M., & Williams, T. (1943). Helvolic acid, an antibiotic produced by *Aspergillus fumigatus*, mut. helvola Yuill. *British Journal of Experimental Pathology, 3*, 108.

Chávez, R., Vaca, I., & García-Estrada, C. (2023). Secondary metabolites produced by the blue-cheese ripening Mold *Penicillium roqueforti*; biosynthesis and regulation mechanisms. *Journal of Fungi, 9*, 459. https://doi.org/10.3390/jof9040459

Chen, F., & Hu, X. (2005). Study on red fermented rice with high concentration of monacolin K and low concentration of citrinin. *International Journal of Food Microbiology, 103*, 331–337. https://doi.org/10.1016/j.ijfoodmicro.2005.03.002

Chen, W., He, Y., Zhou, Y., et al. (2015). Edible filamentous fungi from the species *Monascus*: Early traditional fermentations, modern molecular biology, and future genomics. *Comprehensive Reviews in Food Science and Food Safety, 14*, 555–567. https://doi.org/10.1111/1541-4337.12145

Chou, T. S., Eisenbraun, E. J., & Rapala, R. T. (1967). The chemistry of cephalosporin P1. *Tetrahedron Letters, 8*, 409–414. https://doi.org/10.1016/S0040-4039(00)90960-2

Copetti, M. V. (2019). Fungi as industrial producers of food ingredients. *Current Opinion in Food Science, 25*, 52–56. https://doi.org/10.1016/j.cofs.2019.02.006

Corbu, V. M., Gheorghe-Barbu, I., Ştefania, D. A., et al. (2023). Current insights in fungal importance – A comprehensive review. *Microorganisms, 11*, 1384. https://doi.org/10.3390/microorganisms11061384

Corral-Bobadilla, M., González-Marcos, A., Vergara-González, E. P., & Alba-Elías, F. (2019). Bioremediation of waste water to remove heavy metals using the spent mushroom substrate of *Agaricus bisporus*. *Water, 11*, 454. https://doi.org/10.3390/w11030454

Cripwell, R. A., Rose, S. H., Viljoen-Bloom, M., & Van Zyl, W. H. (2019). Improved raw starch amylase production by *Saccharomyces cerevisiae* using codon optimisation strategies. *FEMS Yeast Research, 19*. https://doi.org/10.1093/femsyr/foy127

Cripwell, R. A., Favaro, L., Viljoen-Bloom, M., & Van Zyl, W. H. (2020). Consolidated bioprocessing of raw starch to ethanol by *Saccharomyces cerevisiae*: Achievements and challenges. *Biotechnology Advances, 42*, 107579. https://doi.org/10.1016/j.biotechadv.2020.107579

Da Cunha, L. P., Casciatori, F. P., Vicente, I. V., et al. (2020). *Metarhizium anisopliae* conidia production in packed-bed bioreactor using rice as substrate in successive cultivations. *Process Biochemistry, 97*, 104–111. https://doi.org/10.1016/j.procbio.2020.07.002

Da S. Pereira, A., Fontes-Sant'Ana, G. C., & Amaral, P. F. F. (2019). Mango agro-industrial wastes for lipase production from *Yarrowia lipolytica* and the potential of the fermented solid as a biocatalyst. *Food and Bioproducts Processing, 115*, 68–77. https://doi.org/10.1016/j.fbp.2019.02.002

Damodaran, D., Vidya Shetty, K., & Raj Mohan, B. (2014). Uptake of certain heavy metals from contaminated soil by mushroom – Galerina vittiformis. *Ecotoxicology and Environmental Safety, 104*, 414–422. https://doi.org/10.1016/j.ecoenv.2013.10.033

Das, S., Naik Deshavath, N., Goud, V. V., & Dasu, V. V. (2019). Bioleaching of Al from spent fluid catalytic cracking catalyst using *Aspergillus* species. *Biotechnology Reports, 23*, e00349. https://doi.org/10.1016/j.btre.2019.e00349

De Carvalho, L. R., Santos, D. R., Dos Santos Lima, C. S., et al. (2023a). Stable polyphenol oxidase produced by *Pleurotus pulmonarius* from fermented peach-palm and cocoa wastes. *Bioenergy Research, 17*, 198–207. https://doi.org/10.1007/s12155-023-10628-0

De Carvalho, M. S., De Menezes, L. H. S., Pimentel, A. B., et al. (2023b). Application of chemometric methods for the optimization secretion of Xylanase by *Aspergillus oryzae* in solid state fermentation and its application in the saccharification of agro-industrial waste. *Waste and Biomass Valorization, 14*, 3183–3193. https://doi.org/10.1007/s12649-022-01832-8

De Oliveira, F., Castellane, T. C. L., De Melo, M. R., & Buzato, J. B. (2022). Preparation of *Aspergillus niger* 426 naringinases for debittering citrus juice utilization of agro-industrial residues. *International Microbiology, 25*, 123–131. https://doi.org/10.1007/s10123-021-00199-5

De Vries, R. P., & Mäkelä, M. R. (2020). Genomic and postgenomic diversity of fungal plant biomass degradation approaches. *Trends in Microbiology, 28*, 487–499. https://doi.org/10.1016/j.tim.2020.01.004

Derossi, A., Caporizzi, R., Azzollini, D., & Severini, C. (2018). Application of 3D printing for customized food. A case on the development of a fruit-based snack for children. *Journal of Food Engineering, 220*, 65–75. https://doi.org/10.1016/j.jfoodeng.2017.05.015

Deshpande, M. V. (1999). Mycopesticide Production by fermentation: Potential and challenges. *Critical Reviews in Microbiology, 25*, 229–243. https://doi.org/10.1080/10408419991299220

Devi, N., Singh, S., Manickam, S., et al. (2022). Itaconic acid and its applications for textile, pharma and agro-industrial purposes. *Sustainability, 14*, 13777. https://doi.org/10.3390/su142113777

Díaz-Urbano, M., Goicoechea, N., Velasco, P., & Poveda, J. (2023). Development of agricultural bio-inoculants based on mycorrhizal fungi and endophytic filamentous fungi: Co-inoculants for improve plant-physiological responses in sustainable agriculture. *Biological Control, 182*, 105223. https://doi.org/10.1016/j.biocontrol.2023.105223

Dikeman, C. L., Bauer, L. L., Flickinger, E. A., & Fahey, G. C. (2005). Effects of stage of maturity and cooking on the chemical composition of select mushroom varieties. *Journal of Agricultural and Food Chemistry, 53*, 1130–1138. https://doi.org/10.1021/jf0485411

Dos Reis, C. B. L., Sobucki, L., Mazutti, M. A., et al. (2018). Production of chitinase from *Metarhizium anisopliae* by solid-state fermentation using sugarcane bagasse as substrate. *Industrial Biotechnology, 14*, 230–234. https://doi.org/10.1089/ind.2017.0031

Dubey, M. K., Zehra, A., Aamir, M., et al. (2017). Improvement strategies, cost effective production, and potential applications of fungal glucose oxidase (GOD): Current updates. *Frontiers in Microbiology, 8*, 1032. https://doi.org/10.3389/fmicb.2017.01032

Dufossé, L. (2018). Red colourants from filamentous fungi: Are they ready for the food industry? *Journal of Food Composition and Analysis, 69*, 156–161. https://doi.org/10.1016/j.jfca.2017.11.002

Dufossé, L., Fouillaud, M., Caro, Y., et al. (2014). Filamentous fungi are large-scale producers of pigments and colorants for the food industry. *Current Opinion in Biotechnology, 26*, 56–61. https://doi.org/10.1016/j.copbio.2013.09.007

Dulay, R. (2012). Aseptic cultivation of Coprinus comatus (O. F. Mull.) gray on various pulp and paper wastes. *Mycosphere, 3*, 392–397. https://doi.org/10.5943/mycosphere/3/3/10

Dusengemungu, L., Kasali, G., Gwanama, C., & Mubemba, B. (2021). Overview of fungal bioleaching of metals. *Environmental Advances, 5*, 100083. https://doi.org/10.1016/j.envadv.2021.100083

Egea, M. B., Dantas, L. A., Sousa, T. L. D., et al. (2023). The potential, strategies, and challenges of *Monascus* pigment for food application. *Frontiers in Sustainable Food Systems, 7*, 1141644. https://doi.org/10.3389/fsufs.2023.1141644

El-Gendi, H., Saleh, A. K., Badierah, R., et al. (2021). A comprehensive insight into fungal enzymes: Structure, classification, and their role in mankind's challenges. *Journal of Fungi, 8*, 23. https://doi.org/10.3390/jof8010023

Elyasi Far, B., Ahmadi, Y., Yari Khosroshahi, A., & Dilmaghani, A. (2020). Microbial alpha-amylase production: progress, challenges and perspectives. *Advanced Pharmaceutical Bulletin, 10*, 350–358. https://doi.org/10.34172/apb.2020.043

Erdoğmuş, S. F., Altıntaş, Ö. E., & Çelik, S. (2023). Production of fungal chitosan and fabrication of fungal chitosan/polycaprolactone electrospun nanofibers for tissue engineering. *Microscopy Research & Technique, 86*, 1309–1321. https://doi.org/10.1002/jemt.24315

Erkan, S. B., Gürler, H. N., Bilgin, D. G., et al. (2020). Production and characterization of tempehs from different sources of legume by *Rhizopus oligosporus*. *LWT, 119*, 108880. https://doi.org/10.1016/j.lwt.2019.108880

Esaki, H., Onozaki, H., Kawakishi, S., & Osawa, T. (1996). New antioxidant isolated from tempeh. *Journal of Agricultural and Food Chemistry, 44*, 696–700. https://doi.org/10.1021/jf950454t

Esawy, M. A., Gamal, A. A., & Kamel, Z. (2022). Optimization of *Aspergillus niger* NRC1ami Pectinase using citrus peel pectin, purification, and thermodynamic characterization of the free and modified enzyme. *Waste and Biomass Valorization, 13*, 4823–4837. https://doi.org/10.1007/s12649-022-01838-2

Escaramboni, B., Garnica, B. C., Abe, M. M., et al. (2022). Food waste as a feedstock for fungal biosynthesis of amylases and proteases. *Waste and Biomass Valorization, 13*, 213–226. https://doi.org/10.1007/s12649-021-01511-0

Fadiji, A. E., & Babalola, O. O. (2020). Exploring the potentialities of beneficial endophytes for improved plant growth. *Saudi Journal of Biological Sciences, 27*, 3622–3633. https://doi.org/10.1016/j.sjbs.2020.08.002

FAO. (2023). World Food and Agriculture – Statistical Yearbook 2023. *Rome*. https://doi.org/10.4060/cc8166en

Feng, K. C., Liu, B. L., & Tzeng, Y. M. (2000). *Verticillium lecanii* spore production in solid-state and liquid-state fermentations. *Bioprocess Engineering, 23*, 25–29. https://doi.org/10.1007/s004499900115

Feng, S.-S., Li, W., Hu, Y.-J., et al. (2022). The biological activity and application of *Monascus* pigments: a mini review. *International Journal of Food Engineering, 18*, 253–266. https://doi.org/10.1515/ijfe-2021-0235

Fermented Foods and Beverages Market Report. (2025). Accessed 8 May 2025. https://www.meticulousresearch.com/product/fermented-foods-and-beverages-market-5469

Ferrara, A., Velotto, S., & Ferranti, P. (2023). Perspective and emerging sources novel foods and ingredients from fungi. In *Sustainable food science – A comprehensive approach* (pp. 220–228). Elsevier.

Finnigan, T. J. A. (2011). Mycoprotein: origins, production and properties. In *Handbook of food proteins* (pp. 335–352). Elsevier.

Finnigan, T. J., Wall, B. T., Wilde, P. J., et al. (2019). Mycoprotein: The future of nutritious non-meat protein, a symposium review. *Current Developments in Nutrition, 3*, nzz021. https://doi.org/10.1093/cdn/nzz021

Fomina, M., Yurieva, O., Pavlychenko, A., et al. (2024). Application of natural fungi in bioconversion of lignocellulosic waste to second-generation ethanol. *Biosystems Diversity, 32*, 45–59. https://doi.org/10.15421/012405

Fraatz, M. A., & Zorn, H. (2011). Fungal flavours. In M. Hofrichter (Ed.), *Industrial applications* (pp. 249–268). Springer.

Fungal Protein Global Market Report. (2025). *The Business research company*. Accessed 8 May 2025. https://www.thebusinessresearchcompany.com/report/fungal-protein-global-market-report

Furey, B., Slingerland, K., Bauter, M. R., et al. (2022). Safety evaluation of Fy Protein™ (Nutritional Fungi Protein), a macroingredient for human consumption. *Food and Chemical Toxicology, 166*, 113005. https://doi.org/10.1016/j.fct.2022.113005

Ghanem, N. B., Yusef, H. H., & Mahrouse, H. K. (2000). Production of Aspergillus terreus xylanase in solid-state cultures: Application of the Plackett–Burman experimental design to evaluate nutritional requirements. *Bioresource Technology, 73*, 113–121. https://doi.org/10.1016/S0960-8524(99)00155-8

Ghosh, S. K., Sen, R., Chanakya, H. N., & Pariatamby, A. (Eds.). (2020). *Bioresource utilization and bioprocess*. Springer.

Ghosh, S., Rusyn, I., Dmytruk, O. V., et al. (2023). Filamentous fungi for sustainable remediation of pharmaceutical compounds, heavy metal and oil hydrocarbons. *Frontiers in Bioengineering and Biotechnology, 11*, 1106973. https://doi.org/10.3389/fbioe.2023.1106973

Ghyadh, B., Al-Ashoor, A., & Ibrahim, Z. (2019). Biotechnology of wastewater treatment with fungi (*Aspergillus niger* and *Rhizopus oligosporium*). *Plant Archives, 19*, 1546–1549.

Gmoser, R., Sintca, C., Taherzadeh, M. J., & Lennartsson, P. R. (2019). Combining submerged and solid state fermentation to convert waste bread into protein and pigment using the edible filamentous fungus N. intermedia. *Waste Management, 97*, 63–70. https://doi.org/10.1016/j.wasman.2019.07.039

Gopaliya, D., Kumar, V., & Khare, S. K. (2021). Recent advances in itaconic acid production from microbial cell factories. *Biocatalysis and Agricultural Biotechnology, 36*, 102130. https://doi.org/10.1016/j.bcab.2021.102130

Guo, M., Song, W., & Buhain, J. (2015). Bioenergy and biofuels: History, status, and perspective. *Renewable and Sustainable Energy Reviews, 42*, 712–725. https://doi.org/10.1016/j.rser.2014.10.013

Haq, I.-U., Ali, S., & Iqbal, J. (2003). Direct production of citric acid from raw starch by Aspergillus niger. *Process Biochemistry, 38*, 921–924. https://doi.org/10.1016/S0032-9592(02)00201-7

Hatzinikolaou, D. G., Hansen, O. C., Macris, B. J., et al. (1996). A new glucose oxidase from *Aspergillus niger*: Characterization and regulation studies of enzyme and gene. *Applied Microbiology and Biotechnology, 46*, 371–381. https://doi.org/10.1007/BF00166232

Hayashida, S., Ohta, K., & Mo, K. (1988). Cellulases of Humicola insolens and Humicola grisea. In *Methods in enzymology* (pp. 323–332). Elsevier.

Hernández Rodríguez, A., Díaz Pacheco, A., Martínez Tolibia, S. E., et al. (2024). Bioprocess of Gibberellic acid by *Fusarium fujikuroi*: The challenge of regulation, raw materials, and product yields. *Jounal of Fungi, 10*, 418. https://doi.org/10.3390/jof10060418

Hittinger, C. T., Steele, J. L., & Ryder, D. S. (2018). Diverse yeasts for diverse fermented beverages and foods. *Current Opinion in Biotechnology, 49*, 199–206. https://doi.org/10.1016/j.copbio.2017.10.004

Hoque, E., & Fritscher, J. (2019). Multimetal bioremediation and biomining by a combination of new aquatic strains of Mucor hiemalis. *Scientific Reports, 9*, 10318. https://doi.org/10.1038/s41598-019-46560-7

Howson, S. J., & Davis, R. P. (1983). Production of phytate-hydrolysing enzyme by some fungi. *Enzyme and Microbial Technology, 5*, 377–382. https://doi.org/10.1016/0141-0229(83)90012-1

Hu, Z.-C., Li, W.-J., Zou, S.-P., et al. (2020). Mutagenesis of echinocandin B overproducing *Aspergillus nidulans* capable of using starch as main carbon source. *Preparative Biochemistry & Biotechnology, 50*, 745–752. https://doi.org/10.1080/10826068.2020.1734940

Huang, W. W., Ge, X. Y., Huang, Y., et al. (2021). High-yield strain of fusidic acid obtained by atmospheric and room temperature plasma mutagenesis and the transcriptional changes involved in improving its production in fungus *Fusidium coccineum*. *Journal of Applied Microbiology, 130*, 405–415. https://doi.org/10.1111/jam.14797

Huerta-Rosas, B., Cano-Rodríguez, I., Gamiño-Arroyo, Z., et al. (2020). Aerobic processes for bioleaching manganese and silver using microorganisms indigenous to mine tailings. *World Journal of Microbiology and Biotechnology, 36*, 124. https://doi.org/10.1007/s11274-020-02902-6

Hyde, K. D., Jeewon, R., Chen, Y.-J., et al. (2020). The numbers of fungi: Is the descriptive curve flattening? *Fungal Diversity, 103*, 219–271. https://doi.org/10.1007/s13225-020-00458-2

Indratmi, D., Sari, J. E., Wachid, M., & Rachmawan, M. D. (2024). Utilization of rice husk waste and gliricidia to pink oyster mushroom production and nutritional content. *BIO Web of Conferences, 104*, 00007. https://doi.org/10.1051/bioconf/202410400007

Jain, A., Yadav, S., Nigam, V. K., et al. (2017). Fungal-Mediated Solid Waste Management: A Review. In: Prasad, R. (eds) Mycoremediation and Environmental Sustainability. *Fungal Biology*. Springer, Cham. https://doi.org/10.1007/978-3-319-68957-9_9

Jebapriya, G., Gnanasalomi, V., & Gnanadoss, J. (2013). Application of mushroom fungi in solid waste management. *International Journal of Computing Algorithm, 2*, 279–285.

Jo, E.-Y., Cheon, J.-L., & Ahn, J.-H. (2013). Effect of food waste compost on the antler-type fruiting body yield of *Ganoderma lucidum*. *Mycobiology, 41*, 42–46. https://doi.org/10.5941/MYCO.2013.41.1.42

Jonathan, S. G., & Fasidi, I. O. (2001). Effect of carbon, nitrogen and mineral sources on growth of *Psathyerella atroumbonata* (Pegler), a Nigerian edible mushroom. *Food Chemistry, 72*, 479–483. https://doi.org/10.1016/S0308-8146(00)00265-X

Kaewchai, S., Soytong, K., & Hyde, K. (2009). Mycofungicides and fungal biofertilizers. *Fungal Diversity, 38*, 25–50.

Keerthana, K., Anukiruthika, T., Moses, J. A., & Anandharamakrishnan, C. (2020). Development of fiber-enriched 3D printed snacks from alternative foods: A study on button mushroom. *Journal of Food Engineering, 287*, 110116. https://doi.org/10.1016/j.jfoodeng.2020.110116

Khan, I., Aftab, M., Shakir, S., et al. (2019a). Mycoremediation of heavy metal (Cd and Cr)–polluted soil through indigenous metallotolerant fungal isolates. *Environmental Monitoring and Assessment, 191*, 585. https://doi.org/10.1007/s10661-019-7769-5

Khan, Y. M., Munir, H., & Anwar, Z. (2019b). Optimization of process variables for enhanced production of urease by indigenous *Aspergillus niger* strains through response surface methodology. *Biocatalysis and Agricultural Biotechnology, 20*, 101202. https://doi.org/10.1016/j.bcab.2019.101202

Kim, J. S., Kim, J., Kim, S. H., & Moon, K.-D. (2025). Development of chickpea tempeh using *Rhizopus oryzae* for dysphagia diet: Effect of fermentation time and heat treatment. *Innovative Food Science & Emerging Technologies, 100*, 103940. https://doi.org/10.1016/j.ifset.2025.103940

Koli, S. H., Suryawanshi, R. K., Mohite, B. V., & Patil, S. V. (2019). Prospective of *Monascus* pigments as an additive to commercial sunscreens. *Natural Product Communications, 14*, 1934578X19894095. https://doi.org/10.1177/1934578X19894095

Koul, B., & Farooq, B. (2020). Mycotechnology: Utility of fungi in food and beverage industries. In *New and future developments in microbial biotechnology and bioengineering* (pp. 133–153). Elsevier.

Kour, R., Singh, S., Sharma, H. B., et al. (2023) Persistence and remote sensing of agri-food wastes in the environment: Current state and perspectives. *Chemosphere, 317*, 137822. https://doi.org/10.1016/j.chemosphere.2023.137822

Kovac, B., & Raspor, P. (1997). The use of the mould *Rhizopus oligosporus* in food production. *Food Technology and Biotechnology, 35*, 69–73.

Králová, M., Frébortová, J., Pěnčík, A., & Frébort, I. (2021). Overexpression of Trp-related genes in Claviceps purpurea leading to increased ergot alkaloid production. *New Biotechnology, 61*, 69–79. https://doi.org/10.1016/j.nbt.2020.11.003

Kuivanen, J., Wang, Y.-M. J., & Richard, P. (2016). Engineering Aspergillus niger for galactaric acid production: elimination of galactaric acid catabolism by using RNA sequencing and CRISPR/Cas9. *Microbial Cell Factories, 15*, 210. https://doi.org/10.1186/s12934-016-0613-5

Kuligowski, M., Pawłowska, K., Jasińska-Kuligowska, I., & Nowak, J. (2016). Isoflavone composition, polyphenols content and antioxidative activity of soybean seeds during tempeh fermentation. *CyTA – Journal of Food*, 1–7. https://doi.org/10.1080/19476337.2016.1197316

Kulshreshtha, S., Mathur, N., Bhatnagar, P., & Jain, B. (2010). Bioremediation of industrial waste through mushroom cultivation. *Journal of Environmental Biology, 31*, 441–444.

Kulshreshtha, S., Mathur, N., & Bhatnagar, P. (2014). Mushroom as a product and their role in mycoremediation. *AMB Express, 4*, 29. https://doi.org/10.1186/s13568-014-0029-8

Kumar, G. R., Singh, S., Gupta, N., & Das, M. D. (2011). Extracellular ribonuclease production from *Aspergillus niger* ATCC 26550: Process optimization by biostatistical analysis. *International Journal of Biotechnology and Biochemistry, 7*, 279–291.

Kumar, D., Singh, P. N., Willer, H., et al. (2023). Fungal Biofertilizers and biopesticides and their roles in sustainable agriculture. In *Applied mycology for agriculture and foods* (1st ed., pp. 165–216). Apple Academic Press.

Kumura, H., Satoh, M., Machiya, T., et al. (2019). Lipase and protease production of dairy *Penicillium* sp. on milk-protein-based solid substrates. *International Journal of Dairy Technology, 72*, 403–408. https://doi.org/10.1111/1471-0307.12597

Lan, T.-Q., Wei, D., Yang, S.-T., & Liu, X. (2013). Enhanced cellulase production by *Trichoderma viride* in a rotating fibrous bed bioreactor. *Bioresource Technology, 133*, 175–182. https://doi.org/10.1016/j.biortech.2013.01.088

Langer, I., Santner, J., Krpata, D., et al. (2012). Ectomycorrhizal impact on Zn accumulation of *Populus tremula* L. grown in metalliferous soil with increasing levels of Zn concentration. *Plant and Soil, 355*, 283–297. https://doi.org/10.1007/s11104-011-1098-y

Lee, J., & Paetzel, M. (2011). Structure of the catalytic domain of glucoamylase from *Aspergillus niger. Acta Crystallographica Section F, 67*, 188–192. https://doi.org/10.1107/S1744309110049390

Lee, K., Lee, S. K., & Lee, B. D. (2006). Aspergillus oryzae as probiotic in poultry – A review. *International Journal of Poultry Science, 5*, 1–3. https://doi.org/10.3923/ijps.2006.1.3

Lee, D., Pan, J. H., Kim, D., et al. (2024). Mycoproteins and their health-promoting properties: *Fusarium* species and beyond. *Comprehensive Reviews in Food Science and Food Safety, 23*, e13365. https://doi.org/10.1111/1541-4337.13365

Lessard, M.-H., Bélanger, G., St-Gelais, D., & Labrie, S. (2012). The composition of Camembert cheese-ripening cultures modulates both mycelial growth and appearance. *Applied and Environmental Microbiology, 78*, 1813–1819. https://doi.org/10.1128/AEM.06645-11

Li, C., Xu, D., Zhao, M., et al. (2014). Production optimization, purification, and characterization of a novel acid protease from a fusant by Aspergillus oryzae and Aspergillus niger. *European Food Research and Technology, 238*, 905–917. https://doi.org/10.1007/s00217-014-2172-5

Li, K., Qiao, K., Xiong, J., et al. (2023). Nutritional values and bio-functional properties of fungal proteins: applications in foods as a sustainable source. *Food, 12*, 4388. https://doi.org/10.3390/foods12244388

Li, X., Sun, L., Chimonas, S., et al. (2024a). Symptom improvements and adverse effects with Reishi mushroom use: A cross-sectional survey of cancer patients. *Integrative Medicine Research, 13*, 101089. https://doi.org/10.1016/j.imr.2024.101089

Li, Y. P., Ahmadi, F., Kariman, K., & Lackner, M. (2024b). Recent advances and challenges in single cell protein (SCP) technologies for food and feed production. *npj Science of Food, 8*, 66. https://doi.org/10.1038/s41538-024-00299-2

Lima, P. C., Karimian, P., Johnston, E., & Hartley, C. J. (2024). The use of Trichoderma spp. for the bioconversion of agro-industrial waste biomass via fermentation: A review. *Fermentation, 10*, 442. https://doi.org/10.3390/fermentation10090442

Lin, Y.-L., Wang, T.-H., Lee, M.-H., & Su, N.-W. (2008). Biologically active components and nutraceuticals in the Monascus-fermented rice: A review. *Applied Microbiology and Biotechnology, 77*, 965–973. https://doi.org/10.1007/s00253-007-1256-6

Lonsane, B. K., Ghildyal, N. P., Budiatman, S., & Ramakrishna, S. V. (1985). Engineering aspects of solid state fermentation. *Enzyme and Microbial Technology, 7*, 258–265. https://doi.org/10.1016/0141-0229(85)90083-3

Lübeck, M., & Lübeck, P. S. (2022). Fungal cell factories for efficient and sustainable production of proteins and peptides. *Microorganisms, 10*, 753. https://doi.org/10.3390/microorganisms10040753

Machida, M., Yamada, O., & Gomi, K. (2008). Genomics of *Aspergillus oryzae*: Learning from the history of koji mold and exploration of its future. *DNA Research, 15*, 173–183. https://doi.org/10.1093/dnares/dsn020

Maddimadugu, R. R., Panyam, S. Y., Battala, V. S. P., et al. (2024). Novel fungal co-culture technique for enhanced bioconversion of agro-industrial waste to amylase. *Current Agriculture Research Journal, 11*, 740–752. https://doi.org/10.12944/CARJ.11.3.05

Maeda, R. N., Barcelos, C. A., Anna, L. M. M. S., & Pereira, N. (2013). Cellulase production by *Penicillium funiculosum* and its application in the hydrolysis of sugar cane bagasse for second generation ethanol production by fed batch operation. *Journal of Biotechnology, 163*, 38–44. https://doi.org/10.1016/j.jbiotec.2012.10.014

Magnuson, J. K., & Lasure, L. L. (2004). Organic acid production by filamentous fungi. In J. S. Tkacz & L. Lange (Eds.), *Advances in fungal biotechnology for industry, agriculture, and medicine* (pp. 307–340). Springer.

Maini Rekdal, V., Villalobos-Escobedo, J. M., Rodriguez-Valeron, N., et al. (2024). *Neurospora intermedia* from a traditional fermented food enables waste-to-food conversion. *Nature Microbiology, 9*, 2666–2683. https://doi.org/10.1038/s41564-024-01799-3

Mäkelä, M. R., Donofrio, N., & De Vries, R. P. (2014). Plant biomass degradation by fungi. *Fungal Genetics and Biology, 72*, 2–9. https://doi.org/10.1016/j.fgb.2014.08.010

Manavalan, A., Adav, S. S., & Sze, S. K. (2011). iTRAQ-based quantitative secretome analysis of Phanerochaete chrysosporium. *Journal of Proteomics, 75*, 642–654. https://doi.org/10.1016/j.jprot.2011.09.001

Martínez-Carrera, D., Aguilar, A., Martínez, W., et al. (2000). Commercial production and marketing of edible mushrooms cultivated on coffee pulp in Mexico. In T. Sera, C. R. Soccol, A. Pandey, & S. Roussos (Eds.), *Coffee biotechnology and quality* (pp. 471–488). Springer.

Mishra, S., Kumar, P., & Malik, A. (2016). Suitability of agricultural by-products as production medium for spore production by *Beauveria bassiana* HQ917687. *International Journal Of Recycling of Organic Waste in Agriculture, 5*, 179–184. https://doi.org/10.1007/s40093-016-0127-5

Mohamadhasani, F., & Rahimi, M. (2022). Growth response and mycoremediation of heavy metals by fungus *Pleurotus* sp. *Scientific Reports, 12*, 19947. https://doi.org/10.1038/s41598-022-24349-5

Moore, D., & Chiu, S. (2001). Fungal products as food. In S. B. Pointing & K. D. Hyde (Eds.), *Bio-exploitation of flamentous fungi* (pp. 223–251). Fungal Diversity Press.

Mushroom Market Report. (2025). *The Business Research Company*. Accessed 8 May 2025. https://www.researchandmarkets.com/reports/5939329/mushroom-market-report

MycoWorks. (2025). *Grow the future of materials* [Online]. https://www.mycoworks.com/

Narwade, J. D., Odaneth, A. A., & Lele, S. S. (2023). Solid-state fermentation in an earthen vessel: *Trichoderma viride* spore-based biopesticide production using corn cobs. *Fungal Biology, 127*, 1146–1156. https://doi.org/10.1016/j.funbio.2023.06.007

Nasseri, A. T., Rasoul-Ami, S., Morowvat, M. H., & Ghasemi, Y. (2011). Single cell protein: Production and process. *American Journal of Food Technology, 6*, 103–116. https://doi.org/10.3923/ajft.2011.103.116

Naveen Kumar, K. J., & Prakash, J. (2021). Bioremoval of different heavy metals in industrial effluent by the resistant fungal strain *Aspergillus niger*. *Nature Environment and Pollution Technology, 20*, 10.46488/NEPT.2021.v20i04.006.

Nguyen, B. V. G., Tran, L. X. T., Ha-Nguyen, A.-T., et al. (2025). Endophytic fungi isolated from Vietnamese nut grass (*Cyperus rotundus* L. Cyperaceae) – A promising solution to mitigate the prime phenomenon of antibiotic resistance. *Heliyon, 11*, e41920. https://doi.org/10.1016/j.heliyon.2025.e41920

Nicolaou, K. C., & Rigol, S. (2018). A brief history of antibiotics and select advances in their synthesis. *Journal of Antibiotics (Tokyo), 71*, 153–184. https://doi.org/10.1038/ja.2017.62

Niego, A. G. T., Lambert, C., Mortimer, P., et al. (2023). The contribution of fungi to the global economy. *Fungal Diversity, 121*, 95–137. https://doi.org/10.1007/s13225-023-00520-9

Nouri, N., Sadeghi, L., & Marefat, A. (2024). Production of alkaline protease by *Aspergillus niger* in a new combinational paper waste culture medium. *Journal of Bioscience and Bioengineering, 137*, 173–178. https://doi.org/10.1016/j.jbiosc.2023.12.010

Nout, M. (2000). Useful role of fungi in food processing. In *Introduction to food-borne fungi* (6th ed., pp. 364–374). Centraalbureau voor Schimmelcultures.

Nout, M. J. R. (2014). Food technologies: Fermentation. In *Encyclopedia of food safety* (pp. 168–177). Elsevier.

Nyamful, A. (2013). *Characterization of glucoamylase produced by Aspergillus niger and Rhizopus* sp. Thesis, Kwame Nkrumah University.

Ohara, A., Soares De Castro, R. J., Goia Nishide, T., et al. (2015). Invertase production by *Aspergillus niger* under solid state fermentation: Focus on physical–chemical parameters, synergistic and antagonistic effects using agro-industrial wastes. *Biocatalysis and Agricultural Biotechnology, 4*, 645–652. https://doi.org/10.1016/j.bcab.2015.06.008

Pan, T.-M., & Hsu, W.-H. (2014). Monascus-fermented products. In *Encyclopedia of food microbiology* (pp. 815–825). Elsevier.

Pandey, A., & Gundevia, H. (2008). Role of the fungus–*Periconiella* sp. in destruction of biomedical waste. *Journal of Environmental Science & Engineering, 50*, 239–240.

Pandey, M., Senthil Kumaran, G., & Vasudeo, G. (2014). Making mushroom production process a zero waste enterprise. *International Journal of Environmental Sciences, 5*, 236–242.

Parapouli, M., Vasileiadi, A., Afendra, A.-S., et al. (2020). *Saccharomyces cerevisiae* and its industrial applications. *AIMS Microbiology, 6*, 1–32. https://doi.org/10.3934/microbiol.2020001

Passoth, V. (2017). Lipids of Yeasts and filamentous fungi and their importance for biotechnology. In A. A. Sibirny (Ed.), *Biotechnology of yeasts and filamentous fungi* (pp. 149–204). Springer.

Paul, M. K., Umeash, B. T., & Mathew, J. (2022). Recent advances and technologies in chitinase production under solid-state fermentation. *Biosciences Biotechnology Research Asia, 19*, 815–825. https://doi.org/10.13005/bbra/3033

Pfaller, M. A., Boyken, L., Hollis, R. J., et al. (2005). In vitro activities of anidulafungin against more than 2,500 clinical isolates of *Candida* spp., including 315 isolates resistant to fluconazole. *Journal of Clinical Microbiology, 43*, 5425–5427. https://doi.org/10.1128/JCM.43.11.5425-5427.2005

Pidosky. (2025). *Mushroom kingdom: Czechia unveils world's first fungi-based house* [Online]. expact.cz. https://www.expats.cz/czech-news/article/mushroom-kingdom-czechia-unveils-world-s-first-fungi-based-house

Pinheiro, V. E., Michelin, M., Vici, A. C., et al. (2020). Trametes versicolor laccase production using agricultural wastes: a comparative study in Erlenmeyer flasks, bioreactor and tray. *Bioprocess and Biosystems Engineering, 43*, 507–514. https://doi.org/10.1007/s00449-019-02245-z

Pletsch, M., De Araujo, B. S., & Charlwood, B. V. (1999). Novel biotechnological approaches in environmental remediation research. *Biotechnology Advances, 17*, 679–687. https://doi.org/10.1016/S0734-9750(99)00028-2

Pochanavanich, P., & Suntornsuk, W. (2002). Fungal chitosan production and its characterization. *Letters in Applied Microbiology, 35*, 17–21. https://doi.org/10.1046/j.1472-765X.2002.01118.x

Polizeli, M. L. T. M., Rizzatti, A. C. S., Monti, R., et al. (2005). Xylanases from fungi: Properties and industrial applications. *Applied Microbiology and Biotechnology, 67*, 577–591. https://doi.org/10.1007/s00253-005-1904-7

Pouris, J., Kolyva, F., Bratakou, S., et al. (2024). The role of fungi in food production and processing. *Applied Sciences, 14*, 5046. https://doi.org/10.3390/app14125046

Prado-Acebo, I., Cubero-Cardoso, J., Lu-Chau, T. A., & Eibes, G. (2024). Integral multi-valorization of agro-industrial wastes: A review. *Waste Management, 183*, 42–52. https://doi.org/10.1016/j.wasman.2024.05.001

Rabot, C., Chen, Y., Lin, S.-Y., et al. (2023). Polystyrene upcycling into fungal natural products and a biocontrol agent. *Journal of the American Chemical Society, 145*, 5222–5230. https://doi.org/10.1021/jacs.2c12285

Raol, G. G., Raol, B. V., Prajapati, V. S., & Bhavsar, N. H. (2015). Utilization of agro-industrial waste for β-galactosidase production under solid state fermentation using halotolerant *Aspergillus tubingensis* GR1 isolate. *3 Biotech, 5*, 411–421. https://doi.org/10.1007/s13205-014-0236-7

Ray, R. R. (2004). β-amylases from various fungal strains. *Acta Microbiologica et Immunologica Hungarica, 51*, 85–95. https://doi.org/10.1556/AMicr.51.2004.1-2.6

Reginatto, C., Rossi, C., Miglioranza, B. G., et al. (2017). Pectinase production by *Aspergillus niger* LB-02-SF is influenced by the culture medium composition and the addition of the enzyme inducer after biomass growth. *Process Biochemistry, 58*, 1–8. https://doi.org/10.1016/j.procbio.2017.04.018

RFA. (2025). *Annual ethanol production* [Online]. Renewable Fuels Association.

Rissoni Toledo, A. G., Reyes Andrade, J. C., Palmieri, M. C., et al. (2021). Innovative method for encapsulating highly pigmented biomass from *Aspergillus nidulans* mutant for copper ions removal and recovery. *PLoS One, 16*, e0259315. https://doi.org/10.1371/journal.pone.0259315

Ritala, A., Häkkinen, S. T., Toivari, M., & Wiebe, M. G. (2017). Single cell protein – State-of-the-art, industrial landscape and patents 2001–2016. *Frontiers in Microbiology, 8*, 2009. https://doi.org/10.3389/fmicb.2017.02009

Roelofsen, P. A., & Talens, A. (1964). Changes in some B Vitamins during molding of soybeans by *Rhizopus oryzae* in the production of Tempeh Kedelee. *Journal of Food Science, 29*, 224–226. https://doi.org/10.1111/j.1365-2621.1964.tb01722.x

Royse, D. J. (2014, November 19–22). *A global perspective on the high five: Agaricus, Pleurotus, Lentinula, Auricularia & Flammulina.* In Proceedings of 8th international conference on mushroom biology and mushroom products (ICMBMP8) (Vol. I & II, pp. 1–6).

Sala, A., Artola, A., Sánchez, A., & Barrena, R. (2020). Rice husk as a source for fungal biopesticide production by solid-state fermentation using *B. bassiana* and *T. harzianum*. *Bioresource Technology, 296*, 122322. https://doi.org/10.1016/j.biortech.2019.122322

Sala, A., Vittone, S., Barrena, R., et al. (2021). Scanning agro-industrial wastes as substrates for fungal biopesticide production: Use of *Beauveria bassiana* and *Trichoderma harzianum* in solid-state fermentation. *Journal of Environmental Management, 295*, 113113. https://doi.org/10.1016/j.jenvman.2021.113113

Sawangwan, T., & Saman, P. (2016). Prebiotic synthesis from rice using *Aspergillus oryzae* with solid state fermentation. *Agriculture and Natural Resources, 50*, 227–231. https://doi.org/10.1016/j.anres.2016.02.004

Schilling, C., & Weiss, S. (2021). A roadmap for industry to harness biotechnology for a more circular economy. *New Biotechnology, 60*, 9–11. https://doi.org/10.1016/j.nbt.2020.08.005

Shah, S. S., Palmieri, M. C., Sponchiado, S. R. P., & Bevilaqua, D. (2020). Enhanced bio-recovery of aluminum from low-grade bauxite using adapted fungal strains. *Brazilian Journal of Microbiology, 51*, 1909–1918. https://doi.org/10.1007/s42770-020-00342-w

Sharma, S., Dhar, M. K., & Kaul, S. (2023). Antagonistic, plant growth promoting and extracellular hydrolytic enzyme activity of fungal endophytes of Dioscorea bulbifera L. *Biocatalysis and Agricultural Biotechnology, 50*, 102694. https://doi.org/10.1016/j.bcab.2023.102694

Show, P. L., Oladele, K. O., Siew, Q. Y., et al. (2015). Overview of citric acid production from *Aspergillus niger*. *Frontiers in Life Science, 8*, 271–283. https://doi.org/10.1080/21553769.2015.1033653

Singh, S., Dhanjal, D. S., Thotapalli, S., et al. (2020). Importance and recent aspects of fungi-based food ingredients. In *New and future developments in microbial biotechnology and bioengineering* (pp. 245–254). Elsevier.

Singhania, R. R., Patel, A. K., Soccol, C. R., & Pandey, A. (2009). Recent advances in solid-state fermentation. *Biochemical Engineering Journal, 44*, 13–18. https://doi.org/10.1016/j.bej.2008.10.019

Sinha, B., Rajendran, P., & Devi, P. S. (2022). Mass production of *Trichoderma* from agricultural waste and its application for plant disease management. In C. Baskar, S. Ramakrishna, S. Baskar, et al. (Eds.), *Handbook of solid waste management* (pp. 619–633). Springer.

Son, S. Y., Lee, S., Singh, D., et al. (2018). Comprehensive secondary metabolite profiling toward delineating the solid and submerged-state fermentation of Aspergillus oryzae KCCM 12698. *Frontiers in Microbiology, 9*, 1076. https://doi.org/10.3389/fmicb.2018.01076

Song, P., Yuan, K., Qin, T., et al. (2018a). Metabolomics profiling reveals the mechanism of increased pneumocandin B0 production by comparing mutant and parent strains. *Journal of Industrial Microbiology and Biotechnology, 45*, 767–780. https://doi.org/10.1007/s10295-018-2047-z

Song, P., Zhang, K., Zhang, S., et al. (2018b). Enhancement of pneumocandin B0 production in Glarea lozoyensis by low-temperature adaptive laboratory evolution. *Frontiers in Microbiology, 9*, 2788. https://doi.org/10.3389/fmicb.2018.02788

Song, P., Zhang, X., Wang, S., et al. (2023). Microbial proteases and their applications. *Frontiers in Microbiology, 14*, 1236368. https://doi.org/10.3389/fmicb.2023.1236368

Sreekantiah, K., Jaleel, S., & Ramachandra Rao, T. (1968). Exocellular catalase produced by strain of *Rhizopus niveus*. *Current Science, 37*, 626.

Stahmann, K.-P. (2011). Production of Vitamin B2 and a polyunsaturated fatty acid by fungi. In M. Hofrichter (Ed.), *Industrial applications* (pp. 235–247). Springer.

Steudler, S., Werner, A., & Walther, T. (2019). It is the mix that matters: substrate-specific enzyme production from filamentous fungi and bacteria through solid-state fermentation. In S. Steudler, A. Werner, & J. J. Cheng (Eds.), *Solid state fermentation* (pp. 51–81). Springer.

Strong, P. J., Self, R., Allikian, K., et al. (2022). Filamentous fungi for future functional food and feed. *Current Opinion in Biotechnology, 76*, 102729. https://doi.org/10.1016/j.copbio.2022.102729

Subash, N., Meenakshisundaram, M., Sasikumar, C., & Unnamalai, N. (2014). Mass cultivation of *Trichoderma harzianum* using agricultural waste as a substrate for the management of damping off disease and growth promotion in chilli plants (*Capsicum annuum* L.). *International Journal of Pharmacy and Pharmaceutical Sciences, 6*, 188–192.

Subedi, S., Kunwar, N., Pandey, K. R., & Joshi, Y. R. (2023). Performance of oyster mushroom (*Pleurotus ostreatus*) on paddy straw, water hyacinth and their combinations. *Heliyon, 9*, e19051. https://doi.org/10.1016/j.heliyon.2023.e19051

Surya, R. (2024). Fermented foods of Southeast Asia other than soybean- or seafood-based ones. *Journal of Ethnic Foods, 11*, 27. https://doi.org/10.1186/s42779-024-00241-7

Suyanto, E., Soetarto, E. S., & Cahyanto, M. N. (2019). Production and optimization of lipase by *Aspergillus niger* using coconut pulp waste in solid state fermentation. *Journal of Physics Conference Series, 1374*, 012005. https://doi.org/10.1088/1742-6596/1374/1/012005

Taneja, K., Gupta, S., & Chander Kuhad, R. (2002). Properties and application of a partially purified alkaline xylanase from an alkalophilic fungus *Aspergillus nidulans* KK-99. *Bioresource Technology, 85*, 39–42. https://doi.org/10.1016/S0960-8524(02)00064-0

Tang, A., Lu, Y., Li, Q., et al. (2021). Simultaneous leaching of multiple heavy metals from a soil column by extracellular polymeric substances of *Aspergillus tubingensis* F12. *Chemosphere, 263*, 127883. https://doi.org/10.1016/j.chemosphere.2020.127883

Tiquia, S. M., Richard, T. L., & Honeyman, M. S. (2002). Carbon, nutrient, and mass loss during composting. *Nutrient Cycling in Agroecosystems, 62*, 15–24. https://doi.org/10.1023/A:1015137922816

Trillas, M. I., Casanova, E., Cotxarrera, L., et al. (2006). Composts from agricultural waste and the *Trichoderma asperellum* strain T-34 suppress *Rhizoctonia solani* in cucumber seedlings. *Biological Control, 39*, 32–38. https://doi.org/10.1016/j.biocontrol.2006.05.007

Troiano, D., Orsat, V., & Dumont, M. J. (2020). Status of filamentous fungi in integrated biorefineries. *Renewable and Sustainable Energy Reviews, 117*, 109472. https://doi.org/10.1016/j.rser.2019.109472

Udagawa, K., & Abe, S. (1961). *Production of Griseofulvin by some strains of the Genus Penicillum.*

Ugalde, U. O., & Castrillo, J. I. (2002). Single cell proteins from fungi and yeasts. In *Applied mycology and biotechnology* (pp. 123–149). Elsevier.

Umsza-Guez, M. A., Díaz, A. B., de Ory, I., et al. (2011). Xylanase production by *Aspergillus awamori* under solid state fermentation conditions on tomato pomace. *Brazilian Journal of Microbiology, 42*, 1585–1597. https://doi.org/10.1590/S1517-83822011000400046

Unuofin, J. O. (2021). The sustainable production of a novel laccase from wheat bran by *Bordetella* sp. JWO16: Toward a total environment. *Catalysts, 11*, 677. https://doi.org/10.3390/catal11060677

Varma, P. K., Uppala, S., Pavuluri, K., et al. (2017). Endophytes: Role and functions in crop health. In D. P. Singh, H. B. Singh, & R. Prabha (Eds.), *Plant-microbe interactions in agro-ecological perspectives* (pp. 291–310). Springer.

Venkata Dasu, V., Panda, T., & Chidambaram, M. (2003). Determination of significant parameters for improved griseofulvin production in a batch bioreactor by Taguchi's method. *Process Biochemistry, 38*, 877–880. https://doi.org/10.1016/S0032-9592(02)00068-7

Venturini Copetti, M. (2019). Yeasts and molds in fermented food production: an ancient bioprocess. *Current Opinion in Food Science, 25*, 57–61. https://doi.org/10.1016/j.cofs.2019.02.014

Vijay, K., & Kavitha, T. (2022). Inoculation of medicinal plant resident *Staphylococcus hominis* subsp. novobiosepticus strain KE6 for improved plant growth and antagonism to pithomyces atro-olivaceous in groundnuts. *Asian Journal of Research in Biosciences*, 12–25.

Vivek, K., & Venkitasamy, C. (2023). Role and Applications of fungi in food and fermentation technology. In I. Singh, V. R. Rajpal, & S. S. Navi (Eds.), *Fungal resources for sustainable economy* (pp. 71–87). Springer.

Volford, B., Varga, M., Szekeres, A., et al. (2021). β-Galactosidase-producing isolates in mucoromycota: Screening, enzyme production, and applications for functional oligosaccharide synthesis. *Journal of Fungi, 7*, 229. https://doi.org/10.3390/jof7030229

Wainwright, M. (1989). Moulds in ancient and more recent medicine. *Mycologist, 3*, 21–23. https://doi.org/10.1016/S0269-915X(89)80010-2

Wainwright, M. (2008). Some highlights in the history of fungi in medicine – A personal journey. *Fungal Biology Reviews, 22*, 97–102. https://doi.org/10.1016/j.fbr.2008.11.001

Wang, T., & Lin, T. (2007). *Monascus* rice products. In *Advances in food and nutrition research* (pp. 123–159). Elsevier.

Wang, J.-J., & Pan, T.-M. (2003). Effect of red mold rice supplements on serum and egg yolk cholesterol levels of laying hens. *Journal of Agricultural and Food Chemistry, 51*, 4824–4829. https://doi.org/10.1021/jf034296e

Wang, F., Xu, L., Zhao, L., et al. (2019). Fungal laccase production from lignocellulosic agricultural wastes by solid-state fermentation: A review. *Microorganisms, 7*, 665. https://doi.org/10.3390/microorganisms7120665

Wang, Y., Zhang, Q., Yousaf, J., et al. (2025). Analysis and comparison of lipids in *Monascus* -fermented cheese from different ripening periods based on UHPLC-QTRAP MS quantitative lipidomics. *Journal of Agricultural and Food Chemistry, 73*, 4385–4395. https://doi.org/10.1021/acs.jafc.4c10355

Waqas, M., Khan, A. L., Hamayun, M., et al. (2015). Endophytic fungi promote plant growth and mitigate the adverse effects of stem rot: an example of *Penicillium citrinum* and *Aspergillus terreus*. *Journal of Plant Interactions, 10*, 280–287. https://doi.org/10.1080/17429145.2015.1079743

Wiebe, M. G. (2004). QuornTM myco-protein – Overview of a successful fungal product. *Mycologist, 18*, 17–20. https://doi.org/10.1017/S0269915X04001089

Wu, B., Hussain, M., Zhang, W., et al. (2019). Current insights into fungal species diversity and perspective on naming the environmental DNA sequences of fungi. *Mycology, 10*, 127–140. https://doi.org/10.1080/21501203.2019.1614106

Yan, Y., Luo, X., Fan, M., et al. (2023). NaCl stress enhances pigment accumulation and synthesis in red rice during the germination stage. *Food Bioscience, 56*, 103224. https://doi.org/10.1016/j.fbio.2023.103224

Yang, W., Guo, F., & Wan, Z. (2013). Yield and size of oyster mushroom grown on rice/wheat straw basal substrate supplemented with cotton seed hull. *Saudi Journal of Biological Sciences, 20*, 333–338. https://doi.org/10.1016/j.sjbs.2013.02.006

Yang, M., Yu, Y., Zhai, H., et al. (2018). Ecological dyeing of silk fabric with *Monascus*. *The Journal of The Textile Institute, 109*, 1329–1334. https://doi.org/10.1080/00405000.2018.1423932

Yélamos, A. M., Marcos, J. F., Manzanares, P., & Garrigues, S. (2025). Harnessing filamentous fungi for enzyme cocktail production through rice bran bioprocessing. *Journal of Fungi, 11*, 106. https://doi.org/10.3390/jof11020106

Yildirim, N., & Yildiz, A. (2009). The effect of Pleurotus eryngii (DC. ex Fr.) Quel. on rice bran and wheat straw under the solid-state bioconversion for ruminant feed. *Journal of Animal & Plant Sciences, 5*, 475–482.

Yin, Z. N., Wu, W. J., Sun, C. Z., et al. (2019). Antioxidant and anti-inflammatory capacity of ferulic acid released from wheat bran by solid-state fermentation of *Aspergillus niger*. *Biomedical and Environmental Sciences, 32*, 11–21. https://doi.org/10.3967/bes2019.002

Yip, D., & Gerriets, V. (2025). *Penicillin*. In: StatPearls [Internet].

Zhang, Y., He, S., & Simpson, B. K. (2018). Enzymes in food bioprocessing – Novel food enzymes, applications, and related techniques. *Current Opinion in Food Science, 19*, 30–35. https://doi.org/10.1016/j.cofs.2017.12.007

Zhang, C., Ali Khan, R. A., Wei, H., et al. (2022a). Rapid and mass production of biopesticide Trichoderma Brev T069 from cassava peels using newly established solid-state fermentation bioreactor system. *Journal of Environmental Management, 313*, 114981. https://doi.org/10.1016/j.jenvman.2022.114981

Zhang, Y., Wei, R., Azi, F., et al. (2022b). Solid-state fermentation with *Rhizopus oligosporus* RT-3 enhanced the nutritional properties of soybeans. *Frontiers in Nutrition, 9*, 972860. https://doi.org/10.3389/fnut.2022.972860

Zhang, Z.-Y., Li, X., Chen, W.-H., et al. (2023). Culturable fungi from urban soils in China II, with the description of 18 novel species in Ascomycota (Dothideomycetes, Eurotiomycetes, Leotiomycetes and Sordariomycetes). *Mycokeys, 98*, 167–220. https://doi.org/10.3897/mycokeys.98.102816

Zhgun, A. A. (2023). Industrial production of antibiotics in fungi: current state, deciphering the molecular basis of classical strain improvement and increasing the production of high-yielding strains by the addition of low-molecular weight inducers. *Fermentation, 9*, 1027. https://doi.org/10.3390/fermentation9121027

Zhou, Y., Li, Z., Xu, C., et al. (2023). Evaluation of corn stalk as a substrate to cultivate king oyster mushroom (*Pleurotus eryngii*). *Horticulturae, 9*, 319. https://doi.org/10.3390/horticulturae9030319

Ziemons, S., Koutsantas, K., Becker, K., et al. (2017). Penicillin production in industrial strain *Penicillium chrysogenum* P2niaD18 is not dependent on the copy number of biosynthesis genes. *BMC Biotechnology, 17*, 16. https://doi.org/10.1186/s12896-017-0335-8

Zurek, M., Hebinck, A., & Selomane, O. (2022). Climate change and the urgency to transform food systems. *Science, 376*, 1416–1421. https://doi.org/10.1126/science.abo2364

Fermented Food Fungal Proteins and Peptides: Past, Present and Future

Amro Abd Al Fattah Amara and Arti Gupta

Abstract In diverse places worldwide, humans have used fungi (intentionally or non-intentionally) for producing fermented food (FF) that can survive (not spoilage). One generic name for those products is FF. Modern science has investigated these products that involve fungi, yeast, and mushrooms. They include soy sauce, miso, tempeh, mold cheeses, beer, and wine. Some FFs contain native fungal proteins and peptides that are produced during the various food fermentation processes (FPs). They are of special importance. Or protein that hydrolyzed because of fungal activity to other simpler forms including active peptides. This chapter is concerned with describing some important native active fungal proteins and peptides that can reach our bodies so that it can perform desired activities through consuming different FFs. The main aim is to highlight the importance of such FFs to their consumers who might not be able to cover the cost of the pharmaceutical-grade proteins or peptides with medicinal purposes. At the same time, it is an important reference for those interested in purifying active protein or peptides from FFs. Some examples of purified proteins and peptides from FFs will be described as well.

Keywords Fermented foods · Fungi · Yeast · Mushrooms · Active proteins · Active peptides

A. A. A. F. Amara (✉)
Protein Research Department, Genetic Engineering and Biotechnology Research Institute [GEBRI], City of Scientific Research and Technological Applications [SRTA-City], Universities and Research Center District, Alex, Egypt
e-mail: aamara@srtacity.sci.eg

A. Gupta
Department of Zoology, Shri Awadh Raj Singh Degree College, Ram Manohar Lohia University, Ayodhya, India

A. Gupta (ed.), *Prospects of Fungal Biotechnologies for Livestock Volume 2*,
Fungal Biology, https://doi.org/10.1007/978-3-032-06478-3_6

Abbreviations

AMC	Artisanal Minas Cheese
BM	Beneficial microbe
GIS	Gastrointestinal system
FF	Fermented food
FM	Fermented milk
FP	Fermentation processes
LAB	Lactic-acid bacteria
PTM	Post-translational modification
SCP	Single-cell protein
TFF	Traditional fermented food

1 Introduction

Humans have learned to ferment food (FF) as a tool for preservation (Hatoum et al., 2012), obtaining new taste (umami taste) (Fan et al., 2025), aroma (Orban et al., 2019), etc. FFs (Anukam & Reid, 2009) are a result of the contact of humans with nature (Kunyeit et al., 2024; Valentino et al., 2024). Some of such foods' FPs are treated as secrets (El Sheikha & Hu, 2020) for those who are discovering them such as kefir (O'Brien et al., 2016). This was a gift from the Arabs to the Caucasians who kept it a secret for hundreds of years (Amara & Shibl, 2015). Why are FFs popular everywhere (Das et al., 2023; Huang et al., 2023; Khorshidian et al., 2020; Paul et al., 2024; Saleem et al., 2024; Wu et al., 2018)? Basically because of their taste and the comfort that they bring during and after their consumption (Amara & Shibl, 2015). Modern science has declared the various components of FFs that are beneficial to humans (Marco et al., 2017). Many of the microbes involved in food fermentation were identified and the genomes of the most interesting ones were sequenced (Pretorius, 2017; Querol & Bond, 2009; Gong et al., 2020; Lan et al., 2024; Yuan et al., 2019; Umemura et al., 2013; Wang et al., 2010). To simplify traditional FFs and differentiate them from the cell-fermented food, it is a food that was traditionally processed by a previously fermented successful food or microbes that naturally existed in native raw substrate (e.g., yeasts on grape surfaces) that react collectively to digest part of their components or their entire component to simple structures or to new compounds interesting and beneficial to humans. This definition could be generic for a traditional fermentation process. The modern definition for the fermentation process is 'a process in which an agent causes an organic substance to break down into simpler substances; in particular, the anaerobic breakdown of sugar into alcohol'. Microbes perform vital functions such as those existing in our gastrointestinal system (GIS) microbiota, where they are involved in changing our foods into simple and beneficial macro and micro molecules that our bodies could consume and benefit from. One should differentiate between two types of microbes:

those that become a part of our GIS and those that do not permanently exist in our GIS but support the digestion process upon eating foods, or fermented foods that contain them, as well as consuming them in pure viable form directly. Nevertheless, the beneficial microbes that are involved in food fermentation perform additional tasks such as taste, aroma, comfort, etc., which technically could be done in our GIS but one could not sense such activity. One should also observe that in most cases the original (non-FFs) such as some proteins are originally inactive and become active after digestion. Such phenomena are highly important as they will be active only in the correct place. As an exception FFs could contain some forms that really turned out to be active such as peptides that are produced from protein hydrolysis. Perhaps the best examples are daily fermented products. In a symbiotic relationship these include fungal/fungal, yeast/yeast (Rahmani et al., 2022), yeast/fungal, yeast/bacterial (Devanthi et al., 2021; Lopitz-Otsoa et al., 2006); fungal/bacterial interactions (Gan et al., 2020; Yang et al., 2025; Furukawa et al., 2013) to create new amazing active protein and peptides. Even each beneficial microbe (BM) can be considered a small factory that does a complicated chemical modification for a certain substrate (the food) during a process named 'fermentation' to produce new products; but a symbiotic relationship of a group of BMs will probably give better results. Traditional fermentation is usually a long and complicated process to reach a certain quality of the final product that is in great demand from the consumer, *de nuovo* products that did not exist before a new product. Each specific condition activates the metabolic power of the microbes in a specific manner and rate to produce specific products. Our aim is to get the benefit from the useful products whose safety has been proven over time by their utilization. Some FFs have attracted the attention of certain civilizations worldwide and become a basic part of everyday meals. Mostly, they vary because of the diverse existing earth and sea and other products from diverse geographic areas. In fact, the process is usually so simple to conduct and did not need so much experience, but rather careful processing and time.

The properties of the microorganisms involved in food fermentation including different forms of fungi and yeast were exploited over the decades. Many of the microbes involved in fermented cheese, bread, wine, and beer, etc., have been identified and well studied. More recently, the possible applications diversified with, for example, the use of natural products of microbial origin, compounds with pharmaceuticals, biocarburants, biopesticides, or bio-surfactant properties.

Although the sight of mold in food is a sign that it is contaminated and should be discarded, humans in ancient times remarkably succeeded in using an intelligent pre-step to grow beneficial fungi first in semi-cooked grains such as soybeans (koji), then using this formula to seed the coming fermentation process. Simple tactics were practiced here and there where either a part of a successful fermentation was used as a seed inoculum (bread processing) or just by using the same container in which a previous successful fermentation process had been carried out (cheese, vinegar, wine, etc.). Humans also observed that some fresh vegetables and fruits could ferment by themselves. A large number of FPs could reach a level of a complete industry existed hundreds of years ago to produce products on a commercial scale such as wine, vinegar, and cheese. In fact, the success of certain fermentation processes was well validated by the color, smell, taste, and sometimes by the visual characterization of the filamentous fungi hypha and color. In some FFs, the presence of visible fungal mycelium is a part of the food dynamics (McGovern, 2007, 2020; Ashaolu & Reale, 2020; Anupma & Tamang, 2016). In diverse parts of the world, there are many different types of FFs, where some have similar traditional technologies that are dependent on the fungi or the activity of the fungi.

Probably FF started with the first use of food that can be fermented if left alone, which is called natural fermentation, as with milk. Meanwhile, there are many reports that describe the beginning of the first fermentation process in may be 13,000 years B.C. Apparently, the best proven date until now started before 7000 B.C. in Iraq, where fermented milk was reported in 7000 B.C., and the Dahi-coagulated sour milk in India in 6000–4000 B.C. Egyptians discovered how to use starter culture containing yeasts to make leavened bread and wine (Ross et al., 2002; Prajapati & Nair, 2003; Ray & Joshi, 2014), as in Fig. 1.

Some of these activities including FFs such as cheese production (e.g., blue cheese (Ropars et al., 2017)); beverages (Ferreira et al., 2022); milk products such as nunu, a Ghanaian fermented milk (FM) (Akabanda et al., 2013); kefir variants

Fig. 1 Timeline of the fermented food throughout the history, updated from Ray and Joshi (2014). (Prepared in this study)

(Rahmani et al., 2022; Youn et al., 2023); fermented tea leaves such as laphet, a traditional fermented tea in Myanmar (Bo et al., 2020); and many other types of FFs such as fermented meat (Kozlovsky et al., 2014; Lo et al., 2023) and fishes (Nie et al., 2025), as well as fungi involved in fermented drinks (Maragkoudakis et al., 2016), single-cell protein (Amara & El-Baky, 2023) and mushrooms (Chi et al., 2025; Sittisart et al., 2024; Sivamaruthi et al., 2024), etc. The fungal FFs and products are common in Asia (Aidoo et al., 2006; Fukami et al., 2021; Rodzi & Lee, 2021), the Middle East (El-Gendy, 1983), and Europe (Tamang & Samuel, 2010; Ashaolu et al., 2025). Examples of Egyptian FFs are different kinds of cheese, salted fish, Mish cheese, and fermented fava beans used in Tameya. Asian FFs include fermented soybeans, rice or barley, natto fermentation, fermented kimchi, kefir grains, and FM. European FFs include the blue cheese different varieties that are dependent on *Penicillium roqueforti*. Blue cheese products include Roquefort, in which kokumi-active glutamyl peptides are produced (Toelstede & Hofmann, 2009). Stilton (Gkatzionis et al., 2014) and Camembert cheese have yeasts with diverse activities included in the fermentation process (Addis et al., 2001; Roostita & Fleet, 1996; Schmidt & Daudin, 1983). Further well-known blue cheeses include Bleu de Bresse and Bleu d'Auvergne (France), Danablu (Denmark), Blue Cheshire (England), in addition to other types (Ropars et al., 2017; Encyclopedia Britannica, 2013a; Siemens & Zawistowski, 1993). Brie and wine (Albergaria et al., 2010) are produced on a large scale dependent on *S. cerevisiae*. *Brettanomyces* is sometimes another yeast strain that is involved in wine fermentation (Branco et al., 2019).

African FFs include ogi (pap), a fermented cereal pudding produced from maize, sorghum, or millet (Fernandes et al., 2021); kenkey, a fermented corn dough product from Ghana; injera, a sourdough flatbread from Ethiopia produced from teff flour (Suraj et al., 2024); togwa, a traditional fermented beverage produced from millet or sorghum, popular in Tanzania (Mugula et al., 2003); uji, a fermented porridge produced from various grains, common in East Africa; and bakalanga (or bikalanga), a fermented fish product from the Democratic Republic of the Congo. North American FFs include sauerkraut, a fermented cabbage, popular in many North American households, especially among those of German descent. Kimchi is a traditional Korean dish that has gained popularity in North America, produced from fermented vegetables, primarily napa cabbage and radishes. Cucumbers fermented in brine or vinegar, are commonly found in North American cuisine. Yogurt, a dairy product made by FM with specific bacterial cultures is consumed worldwide. Beer is an alcoholic beverage produced from fermented grains, primarily barley. Sourdough bread is produced from a naturally fermented starter culture that includes wild yeasts and lactic-acid bacteria (LAB). Fermented hot sauce is produced from fermented chili peppers, gaining popularity in various cuisines. South American FFs include chicha, a traditional fermented beverage produced from maize (corn), often consumed in various forms across South America, particularly in the Andes. Cachaça is a Brazilian spirit produced from fermented sugarcane juice, often used in cocktails such as the caipirinha. Kefir, originally from the Caucasus region, has become popular in South America, particularly in Argentina and Brazil. Fermented cassava (tapioca) is fermented to produce various traditional dishes, such as "farinha" in Brazil, which is a fermented cassava flour. Aji amarillo paste is a fermented chili paste produced from aji amarillo peppers that is commonly used in Peruvian cuisine. Sour corn (choclo) in which corn is fermented to create a sour flavor, is often used in traditional dishes in some regions. Mandioca (manioc) is a fermented manioc used to make traditional dishes in Brazil as well as in other countries, often resulting in a sour flavor. Fermented fish such as "canchal" are used in Peru. Australia has a growing interest in FFs, influenced by both traditional indigenous practices and contemporary culinary trends. Kimchi, originally a Korean dish, has become popular in Australia, often made with local vegetables. Tempeh, a fermented soybean product that originated in Indonesia, is widely produced and consumed in Australia. Sourdough has become increasingly popular in Australia, with many artisanal bakeries focusing on traditional methods.

Fermented fish is popular in some indigenous Australian communities that have traditional practices of fermenting fish that can enhance flavor and preservation.

Nowadays, some diseases have been prevented by consuming FFs. For example, the intake of FFs was reported to be a crucial step toward successful management of Alzheimer's disease (Kumar et al., 2022). The explanation for this effect might be given to the increase in particular nutrients such as protein and vitamins. In contrast, unwanted compounds or those not good for our health such as the antinutritional chemicals including phytates, tannins, and polyphenols may disappear because of food fermentation or at least their amounts are reduced. For more details, refer to Kumar et al. (2022) and the references within (Kumar et al., 2022). In Africa (for example, Mozambique and Uganda), mycelial fungi are used for fermentation as exemplified by their use in detoxifying bitter cassava roots (Kumar et al., 2022).

Fungal FFs and their products are common (Tamang & Samuel, 2010; Bourdichon et al., 2012a). Fungi are involved in developing different dairy products, particularly in the ripening processes of various types of cheese and for enzyme production (Anupma & Tamang, 2016; Chávez et al., 2011; Spinnler & Gripon, 2004). The genus *Aspergillus* is used widely in diverse fermentation processes (Tamang & Samuel, 2010; Hagi et al., 2022; Mogensen et al., 2009; Nout, 2000, 2007).

FermFooDb is a database of bioactive peptides (BAPs) derived from FFs (Chaudhary et al., 2021; Habib et al., 2025; Kumari et al., 2023; Mgbodile & Nwagu, 2023; Rosa et al., 2023). Cheese is a rich source of casein proteins (Yangılar & Oğuzhan Yıldız, 2016) that can be broken down into BAPs during fermentation (Behnke & Schalinatus, 1975; Iwaniak et al., 2022). These peptides can have various health benefits, including antihypertensive and antimicrobial effects (Bravo et al., 2022). Molds such as *Penicillium nalgiovense* are used in fermented dry-cured sausages (Emri et al., 2013; Färber & Geisen, 1994; Moavro et al., 2022). Fungi can produce various enzymes, including amylases (Sandiford et al., 1994), proteases (el-Maghrabi et al., 1990), lipases (Chang et al., 2006), cellulases (Amaike Campen et al., 2017), pectinases (Paup et al., 2023), tannase (Zhong et al., 2004), and asparaginase (Chung et al., 2021). These enzymes are used in industries such as food (e.g., brewing, baking), textiles, and biofuels for breaking down complex substrates. Fungi can be engineered to produce therapeutic proteins, peptides, polypeptide hormone (e.g., insulin), vaccines, etc. Fungi can stand alone to produce food, for example, *Fusarium venenatum*, *Neurospora crassa*, and *Ischnoderma benzoinum* are used to produce mycoprotein (Bartholomai et al., 2022; Liang et al., 2023), a protein meat substitute marketed as Quorn. Mycoprotein is a sustainable protein source with a lower environmental impact than traditional livestock. This chapter is concerned with highlighting the importance of FFs as an open-source treasure for anyone, and their important products, especially protein and peptides.

2 The Folk Myth and Belief

Some of the FFs are very popular (Das et al., 2023; Huang et al., 2023; Khorshidian et al., 2020; Paul et al., 2024; Saleem et al., 2024; Wu et al., 2018; Brüssow, 2020). In the past, humans mainly fermented food to save foods from season to season. But they were also used for their other positive properties (Amara & Shibl, 2015; Amara & El-Baky, 2023). On the one hand they are tasty, appetizing, healthy (if well prepared), supportive to the digestive system, etc.; on the other they are ready to be eaten directly and at any time. Nowadays, they are a source of profit. Many companies gain billions of dollars each year from various FF products such as beers, wine, and other drinks; cheese, yogurt (Nielsen et al., 2021), and other fermented dairy products (Cosentino et al., 2001); cookies, sausages, fish, as well as other fermented types of meats. The market margin is really huge (Toro-Funes et al., 2015; Xu et al., 2015) considering that the products are produced by small producers alongside the mega companies in the field. FFs gain their popularity not only from their producer but from their consumers as well (Kim et al., 2010). It is a remarkable question after the huge improvement and spread of the international industrial food products in nearly every country why traditional FFs are still popular and have a wide range of consumers from varying economic levels. Consumers trust in FFs because of their comfort effect. Many try to produce the traditional FFs that were originally prepared at home by using a pre-isolated and purified microbe, but the products did not gain the same quality. It is a slow, precise process, an ancestral legacy, folk belief, and trust and when trust is gained for a certain name, the producer will start to grow, as well as the number of consumers.

So, the best practice to get the same quality is to use the old technique, particularly the seeding process that guarantees either maintaining the original naturally grown microbes on the used food (e.g., milk, grapes, etc.) or seeding the new batch with an old sample obtained from a successfully produced FF batch that contains all the needed microbes. Alternatively, one could observe that in some places such as Egypt the newly produced cheeses that are going to be aged are just put in the same place that contained a previously successfully aged batch or in a vessel that still contains the residue of a successfully aged batch.

Nowadays, the FFs have gone beyond being useful to become a sign of place, history and civilization, where each population has their own choices from the FF dishes that transfer from generation to generation. The respect for the national FFs can be observed in the various published articles (Amara & Shibl, 2015; Amara & El-Baky, 2023; Kothe et al., 2022; Zheng et al., 2018; Chourasia et al., 2025). These days, the FF attracts the attention of sectors other than those with food-related interests, such as the medicinal sector. Perhaps FFs such as kefir or yogurt are responsible for the first scientific interest but not for the first human belief and practice of using FFs to promote health (Amara & Shibl, 2015). The domain of probiotics (Agarbati et al., 2020; de Miranda et al., 2023; Fernández-Pacheco et al., 2023) and prebiotics (Ballini et al., 2023; Cheng et al., 2025; Oliveira et al., 2012) that were previously discussed only by scientists have nowadays become a common public interest. The probiotic has become a popular noun and a sign of the product's quality thanks to the pioneers in this field and those who make the effort to highlight the benefits of FFs.

Apparently, nearly all old civilizations in Asia and Middle East as well as in Old Europe were aware of the benefits of FFs. Nowadays, the beneficial microbes of a few FFs out of many were identified and some are considered as a miracle for health promotion and protection against illness. The skin condition eczema was known by many who observed it to be cured after drinking FM (Amara & Shibl, 2015; Amara & El-Baky, 2023). An old belief was that healthy persons' saliva could cure illness and improve health. Such practices are used in many Arabic countries, but to our knowledge it is used in Egypt and Morocco (personal communication). Kefir is believed to follow the same concept. There are many wonderful practices where a scientist might have difficulty explaining it. Another Arab custom is to give a date to a healthy person to chew it and then put a piece of it in the mouth of the newborn, believing that the newborn will become healthy like him. This may be explained by the transfer of some beneficial microbes from the mouth of the healthy person to the newborn. Virus C-infected people have been known to be cured by camel milk. New science has proven the truth of some of these practices and has discovered the scientific explanation for many. Only recently has modern science proven the existence of active peptides in milk that could fight against microbes and viruses. FFs free more and more forms of peptides. That might put a question mark as to how were old civilizations aware of many scientific issues. The use of honey and milk, milk (or FM) and olive oil, or all of them together is a popular mixture for improving the digestive system. Different kinds of FM, such as yogurt, different types of cheeses, kefir, sour milk, etc., are rich with diverse prokaryotic and eukaryotic BMs, some of which or many of them have probiotic properties as well as peptides, enzymes, active proteins, etc. A really huge number of recipes containing milk, milk products, and FM for treating illnesses were used in the folk remedies of nearly all unknown symptoms. The use of a combined compound at once in the form of crude extract mixed with FFs for treatment is at the core of the medicine of old civilizations, which believed that what has come from the nature should be better than any synthetic structures or any additives that might change its original structures.

3 Filamentous Fungi Beneficial to Fermented Food

Filamentous fungi have unique properties that make them the best candidates for use in food cell factories for producing proteins and peptides. In fact, the diverse strains of fungi wild types are doing that spontaneously during the different types of processes involved in food fermentation. Collectively, fungi as well as bacteria and other microbial forms could collaborate in food fermentation. The FP converts complex food or raw material into simple gradients or partially changes their form into a new useful product (e.g., milk fermented to cheese). That can happen owing to the activity of various enzymes and active converted proteins that could degrade macromolecules to simpler forms (e.g., peptides or amino acids [monomeric forms]). For example, proteins can be decomposed to amino acids, peptides, etc. During this metabolism additional natural products, aromas, and so on, are produced. Fungi have unique metabolic capabilities, some of them have rapid growth rates, and they have a developed an efficient secretory system (Schwentke et al., 2014) able to secrete macromolecules including proteins into the culture medium.

Fungi have other privileges while they are eukaryotes, which means that they have cellular machinery similar or close to higher organisms including humans. This allows processing of many important eukaryotic products, particularly active peptides and proteins such as insulin (Baeshen et al., 2014). Even eukaryotic fungi are good candidates for production of a complicated macromolecule such as insulin or proinsulin, but in many cases the strain used is subjected to different modifications to be a suitable cell factory (Baeshen et al., 2014). Fungi also have a well-differentiated and developed gene translation system higher than prokaryotic cells. They also have an advanced system for protein and peptides post-translation modification (PTM; e.g., glycosylation) that is essential for the functionality of many proteins and peptides, as well as advanced signaling and transferring structures. Fungi can grow quickly in minimal water or moisture content and can be cultivated in various conditions, making them suitable for large-scale production. Amazing so old forms of seed culture (e.g., koji) contain different fungal species as a starter culture each can be used for a specific fermentation process. Koji, baker yeast starters, Mish cheese aging, traditional wine fermentation, etc., are all examples of precise steps to successful fermentation practiced before the advent of microbiology. The ability of the fungi to secrete proteins directly into the culture medium is an interesting feature for producing enzymes and other extracellular active proteins and macro and micromolecules, simplifying the downstream processing and purifying the desired products. Fungi possess a wide range of metabolic pathways that enable various compounds to be produced, including enzymes, BAPs, and therapeutic proteins. Advances in molecular biology, genetic engineering techniques, bioinformatics, gene, protein, and peptide databases, etc., allow modification of genes to engineer proteins or conduct strain improvements to enhance protein production, increase the secretion and the overall yield, etc. Nearly all techniques applied to prokaryotic cells or procaryotic genes and proteins are used within the eucaryotes including classical mutagenesis, random mutagenesis, site-directed mutagenesis, genetic engineering, protein engineering, strain engineering, strain improvement, etc.

3.1 Yeast

The most famous yeast (out of hundreds) involved in food fermentation is *Saccharomyces cerevisiae*, which is commonly known as Baker's yeast or brewer's yeast. Bread production is a real fermentation process including all the properties of successful fermentation including producing useful protein, peptides, CO_2, alcohol, degrading undigestible products to a simpler form, etc. The process needs nothing but water and a place for processing. During the bread production *S. cerevisiae* ferments sugars to produce carbon dioxide and alcohol, contributing to the leavening of the bread and producing alcoholic beverages. During this process different types of protein, peptides and amino acids are produced, as well as many other useful products. One could develop a fermentation process without adding a seed culture (amount of yeast) whereas in bread, barley and other forms used in bread production contain natural yeasts that could start the process. In fact, in a small farm, strains like yeast or a bacteria like Lactobacillus could be simply transferred from one place to others just through handling. At least in the place(s) where bread is prepared, milk is fermented, cheeses are aged, etc.

Before the advent of the science of microbiology, there was no preprepared wet or dry yeast. Wet, dry, and fast-dry yeast can be found in nearly any market. Many companies worldwide are active in the production of starter culture that contains yeast or mold (Mizuochi & Betts, 2022). Nowadays, besides the bread making and alcoholic drink making *S. cerevisiae* is widely used in diverse other applications such as production of single-cell proteins (SCPs) (Amara & El-Baky, 2023; Do et al., 2024), proteins, and peptide production, etc. *S. cerevisiae*, in addition to other yeasts, collectively increase the protein content in bread and in any FFs. Yeast can be found and isolated from diverse environments, particularly from fruits and any sugary environment. There are a respective number of FFs that contain yeast (Anupma & Tamang, 2016; Tamang et al., 2015). *Saccharomyces* yeasts also form a symbiotic relation with bacteria to ferment a particular type of food (Tamang et al., 2015; Witthuhn et al., 2005). Another famous strain is the *P. pastoris*, a methylotrophic yeast that has become popular for producing respective levels of recombinant proteins. It can utilize methanol as a carbon source, allowing for high-density fermentation. It used in producing enzymes, antibodies, and therapeutic proteins such as human growth hormone. *P. pastoris* is often used in biotechnology for protein expression (Vijayakumar & Venkataraman, 2024; Zhong et al., 2014) and can also be cultivated for SCP. *Kluyveromyces lactis* yeast is used for protein production, especially for proteins that require PTMs. *Kluyveromyces marxianus* is able to ferment lactose (Diniz et al., 2012), and is used to produce SCP (Carrasco de Mendoza et al., 1991; Yadav, Bezawada, Ajila, et al., 2014a; b) from dairy waste used in producing lactate dehydrogenase in addition to other enzymes. *S. cerevisiae* produces a 5–10 kDa peptidic fraction responsible for inhibiting malolactic fermentation (Rizk et al., 2016; Rizk et al., 2018). SCP derived from yeast is a highly nutritious protein source that is gaining attention for its potential in food and feed applications. *Candida utilis*, also known as torula yeast, can utilize various carbon sources, including waste products from the food industry, making it a sustainable option for SCP production (Buitrago Mora et al., 2019). Yeast SCP typically contains 40–60% protein by dry weight, making it a rich protein source. Yeast protein is generally rich in essential amino acids, particularly lysine, which is often limited in plant-based proteins. Yeast is a good source of B vitamins (such as B1, B2, B3, B6, and B12) and minerals such as selenium, zinc, and iron. Yeast SCP is widely used in animal nutrition, particularly in poultry, aquaculture, and livestock feeds, as a protein supplement. Yeast-extract powders are used in various food products, including protein bars, snacks, and dietary supplements. They can also enhance flavor because of their umami properties. Yeast SCP can be incorporated into meat analogs in addition to other protein-rich food products, providing texture and nutritional benefits. Yeast is used in FPs for producing biofuels, beside other byproducts, contributing to sustainable practices. *Candida milleri* yeast is often used in sourdough bread fermentation. It works alongside LAB to produce the characteristic sour flavor and texture of sourdough.

3.2 Filamentous Fungi

Filamentous fungi are used extensively in the processing of traditional FF and in producing different types of starter cultures in Asia (e.g., koji). They make several contributions, such as saccharification (Li et al., 2022), liquefaction (Suesse et al., 2016), and as a part of ethanol production and product refining to produce different types of low-alcohol beverages and high-alcohol distilled liquor (Tamang & Samuel, 2010; Suesse et al., 2016; Chen et al., 2014; Lee et al., 2014; Thapa & Tamang, 2004). In Europe, they are used in developing different dairy products, particularly in the ripening processes of various types of cheese (such as Roquefort, Camembert) and for enzyme production (Anupma & Tamang, 2016; Chávez et al., 2011; Spinnler & Gripon, 2004). Several species of filamentous fungi become the vehicle of choice to produce essential microbiota for alcohol-producing dry starters, FM, and cheeses (Tamang & Samuel, 2010; Bourdichon et al., 2012a, b). There are a significant number of the fungal genera involved in FF production. For more details, refer to McGovern (2007), Ashaolu and Reale (2020), Anupma and Tamang (2016)), and the references within. Many of these fungi have diffused to other parts of the world, or local similar or new strains are isolated. Within the genus *Aspergillus* (Hagi et al., 2022), there are important species such as *Aspergillus acidus*, *A. oryzae*, *A. niger*, *A. sojae*, *A. sydowii*, *A. versicolor*, and *A. flavus*, which are used in traditional fermented food (TFF) production such as soy sauce fermentations, miso, sake, awamori liquors, and pu-erh tea (Tamang & Samuel, 2010; Mogensen et al., 2009; Nout, 2000, 2007). The genus *Fusarium* currently contains nearly 200 accepted species, with the most common in the food industry (cheese fermentations), *Fusarium domesticum* (Bachmann et al., 2005; Savary et al., 2023), *F. solani*, and *F. venenatum* (Tong, Chen, et al., 2024a; b). *F. venenatum* is used to produce mycoprotein, the main ingredient in Quorn, a meat substitute. It is cultivated in FPs to create a high-protein, low-fat food product. Species from the *Fusarium* (Savary et al., 2023) genus are recognized as promising sources of several enzymes for industrial applications, as well as "helpers" that promote production of aromas and pigments to the final products. Some of the secondary metabolites produced by *Fusarium* spp. strains are mycotoxins that affect human and animal health, but studies have demonstrated that producing mycotoxins stops under industrial conditions (Anupma & Tamang, 2016; Thrane, 2007). Within the genus *Rhizopus* (Ibarruri & Hernández, 2019), *Rhizopus oligosporus* (Wainaina et al., 2020), *R. microspores* (da Silva et al., 2020), *R. stolonifer*, *R. arrhizus*, *R. delemar* (Janssen et al., 2006), *Rhizopus oryzae* (Sar et al., 2020), and *R. delemar* (Janssen et al., 2006). *Rhizopus* sp., strains were detected in yaoqu from China and banhmen from Vietnam, which are strong amylase producers.(Tamang & Samuel, 2010; Gryganskyi et al., 2018; Thanh et al., 2008).

Aspergillus oryzae, is essential in producing traditional Asian FFs such as soy sauce, miso, and sake. It helps in breaking down starches and proteins into sugars and amino acids. It produces enzymes such as amylases and proteases, which facilitate the food fermentation. It is also used in fermenting a typical traditional product such as tempeh, a fermented soybean food. *R. oryzae* is considered to be a Generally Regarded As Safe filamentous fungus (Anupma & Tamang, 2016; Tamang & Samuel, 2010; Tamang et al., 2015; Ly et al., 2018; Lv et al., 2012; Bora et al., 2016). Within the genus *Penicillium*(Engel et al., 1982), *Penicillium camemberti* (Le Dréan et al., 2010), *P. biforme*, *P. fuscoglaucum*, *P. palitans*, *P. commune* (Jurado & Vicente, 2020) (as the wild-type "ancestor" of *P. camemberti*), *P. caseifulvum* (Suhr et al., 2002), *P. nalgiovense* (Färber & Geisen, 1994), *P. chrysogenum* (Hidalgo et al., 2014), *P. solitum*, *P. nordicum*, *P. expansum*, and *P. roqueforti* are the most common species involved in the microbiota of several types of cheeses, such as Camembert, Brie, and "blue cheeses," as well as fermented mild-ripened sausages (such as salami) and raw dry- or smoke-cured meat products (such as ham) (Pitt et al., 1986; Polonelli et al., 1987; Giraud et al., 2010). *Penicillium camemberti* is used in cheese production (e.g., Camembert and Roquefort cheeses). These molds participate in the development of flavor, aroma, and texture in cheeses through their enzymatic activity. Other different fungal genera exist in food, such as *Actinomucor* (Yin et al., 2021; Zhang et al., 2022), *Amylomyces*, *Bjerkandera* (Feijoo et al., 1999), *Brettanomyces* (Péter et al., 2017), *Candida* (Dishan et al., 2025), *Cryptococcus* (Castanha et al., 2014), *Cyberlindnera* (van Rijswijck et al., 2017), *Cystofilobasidium*, *Debaryomyces* (Leclercq-Perlat et al., 2000), *Dekkera* (Branco et al., 2019), *Galactomyces* (Grygier et al., 2020), *Geotrichum* (Grygier et al., 2020; Lessard et al., 2014), *Cladosporium* (Messini et al., 2017), *Guehomyces*, *Hanseniaspora* (Branco et al., 2015), *Hansenula* (Cui et al., 2020), *Hyphopichia*, *Issatchenkia* (Miao et al., 2018), *Kazachstania*, *Kluyveromyces* (Arfi et al., 2002), *Lecanicillium*, *Lachancea* (Bellut et al., 2020), *Metschnikowia* (Wu, Li, et al., 2025a), *Monascus* (Xia et al., 2020), *Mucor* (Hermet et al., 2012), *Neurospora* (Poltorak et al., 2024), *Pichia* (Kahve, 2023), *Rhodotorula* (Ma et al., 2025), *Rhodosporidium* (Liu et al., 2021), *Saccharomyces* (Hansen & Jakobsen, 2001),

Saccharomycodes, *Saccharomycopsis* (Yu et al., 2013), *Schizosaccharomyces* (Huh et al., 2018), *Schwanniomyces*, *Sporobolomyces* (Dagher et al., 2013), *Scopulariopsis*, *Sporendonema*, *Starmerella*, *Torulaspora* (Andrade et al., 2019), *Trametes* (Wagner et al., 2024), *Trichosporon* (Vorbeck & Cone, 1963), *Trigonopsis*, *Torulopsis*, *Ustilago*, *Wickerhamomyces*, *Yarrowia* (Suzzi et al., 2001), *Zygosaccharomyces* (Ogata & Kuroki, 2021), and *Zygotorulaspora* (McGovern, 2007, 2020; Ashaolu & Reale, 2020; Anupma & Tamang, 2016).

As above, many fungal genus and strains are involved in diverse FPs. Some are common in a particular product because of the condition, the food origin and type, the seed culture, etc. But there are a number of other fungi, yeasts, and bacteria that exist as well. One should highlight that some strains exist under the control of others or in a symbiotic relationship. Other forms of relations that govern the existence of different microbes and different strains are also reported. Using a single strain will not give the same result for any of the FFs that need a consortium of strains that are synergistic, symbiotic, or antagonistic to each other. To obtain the proper taste, aroma, quality, etc., a seed culture from the original FF is probably used. One can do that from obtaining FFs from local markets, whereas most industrially produced products are passed through a strain deactivation process, in addition to other used strategies, which are applied to guarantee blocking of microbial activity, including their reproduction.

The ancient microbiological tactics in many cases prove to be efficient. In fact, those tactics are optimized through the ages by the producers. The techniques are transferred from hand to hand, from generation to another and so on. For more details one should read about the history of some of those products such as Kefir, different cheeses, fermented milk, soybeans, etc. Each country has unique food resources. For that the opportunity of each country to produce their own special fermented products is high. In addition, some countries nowadays build their economy on cheese products as a solid part of its resources. Moreover, governments must support small producers to be a unique source of a special kind of different FFs or at least make efforts to document this art carefully before it disappears.

Neurospora sitophila is used in producing certain FFs, such as the traditional Chinese fermented rice dish known as "jiuniang." It contributes to food fermentation by breaking down starches and sugars. The coexistence of some bacterial species could regulate some fungal species that might be harmful. This highlights the usefulness of using the microbiota of a safely produced traditional food over using a single isolated microbe. For example, *A. flavu*, produces aflatoxin and *A. versicolor* toxic metabolites. The coexistence of LAB may play an antagonistic role against *A. species* inside the structure of the food matrix that potentially leads to reducing aflatoxin production (Bachmann et al., 2005; Adebo et al., 2019; Shakeel et al., 2018; Grafenham et al., 2011). In fact, this point still needs more investigation. Anyhow, FF is safe depending on the activity of LAB and most probably only safe yeast strains that coexist in large quantities whereas numbers of other microbes are significantly reduced.

4 Fermented Cheese Products

Cheese is a dairy product produced from the curd of milk, which is formed by coagulating milk proteins (casein), which are then separated from the liquid whey. The milk used can be from cows, goats, sheep, and buffalo. Some cheese such as Kharish cheese does not include any additives to coagulate the milk protein. No rennet or acid is added to the milk to cause it to curdle, forming curds and whey. In some types of cheese curds are cut into smaller pieces to facilitate whey drainage. Others are cooked or washed to achieve the desired texture and moisture content. The whey is drained off, leaving behind the curds. Salt is added to preserve the cheese or in another word to incapacitate the growth of any undesired microbes. Also, salt gives a taste to the cheese. Cheeses are aged for varying periods and could be injected or covered by some microbes to give a special taste, aroma, flavor, color, etc. Hundreds of varieties of cheese are produced from the milk of cows, goats, sheep, water buffalo, horses, llamas, and yaks. Products vary according to the selection and treatment of the milk; adjustment of its fat content; heating or pasteurizing; and adding enzymes or cultures of bacteria, molds, or yeast. Curd formation varies with changes in temperature, time, acidity for coagulation, proportions of rennet and acid, and the speed and extent of removal of the whey. Ripening and curing consist of biological and chemical changes that occur in the cheese and are affected by the moisture content, acidity, texture, shape, size, and microorganisms in the cheese. These changes alter the consistency as well as the flavor of the cheese. Before ripening, cheese is said to be fresh or green; after ripening it is called cured, aged, or ripened. The chemical changes may be classified broadly as the breakdown of fats to fatty acids; proteins to amino acids; and lactose to products such as lactic, acetic, and propionic acids, diacetyl, and carbon dioxide. Flavorful products of ripening include volatile fatty acids, ketones, esters, alcohols, peptides (Peres Fabbri et al., 2024), amino acids, ammonia, and hydrogen sulfide. In modern factories, cheese is mass produced according to standardized recipes and techniques that result in a more uniform product. Diverse production was of primary economic importance for hundreds of years in the Netherlands, France (Colombo et al., 2021), Italy (Anelli et al., 2018), and Switzerland (De Respinis et al., 2023). Certain cheeses are historically associated with particular areas. Maintaining quality is enhanced by a variety of procedures, including heating and kneading (Italian provolone) and soaking in brine (Egyptian Domiati). The loosely knit structure of French Roquefort develops blue mold only in a cool, wet atmosphere such as occurs naturally in the caves of Roquefort (Encyclopædia Britannica, 2013a).

4.1 Blue Cheese

Classic cow's-milk cheese of Normandy (Desmasures et al., 1998) is named for a village in that region; its characteristic creamy, ivory-colored interior and downy white surface, resembling that of Brie, result from the *Penicillium camemberti* mold with which the curd is treated. Camembert curd is customarily shaped in disks; by the action of the mold, it ripens within 6–8 weeks. The flavor varies from mild to strong as the cheese ripens. Camembert is exported worldwide and reproduced in many countries, although the prototype from Normandy. It dates to the late seventeenth century, and remains unparalleled (Pitt et al., 1986; Polonelli et al., 1987; Encyclopædia Britannica, 2013a; Mei et al., 2014). The release and migrating taste-active compounds during ripening have been studied (Engel et al., 2001). Yeast and bacteria exist in the Camembert cheese microbiota (Addis et al., 2001; Roostita & Fleet, 1996; Schmidt & Daudin, 1983). *Penicillium camemberti* has antigenic characteristics against common cheese contaminants (Polonelli et al., 1987).

Roquefort is generally formed in cylinders. The white paste of the interior is marbled with blue mold; the cheese is wrapped in foil (Ropars et al., 2017; Siemens & Zawistowski, 1993; Encyclopædia Britannica, 2013a; Engel and Prokopek 1979, 1980; Vallone et al., 2018). All blue mold cheeses are always made with *P. Roquefort*, which is inoculated and grows inside the product. *P. camemberti* is used in mold-ripened soft cheeses such as Camembert and Brie. It is inoculated together with a lactic starter culture or, alternatively, sprayed on the product and grows on the surface of the cheese piece (Ropars et al., 2017). *P. roqueforti* can be produced as secondary metabolite components that fall within the category of mycotoxins (Vallone et al., 2018). Today, safe starter cultures are available that do not have the ability to produce most of these mycotoxins, and the presence of roquefortine C and isofumigaclavines in the amounts usually found on cheeses does not seem to represent a risk for human health (Siemens & Zawistowski, 1993; Engel and Prokopek 1979, 1980; Hymery et al., 2014).

Stilton is a classic English blue cheese produced from cow's milk, named for the village in Huntingdonshire where, according to tradition, it was first sold in the late eighteenth century. Originally made exclusively from the milk of Shorthorn cows, as some still is, Stilton has a rich, intense flavor that is less salty than that of Roquefort. The texture is semisoft and moist or creamy, as the cheese is not pressed, and it is ivory yellow in color, streaked with bluish-green veins of *Penicillium glaucum* mold. The beige rind is wrinkled and crusty. Stilton is aged for 4–8 months, sometimes longer (Encyclopædia Britannica, 2013a). Strains from fungus (*P. roqueforti*), yeast (*Yarrowia lipolytica, Debaryomyces hansenii, Trichosporon ovoides, and Candida catenulate*) are existed (Gkatzionis et al., 2014).

Gorgonzola cheese takes its name from the town of Gorgonzola, in the Lombardy region, northern Italy, northeast of Milan. This cheese is soft when freshly made; after being drained twice, it is then oven dried for 20 days and pierced with copper needles to promote the internal formation of the characteristic greenish-blue mold

(*P. roqueforti*). This cheese is also made in other parts of Lombardy and in Piedmont and Emilia-Romagna (Encyclopædia Britannica, 2013a; Vallone et al., 2018).

Brie cheese is a soft-ripened cow's-milk cheese named for the district in northeastern France in which it is made. The preliminary soft curd of Brie is molded in flat, pancake-like rounds. After the curd becomes firm, the cheese is sprayed with spores of *Penicillium candidum*, a mold that forms a thin, downy white crust. The cheese ripens within 3–4 weeks. The interior is an ivory-colored paste that has a semisoft, chalky center when young but becomes soft and creamy throughout when fully ripened. Highly perishable, it remains in this state only briefly; at advanced stages of ripeness ammoniated odors develop, and the cheese begins to dry out and harden. Brie originated near Paris, where unpasteurized, farm-produced Brie de Meaux and Brie de Melun are the most highly esteemed versions. Most Brie today is made in factories from pasteurized milk, which prolongs the life of the cheese but weakens its flavor. Brie is now widely imitated in many cheese-producing countries (Encyclopædia Britannica, 2013a).

4.2 Cottage

Cottage cheese is a fresh, soft, unripened cheese consisting of curds of varying sizes, usually mixed with some whey or cream. It is white and mild but faintly sour in taste. In commercial diverse production, the curds are derived from pasteurized skim milk or reconstituted, low-fat milk products. The whey is drained but not pressed from the curds, thus leaving a certain amount of liquid. In this form, cottage cheese has a low-fat content and is a popular food in low-fat diets. If cream is added and the product contains 4% or more fat, it is sold as creamed cottage cheese (Encyclopædia Britannica, 2013a). *Galactomyces geotrichum* mold isolated from a traditional fried cottage cheese produced omega-3 fatty acids (Grygier et al., 2020). Regulations are applied to keep the amount of yeast and fungi in cottage cheese at acceptable levels (Powers et al., 1971).

4.3 Cheddar

Cheddar is a hard cow's-milk cheese named for the district of its origin in the southwestern county of Somerset, England. Cheddar is one of England's oldest cheeses. The original, so-called farmhouse variety remains in limited production in modern times. In the traditional method of cheddar manufacturers, the firm curd is cut, or "cheddared," into small bits to drain the whey and then pressed firmly into cylinders. The cheese, a light orange-yellow in color, is wrapped in thin muslin and coated with wax. It is aged for a minimum of 3–6 months, preferably 1.5–2 years.

Its mellow, rich flavor becomes more intense with age but should not be bitter. Factory-made cheddar, now by far the norm, attains a certain standard in England and in some North American versions, notably Canadian, Vermont, New York, and Wisconsin (Encyclopædia Britannica, 2013b). Yeasts belonging to *Debaryomyces* and *Candida*, *Cryptococcus*, *Rhodotorula*, *Yarrowia*, *Pichia*, *Trichosporon*, *Torulaspora*, *Issatchenkia*, *Saccharomyces*, and *Zygosaccharomyces* have been reported (Ferreira & Viljoen, 2003; Shakeel Ur et al., 2003; Viljoen & Greyling, 1995).

4.4 Feta

Feta is a fresh, white, soft, or semisoft cheese of Greece, originally made exclusively from goat's or sheep's milk but in modern times containing cow's milk. Feta is not cooked or pressed but is cured briefly in a brine solution that adds a salty flavor to the sharp tang of goat's or sheep's milk. The making, or so-called pickling, of feta is an ancient practice. It is considered the national cheese of Greece, but feta is now made in many parts of the world. The best versions adhere to the original ingredients. Exported feta is usually pasteurized, but in Greece the cheese is still made in the traditional manner by mountain farmers. When stored in a liquid bath of brine, water, or milk, it will keep for months; otherwise it dries out quickly (Encyclopædia Britannica, 2013a). Fungi is part of the cheese biota (Belessi et al., 2008) and the yeast as well (Bintsis et al., 2003; Fadda et al., 2001; Psomas et al., 2001, 2003; Vivier et al., 1994).

4.5 Gouda

Gouda cheese is a semisoft cow's-milk cheese of the Netherlands, named for the town of its origin. Gouda is traditionally made in flat wheels, each with a thin natural rind coated in yellow paraffin. A smaller variant is called baby Gouda. Gouda has a smooth-textured interior of a pale ivory color. Flavors are bland and creamy, except for aged Gouda, which is darker gold in color, stronger and saltier in flavor, and harder in texture. Low-fat Goudas are also produced. Gouda is one of the oldest of the European cheeses, probably dating back to the twelfth century, and is widely imitated in other cheese-producing countries. True Gouda has "Holland" stamped on the rind; farm-made Gouda bears an imprint of the word "Boerenkaas," which means "farmer's cheese."(Encyclopædia Britannica, 2013a). Yeast (Viljoen & Greyling, 1995) such as *S. cerevisiae* and *Yarrowia lipolytica*, *Kluyveromyces marxianus*, *Torulaspora delbrueckii*, *Rhodotorula glutinis*, *Cryptococcus albidus*, and *Candida catenulate* are part of Gouda cheese microbiota (Welthagen & Viljoen, 1998).

4.6 Cave Cheese

Apulian cave cheese was originally produced in Italy from raw milk of sheep or cows, a semi-hard cheese, aged in natural cellars or tufa caves and *Penicillium gravinicasei* is part of its microbiota (Anelli et al., 2018; Martin & Cotter, 2023). Also known as "Caciocavallo Podolico," it originates from the Apulia region (Puglia) in southern Italy. Caciocavallo Podolico produced from the milk of Podolica cows, a breed native to the region. The cheese has a rich, buttery flavor with a slightly tangy and nutty profile. The taste can vary depending on the aging step. It has a firm and elastic texture, which becomes crumblier as it ages. Typically, it is shaped like a teardrop or a gourd and is often tied with a rope for hanging during the aging step. The cows graze on the rich pastures of the Apulian countryside, which contributes to the cheese's unique flavor. The milk is warmed and then curdled using natural rennet. The curds are drained and then placed into molds to form the characteristic shape. The cheese is aged in caves or cellars that provide a controlled environment that enhances its flavor and texture. Aging can last from a few months to over a year. Producing Apulian cave cheese is deeply rooted in local traditions and usually made by small-scale artisanal producers (Vacca et al., 2024).

Italy cave cheese is produced from raw milk of sheep or cow, semi-hard cheese, ripened in natural caves where strains like *Aspergillus westerdijkiae, Penicillium biforme, Penicillium chrysogenum, Aspergillus puulaauensis* can be grown well. (Martin & Cotter, 2023; Anelli et al., 2019). Cave cheese refers to a type of cheese that is aged in natural caves or cave-like environments, which can significantly influence its flavor, texture, and overall characteristics. The unique conditions of caves, such as temperature, humidity, and airflow, create an ideal environment for aging cheese. These conditions can promote the growth of specific molds and bacteria that participate in the cheese's flavor profile. Cave-aged cheeses often have complex and robust flavors that can include earthy, nutty, and sometimes tangy notes. The specific flavor will depend on the type of cheese and the conditions of the cave. The texture of cave cheese can vary widely, from creamy and soft to hard and crumbly, depend on the cheese variety and aging process. Cave cheeses may have a natural rind that develops during the aging step, often featuring a unique coloration and texture because of the molds and bacteria present in the cave environment. Cave cheeses can be produced from various types of milk, including cow, goat, or sheep milk, depending on the cheese variety. The cheese is made using traditional methods, where the milk is curdled, drained, and formed into wheels or blocks. After the initial production, the cheese is moved to a cave or cave-like aging facility. Aging can last from several weeks to several months or even years, allowing the cheese to develop its unique characteristics. Many cave cheeses are produced using traditional methods that were passed down through generations. The use of caves for aging is usually rooted in local customs and practices. Different regions may have their own specific types of cave cheese, each with distinct flavors and characteristics that depend on local milk sources and cave conditions. A famous blue cheese from France is aged in the natural caves of the Roquefort-sur-Soulzon region. It has a nutty and complex flavor profile and characteristic blue veins. Divle cave cheese is produced in Turkey from raw milk ripened in caves in goat skin bags *Penicillium* spp., *Mucor racemosus, Trichoderma* spp., *Yarrowia lipolytica* can be found in cave chees (Martin & Cotter, 2023; Ozturkoglu-Budak et al., 2016).

4.7 Artisanal Minas

Artisanal Minas Cheese (AMC) is made from raw cows' milk, without previous heat treatment, rennet, and endogenous culture (*Debaryomyces hansenii, Yarrowia lipolytica, Aspergillus fumigatus, P. roqueforti, Penicillium* spp., *Scopulariopsis* spp.) (Martin & Cotter, 2023; de Souza et al., 2021). AMC is also known as "Queijo Minas Artesanal," and is a traditional cheese from the Minas Gerais state in Brazil. Minas cheese is a fresh, soft cheese that can be produced from cow's milk, goat's milk, or a combination. The artisanal version is typically made using traditional methods. It has a mild, slightly tangy flavor that can vary depending on the milk source and the specific production techniques used. The taste is usually described as creamy and slightly sweet. The cheese is soft and moist, with a smooth texture. It can be found in various forms, including fresh (unaged) and aged varieties, with the aged versions being firmer and crumblier. AMC is usually white or pale yellow and usually sold in round or rectangular shapes. The cheese is traditionally produced from raw milk from local dairy farms, which contributes to its unique flavor profile. The milk is heated and then curdled using rennet or natural acids. The curds are drained to remove excess whey (Aragão et al., 2025; Martins et al., 2015).

4.8 Cotija

Cotija cheese is produced in Mexico from raw milk, and is a semi-hard cheese where yeasts like *Scheffersomyces*, *Saccharomyces*, *Candida*, and *Galactomyces* can be isolated (Martin & Cotter, 2023; Murugesan et al., 2018). Cotija cheese has a crumbly texture and salty flavor. Cotija is a fresh, crumbly cheese made primarily from cow's milk. It is usually compared with feta cheese because of its similar texture and saltiness. The cheese has a strong, salty flavor that can vary in intensity depending on the aging step. Fresh Cotija is milder, whereas aged Cotija has a more pronounced taste. Cotija's typical crumbliness means that it can be easily crumbled or grated, making it a versatile ingredient in various dishes. The cheese is usually white and can be found in blocks or crumbled form. Aged Cotija may have a slightly firmer texture. Although Cotija is traditionally produced from cow's milk, some variations may use goat's milk. The milk is heated and curdled using rennet or natural acids, forming curds. The curds are drained and may be pressed lightly to remove excess whey, resulting in a moist cheese. Cotija can be consumed fresh or aged for a few months. Fresh Cotija is typically used immediately. Cotija cheese is commonly used as a topping for various Mexican dishes, such as tacos, enchiladas, and salads. It adds a salty and creamy element to these dishes. It is usually sprinkled over elote (Mexican street corn), used in quesadillas, or incorporated into salsas and dips. If Cotija is unavailable, feta cheese is used as a substitute because of its similar texture and flavor profile. Cotija cheese is named after the town of Cotija in the state of Michoacán, Mexico, where it was produced for centuries. It is a staple in Mexican cuisine and usually used in traditional dishes, reflecting the culinary heritage of the region (Martin & Cotter, 2023; Murugesan et al., 2018)..

5 Kefir

Kefir is a fermented dairy product. It was coined in Caucasus in the 1900s. It has a tangy flavor. It is made by FM with kefir grains that are a combination of bacteria and yeast that form a gelatinous structure. The most common type is produced from cow's, goat's, or sheep's milk. It is creamy, tangy, and slightly effervescent. It is consumed as a drink, used in smoothies, salad dressings, or as a base for sauces. After reaching the USA and South America, a nondairy version made by fermented sugar water, fruit juice, or coconut water with water kefir grains was established. The product is sweet and slightly fizzy, with a fruity taste depending on the added ingredients. It is a refreshing beverage, often flavored with fruits or herbs. A thicker version of milk, kefir is similar to sour cream or yogurt. It is used as a topping, in dips, or as a base for dressings. It contains a symbiotic culture of bacteria and yeast named symbiotic culture of bacteria and yeast that ferments the milk or sugar water. Depending on the type of the kefir being made, the milk type is selected and is combined with kefir grains and left to ferment at room temperature for 12 to 48 h. The fermentation time affects the flavor and thickness. Sugar water is mixed with water kefir grains and fermented for 24–48 h. Flavorings like fruits or herbs can be added during fermentation. After fermentation, the kefir is strained to remove the grains, which can be reused for the next batch. Kefir is rich in beneficial bacteria and yeast that can support gut health and improve digestion. It is a good source of protein, calcium, B vitamins, in addition to other essential nutrients. The fermentation breaks down lactose, making kefir easier to digest for those who are lactose intolerant. The probiotics in kefir may help to enhance the immune system and reduce the risk of infections. Some studies suggest that kefir might have anti-inflammatory effects and contribute to improving the overall health quality. Kefir has many medicinal properties. It can inhibit angiotensin-converting enzyme activity and reduce systolic arterial pressure in hypertensive rats (Amorim et al., 2019; González-Orozco et al., 2024). The grains are composed of colonies of living lactobacilli (Hong et al., 2010) and yeast (Romanin et al., 2010; Lu et al., 2014; Gientka et al., 2017; Lo et al., 2021). Yeasts found in kefir include *Candida kefir*, *S. cerevisiae*. *Kluyveromyces marxianus* (González-Orozco et al., 2024; Lo et al., 2021), *Saccharomyces fragilis*, *Torulaspora delbrueckii*, and *Kazachstania unispora* (Guzel-Seydim et al., 2011). In one study 21 yeast strains from the genera

Saccharomyces, *Kluyveromyces*, and *Issatchenkia* were found (Romanin et al., 2010). The kefir grains are initially created by auto-aggregations of *Lactobacillus kefiranofaciens* (Vancanneyt et al., 2004) and *Saccharomyces turicensis* (Wyder et al., 1999), *Kluyveromyces marxianus* (Lo et al., 2021), and others (Lu et al., 2014). They are biofilm producers. In addition, they can adhere multiply to the surface to become a three-dimensional micro-colony (Wyder et al., 1999; Wang et al., 2012). Milk used in kefir preparation includes milk from cows (García Fontán et al., 2006; Pires et al., 2021), goats (Wang et al., 2021; Sun et al., 2025), or sheep. Kefir grains (Wang et al., 2021) are similar to a thin yogurt or ayran (Ferreira et al., 2010; Dimitrellou et al., 2015; Dallas et al., 2016). They live together in a symbiotic relationship and ferment milk at room temperature. Traditional kefir was made in goatskin bags that were hung near a doorway. The bag was knocked by anyone passing by to keep the milk and kefir grains well mixed. Such a practice is usually also done by traveler Bedouin in the Middle East. During food fermentation, the lactose is digested, which makes kefir ideal for those who are lactose intolerant. Kefir possesses natural antibiotics and is rich in vitamins B1 and B12, calcium, folic acid, phosphor, and vitamin K. Peptidomic analysis reveals proteolytic activity of kefir microorganisms on bovine milk proteins (Dallas et al., 2016) (proteolysis) (Ferreira et al., 2010).

6 Fermented Soybeans

Soybeans can be examined as an example of the early recognition of the benefits of fungi that are involved in traditional food. Fermented soy foods are associated with decreased low-density lipoprotein and total blood cholesterol levels. It also reduces the risk of the other vascular effects associated with reduced risk of coronary heart disease. Tofu, tempeh, miso, soy flour, soy milk, and soy nuts are examples of soy foods. These foods contain isoflavones, estrogen-like compounds. Isoflavones are thought to be responsible for beneficial cardiovascular effects (Cabezudo et al., 2022).

6.1 Tempeh or Tempe

Tempeh is produced from fermented cooked soybeans with the mold *Rhizopus oligosporus* or the *Rhizopus oryzae* and is used in FP (known like tempeh starter) (Moreno et al., 2002; Supriyanto et al., 1994). The food fermentation binds the soybeans into a firm cake, enhancing its nutritional value and flavor. The place of production is in Indonesia for thousands of years. It remains a major food staple and an inexpensive source of dietary protein. It is being spread to other countries such as Malaysia and the Netherlands. Fermented soybeans in tempeh contain vitamin B12, a byproduct of the FP. B12 is important, especially for vegetarians and in particular for pregnant and lactating women. That can lead to anemia and irreversible nerve damage. B12 is also a necessary vitamin for pregnant and lactating women. Tempeh enhances the body absorbing isoflavones. Tempeh is high in fiber, enzymes, manganese, and beneficial bacteria. It is a good digestible protein source and it is a great meat substitute (high in protein). It is an excellent source of calcium and contains all essential amino acids. It can contain a blend of grains, beans, or other vegetables. *Rhizopus oligosporus* is used as tempeh starter (Moreno et al., 2002; Supriyanto et al., 1994). The food fermentation process also removes the enzyme inhibitors that occur naturally in soybeans. Tempeh-producing fungi are *Fusarium* sp. BLB (Sugimoto et al., 2007). It is high in dietary fiber, which supports digestive health. Tempeh may help to lower cholesterol levels and improve heart health because of its healthy fat content and fiber. The calcium and vitamin K content in tempeh can contribute to bone health. High in protein and fiber, tempeh can help to promote satiety and support weight management. Tempeh can be sliced or cubed and added to stir-fries for a protein boost. It can be crumbled or diced and used in salads for added texture and flavor. Tempeh is rich in protein (Wu, Wee, et al., 2025b), and food fermentation increases the bioavailability of amino acids. It also produces BAPs, which may have health benefits, such as anti-inflammatory properties (Wu, Wee, et al., 2025b).

6.2 Tofu

Tofu is a soft, bland, custard-like food product produced from soybeans. It is an important source of protein in the cuisines of China, Japan, Korea, and Southeast Asia. Tofu is believed to date from the Han Dynasty (206 BC–AD 220). Tofu is produced from dried soybeans that are soaked in water, crushed, and boiled. The mixture is separated into solid pulp (okara) and soy "milk." Coagulants such as calcium and magnesium chlorides and sulfates are added to the soy milk to separate the curds from the whey. The soy milk is poured into molds to allow the carbohydrate-laden whey to drain off. The resultant soft cakes are cut into squares and stored under water until sold, in bulk or in individual water-filled tubs. Tofu is 6–8% protein and high in calcium, potassium, and iron. Okara and soy whey, the byproducts of tofu manufacture, are also eaten. The skin that forms on top of the soy milk as it stands is removed and dried in sheets for use in vegetarian dishes. Tofu may be grilled, deep-fried, simmered, stir-fried, steamed, or eaten fresh (Encyclopædia Britannica, 2013c).

Critical Points Extracted from Fermented Food

Some critical points concerning the fermented food could highlight how one could benefit from them.

1. FFs have beneficial microbes. Some of them can be considered as probiotic microbes.

2. FFs are so variant where any change of the chemical, physical, and biological condition could lead to a different product. For those who search for a succesful business, FFs consider as an open source for different consumers' accepted products. The huge number of fermented cheeses are a good example.
3. Changes that happen during FP have a double benefit. The first is that the food becomes more digestible; the second is the beneficial products and microbes that are produced during the FP.
4. Fungi, yeasts, and mushrooms are involved in the production of many compounds, peptides, active proteins, amino acids, etc.
5. Each ethnic group, tribe, country, etc. has their specific types of FFs, which are probably homogenous with their beliefs, morale, lifestyle, etc.
6. There are some microbial species that are usually beneficial and can be used at any time for an FP. *Saccharomyces cerevisiae* is an example.
7. If an FP fails, that means a problem existence in the process and did not mean the inability of the microbe(s) to ferment food.
8. The best way to get the same quality of a TFF is to use a starter seed culture from a previously successful sample. In addition, the same conditions should be followed.
9. Different beneficial microbes from various prokaryotic and eukaryotic species could be involved in processing a single FF. Some microbes will control others whereas others will be in a symbiotic relationship.
10. A huge number of enzymes, peptides, active protein vitamins, compounds, etc., were produced through FFs.
11. The most successful BMs are used to conduct a new single-cell-based FP, or as an SCP or mycoprotein.
12. FFs are treasures in the hand of any of us and each formula considers an additional product. This formula could exist in any place around the world but could benefit everyone here and there.

7 Conclusion

Fungi are used as powerful platforms for producing imported proteins and peptides, with applications in various fields, including biotechnology, the pharmaceutical industry, and food production. Research and advances in genetic technology and fermentation continue to increase the effectiveness of fungi to stand alone as a cell factory. FFs had been invented, developed, distributed, through several successive generations who had been interested in preserving their products and developing new dishes. Collectively, across diverse places throughout the world, one could observe the existence of several types and varieties of FFs. They are currently many and various. They are related to all food products such as milk products, for example, various types of cheese; kefir, sour milk, yogurt, etc.; marine products such as fishes, Crustacea, etc.; agriculture products such as different types of meat products, fruits, vegetables, seeds, etc.; other food and drink variants are also included. Food fermentation breaks down proteins into amino acids (Jia et al., 2011; Kasperek et al., 2024), including glutamate, which contributes to the umami flavor (Kim et al., 2017). BAPs with antioxidant properties may also be produced. FFs are collectively useful. Inspired by the negative side effects of alcoholic products, they have numerous positive effects on health. They support the GIS with permanent probiotic microbes or with microbes that could support the digestion of any meal. Nowadays, FF has entered a new horizon: medicinal and pharmaceutical products. Many useful peptides and active proteins as well as other structures are identified. The margin of the FFs market is huge. Governors are invited to support any local fermented products. This is the best way to preserve all local FFs.

References

Addis, E., Fleet, G. H., Cox, J. M., Kolak, D., & Leung, T. (2001). The growth, properties and interactions of yeasts and bacteria associated with the maturation of Camembert and blue-veined cheeses. *International Journal of Food Microbiology, 69*, 25–36. https://doi.org/10.1016/s0168-1605(01)00569-4

Adebo, O., Kayitesi, E., & Njobeh, P. (2019). Reduction of mycotoxins during fermentation of whole grain sorghum to whole grain ting (a southern African food). *Toxins, 11*, 180. https://doi.org/10.3390/toxins11030180

Agarbati, A., Canonico, L., Marini, E., Zannini, E., Ciani, M., & Comitini, F. (2020). Potential probiotic yeasts sourced from natural environmental and spontaneous processed foods. *Foods, 9*, 287. https://doi.org/10.3390/foods9030287

Aidoo, K. E., Nout, M. J., & Sarkar, P. K. (2006). Occurrence and function of yeasts in Asian indigenous fermented foods. *FEMS Yeast Research, 6*, 30–39. https://doi.org/10.1111/j.1567-1364.2005.00015.x

Akabanda, F., Owusu-Kwarteng, J., Tano-Debrah, K., Glover, R. L., Nielsen, D. S., & Jespersen, L. (2013). Taxonomic and molecular characterization of lactic acid bacteria and yeasts in nunu, a Ghanaian fermented milk product. *Food Microbiology, 34*, 277–283. https://doi.org/10.1016/j.fm.2012.09.025

Albergaria, H., Francisco, D., Gori, K., Arneborg, N., & Gírio, F. (2010). Saccharomyces cerevisiae CCMI 885 secretes peptides that inhibit the growth of some non-Saccharomyces wine-related strains. *Applied Microbiology and Biotechnology, 86*, 965–972. https://doi.org/10.1007/s00253-009-2409-6

Amaike Campen, S., Lynn, J., Sibert, S. J., Srikrishnan, S., Phatale, P., Feldman, T., Guenther, J. M., Hiras, J., Tran, Y. T. A., Singer, S. W., et al. (2017). Expression of naturally ionic liquid-tolerant thermophilic cellulases in Aspergillus Niger. *PLoS One, 12*, e0189604. https://doi.org/10.1371/journal.pone.0189604

Amara, A. A., & El-Baky, N. A. (2023). Fungi as a source of edible proteins and animal feed. *Journal of Fungi, 9*. https://doi.org/10.3390/jof9010073

Amara, A. A., & Shibl, A. (2015). Role of probiotics in health improvement, infection control and disease treatment and management. *Saudi Pharmaceutical Journal, 23*, 107–114. https://doi.org/10.1016/j.jsps.2013.07.001

Amorim, F. G., Coitinho, L. B., Dias, A. T., AGF, F., Monteiro, B. L., LCD, R., TMC, P., Campagnaro, B. P., De Pauw, E., Vasquez, E. C., & Quinton, L. (2019). Identification of new bioactive peptides from kefir milk through proteopeptidomics: Bioprospection of antihypertensive molecules. *Food Chemistry, 282*, 109–119. https://doi.org/10.1016/j.foodchem.2019.01.010

Andrade, R. P., Oliveira, D. R., Lopes, A. C. A., de Abreu, L. R., & Duarte, W. F. (2019). Survival of Kluyveromyces lactis and Torulaspora delbrueckii to simulated gastrointestinal conditions and their use as single and mixed inoculum for cheese production. *Food Research International, 125*, 108620. https://doi.org/10.1016/j.foodres.2019.108620

Anelli, P., Peterson, S. W., Haidukowski, M., Logrieco, A. F., Moretti, A., Epifani, F., & Susca, A. (2018). Penicillium gravinicasei, a new species isolated from cave cheese in Apulia, Italy. *International Journal of Food Microbiology, 282*, 66–70. https://doi.org/10.1016/j.ijfoodmicro.2018.06.006

Anelli, P., Haidukowski, M., Epifani, F., Cimmarusti, M. T., Moretti, A., Logrieco, A., & Susca, A. (2019). Fungal mycobiota and mycotoxin risk for traditional artisan Italian cave cheese. *Food Microbiology, 78*, 62–72. https://doi.org/10.1016/j.fm.2018.09.014

Anukam, K. C., & Reid, G. (2009). African traditional fermented foods and probiotics. *Journal of Medicinal Food, 12*, 1177–1184. https://doi.org/10.1089/jmf.2008.0163

Anupma, A., & Tamang, J. P. (2016). Diversity of filamentous fungi isolated from some amylase and alcohol-producing starters of India. *Frontiers in Microbiology, 11*. https://doi.org/10.3389/fmicb.2020.00905

Aragão, M. O. P., Lima, F. R., Passamani, F. R. F., Santos, M. A. A., Rezende, J. P., & Batista, L. R. (2025). Fungal and bacterial diversity present on the rind and core of Natural Bloomy Rind Artisanal Minas Cheese from the Canastra region, Brazil. *Food Research International, 202*, 115724. https://doi.org/10.1016/j.foodres.2025.115724

Arfi, K., Spinnler, H. E., Tache, R., & Bonnarme, P. (2002). Production of volatile compounds by cheese-ripening yeasts: Requirement for a methanethiol donor for S-methyl thioacetate synthesis by Kluyveromyces lactis. *Applied Microbiology and Biotechnology, 58*, 503–510. https://doi.org/10.1007/s00253-001-0925-0

Ashaolu, T., & Reale, A. (2020). A holistic review on Euro-Asian lactic acid bacteria fermented cereals and vegetables. *Microorganisms, 8*, 1176. https://doi.org/10.3390/microorganisms8081176

Ashaolu, T. J., Varga, L., & Greff, B. (2025). Nutritional and functional aspects of European cereal-based fermented foods and beverages. *Food Research International, 209*, 116221. https://doi.org/10.1016/j.foodres.2025.116221

Bachmann, H. P., Bobst, C., Bütikofer, U., Casey, M. G., Dalla Torre, M., Fröhlich-Wyder, M. T., & Fürst, M. (2005). Occurrence and significance of Fusarium domesticum alias Anticollanti on smear-ripened cheeses. *LWT - Food Science and Technology, 38*, 399–407. https://doi.org/10.1016/j.lwt.2004.05.018

Baeshen, N. A., Baeshen, M. N., Sheikh, A., Bora, R. S., Ahmed, M. M., Ramadan, H. A., Saini, K. S., & Redwan, E. M. (2014). Cell factories for insulin production. *Microbial Cell Factories, 13*, 141. https://doi.org/10.1186/s12934-014-0141-0

Ballini, A., Charitos, I. A., Cantore, S., Topi, S., Bottalico, L., & Santacroce, L. (2023). About functional foods: The probiotics and prebiotics state of art. *Antibiotics (Basel), 12*. https://doi.org/10.3390/antibiotics12040635

Bartholomai, B. M., Ruwe, K. M., Thurston, J., Jha, P., Scaife, K., Simon, R., Abdelmoteleb, M., Goodman, R. E., & Farhi, M. (2022). Safety evaluation of Neurospora crassa mycoprotein for use as a novel meat alternative and enhancer. *Food and Chemical Toxicology, 168*, 113342. https://doi.org/10.1016/j.fct.2022.113342

Behnke, U., & Schalinatus, E. (1975). Formation of bitter peptides in cheese and from casein. *Nahrung, 19*, 839–843.

Belessi, C. I., Papanikolaou, S., Drosinos, E. H., & Skandamis, P. N. (2008). Survival and acid resistance of Listeria innocua in Feta cheese and yogurt, in the presence or absence of fungi. *Journal of Food Protection, 71*, 742–749. https://doi.org/10.4315/0362-028x-71.4.742

Bellut, K., Krogerus, K., & Arendt, E. K. (2020). Lachancea fermentati strains isolated from Kombucha: Fundamental insights, and practical application in low alcohol beer brewing. *Frontiers in Microbiology, 11*, 764. https://doi.org/10.3389/fmicb.2020.00764

Bintsis, T., Vafopoulou-Mastrojiannaki, A., Litopoulou-Tzanetaki, E., & Robinson, R. K. (2003). Protease, peptidase and esterase activities by lactobacilli and yeast isolates from Feta cheese brine. *Journal of Applied Microbiology, 95*, 68–77. https://doi.org/10.1046/j.1365-2672.2003.01980.x

Bo, B., Kim, S. A., & Han, N. S. (2020). Bacterial and fungal diversity in Laphet, traditional fermented tea leaves in Myanmar, analyzed by culturing, DNA amplicon-based sequencing, and PCR-DGGE methods. *International Journal of Food Microbiology, 320*, 108508. https://doi.org/10.1016/j.ijfoodmicro.2020.108508

Bora, S. S., Keot, J., Das, S., Sarma, K., & Barooah, M. (2016). Metagenomics analysis of microbial communities associated with a traditional rice wine starter culture (Xaj-pitha) of Assam, India. *3 Biotech, 6*. https://doi.org/10.1007/s13205-016-0471-1

Bourdichon, F., Casaregola, S., Farrokh, C., Frisvad, J. C., Gerds, M. L., Hammes, W. P., Harnett, J., Huys, G., Laulund, S., Ouwehand, A., et al. (2012a). Food fermentations: Microorganisms with technological beneficial use. *International Journal of Food Microbiology, 154*, 87–97. https://doi.org/10.1016/j.ijfoodmicro.2011.12.030

Bourdichon, F., Casaregola, S., Farrokh, C., Frisvad, J. C., Gerds, M. L., Hammes, W. P., Harnett, J., Huys, G., Laulund, S., Ouwehand, A., et al. (2012b). Erratum to "Food fermentations: microorganisms with technological beneficial use" [International Journal of Food Microbiology 154 (2012) 87–97]. *International Journal of Food Microbiology, 156*, 301. https://doi.org/10.1016/j.ijfoodmicro.2012.04.003

Branco, P., Viana, T., Albergaria, H., & Arneborg, N. (2015). Antimicrobial peptides (AMPs) produced by Saccharomyces cerevisiae induce alterations in the intracellular pH, membrane permeability and culturability of Hanseniaspora guilliermondii cells. *International Journal of Food Microbiology, 205*, 112–118. https://doi.org/10.1016/j.ijfoodmicro.2015.04.015

Branco, P., Sabir, F., Diniz, M., Carvalho, L., Albergaria, H., & Prista, C. (2019). Biocontrol of Brettanomyces/Dekkera bruxellensis in alcoholic fermentations using saccharomycin-overproducing Saccharomyces cerevisiae strains. *Applied Microbiology and Biotechnology, 103*, 3073–3083. https://doi.org/10.1007/s00253-019-09657-7

Bravo, F. I., Mas-Capdevila, A., López-Fernández-Sobrino, R., Torres-Fuentes, C., Mulero, M., Alcaide-Hidalgo, J. M., & Muguerza, B. (2022). Identification of novel antihypertensive peptides from wine lees hydrolysate. *Food Chemistry, 366*, 130690. https://doi.org/10.1016/j.foodchem.2021.130690

Brüssow, H. (2020). Bioarchaeology: A profitable dialogue between microbiology and archaeology. *Microbial Biotechnology, 13*, 406–409. https://doi.org/10.1111/1751-7915.13527

Buitrago Mora, H. M., Piñeros, M. A., Espinosa Moreno, D., Restrepo Restrepo, S., Cardona Jaramillo, J. E. C., Álvarez Solano, Ó. A., Fernandez-Niño, M., & González Barrios, A. F. (2019). Multiscale design of a dairy beverage model composed of Candida utilis single cell protein supplemented with oleic acid. *Journal of Dairy Science, 102*, 9749–9762. https://doi.org/10.3168/jds.2019-16729

Cabezudo, I., Galetto, C. S., Romanini, D., Furlán, R. L. E., & Meini, M. R. (2022). Production of gallic acid and relevant enzymes by Aspergillus Niger and Aspergillus oryzae in solid-state fermentation of soybean hull and grape pomace. *Biomass Conversion and Biorefinery, 13*, 14939–14947. https://doi.org/10.1007/s13399-022-03435-8

Carrasco de Mendoza, M. S., Basílico, J. C., Umansky, G., & Scarinci, H. E. (1991). Biological value of the unicellular protein of Kluyveromyces marxianus var. lactis. *Archivos Latinoamericanos de Nutrición, 41*, 72–78.

Castanha, R. F., Mariano, A. P., de Morais, L. A., Scramin, S., & Monteiro, R. T. (2014). Optimization of lipids production by Cryptococcus laurentii 11 using cheese whey with molasses. *Brazilian Journal of Microbiology, 45*, 379–387. https://doi.org/10.1590/s1517-83822014000200003

Chang, S. W., Lee, G. C., & Shaw, J. F. (2006). Efficient production of active recombinant Candida rugosa LIP3 lipase in Pichia pastoris and biochemical characterization of the purified enzyme. *Journal of Agricultural and Food Chemistry, 54*, 5831–5838. https://doi.org/10.1021/jf060835e

Chaudhary, A., Bhalla, S., Patiyal, S., Raghava, G. P. S., & Sahni, G. (2021). FermFooDb: A database of bioactive peptides derived from fermented foods. *Heliyon, 7*, e06668. https://doi.org/10.1016/j.heliyon.2021.e06668

Chávez, R., Fierro, F., García-Rico, R. O., & Laich, F. (2011). Mold-fermented foods: Penicillium spp. as ripening agents in the elaboration of cheese and meat products. *Mycofactories*, 73–98. https://doi.org/10.2174/978160805223311101010073

Chen, B., Wu, Q., & Xu, Y. (2014). Filamentous fungal diversity and community structure associated with the solid state fermentation of Chinese Maotai-flavor liquor. *International Journal of Food Microbiology, 179*, 80–84. https://doi.org/10.1016/j.ijfoodmicro.2014.03.011

Cheng, L., Wang, F., Guo, Y., Du, Q., Zeng, X., Wu, Z., Guo, Y., Tu, M., & Pan, D. (2025). Potential prebiotic properties and proliferation mechanism of fermented milk-derived polypeptides. *Food Chemistry, 463*, 141335. https://doi.org/10.1016/j.foodchem.2024.141335

Chi, X., Zhang, Y., Yang, Q., Zhang, J., Sun, B., & Ai, N. (2025). An insight into specific flavor sensation in fermented milk: Linalool and mushroom alcohol. *Journal of Dairy Science, 108*, 5741–5753. https://doi.org/10.3168/jds.2025-26449

Chourasia, R., Abedin, M. M., Phukon, L. C., Sarkar, P., Sharma, S., Sahoo, D., Singh, S. P., & Kumar Rai, A. (2025). Unearthing novel and multifunctional peptides in peptidome of fermented chhurpi cheese of Indian Himalayan region. *Food Research International, 201*, 115651. https://doi.org/10.1016/j.foodres.2024.115651

Chung, D., Kwon, Y. M., & Yang, Y. (2021). Telomere-to-telomere genome assembly of asparaginase-producing *Trichoderma simmonsii*. *BMC Genomics, 22*, 830. https://doi.org/10.1186/s12864-021-08162-4

Colombo, M. L., Cimino, C. V., Bruno, M. A., Hugo, A., Liggieri, C., Fernández, A., & Vairo-Cavalli, S. (2021). Artichoke cv. Francés flower extract as a rennet substitute: Effect on textural, microstructural, microbiological, antioxidant properties, and sensory acceptance of miniature cheeses. *Journal of the Science of Food and Agriculture, 101*, 1382–1388. https://doi.org/10.1002/jsfa.10749

Cosentino, S., Fadda, M. E., Deplano, M., Mulargia, A. F., & Palmas, F. (2001). Yeasts associated with Sardinian ewe's dairy products. *International Journal of Food Microbiology, 69*, 53–58. https://doi.org/10.1016/s0168-1605(01)00572-4

Cui, J., Xia, P., Zhang, L., Hu, Y., Xie, Q., & Xiang, H. (2020). A novel fermented soybean, inoculated with selected Bacillus, Lactobacillus and Hansenula strains, showed strong antioxidant and anti-fatigue potential activity. *Food Chemistry, 333*, 127527. https://doi.org/10.1016/j.foodchem.2020.127527

da Silva, R. R., Souto, T. B., Gonsales da Rosa, N., de Oliveira, L. C. G., Juliano, M. A., Juliano, L., Rosa, J. C., & Cabral, H. (2020). Evaluation of the milk clotting properties of an aspartic peptidase secreted by Rhizopus microsporus. *Preparative Biochemistry & Biotechnology, 50*, 226–233. https://doi.org/10.1080/10826068.2019.1683861

Dagher, S. F., Azcarate-Peril, M. A., & Bruno-Bárcena, J. M. (2013). Heterologous expression of a bioactive β-Hexosyltransferase, an enzyme producer of prebiotics, from *Sporobolomyces singularis*. *Applied and Environmental Microbiology, 79*, 1241–1249. https://doi.org/10.1128/AEM.03491-12

Dallas, D. C., Citerne, F., Tian, T., Silva, V. L., Kalanetra, K. M., Frese, S. A., Robinson, R. C., Mills, D. A., & Barile, D. (2016). Peptidomic analysis reveals proteolytic activity of kefir microorganisms on bovine milk proteins. *Food Chemistry, 197*, 273–284. https://doi.org/10.1016/j.foodchem.2015.10.116

Das, D., Afzal, N. U., Wann, S. B., Kalita, J., & Manna, P. (2023). A ~24 kDa protein isolated from protein isolates of Hawaijar, popular fermented soy food of North-East India exhibited promising antidiabetic potential via stimulating PI3K/AKT/GLUT4 signaling pathway of muscle glucose metabolism. *International Journal of Biological Macromolecules, 224*, 1025–1039. https://doi.org/10.1016/j.ijbiomac.2022.10.187

de Miranda, N. M. Z., de Souza, A. C., de Souza Costa Sobrinho, P., Dias, D. R., Schwan, R. F., & Ramos, C. L. (2023). Novel yeasts with potential probiotic characteristics isolated from the endogenous ferment of artisanal Minas cheese. *Brazilian Journal of Microbiology, 54*, 1021–1033. https://doi.org/10.1007/s42770-023-01002-5

De Respinis, S., Caminada, A., Pianta, E., Buetti-Dinh, A., Riva Scettrini, P., Petrini, L., Tonolla, M., & Petrini, O. (2023). Fungal communities on alpine cheese rinds in southern Switzerland. *Botanical Studies, 64*, 6. https://doi.org/10.1186/s40529-023-00371-2

de Souza, T. P., Evangelista, S. R., Passamani, F. R. F., Bertechini, R., de Abreu, L. R., & Batista, L. R. (2021). Mycobiota of Minas artisanal cheese: Safety and quality. *International Dairy Journal, 120*, 105085. https://doi.org/10.1016/j.idairyj.2021.105085

Desmasures, N., Mangin, I., Corroler, D., & Guéguen, M. (1998). Characterization of lactococci isolated from milk produced in the Camembert region of Normandy. *Journal of Applied Microbiology, 85*, 999–1005. https://doi.org/10.1111/j.1365-2672.1998.tb05264.x

Devanthi, P. V. P., Kho, K., Nurdiansyah, R., Briot, A., Taherzadeh, M. J., & Aslanzadeh, S. (2021). Do Kombucha symbiotic cultures of bacteria and yeast affect bacterial cellulose yield in molasses? *Journal of Fungi (Basel), 7*, 705. https://doi.org/10.3390/jof7090705

Dimitrellou, D., Kandylis, P., Kourkoutas, Y., Koutinas, A. A., & Kanellaki, M. (2015). Cheese production using kefir culture entrapped in milk proteins. *Applied Biochemistry and Biotechnology, 176*, 213–230. https://doi.org/10.1007/s12010-015-1568-4

Diniz, R. H., Silveira, W. B., Fietto, L. G., & Passos, F. M. (2012). The high fermentative metabolism of Kluyveromyces marxianus UFV-3 relies on the increased expression of key lactose metabolic enzymes. *Antonie Van Leeuwenhoek, 101*, 541–550. https://doi.org/10.1007/s10482-011-9668-9

Dishan, A., Ozkaya, Y., Temizkan, M. C., Barel, M., & Gonulalan, Z. (2025). Candida species covered from traditional cheeses: Characterization of C. albicans regarding virulence factors, biofilm formation, caseinase activity, antifungal resistance and phylogeny. *Food Microbiology, 127*, 104679. https://doi.org/10.1016/j.fm.2024.104679

Do, S. H., Lee, T. G., & Kim, S. K. (2024). Enhancing protein content in wild-type *Saccharomyces cerevisiae* via random mutagenesis and optimized fermentation conditions. *Journal of Microbiology and Biotechnology, 34*, 1912–1918. https://doi.org/10.4014/jmb.2405.05027

El Sheikha, A. F., & Hu, D. M. (2020). Molecular techniques reveal more secrets of fermented foods. *Critical Reviews in Food Science and Nutrition, 60*, 11–32. https://doi.org/10.1080/10408398.2018.1506906

El-Gendy, S. M. (1983). Fermented foods of Egypt and the Middle East. *Journal of Food Protection, 46*, 358–367. https://doi.org/10.4315/0362-028x-46.4.358

el-Maghrabi, E. A., Dixon, D. M., & Burnett, J. W. (1990). Characterization of Candida albicans epidermolytic proteases and their role in yeast-cell adherence to keratinocytes. *Clinical and Experimental Dermatology, 15*, 183–191. https://doi.org/10.1111/j.1365-2230.1990.tb02069.x

Emri, T., Tóth, V., Nagy, C. T., Nagy, G., Pócsi, I., Gyémánt, G., Antal, K., Balla, J., Balla, G., Román, G., et al. (2013). Towards high-siderophore-content foods: Optimisation of coprogen production in submerged cultures of Penicillium nalgiovense. *Journal of the Science of Food and Agriculture, 93*, 2221–2228. https://doi.org/10.1002/jsfa.6029

Encyclopædia Britannica. (2013a). *Encyclopædia Britannica student and home edition.* Encyclopædia Britannica.

Encyclopædia Britannica. (2013b). *Cheddar-Encyclopædia Britannica student and home edition.* Encyclopædia Britannica.

Encyclopædia Britannica. (2013c). *Tofu-Encyclopædia Britannica student and home edition.* Encyclopædia Britannica.

Engel, G., & Prokopek, D. (1979). KeinNachweis von Penicillium roqueforti-Toxin in Kase. *Milschwissenschaft, 34*, 272–274.

Engel, G., & Prokopek, D. (1980). KeinNachweis von Patulin and Penicillinsaure in mit Patulin- und Penicillinsaure-bildenden Penicillium roqueforti-Stammenhergestell-ten Kasen. *Milchwissenschaft, 35*, 218–220.

Engel, G., von Milczewski, K. E., Prokopek, D., & Teuber, M. (1982). Strain-specific synthesis of mycophenolic acid by Penicillium roqueforti in blue-veined cheese. *Applied and Environmental Microbiology, 43*, 1034–1040. https://doi.org/10.1128/aem.43.5.1034-1040.1982

Engel, E., Tournier, C., Salles, C., & Le Quéré, J. L. (2001). Evolution of the composition of a selected bitter Camembert cheese during ripening: Release and migration of taste-active compounds. *Journal of Agricultural and Food Chemistry, 49*, 2940–2947. https://doi.org/10.1021/jf000966u

Fadda, M. E., Cosentino, S., Deplano, M., & Palmas, F. (2001). Yeast populations in Sardinian feta cheese. *International Journal of Food Microbiology, 69*, 153–156. https://doi.org/10.1016/s0168-1605(01)00586-4

Fan, Y., Gan, R., Zhang, Z., Xu, J., Liu, S., Bu, Y., Cao, C., Liu, Q., Xia, X., Kong, B., et al. (2025). Flavor effect, application status, and research trend of umami peptides based on microbial fermentation in food. *Food Microbiology, 130*, 104769. https://doi.org/10.1016/j.fm.2025.104769

Färber, P., & Geisen, R. (1994). Antagonistic activity of the food-related filamentous fungus Penicillium nalgiovense by the production of penicillin. *Applied and Environmental Microbiology, 60*, 3401–3404. https://doi.org/10.1128/aem.60.9.3401-3404.1994

Feijoo, G., Moreira, M., Roca, E., & Lema, J. (1999). Use of cheese whey as a substrate to produce manganese peroxidase by Bjerkandera sp BOS55. *Journal of Industrial Microbiology & Biotechnology, 23*, 86–90. https://doi.org/10.1038/sj.jim.2900691

Fernandes, Â., Bancessi, A., Pinela, J., Dias, M. I., Liberal, Â., Calhelha, R. C., Ćirić, A., Soković, M., Catarino, L., Ferreira, I., et al. (2021). Nutritional and phytochemical profiles and biological activities of Moringa oleifera Lam. edible parts from Guinea-Bissau (West Africa). *Food Chemistry, 341*, 128229. https://doi.org/10.1016/j.foodchem.2020.128229

Fernández-Pacheco, P., Ramos Monge, I. M., Poveda, J. M., Díaz-Maroto, M. C., & Arévalo-Villena, M. (2023). Use of probiotic yeasts with biocontrol activity for fermentation of ewe's milk. *Journal of the Science of Food and Agriculture, 103*, 4107–4118. https://doi.org/10.1002/jsfa.12394

Ferreira, A. D., & Viljoen, B. C. (2003). Yeasts as adjunct starters in matured Cheddar cheese. *International Journal of Food Microbiology, 86*, 131–140. https://doi.org/10.1016/s0168-1605(03)00252-6

Ferreira, I. M., Pinho, O., Monteiro, D., Faria, S., Cruz, S., Perreira, A., Roque, A. C., & Tavares, P. (2010). Short communication: Effect of kefir grains on proteolysis of major milk proteins. *Journal of Dairy Science, 93*, 27–31. https://doi.org/10.3168/jds.2009-2501

Ferreira, I., de Sousa Melo, D., Menezes, A. G. T., Fonseca, H. C., de Assis, B. B. T., Ramos, C. L., Magnani, M., Dias, D. R., & Schwan, R. F. (2022). Evaluation of potentially probiotic yeasts and Lactiplantibacillus plantarum in co-culture for the elaboration of a functional plant-based fermented beverage. *Food Research International, 160*, 111697. https://doi.org/10.1016/j.foodres.2022.111697

Fukami, H., Higa, Y., Hisano, T., Asano, K., Hirata, T., & Nishibe, S. (2021). A review of red yeast rice, a traditional fermented food in Japan and East Asia: Its characteristic ingredients and application in the maintenance and improvement of health in lipid metabolism and the circulatory system. *Molecules, 26*. https://doi.org/10.3390/molecules26061619

Furukawa, S., Watanabe, T., Toyama, H., & Morinaga, Y. (2013). Significance of microbial symbiotic coexistence in traditional fermentation. *Journal of Bioscience and Bioengineering, 116*, 533–539. https://doi.org/10.1016/j.jbiosc.2013.05.017

Gan, X., Cao, D., Zhang, Z., Cheng, S., Wei, L., Li, S., & Liu, B. (2020). Draft genome assembly of Floccularia luteovirens, an edible and symbiotic mushroom on Qinghai-Tibet Plateau. *G3 (Bethesda), 10*, 1167–1173. https://doi.org/10.1534/g3.120.401037

García Fontán, M. C., Martínez, S., Franco, I., & Carballo, J. (2006). Microbiological and chemical changes during the manufacture of Kefir made from cows' milk, using a commercial starter culture. *International Dairy Journal, 16*, 762–767. https://doi.org/10.1016/j.idairyj.2005.07.004

Gientka, I., Kieliszek, M., Jermacz, K., & Błażejak, S. (2017). Identification and characterization of oleaginous yeast isolated from kefir and its ability to accumulate intracellular fats in deproteinated potato wastewater with different carbon sources. *BioMed Research International, 2017*, 6061042. https://doi.org/10.1155/2017/6061042

Giraud, F., Giraud, T., Aguileta, G., Fournier, E., Samson, R., Cruaud, C., Lacoste, S., Ropars, J., Tellier, A., & Dupont, J. (2010). Microsatellite loci to recognize species for the cheese starter and contaminating strains associated with cheese manufacturing. *International Journal of Food Microbiology, 137*, 204–213. https://doi.org/10.1016/j.ijfoodmicro.2009.11.014

Gkatzionis, K., Yunita, D., Linforth, R. S., Dickinson, M., & Dodd, C. E. (2014). Diversity and activities of yeasts from different parts of a Stilton cheese. *International Journal of Food Microbiology, 177*, 109–116. https://doi.org/10.1016/j.ijfoodmicro.2014.02.016

Gong, W., Wang, Y., Xie, C., Zhou, Y., Zhu, Z., & Peng, Y. (2020). Whole genome sequence of an edible and medicinal mushroom, Hericium erinaceus (Basidiomycota, fungi). *Genomics, 112*, 2393–2399. https://doi.org/10.1016/j.ygeno.2020.01.011

González-Orozco, B. D., McGovern, C. J., Barringer, S. A., Simons, C., Jiménez-Flores, R., & Alvarez, V. B. (2024). Development of probiotic yogurt products incorporated with lactobacillus kefiranofaciens OSU-BSGOA1 in mono- and coculture with Kluyveromyces marxianus. *Journal of Dairy Science, 107*, 7718–7733. https://doi.org/10.3168/jds.2024-24756

Grafenham, T., Schroers, H. J., Nirenberg, H. I., & Seifert, K. A. (2011). An overview of the taxonomy, phylogeny, and typification of nectriaceous fungi in Cosmospora, Acremonium, Fusarium, Stilbella, and Volutella. *Studies in Mycology, 68*, 79–113. https://doi.org/10.3114/sim.2011.68.04

Gryganskyi, A. P., Golan, J., Dolatabadi, S., Mondo, S., Robb, S., Idnurm, A., Muszewska, A., Steczkiewicz, K., Masonjones, S., Liao, H.-L., et al. (2018). Phylogenetic and phylogenomic definition of rhizopus species. *G3 Genes|Genomes|Genetics, 8*, 2007–2018. https://doi.org/10.1534/g3.118.200235

Grygier, A., Myszka, K., Juzwa, W., Białas, W., & Rudzińska, M. (2020). Galactomyces geotrichum mold isolated from a traditional fried cottage cheese produced omega-3 fatty acids. *International Journal of Food Microbiology, 319*, 108503. https://doi.org/10.1016/j.ijfoodmicro.2019.108503

Guzel-Seydim, Z. B., Kok-Tas, T., Greene, A. K., & Seydim, A. C. (2011). Review: Functional properties of kefir. *Critical Reviews in Food Science and Nutrition, 51*, 261–268. https://doi.org/10.1080/10408390903579029

Habib, H. M., Ismail, R., Agami, M., & El-Yazbi, A. F. (2025). Exploring the impact of bioactive peptides from fermented Milk proteins: A review with emphasis on health implications and artificial intelligence integration. *Food Chemistry, 481*, 144047. https://doi.org/10.1016/j.foodchem.2025.144047

Hagi, T., Kurahashi, A., Oguro, Y., Kodaira, K., Kobayashi, M., Hayashida, S., Yamashita, H., Arakawa, Y., Miura, T., Sato, K., et al. (2022). Effect of sake lees on cheese components in cheese ripened by Aspergillus oryzae and lactic acid bacteria. *Journal of Dairy Science, 105*, 4868–4881. https://doi.org/10.3168/jds.2021-21721

Hansen, T. K., & Jakobsen, M. (2001). Taxonomical and technological characteristics of Saccharomyces spp. associated with blue veined cheese. *International Journal of Food Microbiology, 69*, 59–68. https://doi.org/10.1016/s0168-1605(01)00573-6

Hatoum, R., Labrie, S., & Fliss, I. (2012). Antimicrobial and probiotic properties of yeasts: From fundamental to novel applications. *Frontiers in Microbiology, 3*, 421. https://doi.org/10.3389/fmicb.2012.00421

Hermet, A., Méheust, D., Mounier, J., Barbier, G., & Jany, J. L. (2012). Molecular systematics in the genus Mucor with special regards to species encountered in cheese. *Fungal Biology, 116*, 692–705. https://doi.org/10.1016/j.funbio.2012.04.002

Hidalgo, P. I., Ullán, R. V., Albillos, S. M., Montero, O., Fernández-Bodega, M., García-Estrada, C., Fernández-Aguado, M., & Martín, J. F. (2014). Molecular characterization of the PR-toxin gene cluster in Penicillium roqueforti and Penicillium chrysogenum: Cross talk of secondary metabolite pathways. *Fungal Genetics and Biology, 62*, 11–24. https://doi.org/10.1016/j.fgb.2013.10.009

Hong, W. S., Chen, Y. P., & Chen, M. J. (2010). The antiallergic effect of kefir Lactobacilli. *Journal of Food Science, 75*, H244–H253. https://doi.org/10.1111/j.1750-3841.2010.01787.x

Huang, L., Tang, Y., Zheng, J., Kan, J., Wu, Y., Wu, Y., Awad, S., Ibrahim, A., & Du, M. (2023). Relationship between the dynamics of flavor compounds and microbial succession in the natural fermentation of Zhalajiao, a popular traditional Chinese fermented chili paste. *Food, 12*. https://doi.org/10.3390/foods12203849

Huh, E., Lim, S., Kim, H. G., Ha, S. K., Park, H. Y., Huh, Y., & Oh, M. S. (2018). Ginger fermented with Schizosaccharomyces pombe alleviates memory impairment via protecting hippocampal neuronal cells in amyloid beta(1-42) plaque injected mice. *Food & Function, 9*, 171–178. https://doi.org/10.1039/c7fo01149k

Hymery, N., Vasseur, V., Coton, M., Mounier, J., Jany, J. L., Barbier, G., & Coton, E. (2014). Filamentous fungi and mycotoxins in cheese: A review. *Comprehensive Reviews in Food Science and Food Safety, 13*, 437–456. https://doi.org/10.1111/1541-4337.12069

Ibarruri, J., & Hernández, I. (2019). Valorization of cheese whey and orange molasses for fungal biomass production by submerged fermentation with Rhizopus sp. *Bioprocess and Biosystems Engineering, 42*, 1285–1300. https://doi.org/10.1007/s00449-019-02127-4

Iwaniak, A., Mogut, D., Minkiewicz, P., Żulewska, J., & Darewicz, M. (2022). An integrated approach to the analysis of antioxidative peptides derived from Gouda cheese with a modified β-casein content. *Scientific Reports, 12*, 13314. https://doi.org/10.1038/s41598-022-17641-x

Janssen, H. G., Hrncirík, K., Szórádi, A., & Leijten, M. (2006). An improved method for sn-2 position analysis of triacylglycerols in edible oils and fats based on immobilised lipase D (Rhizopus delemar). *Journal of Chromatography A, 1112*, 141–147. https://doi.org/10.1016/j.chroma.2005.11.097

Jia, S., Kang, Y. P., Park, J. H., Lee, J., & Kwon, S. W. (2011). Simultaneous determination of 23 amino acids and 7 biogenic amines in fermented food samples by liquid chromatography/quadrupole time-of-flight mass spectrometry. *Journal of Chromatography A, 1218*, 9174–9182. https://doi.org/10.1016/j.chroma.2011.10.040

Jurado, M., & Vicente, C. J. (2020). Penicillium commune affects textural properties and water distribution of hard and extra-hard cheeses. *The Journal of Dairy Research, 87*, 117–122. https://doi.org/10.1017/s0022029919000906

Kahve, H. (2023). In vitro evaluation of the technological and probiotic potential of Pichia kudriavzevii strains isolated from traditional fermented foods. *Current Microbiology, 80*, 379. https://doi.org/10.1007/s00284-023-03505-8

Kasperek, M. C., Velasquez Galeas, A., Caetano-Silva, M. E., Xie, Z., Ulanov, A., La Frano, M., Devkota, S., Miller, M. J., & Allen, J. M. (2024). Microbial aromatic amino acid metabolism is modifiable in fermented food matrices to promote bioactivity. *Food Chemistry, 454*, 139798. https://doi.org/10.1016/j.foodchem.2024.139798

Khorshidian, N., Yousefi, M., & Mortazavian, A. M. (2020). Fermented milk: The most popular probiotic food carrier. *Advances in Food and Nutrition Research, 94*, 91–114. https://doi.org/10.1016/bs.afnr.2020.06.007

Kim, H. G., Hong, J. H., Song, C. K., Shin, H. W., & Kim, K. O. (2010). Sensory characteristics and consumer acceptability of fermented soybean paste (Doenjang). *Journal of Food Science, 75*, S375–S383. https://doi.org/10.1111/j.1750-3841.2010.01771.x

Kim, Y., Kim, E. Y., Son, H. J., Lee, J. J., Choi, Y. H., & Rhyu, M. R. (2017). Identification of a key umami-active fraction in modernized Korean soy sauce and the impact thereof on bitter-masking. *Food Chemistry, 233*, 256–262. https://doi.org/10.1016/j.foodchem.2017.04.123

Kothe, C. I., Mohellibi, N., & Renault, P. (2022). Revealing the microbial heritage of traditional Brazilian cheeses through metagenomics. *Food Research International, 157*, 111265. https://doi.org/10.1016/j.foodres.2022.111265

Kozlovsky, A. G., Zhelifonova, V. P., Antipova, T. V., Baskunov, B. P., Ivanushkina, N. E., & Ozerskaya, S. M. (2014). Exo-metabolites of mycelial fungi isolated in production premises of cheese-making and meat-processing plants. *Food Additives & Contaminants. Part A, Chemistry, Analysis, Control, Exposure & Risk Assessment, 31*, 300–306. https://doi.org/10.1080/19440049.2013.862350

Kumar, M. R., Azizi, N. F., Yeap, S. K., Abdullah, J. O., Khalid, M., Omar, A. R., Osman, M. A., Leow, A. T. C., Mortadza, S. A. S., & Alitheen, N. B. (2022). Clinical and preclinical studies of fermented foods and their effects on Alzheimer's disease. *Antioxidants (Basel), 11*, 883. https://doi.org/10.3390/antiox11050883

Kumari, R., Sharma, N., Sharma, S., Samurailatpam, S., Padhi, S., Singh, S. P., & Kumar Rai, A. (2023). Production and characterization of bioactive peptides in fermented soybean meal produced using proteolytic Bacillus species isolated from kinema. *Food Chemistry, 421*, 136130. https://doi.org/10.1016/j.foodchem.2023.136130

Kunyeit, L., Rao, R. P., & Anu-Appaiah, K. A. (2024). Yeasts originating from fermented foods, their potential as probiotics and therapeutic implication for human health and disease. *Critical Reviews in Food Science and Nutrition, 64*, 6660–6671. https://doi.org/10.1080/10408398.2023.2172546

Lan, Y., Cong, Q., Yu, Q., Liu, L., Cui, X., Li, X., Wang, Q., Yang, S., Yu, H., & Kong, Y. (2024). Genome sequencing of three pathogenic fungi provides insights into the evolution and pathogenic mechanisms of the cobweb disease on cultivated mushrooms. *Food, 13*. https://doi.org/10.3390/foods13172779

Le Dréan, G., Mounier, J., Vasseur, V., Arzur, D., Habrylo, O., & Barbier, G. (2010). Quantification of Penicillium camemberti and P. roqueforti mycelium by real-time PCR to assess their growth dynamics during ripening cheese. *International Journal of Food Microbiology, 138*, 100–107. https://doi.org/10.1016/j.ijfoodmicro.2009.12.013

Leclercq-Perlat, M. N., Oumer, A., Bergere, J. L., Spinnler, H. E., & Corrieu, G. (2000). Behavior of Brevibacterium linens and Debaryomyces hansenii as ripening flora in controlled production of smear soft cheese from reconstituted milk: Growth and substrate consumption dairy foods. *Journal of Dairy Science, 83*, 1665–1673. https://doi.org/10.3168/jds.s0022-0302(00)75035-1

Lee, S. C., Billmyre, R. B., Li, A., Carson, S., Sykes, S. M., Huh, E. Y., Mieczkowski, P., Ko, D. C., Cuomo, C. A., & Heitman, J. (2014). Analysis of a food-borne fungal pathogen outbreak: virulence and genome of a Mucor circinelloides isolate from yogurt. *mBio, 5*. https://doi.org/10.1128/mbio.01390-14

Lessard, M. H., Viel, C., Boyle, B., St-Gelais, D., & Labrie, S. (2014). Metatranscriptome analysis of fungal strains Penicillium camemberti and Geotrichum candidum reveal cheese matrix breakdown and potential development of sensory properties of ripened Camembert-type cheese. *BMC Genomics, 15*, 235. https://doi.org/10.1186/1471-2164-15-235

Li, Y., Li, T., Guo, J., Wang, G., & Chen, G. (2022). Expression and characterization of a novel lytic polysaccharide monooxygenase, PdLPMO9A, from the edible fungus Pleurotus djamor and its synergistic interactions with cellulase in corn straw biomass saccharification. *Bioresource Technology, 348*, 126792. https://doi.org/10.1016/j.biortech.2022.126792

Liang, J., Xu, N., Nedele, A. K., Rigling, M., Zhu, L., Zhang, Y., Stöppelmann, F., Hannemann, L., Heimbach, J., Kohlus, R., et al. (2023). Upcycling of Soy Whey with Ischnoderma benzoinum toward production of bioflavors and mycoprotein. *Journal of Agricultural and Food Chemistry, 71*, 9070–9079. https://doi.org/10.1021/acs.jafc.3c01169

Liu, M., Zhang, W., Yao, J., & Niu, J. (2021). Production, purification, characterization, and biological properties of Rhodosporidium paludigenum polysaccharide. *PLoS One, 16*, e0246148. https://doi.org/10.1371/journal.pone.0246148

Lo, S. C., Yang, C. Y., Mathew, D. C., & Huang, C. C. (2021). Growth and autolysis of the kefir yeast Kluyveromyces marxianus in lactate culture. *Scientific Reports, 11*, 14552. https://doi.org/10.1038/s41598-021-94101-y

Lo, Y. C., Bruxaux, J., Rodríguez de la Vega, R. C., O'Donnell, S., Snirc, A., Coton, M., Le Piver, M., Le Prieur, S., Roueyre, D., Dupont, J., et al. (2023). Domestication in dry-cured meat Penicillium fungi: Convergent specific phenotypes and horizontal gene transfers without strong genetic subdivision. *Evolutionary Applications, 16*, 1637–1660. https://doi.org/10.1111/eva.13591

Lopitz-Otsoa, F., Rementeria, A., Elguezabal, N., & Garaizar, J. (2006). Kefir: A symbiotic yeasts-bacteria community with alleged healthy capabilities. *Revista Iberoamericana de Micología, 23*, 67–74. https://doi.org/10.1016/s1130-1406(06)70016-x

Lu, M., Wang, X., Sun, G., Qin, B., Xiao, J., Yan, S., Pan, Y., & Wang, Y. (2014). Fine structure of Tibetan kefir grains and their yeast distribution, diversity, and shift. *PLoS One, 9*, e101387. https://doi.org/10.1371/journal.pone.0101387

Lv, X.-C., Huang, Z.-Q., Zhang, W., Rao, P.-F., & Ni, L. (2012). Identification and characterization of filamentous fungi isolated from fermentation starters for Hong Qu glutinous rice wine brewing. *The Journal of General and Applied Microbiology, 58*, 33–42. https://doi.org/10.2323/jgam.58.33

Ly, S., Mith, H., Tarayre, C., Taminiau, B., Daube, G., Fauconnier, M.-L., & Delvigne, F. (2018). Impact of microbial composition of Cambodian traditional dried starters (Dombea) on flavor compounds of rice wine: Combining amplicon sequencing with HP-SPME-GCMS. *Frontiers in Microbiology, 9*. https://doi.org/10.3389/fmicb.2018.00894

Ma, X., Zhou, B., Jiang, L., Xie, M., Rong, Z., Yin, S., Wang, F., Liu, Y., & Li, X. (2025). Microbial interactions between Lactoplantibacillus plantarum and Rhodotorula mucilaginosa in the fermented fish juice system. *Food Research International, 208*, 116166. https://doi.org/10.1016/j.foodres.2025.116166

Maragkoudakis, P., Vendramin, V., Bovo, B., Treu, L., Corich, V., & Giacomini, A. (2016). Potential use of scotta, the by-product of the ricotta cheese manufacturing process, for the production of fermented drinks. *The Journal of Dairy Research, 83*, 104–108. https://doi.org/10.1017/s002202991500059x

Marco, M. L., Heeney, D., Binda, S., Cifelli, C. J., Cotter, P. D., Foligné, B., Gänzle, M., Kort, R., Pasin, G., Pihlanto, A., et al. (2017). Health benefits of fermented foods: Microbiota and beyond. *Current Opinion in Biotechnology, 44*, 94–102. https://doi.org/10.1016/j.copbio.2016.11.010

Martin, J. G. P., & Cotter, P. D. (2023). Filamentous fungi in artisanal cheeses: A problem to be avoided or a market opportunity? *Heliyon, 9*, e15110. https://doi.org/10.1016/j.heliyon.2023.e15110

Martins, J. M., Galinari, É., Pimentel-Filho, N. J., Ribeiro, J. I., Jr., Furtado, M. M., & Ferreira, C. L. (2015). Determining the minimum ripening time of artisanal Minas cheese, a traditional Brazilian cheese. *Brazilian Journal of Microbiology, 46*, 219–230. https://doi.org/10.1590/s1517-838246120131003

McGovern, P. E. (2007). *Ancient wine: The search for the origins of viniculture*. Princeton University Press.

McGovern, P. E. (2020). *Ancient wine: The search for the origins of viniculture*. Princeton University Press.

Mei, J., Guo, Q., Wu, Y., & Li, Y. (2014). Microbial diversity of a Camembert-type cheese using freeze-dried Tibetan kefir coculture as starter culture by culture-dependent and culture-independent methods. *PLoS One, 9*, e111648. https://doi.org/10.1371/journal.pone.0111648

Messini, A., Buccioni, A., Minieri, S., Mannelli, F., Mugnai, L., Comparini, C., Venturi, M., Viti, C., Pezzati, A., & Rapaccini, S. (2017). Effect of chestnut tannin extract (Castanea sativa Miller) on the proliferation of Cladosporium cladosporioides on sheep cheese rind during the ripening. *International Dairy Journal, 66*, 6–12. https://doi.org/10.1016/j.idairyj.2016.10.012

Mgbodile, F. C., & Nwagu, T. N. T. (2023). Probiotic therapy, African fermented foods and food-derived bioactive peptides in the management of SARS-CoV-2 cases and other viral infections. *Biotechnology Reports (Amsterdam, Netherlands), 38*, e00795. https://doi.org/10.1016/j.btre.2023.e00795

Miao, Y., Xiong, G., Li, R., Wu, Z., Zhang, X., & Weng, P. (2018). Transcriptome profiling of Issatchenkia orientalis under ethanol stress. *AMB Express, 8*, 39. https://doi.org/10.1186/s13568-018-0568-5

Mizuochi, S., & Betts, G. (2022). Validation of the CompactDry "Nissui" YMR for enumeration of yeasts and molds in a variety of foods: AOAC performance tested method SM 092002. *Journal of AOAC International, 105*, 1408–1417. https://doi.org/10.1093/jaoacint/qsac028

Moavro, A., Pino, F., Sanchez-Díaz, M., Delfederico, L., & Ludemann, V. (2022). Sensory analysis for stuffed cheese with Penicillium nalgiovense superficial growth. *Food Science and Technology International, 28*, 502–513. https://doi.org/10.1177/10820132211023673

Mogensen, J. M., Varga, J., Thrane, U., & Frisvad, J. C. (2009). *Aspergillus acidus* from Puerh tea and black tea does not produce ochratoxin A and fumonisin B2. *International Journal of Food Microbiology, 132*, 141–144. https://doi.org/10.1016/j.ijfoodmicro.2009.04.011

Moreno, M. R. F., Leisner, J. J., Tee, L. K., Ley, C., Radu, S., Rusul, G., Vancanneyt, M., & De Vuyst, L. (2002). Microbial analysis of Malaysian tempeh, and characterization of two bacteriocins produced by isolates of Enterococcus faecium. *Journal of Applied Microbiology, 92*, 147–157. https://doi.org/10.1046/j.1365-2672.2002.01509.x

Mugula, J. K., Sørhaug, T., & Stepaniak, L. (2003). Proteolytic activities in togwa, a Tanzanian fermented food. *International Journal of Food Microbiology, 84*, 1–12. https://doi.org/10.1016/s0168-1605(02)00387-2

Murugesan, S., Reyes-Mata, M. P., Nirmalkar, K., Chavez-Carbajal, A., Juárez-Hernández, J. I., Torres-Gómez, R. E., Piña-Escobedo, A., Maya, O., Hoyo-Vadillo, C., Ramos-Ramírez, E. G., et al. (2018). Profiling of bacterial and fungal communities of Mexican cheeses by high throughput DNA sequencing. *Food Research International, 113*, 371–381. https://doi.org/10.1016/j.foodres.2018.07.023

Nie, J., Zhu, S., Zhang, X., Wu, D., Li, X., & Huang, Q. (2025). Effects of starters on the quality of fermented fish (Zaoyu): Key microorganisms for coloring, softening, and improving flavor. *Food Chemistry, 465*, 142087. https://doi.org/10.1016/j.foodchem.2024.142087

Nielsen, L., Rolighed, M., Buehler, A., Knøchel, S., Wiedmann, M., & Marvig, C. (2021). Development of predictive models evaluating the spoilage-delaying effect of a bioprotective culture on different yeast species in yogurt. *Journal of Dairy Science, 104*, 9570–9582. https://doi.org/10.3168/jds.2020-20076

Nout, R. (2000). Useful role of fungi in food processing. In R. A. Samson, E. S. Hoekstra, J. C. Frisvad, & O. Filtenborg (Eds.), *Introduction to food- and airborne fungi* (6th ed.). Centraal Bureau Voor Schimmel Cultures.

Nout, R. (2007). The colonizing fungus as a food provider. In J. Dijksterhuis & R. A. Samson (Eds.), *Food mycology: A multifaceted approach to fungi and food* (Mycology) (pp. 335–352). https://doi.org/10.1201/9781420020984.ch17

O'Brien, K. V., Aryana, K. J., Prinyawiwatkul, W., Ordonez, K. M. C., & Boeneke, C. A. (2016). Short communication: The effects of frozen storage on the survival of probiotic microorganisms found in traditionally and commercially manufactured kefir. *Journal of Dairy Science, 99*, 7043–7048. https://doi.org/10.3168/jds.2015-10284

Ogata, T., & Kuroki, K. (2021). Regulation of mating and mating-type-specific genes in Zygosaccharomyces sp. yeast. *Yeast, 38*, 471–479. https://doi.org/10.1002/yea.3561

Oliveira, D. L., Costabile, A., Wilbey, R. A., Grandison, A. S., Duarte, L. C., & Roseiro, L. B. (2012). In vitro evaluation of the fermentation properties and potential prebiotic activity of caprine cheese whey oligosaccharides in batch culture systems. *BioFactors, 38*, 440–449. https://doi.org/10.1002/biof.1043

Orban, A., Fraatz, M. A., & Rühl, M. (2019). Aroma profile analyses of filamentous fungi cultivated on solid substrates. *Advances in Biochemical Engineering/Biotechnology, 169*, 85–107. https://doi.org/10.1007/10_2019_87

Ozturkoglu-Budak, S., Wiebenga, A., Bron, P. A., & de Vries, R. P. (2016). Protease and lipase activities of fungal and bacterial strains derived from an artisanal raw ewe's milk cheese. *International Journal of Food Microbiology, 237*, 17–27. https://doi.org/10.1016/j.ijfoodmicro.2016.08.007

Paul, S., Hossain, T. J., Ali, F., Hossain, M. E., Chowdhury, T., Faisal, I. K., & Ferdouse, J. (2024). Assessment of the in-vitro probiotic efficacy and safety of Pediococcus pentosaceus L1 and Streptococcus thermophilus L3 isolated from Laban, a popular fermented milk product. *Archives of Microbiology, 206*, 82. https://doi.org/10.1007/s00203-023-03812-5

Paup, V. D., Aplin, J. J., Potter, R. I., Edwards, C. G., Lee, J., & Ross, C. F. (2023). Sensory properties of 6- and 18-month-stored wines made with pectinase-producing non-Saccharomyces yeasts. *Journal of Food Science, 88*, 462–476. https://doi.org/10.1111/1750-3841.16418

Peres Fabbri, L., Cavallero, A., Vidotto, F., & Gabriele, M. (2024). Bioactive peptides from fermented foods: Production approaches, sources, and potential health benefits. *Foods (Basel, Switzerland), 13*, 3369. https://doi.org/10.3390/foods13213369

Péter, G., Dlauchy, D., Tóbiás, A., Fülöp, L., Podgoršek, M., & Čadež, N. (2017). Brettanomyces acidodurans sp. nov., a new acetic acid producing yeast species from olive oil. *Antonie Van Leeuwenhoek, 110*, 657–664. https://doi.org/10.1007/s10482-017-0832-8

Pires, A. F., Marnotes, N. G., Bella, A., Viegas, J., Gomes, D. M., Henriques, M. H. F., & Pereira, C. J. D. (2021). Use of ultrafiltrated cow's whey for the production of whey cheese with Kefir or probiotics. *Journal of the Science of Food and Agriculture, 101*, 555–563. https://doi.org/10.1002/jsfa.10667

Pitt, J. I., Cruickshank, R. H., & Leistner, L. (1986). Penicillium commune, P. camembertii, the origin of white cheese moulds, and the production of cyclopiazonic acid. *Food Microbiology, 3*, 363–371. https://doi.org/10.1016/0740-0020(86)90022-5

Polonelli, L., Morace, G., Rosa, R., Castagnola, M., & Frisvad, J. C. (1987). Antigenic characterization of Penicillium camemberti and related common cheese contaminants. *Applied and Environmental Microbiology, 53*, 872–878. https://doi.org/10.1128/aem.53.4.872-878.1987

Poltorak, A., Zhou, X., Kasuga, T., Xu, Y., & Fan, Z. (2024). Conversion of deproteinized cheese whey to lactobionate by an engineered neurospora crassa strain F5. *Applied Biochemistry and Biotechnology, 196*, 1292–1303. https://doi.org/10.1007/s12010-023-04583-x

Powers, E. M., Ay, C. C., Eckfeldt, G. A., & Rowley, D. B. (1971). Modification of the American Public Health Association procedure for counting yeast and mold in cottage cheese. *Applied Microbiology, 21*, 155–156. https://doi.org/10.1128/am.21.1.155-156.1971

Prajapati, J. B., & Nair, B. M. (2003). *The history of fermented foods* (E. R. Farnworth, Ed.). CRC Press.

Pretorius, I. S. (2017). Synthetic genome engineering forging new frontiers for wine yeast. *Critical Reviews in Biotechnology, 37*, 112–136. https://doi.org/10.1080/07388551.2016.1214945

Psomas, E., Andrighetto, C., Litopoulou-Tzanetaki, E., Lombardi, A., & Tzanetakis, N. (2001). Some probiotic properties of yeast isolates from infant faeces and feta cheese. *International Journal of Food Microbiology, 69*, 125–133. https://doi.org/10.1016/s0168-1605(01)00580-3

Psomas, E. I., Fletouris, D. J., Litopoulou-Tzanetaki, E., & Tzanetakis, N. (2003). Assimilation of cholesterol by yeast strains isolated from infant feces and feta cheese. *Journal of Dairy Science, 86*, 3416–3422. https://doi.org/10.3168/jds.S0022-0302(03)73945-9

Querol, A., & Bond, U. (2009). The complex and dynamic genomes of industrial yeasts. *FEMS Microbiology Letters, 293*, 1–10. https://doi.org/10.1111/j.1574-6968.2008.01480.x

Rahmani, B., Alimadadi, N., Attaran, B., & Nasr, S. (2022). Yeasts from Iranian traditional milk kefir samples: Isolation, molecular identification and their potential probiotic properties. *Letters in Applied Microbiology, 75*, 1264–1274. https://doi.org/10.1111/lam.13794

Ray, R. C., & Joshi, V. K. (2014). Fermented foods: Past, present and future. In *Microorganisms and fermentation of traditional foods* (pp. 1–36).

Rizk, Z., El Rayess, Y., Ghanem, C., Mathieu, F., Taillandier, P., & Nehme, N. (2016). Impact of inhibitory peptides released by Saccharomyces cerevisiae BDX on the malolactic fermentation performed by Oenococcus oeni Vitilactic F. *International Journal of Food Microbiology, 233*, 90–96. https://doi.org/10.1016/j.ijfoodmicro.2016.06.018

Rizk, Z., Rayess, Y. E., Ghanem, C., Mathieu, F., Taillandier, P., & Nehme, N. (2018). Identification of multiple-derived peptides produced by Saccharomyces cerevisiae involved in malolactic fermentation inhibition. *FEMS Yeast Research, 18*. https://doi.org/10.1093/femsyr/foy080

Rodzi, N., & Lee, L. K. (2021). Traditional fermented foods as vehicle of non-dairy probiotics: Perspectives in South East Asia countries. *Food Research International, 150*, 110814. https://doi.org/10.1016/j.foodres.2021.110814

Romanin, D., Serradell, M., González Maciel, D., Lausada, N., Garrote, G. L., & Rumbo, M. (2010). Down-regulation of intestinal epithelial innate response by probiotic yeasts isolated from kefir. *International Journal of Food Microbiology, 140*, 102–108. https://doi.org/10.1016/j.ijfoodmicro.2010.04.014

Roostita, R., & Fleet, G. H. (1996). The occurrence and growth of yeasts in Camembert and blue-veined cheeses. *International Journal of Food Microbiology, 28*, 393–404. https://doi.org/10.1016/0168-1605(95)00018-6

Ropars, J., López-Villavicencio, M., Snirc, A., Lacoste, S., & Giraud, T. (2017). Blue cheese-making has shaped the population genetic structure of the mould Penicillium roqueforti. *PLoS One, 12*, e0171387. https://doi.org/10.1371/journal.pone.0171387

Rosa, L. S., Santos, M. L., Abreu, J. P., Rocha, R. S., Esmerino, E. A., Freitas, M. Q., Mársico, E. T., Campelo, P. H., Pimentel, T. C., Cristina Silva, M., et al. (2023). Probiotic fermented whey-milk beverages: Effect of different probiotic strains on the physicochemical characteristics, biological activity, and bioactive peptides. *Food Research International, 164*, 112396. https://doi.org/10.1016/j.foodres.2022.112396

Ross, R. P., Morgan, S., & Hill, C. (2002). Preservation and fermentation: Past, present and future. *International Journal of Food Microbiology, 79*, 3–16. https://doi.org/10.1016/s0168-1605(02)00174-5

Saleem, G. N., Gu, R., Qu, H., Bahar Khaskheli, G., Rashid Rajput, I., Qasim, M., & Chen, X. (2024). Therapeutic potential of popular fermented dairy products and its benefits on human health. *Frontiers in Nutrition, 11*, 1328620. https://doi.org/10.3389/fnut.2024.1328620

Sandiford, C. P., Tee, R. D., & Taylor, A. J. (1994). The role of cereal and fungal amylases in cereal flour hypersensitivity. *Clinical and Experimental Allergy, 24*, 549–557. https://doi.org/10.1111/j.1365-2222.1994.tb00952.x

Sar, T., Ferreira, J. A., & Taherzadeh, M. J. (2020). Bioprocessing strategies to increase the protein fraction of Rhizopus oryzae biomass using fish industry sidestreams. *Waste Management, 113*, 261–269. https://doi.org/10.1016/j.wasman.2020.06.005

Savary, O., Coton, E., Maillard, M. B., Gaucheron, F., Le Meur, C., Frisvad, J., Thierry, A., Jany, J. L., & Coton, M. (2023). Functional diversity of Bisifusarium domesticum and the newly described Nectriaceae cheese-associated species. *Food Research International, 168*, 112691. https://doi.org/10.1016/j.foodres.2023.112691

Schmidt, J. L., & Daudin, J. J. (1983). The homogeneity of a population of yeasts from Camembert cheeses. *Annals of Microbiology (Paris), 134a*, 399–409.

Schwentke, J., Sabel, A., Petri, A., König, H., & Claus, H. (2014). The yeast Wickerhamomyces anomalus AS1 secretes a multifunctional exo-β-1,3-glucanase with implications for winemaking. *Yeast, 31*, 349–359. https://doi.org/10.1002/yea.3029

Shakeel Ur, R., Farkye, N. Y., Vedamuthu, E. R., & Drake, M. A. (2003). A preliminary study on the effect of adding yeast extract to cheese curd on proteolysis and flavour development of reduced-fat Cheddar. *The Journal of Dairy Research, 70*, 99–103. https://doi.org/10.1017/s0022029902005861

Shakeel, Q., Lyu, A., Zhang, J., Wu, M., Li, G., Hsiang, T., & Yang, L. (2018). Biocontrol of Aspergillus flavus on Peanut Kernels using Streptomyces yanglinensis 3-10. *Frontiers in Microbiology, 9*. https://doi.org/10.3389/fmicb.2018.01049

Siemens, K., & Zawistowski, J. (1993). Occurrence of PR Imine, a metabolite of Penicillium roqueforti, in blue cheese. *Journal of Food Protection, 56*, 317–319. https://doi.org/10.4315/0362-028x-56.4.317

Sittisart, P., Mahidsanan, T., Yuvanatemiya, V., & Srinamngoen, P. (2024). Technological quality and fungal community of Kombucha fermented with hemp leaves and milky mushroom flour (Calocybe indica). *PeerJ, 12*, e18116. https://doi.org/10.7717/peerj.18116

Sivamaruthi, B. S., Sisubalan, N., Kesika, P., Sureka, I., & Chaiyasut, C. (2024). A concise review of the nutritional profiles, microbial dynamics, and health impacts of fermented mushrooms. *Journal of Food Science, 89*, 3973–3994. https://doi.org/10.1111/1750-3841.17172

Spinnler, H. E., & Gripon, J. C. (2004). Surface mould-ripened cheeses. *Cheese: Chemistry, Physics and Microbiology, 2*, 157–174. https://doi.org/10.1016/s1874-558x(04)80043-5

Suesse, A. R., Norton, G. A., & van Leeuwen, J. (2016). Pilot-scale continuous-flow hydrothermal liquefaction of filamentous fungi. *Energy & Fuels, 30*, 7379–7386. https://doi.org/10.1021/acs.energyfuels.6b01229

Sugimoto, S., Fujii, T., Morimiya, T., Johdo, O., & Nakamura, T. (2007). The fibrinolytic activity of a novel protease derived from a tempeh producing fungus, Fusarium sp. BLB. *Bioscience, Biotechnology, and Biochemistry, 71*, 2184–2189. https://doi.org/10.1271/bbb.70153

Suhr, K. I., Haasum, I., Steenstrup, L. D., & Larsen, T. O. (2002). Factors affecting growth and pigmentation of Penicillium caseifulvum. *Journal of Dairy Science, 85*, 2786–2794. https://doi.org/10.3168/jds.S0022-0302(02)74365-8

Sun, Y., Zhao, H., Chang, M., Yue, T., Yuan, Y., & Shi, Y. (2025). Prophylactic effects of Tibetan goat kefir on depression-like behaviors in chronic unpredictable stress model through the gut-brain axis. *Journal of the Science of Food and Agriculture, 105*, 266–275. https://doi.org/10.1002/jsfa.13825

Supriyanto, Fujio, Y., & Hayakawa, I. (1994). Statistical characterization of tempeh starter from the aroma components of soybean tempeh. In *Developments in food engineering* (pp. 522–524). https://doi.org/10.1007/978-1-4615-2674-2_167

Suraj, M., Abewaa, M., Mengistu, A., Bultosa, G., & Bussa, N. (2024). Influence of fermentation conditions, and the blends of sorghum and carrot pulp supplementation on the nutritional and sensory quality of tef injera. *Scientific Reports, 14*, 12819. https://doi.org/10.1038/s41598-024-62420-5

Suzzi, G., Lanorte, M. T., Galgano, F., Andrighetto, C., Lombardi, A., Lanciotti, R., & Guerzoni, M. E. (2001). Proteolytic, lipolytic and molecular characterisation of Yarrowia lipolytica isolated from cheese. *International Journal of Food Microbiology, 69*, 69–77. https://doi.org/10.1016/s0168-1605(01)00574-8

Tamang, J., & Samuel, D. (2010). Dietary cultures and antiquity of fermented foods and beverages. In *Fermented foods and beverages of the world* (pp. 1–40). https://doi.org/10.1201/ebk1420094954-c1

Tamang, J., Thapa, N., Tamang, B., Rai, A., & Chettri, R. (2015). Microorganisms in fermented foods and beverages. In *Health benefits of fermented foods and beverages* (pp. 1–110). https://doi.org/10.1201/b18279-2

Thanh, V. N., Mai, L. T., & Tuan, D. A. (2008). Microbial diversity of traditional Vietnamese alcohol fermentation starters (banh men) as determined by PCR-mediated DGGE. *International Journal of Food Microbiology, 128*, 268–273. https://doi.org/10.1016/j.ijfoodmicro.2008.08.020

Thapa, S., & Tamang, J. P. (2004). Product characterization of kodo ko jaanr: Fermented finger millet beverage of the Himalayas. *Food Microbiology, 21*, 617–622. https://doi.org/10.1016/j.fm.2004.01.004

Thrane, U. (2007). Fungal protein for food. In J. Dijksterhuis & R. A. Samson (Eds.), *Food mycology: A multifaceted approach to fungi and food* (pp. 353–360). CRC Press.

Toelstede, S., & Hofmann, T. (2009). Kokumi-active glutamyl peptides in cheeses and their biogeneration by Penicillium roquefortii. *Journal of Agricultural and Food Chemistry, 57*, 3738–3748. https://doi.org/10.1021/jf900280j

Tong, S., Chen, W., Hong, R., Chai, M., Sun, Y., Wang, Q., & Li, D. (2024a). Efficient mycoprotein production with low CO(2) emissions through metabolic engineering and fermentation optimization of Fusarium venenatum. *Journal of Agricultural and Food Chemistry, 72*, 604–612. https://doi.org/10.1021/acs.jafc.3c08509

Tong, S., Hong, R., Chen, W., Chai, M., Zhang, Y., Sun, Y., Wang, Q., & Li, D. (2024b). Synchronous bioproduction of betanin and mycoprotein in the engineered edible fungus Fusarium venenatum. *Journal of Agricultural and Food Chemistry, 72*, 19462–19469. https://doi.org/10.1021/acs.jafc.4c06071

Toro-Funes, N., Bosch-Fuste, J., Latorre-Moratalla, M. L., Veciana-Nogués, M. T., & Vidal-Carou, M. C. (2015). Biologically active amines in fermented and non-fermented commercial soybean products from the Spanish market. *Food Chemistry, 173*, 1119–1124. https://doi.org/10.1016/j.foodchem.2014.10.118

Umemura, M., Koyama, Y., Takeda, I., Hagiwara, H., Ikegami, T., Koike, H., & Machida, M. (2013). Fine *de novo* sequencing of a fungal genome using only SOLiD short read data: Verification on *Aspergillus oryzae* RIB40. *PLoS One, 8*, e63673–e63673. https://doi.org/10.1371/journal.pone.0063673

Vacca, M., Celano, G., Serale, N., Costantino, G., Calabrese, F. M., Calasso, M., & De Angelis, M. (2024). Dynamic microbial and metabolic changes during Apulian Caciocavallo cheese-making and ripening produced according to a standardized protocol. *Journal of Dairy Science, 107*, 6541–6557. https://doi.org/10.3168/jds.2023-24049

Valentino, V., Magliulo, R., Farsi, D., Cotter, P. D., O'Sullivan, O., Ercolini, D., & De Filippis, F. (2024). Fermented foods, their microbiome and its potential in boosting human health. *Microbial Biotechnology, 17*, e14428. https://doi.org/10.1111/1751-7915.14428

Vallone, L., Giardini, A., & Soncini, G. (2018). Secondary metabolites from Penicillium roqueforti, a starter for the production of Gorgonzola cheese. *Italian Journal of Food Safety, 3*. https://doi.org/10.4081/ijfs.2014.2118

van Rijswijck, I. M. H., Derks, M. F. L., Abee, T., de Ridder, D., & Smid, E. J. (2017). Genome sequences of Cyberlindnera fabianii 65, Pichia kudriavzevii 129, and Saccharomyces cerevisiae 131 isolated from fermented Masau fruits in Zimbabwe. *Genome Announcements, 5*, e00064–e00017. https://doi.org/10.1128/genomeA.00064-17

Vancanneyt, M., Mengaud, J., Cleenwerck, I., Vanhonacker, K., Hoste, B., Dawyndt, P., Degivry, M. C., Ringuet, D., Janssens, D., & Swings, J. (2004). Reclassification of Lactobacillus kefirgranum Takizawa et al. 1994 as Lactobacillus kefiranofaciens subsp. kefirgranum subsp. nov. and emended description of L. kefiranofaciens Fujisawa et al. 1988. *International Journal of Systematic and Evolutionary Microbiology, 54*, 551–556. https://doi.org/10.1099/ijs.0.02912-0

Vijayakumar, V. E., & Venkataraman, K. (2024). A systematic review of the potential of *Pichia pastoris* (Komagataella phaffii) as an alternative host for biologics production. *Molecular Biotechnology, 66*, 1621–1639. https://doi.org/10.1007/s12033-023-00803-1

Viljoen, B. C., & Greyling, T. (1995). Yeasts associated with Cheddar and Gouda making. *International Journal of Food Microbiology, 28*, 79–88. https://doi.org/10.1016/0168-1605(94)00114-1

Vivier, D., Rivemale, M., Reverbel, J. P., Ratomahenina, R., & Galzy, P. (1994). Study of the growth of yeasts from feta cheese. *International Journal of Food Microbiology, 22*, 207–215. https://doi.org/10.1016/0168-1605(94)90143-0

Vorbeck, M. L., & Cone, J. F. (1963). Characteristics of an intracellullar proteinase system of a Trichosporon species isolated from Trappist-type cheese. *Applied Microbiology, 11*, 23–27. https://doi.org/10.1128/am.11.1.23-27.1963

Wagner, T., Pfeifle, H., Hildebrand, G., & Zhang, Y. (2024). Production of a cheese-like aroma via fermentation of plant proteins and coconut oil with the basidiomycetes Cyclocybe aegerita and Trametes versicolor. *Journal of Agricultural and Food Chemistry, 72*, 6544–6553. https://doi.org/10.1021/acs.jafc.4c00219

Wainaina, S., Kisworini, A. D., Fanani, M., Wikandari, R., Millati, R., Niklasson, C., & Taherzadeh, M. J. (2020). Utilization of food waste-derived volatile fatty acids for production of edible Rhizopus oligosporus fungal biomass. *Bioresource Technology, 310*, 123444. https://doi.org/10.1016/j.biortech.2020.123444

Wang, B., Guo, G., Wang, C., Lin, Y., Wang, X., Zhao, M., Guo, Y., He, M., Zhang, Y., & Pan, L. (2010). Survey of the transcriptome of Aspergillus oryzae via massively parallel mRNA sequencing. *Nucleic Acids Research, 38*, 5075–5087. https://doi.org/10.1093/nar/gkq256

Wang, S.-Y., Chen, K.-N., Lo, Y.-M., Chiang, M.-L., Chen, H.-C., Liu, J.-R., & Chen, M.-J. (2012). Investigation of microorganisms involved in biosynthesis of the kefir grain. *Food Microbiology, 32*, 274–285. https://doi.org/10.1016/j.fm.2012.07.001

Wang, H., Sun, X., Song, X., & Guo, M. (2021). Effects of kefir grains from different origins on proteolysis and volatile profile of goat milk kefir. *Food Chemistry, 339*, 128099. https://doi.org/10.1016/j.foodchem.2020.128099

Welthagen, J. J., & Viljoen, B. C. (1998). Yeast profile in Gouda cheese during processing and ripening. *International Journal of Food Microbiology, 41*, 185–194. https://doi.org/10.1016/s0168-1605(98)00042-7

Witthuhn, R. C., Schoeman, T., & Britz, T. J. (2005). Characterisation of the microbial population at different stages of Kefir production and Kefir grain mass cultivation. *International Dairy Journal, 15*, 383–389. https://doi.org/10.1016/j.idairyj.2004.07.016

Wu, J., Tian, T., Liu, Y., Shi, Y., Tao, D., Wu, R., & Yue, X. (2018). The dynamic changes of chemical components and microbiota during the natural fermentation process in Da-Jiang, a Chinese popular traditional fermented condiment. *Food Research International, 112*, 457–467. https://doi.org/10.1016/j.foodres.2018.06.021

Wu, Y., Li, Y., Liang, H., Zhang, S., Lin, X., & Ji, C. (2025a). Enhancing cider quality through co-fermentation with acid protease and esterase-producing Metschnikowia species and Saccharomyces cerevisiae. *Journal of the Science of Food and Agriculture, 105*, 1003–1011. https://doi.org/10.1002/jsfa.13891

Wu, J. Y., Wee, S., Ler, S. G., Henry, C. J., & Gunaratne, J. (2025b). Unraveling the impact of tempeh fermentation on protein nutrients: An in vitro proteomics and peptidomics approach. *Food Chemistry, 474*, 143154. https://doi.org/10.1016/j.foodchem.2025.143154

Wyder, M.-T., Meile, L., & Teuber, M. (1999). Description of Saccharomyces turicensis sp. nov., a new species from Kefyr. *Systematic and Applied Microbiology, 22*, 420–425. https://doi.org/10.1016/s0723-2020(99)80051-4

Xia, Y., Yuan, R., Weng, S., Wang, G., Xiong, Z., Zhang, H., Song, X., Liu, W., & Ai, L. (2020). Proteolysis, lipolysis, texture and sensory properties of cheese ripened by Monascus fumeus. *Food Research International, 137*, 109657. https://doi.org/10.1016/j.foodres.2020.109657

Xu, L., Du, B., & Xu, B. (2015). A systematic, comparative study on the beneficial health components and antioxidant activities of commercially fermented soy products marketed in China. *Food Chemistry, 174*, 202–213. https://doi.org/10.1016/j.foodchem.2014.11.014

Yadav, J. S., Bezawada, J., Ajila, C. M., Yan, S., Tyagi, R. D., & Surampalli, R. Y. (2014a). Mixed culture of Kluyveromyces marxianus and Candida krusei for single-cell protein production and organic load removal from whey. *Bioresource Technology, 164*, 119–127. https://doi.org/10.1016/j.biortech.2014.04.069

Yadav, J. S., Bezawada, J., Elharche, S., Yan, S., Tyagi, R. D., & Surampalli, R. Y. (2014b). Simultaneous single-cell protein production and COD removal with characterization of residual protein and intermediate metabolites during whey fermentation by K. marxianus. *Bioprocess and Biosystems Engineering, 37*, 1017–1029. https://doi.org/10.1007/s00449-013-1072-6

Yang, L., Zhang, J., Chen, Z., Chen, Y., Wang, C., Yu, H., Zuo, F., & Huang, W. (2025). Probiotic-enzyme synergy regulates fermentation of distiller's grains by modifying microbiome structures and symbiotic relationships. *Journal of Agricultural and Food Chemistry, 73*, 5363–5375. https://doi.org/10.1021/acs.jafc.4c11539

Yangılar, F., & Oğuzhan Yıldız, P. (2016). Casein/natamycin edible films efficiency for controlling mould growth and on microbiological, chemical and sensory properties during the ripening of Kashar cheese. *Journal of the Science of Food and Agriculture, 96*, 2328–2336. https://doi.org/10.1002/jsfa.7348

Yin, L., Zhang, Y., Azi, F., Tekliye, M., Zhou, J., Liu, X., Dong, M., & Xia, X. (2021). Neuroprotective potency of tofu bio-processed using *Actinomucor elegans* against hypoxic injury induced by cobalt chloride in PC12 cells. *Molecules (Basel, Switzerland), 26*, 2983.

Youn, H. Y., Kim, D. H., Kim, H. J., Jang, Y. S., Song, K. Y., Bae, D., Kim, H., & Seo, K. H. (2023). A combined in vitro and in vivo assessment of the safety of the yeast strains Kluyveromyces marxianus A4 and A5 isolated from Korean Kefir. *Probiotics and Antimicrobial Proteins, 15*, 129–138. https://doi.org/10.1007/s12602-021-09872-7

Yu, X. J., Chi, Z., Wang, F., Li, J., Chi, Z. M., & Madzak, C. (2013). Expression of the acid protease gene from Saccharomycopsis fibuligera in the marine-derived Yarrowia lipolytica for both milk clotting and single cell protein production. *Applied Biochemistry and Biotechnology, 169*, 1993–2003. https://doi.org/10.1007/s12010-013-0118-1

Yuan, Y., Wu, F., Si, J., Zhao, Y. F., & Dai, Y. C. (2019). Whole genome sequence of Auricularia heimuer (Basidiomycota, fungi), the third most important cultivated mushroom worldwide. *Genomics, 111*, 50–58. https://doi.org/10.1016/j.ygeno.2017.12.013

Zhang, W., Shi, K., Han, Y., Wang, J., Yang, C., Xu, X., & Li, B. (2022). Characterization of Pleurotus citrinopileatus hydrolysates obtained from Actinomucor elegans proteases compared with that by commercial proteases. *Journal of Food Science, 87*, 3737–3751. https://doi.org/10.1111/1750-3841.16256

Zheng, X., Liu, F., Li, K., Shi, X., Ni, Y., Li, B., & Zhuge, B. (2018). Evaluating the microbial ecology and metabolite profile in Kazak artisanal cheeses from Xinjiang, China. *Food Research International, 111*, 130–136. https://doi.org/10.1016/j.foodres.2018.05.019

Zhong, X., Peng, L., Zheng, S., Sun, Z., Ren, Y., Dong, M., & Xu, A. (2004). Secretion, purification, and characterization of a recombinant Aspergillus oryzae tannase in Pichia pastoris. *Protein Expression and Purification, 36*, 165–169. https://doi.org/10.1016/j.pep.2004.04.016

Zhong, Y., Yang, L., Guo, Y., Fang, F., Wang, D., Li, R., Jiang, M., Kang, W., Ma, J., Sun, J., et al. (2014). High-temperature cultivation of recombinant Pichia pastoris increases endoplasmic reticulum stress and decreases production of human interleukin-10. *Microbial Cell Factories, 13*, 163. https://doi.org/10.1186/s12934-014-0163-7

Recent Innovations in Myco-biotechnology: Valourisation of Fungal Resources for a Sustainable and a Circular Economy

S. Ramalakshmi, J. Helan Chandra, K. Sivasubramanian, R. S. Arvind Bharani, R. Lekhavani, and K. Suguna

Abstract Food insecurity, environmental degradation, and climate change are among the persistent global challenges facing humanity, and fungi offer unique and underexplored opportunities to address them. *Resilient food sources, animal feed, biochemicals, biofuels, textiles, and raw materials for the automotive, construction, and transportation sectors are among the many valuable products that these adaptable organisms can effectively derive from organic materials. Their exceptional metabolic diversity and adaptability make them excellent candidates for industrial applications.* The global transition from a petroleum-based economy to a sustainable, bio-based circular economy is being accelerated by advances in fungal biotechnology. *Fungal systems contribute to reducing greenhouse gas emissions and environmental pollution by enabling the production of eco-friendly alternatives and promoting efficient resource utilisation. Additionally, by stabilising agricultural systems and enhancing nutrient availability for a growing global population, fungi offer promising strategies to improve food security.* Developments in fungal biotechnology, along with the rational design and optimisation of both existing and novel fungal cell factories, hold significant potential to mitigate the effects of climate change. These initiatives are aligned with the United Nations Sustainable Development Goals (SDGs), particularly those related to responsible consumption, climate action, and zero hunger. This chapter summarises current prospects and research challenges in the utilisation of fungal resources for a sustainable and

---

S. Ramalakshmi
Department of Bioengineering, Vels Institute of Science, Technology and Advanced Studies, Pallavaram, Chennai, India

J. Helan Chandra (✉)
Ampersand Academy, Ashok Nagar, Chennai, Tamil Nadu, India

K. Sivasubramanian · R. Lekhavani · K. Suguna
Centre for Materials Engineering and Regenerative Medicine, Bharath Institute of Higher Education and Research, Selaiyur, Chennai, Tamil Nadu, India

R. S. Arvind Bharani
Institute of Obstetrics and Gynaecology, Madras Medical College, Egmore, Chennai, Tamil Nadu, India

A. Gupta (ed.), *Prospects of Fungal Biotechnologies for Livestock Volume 2*, Fungal Biology, https://doi.org/10.1007/978-3-032-06478-3_7

circular economy, aiming to raise awareness among academicians, researchers, industry stakeholders, and policymakers about the ongoing fungal biotechnology revolution.

Keywords Biorefinery · Bioremediation · Circular economy · Fungal biotechnology · Fungi-based nutraceuticals · Sustainable agriculture

1 Introduction to Fungal Resources

Fungi are eukaryotic heterotrophs that obtain their nourishment from other living organisms. Broader but underutilised fungal biodiversity can help achieve the Sustainable Development Goals (SDGs) proposed by the United Nations (UN) (Monclaro et al., 2024). Fungal biotechnology is experiencing resurgence 100 years after its emergence, providing sustainable solutions across different markets and industries. The development of fungal biotechnology holds promise for mitigating global warming and advancing SDG reforms. Additionally, producing different metabolites, enzymes, and materials is another advantageous characteristic of fungal resources. Consequently, fungi are a potential food resource that could be consumed directly. Similar to probiotics, fungi have excellent anti-inflammatory and antioxidant properties that enhance animal health (Garg et al., 2024). Antibiotics have noticeably enhanced the likelihood of individuals leading healthy, disease-free lives. Several fungi produce antibiotic compounds to protect plants from soil-borne pests and diseases. Additionally, *Saccharomyces cerevisiae*, *Pichia pastoris*, *Yarrowia lipolytica*, and edible mushrooms (Ascomycota and Basidiomycota) have been extensively used in various commercial applications (Aggarwal & Kumari, 2024). The worldwide assessed mushroom market size was 50.3 billion USD in 2021 and is expected to multiply at a Compound Annual Growth Rate (CAGR) of 9.7% from 2022 to 2030 (Sharma et al., 2024). The contribution of fungi in achieving the United Nations SDGs is presented in Fig. 1. This chapter explores the multifaceted applications of fungi that advance environmental sustainability and improve the circular economy.

2 Fungal Applications in Sustainable Agriculture

As the worldwide population continues to grow, it is highly required to identify sustainable and effective methods to grow our food. Hence, there is a relentless need for farmers to grow more food on less land. It is now common practice to use plant-beneficial microorganisms, such as probiotic and postbiotic combinations, to manage and stimulate the beneficial microbiome that already exists in order to mitigate

Fig. 1 The 17 Sustainable Development Goals (SDGs) outlined by the United Nations. This visual presents a circular arrangement of 17 colour-coded icons, each representing a distinct goal—from poverty eradication and clean water access to climate action and global partnerships. At the centre, the phrase "Sustainable Development Goals" is encircled by a multicoloured ring, symbolising the interconnected nature of these global priorities aimed at achieving peace and prosperity by 2030

all of these issues in plant production. In agriculture and the soil environment, fungal communities are among the crucial soil microbes. Since many fungal microorganisms can produce useful bioactive compounds to promote plant growth and act as bio-control agents, they are known to have potential biotechnological applications in agriculture (Khan et al., 2024).

2.1 *Fungal-Based Pesticides and Fertilisers*

Biopesticides and biofertilisers derived from fungi have demonstrated efficacy in improving plant growth and health while managing pests and diseases in a sustainable manner. Secondary fungal metabolites have recently drawn attention because of their dual function in increasing productivity and pest-controlling effect. Besides, *Trichoderma* and arbuscular mycorrhizal fungi (AMF) are the two important fungi that act as biostimulants in agriculture. These have an impact on plant productivity due to AMFs' capacity to act as bio-fertilisers, bio-protectors, or bio-degraders. AMF is crucial in shielding plants from fungi and bacteria that cause phytopathogenicity. A recent study found several AMF isolates' ability to increase tomato growth and reduce biomass losses caused by infections with significant root disease. According to the findings, AMFs have a great deal of potential to boost crop productivity and resilience. They have also created new opportunities for using different plant and AMF species combinations (Abarca et al., 2024). The role of AMF in soil remediation is presented in Fig. 2.

Biofertilisers that comprise phosphate-solubilising bacteria (PSB) and AMF augment the ability of plants to endure drought and stress. In order to track the growth of cabbage plants and the soil fungus population, a study was carried out in which the soil was modified using fertiliser that included *Trichoderma guizhouense*. It was found that the biofertilisers amended using *Trichoderma* prominently increased the growth of the cabbage plant. This was reasoned due to the changes in the fungal community, such as the rise in the number and relative abundance of native fungal species that promote soil growth. The limitations imposed by their required biotrophic conditions and the absence of suitable equipment for large-scale production represent a few of the reasons why AMF-rhizosphere microflora interactions in horticulture, forestry, and agriculture are still neglected despite these advantages (Yağmur et al., 2024).

Mycorrhiza has lately gained attention due to its use as a biofertiliser in ecosystems and forestry (Wang et al., 2023). *Ectomycorrhiza* (ECM) is a fungal

Fig. 2 Benefits of arbuscular mycorrhiza fungi in agriculture and soil health. The infographic illustrates five key advantages of arbuscular mycorrhiza fungi, including their roles as biofertilisers, bioprotectors, and biostimulators. These fungi enhance water uptake, improve soil quality, and support plant growth—each benefit visually represented with distinct icons branching from a central fungal illustration

association with higher plant roots that benefits both species equally and seems to be crucial for both parties' survival. This association undoubtedly provided significant advantages in stabilising and restoring forest and ecosystem soil aggregation. ECM fungi also improve the nutrient cycle, soil structure, and phytohormone production, boosting seedling growth and survival, increasing the rate at which plants photosynthesise, and maintaining the plants' tolerance level to environmental stresses in forest nurseries—additionally, *Ectomycorrhiza* lowers fertiliser cost in a sustainable way (Yang et al., 2022). The global market of biofertilisers is classified into *Azotobacter* (17%), *Azospirillum* (16%), *Rhizobium* (18%), phosphate-solubilising bacteria (15%), mycorrhiza (17%), blue-green algae (13%), and others (4%). The previous are the main microorganisms that have been used by the top global Agro-companies (Bayer CropScience AG, Mapleton Agri Biotech Pty Limited, Lallemand Inc., Symborg SL, CBF China Bio-Fertilizer AG, National Fertilizer Ltd., among others) (Sansinenea, 2021). Despite the increasing demand, biopesticides production represents only 5% of the global crop production products market (Ramírez-Guzman' et al., 2021), while biofertilisers have a much higher demand percentage, with 10.9% of global Comprehensive Annual Financial Report (Chaparro et al., 2021). These numbers indicate that there is growing interest worldwide in integrating this environmentally beneficial option into the mass food production system.

3 Fungal-Based Biorefinery

Petroleum and fossil fuels have propelled industrial growth in contemporary society for the past 200 years and are about to go extinct. Sustainable biomass processing into various marketable goods and energy is a key component of a biorefinery. Biorefineries combine upstream, midstream, and downstream biomass processing to provide various value-added products. Large-scale biorefineries are anticipated to help nations become more competitive and prosperous in the future by meeting the demand for a variety of bio-based goods and energy in a manner that is economically, socially, and environmentally sustainable. Fungi are crucial in addressing significant global issues (Hassan, 2025). The use of fungal enzymes in industries including textiles, leather, paper, and pulp has allowed for a transition from chemical to biological processing, which has greatly lessened the adverse effects on the environment.

3.1 Fungal Applications in Biofuel and Biogas Production

One of the best alternatives to fossil fuels produced from the depleting petroleum resources is biofuel. Agricultural wastes could be converted into bioethanol, supplying natural resources at a low cost. With different suspended microorganisms, lignocellulosic hydrolysate has been converted to bioethanol using various techniques (Periyasamy et al., 2023). In a continuous bioethanol production method, a study was conducted to encapsulate three microorganisms—*S. cerevisiae*, *T. reesei*, and *P. stipitis*. These microbes were placed in small bioreactor platform (SBP) capsules and studied for bioethanol generation, biological activity, and post-encapsulation survivability.

The encapsulation of microbes in SBP capsules guaranteed their long-term activity and flourishing, simplified their acclimatisation, and gave them a contained, protected environment. The simultaneous availability of saccharification and fermentation for an extended period, roughly 2.5–3 months, without the need to replace the cells or encapsulating matrices, was another important advantage of using SBP capsules. It was witnessed that the effectiveness of the systems ranged from 60% to 70%, indicating the microbes' sustained metabolic activity and long-term prosperity (Rahamim et al., 2022). As fungi break down organic waste and produce enzymes that aid in fermentation, they can be employed to enhance biogas production from that waste.

3.2 The Role of Fungi in Eco-friendly Textile Processing

Fungi are significant in the textile industry's biorefinery. Food waste can be converted into fungal monofilaments by fungal biorefineries, which can then be used in textiles. The fungal cell wall produced by alkali treating the fungal biomass when *Rhizopus delemar* was grown on bread waste included 23% and 36% N-acetyl glucosamine, and glucosamine, respectively, which stand for chitosan and chitin. Acetic or lactic acid was used to protonate the chitosan's amino groups. As a result, a homogenous hydrogel of fungal microfibres formed. The wet-spun monofilaments exhibited a Young's modulus of 4.97 GPa and tensile strengths of up to 69.5 MPa. This work makes the development of sustainable fungal textiles possible, which shows how to create renewable fibres from fungal cell walls grown on an abundance of food waste in an environmentally safe manner (Svensson et al., 2021). A novel biorefinery technology for creating environmentally beneficial materials within the circular economy is the manufacturing of mycelium-based composites, which excel in biodegradability and profit from the upcycling of waste and industrial byproducts. The study involved the incorporation of local textile waste fibres as reinforcements in these mycelium-based composites. In comparison to samples without reinforcement, the hybridisation technique of adding textile waste fibres resulted in distributed cracking in higher deformation testing (30%), reduced integrity loss, and enhanced compressive resistance. Therefore, by using low-carbon footprint techniques, fungus mycelium, and materials made from recycled textile and agricultural waste, this study helped in the decarbonisation of the environment (Voutetaki & Mpalaskas, 2024). Fungi are also used as a versatile biocatalytic tool for treating textile wastewater effluents. *Geotrichum candidum* effectively decolourises textile effluent in 6 h, removing 85.5% and 98.5% of the dye, resulting in increased efficacy in immobilising the fungus on coconut fibres, which improves its retention in polluted areas. The analysis revealed that a major contributing factor to the improved performance of immobilised *G. candidum* was a laccase-driven breakdown of contaminants (Rajhans et al., 2021). Various applications of fungi in the biorefinery sector are presented in Table 1. Despite the demerits of fungi, such as slow growth rate, scaling-up challenges, and energy-intensive processing, the potential of fungi in the biorefinery sector continues to expand, offering innovative solutions for achieving SDGs.

4 Fungal-Based Sustainable Materials

Buildings are increased annually to meet the lifestyles of the community, but they also generate building trash such as glassware, wood, concrete, steel, bricks, and plastics. They affect environmental sanitation and individuals. Scientists seek ways to employ biological resources instead of non-renewable ones. Bio-composite and bio-based materials were created to meet the growing need for "green" materials and productive processing (Carvalho et al., 2024). Fungi-based biomaterials are being exploited to create bio-based innovations (Srivastava et al., 2025). Researchers in several sectors have recognised the use of farm wastes (Shinde et al., 2022). Coconut, rice, rice straw, and coffee husks were used to grow mushrooms. Rice husk compost, when integrated with soil, enhances agricultural yields and acts as a fertiliser during rice irrigation (Rashad et al., 2021). Though mushrooms are incorporated into diet and medicine, they are also used to produce bioproducts.

Mycelium-based composites, which employ natural fibres and mycelium as a binding ingredient, have substituted non-renewable resources. Fungi, like mushrooms, produce mycelium. Composites are made from many mushrooms. *Pleurotus ostreatus* mycelium was cultivated on wheat bran with additives for concrete and verified for compressive strength (Alemu et al., 2022). These remarkable properties of agricultural wastes and mushrooms lead to bio-composites made from rice and coconut husks and indigenous mushrooms like mycelia, *Pleurotus ostreatus*, *Ganoderma*. Construction is a key driver of economic growth, yet wastes hurt the environment and society. This challenge inspires researchers to create biomaterials from fungus to lessen environmental effects. Mycelium-based material is another new bioproduct technology that replaces non-renewable products and reduces waste. Mycelium-based composites are biomaterials made from mushrooms grown on natural fibres (Elhalis, 2025). Architectural and industrial design may use mycelium-based materials as biodegradable alternative resources to produce bricks and walls. Bio-composite is made of biodegradable natural fibres (Patel & Sharma, 2023). Bio-composite compactness, modulus, natural fibre composites, toughness, thermal conductivity, high temperature, and other qualities have been studied. Agricultural waste with mycelium creates 3D bio-composites like mycelium bricks. They may be strong-flexing, non-flammable construction materials that generate less smoke and CO_2 (Sharma & Sumbria, 2022). The architecture and building

Table 1 Fungi in nutraceutical industry

S No	Fungi	Nutritional property	Therapeutic property	References
1.	*Pleurotus ostreatus*	Higher fibre content and lower sugars compared to commercial cookies	Antioxidant property	Espinosa-Páez et al. (2021)
2.	*Cordyceps militaris*	Cordycepin, cordy glucans, cordymin, monosaccharide saponins, and the exopolysaccharide fraction	Antitumour, blood sugar-lowering, diabetes-fighting, inflammation-reducing, infection-fighting, flu-fighting, seizure-preventing, performance-enhancing, and fatigue-fighting properties	Mayirnao et al. (2025)
3.	*Calocybe indica*	Calocyban, lectin, fucogalactoglucan	Cytotoxic, antibacterial, anticoagulant, anti-inflammatory, immunostimulant, anti-obesity, and anti-ageing properties	Mayirnao et al. (2025)
4.	*Flammulina velutipes*	Hetero-galactan (WFVP-N-b1), enokipodin, sterpurol, 2, 5-cuparadiene-1,4-dione, sterpuric acid, flammulinol, flammulin, flammulina polysaccharide protein (FVP), flammulinolide, proflamin, flammuspirone, flamvelutpenoid…	Immune-stimulatory, anticancer, antibacterial, antiaging, neuroprotective, and antihyperlipidemic properties	Sharma et al. (2022)
5.	*Pleurotus ostreatus*	(6S,7S)-6,7-dihydroxy-3,6-dimethyl-2-isovaleroyl-4,5,6,7-tetrahydrobenzofuran, lectin, pleurospiroketal, lovastatin, and pleuran	Angiotensin-converting enzyme inhibitory, immunomodulatory, antibacterial, antitumor, hypolipidemic, hypoglycaemic, and anti-arthritic properties	Kumar (2020)

(continued)

Table 1 (continued)

S No	Fungi	Nutritional property	Therapeutic property	References
6.	*Ganoderma lucidum*	Ganoderone, lectin, methyl ganoderate, ganoderol, ganoboninketal, n-butyl lucidenate, Ganoderic acid, lucidumol, ganorbiformin, lucidenic acid, 7-O-methyl-ganoderic acid, ganodermadiol, ganoderiol, ganodermanondiol, butyl ganoderate, 7-oxo-ganoderic acid, lucialdehyde, 15-hydroxy-ganoderic acid, ganoderal, ganodermic acid, ganodermanonolganodermacetal, ganodermanontriol, ganodermatriol, applanoxidic acid, lucidadiol, methyl lucidenate, colossolactone, fornicatin, ganosporelactone	Antiviral, anticholinesterase, antitubercular, antimalarial, antiplasmodial, antiangiogenic, immunomodulatory, antitumor, pro-apoptotic, antiatherosclerotic, antifibrotic, chemopreventive, inhibition of cholesterol synthesis, immunomodulatory, antihypertension, inhibition of platelet aggregation, antiherpetic, antiosteoporotic, anti-HIV, antiarthritic, antiaging, hepatoprotective, antiulcer, and hypoglycaemic properties	Gharib et al. (2022)
7.	*Agaricus bisporus*	Agaritine, fucogalactan, lectin, ABP-W1 heteroglycan	Hepatoprotective, anti-inflammatory, anticoagulant, anti-ageing, antimicrobial, anti-cytotoxic, anti-obesity, anti-inflammatory, antinociceptive, and immunomodulatory properties	Usman et al. (2021)

(continued)

Table 1 (continued)

S No	Fungi	Nutritional property	Therapeutic property	References
8.	Coprophilous fungi	Mycotoxins, antibiotics, and substances that promote plant development	Low-cariogenic sugar alternatives that are safe for diabetics as well as probiotics, prebiotics, and nutraceuticals. These oligosaccharides are also used as antioxidants and anticancer agents, and they help with immunological modulation, lipid metabolism, mineral absorption, and more	Ojwach et al. (2022)
9.	Mushroom	Polysaccharides, polysaccharide-protein complexes, peptides and proteins, terpenoids and phenolic compounds	Anti-inflammatory, prebiotic, antidiabetic, antibacterial, anticancer, and antioxidant properties	Kour et al. (2022)
10.	*Agaricus bisporus*	Flavonoids, phenolic content, free amino acids, saponins, alkaloids, tannins, vitamin C, potassium, magnesium, phosphorous, and calcium	Hepato-protective activity, antiaging activity, antitumor activity, and antioxidant activity	Krishnamoorthi et al. (2022)

sector has recently been interested in mycelium bricks as clay brick alternatives. Unlike clay bricks, which are made from deforested dirty clay, they are ecologically beneficial and detrimental to humans and the environment (Silvestri, 2022).

Finding biomaterials to replace plastic and recycle agricultural waste is a hot topic. Mycelium-based composites have been produced using bio-fabrication technology that grows fungus on agricultural wastes for 20 years. This study summarises the key features of this process, including fungal species, substrate type, and how each parameter affects mycelium-based composite characteristics and prospective uses. As packaging, fashion, and architecture alternatives, mycelium-based composites are likely to grow. Mycelium-based materials offer various benefits over standard synthetic materials, including cheap cost, safety, biodegradability, and environmental impact. Low mechanical characteristics, significant water absorption, and the lack of standard mycelium-based materials production and testing techniques remain important difficulties that must be solved in the future. An enlarged set of parameters must be established and utilised to understand substrate component interactions and their effects on ultimate material quality. More advanced analytical approaches should be used to assess the appropriateness of mycelium-based materials for each application properly.

5 Fungal-Based Nutraceutical Applications

Fungi, particularly edible mushrooms, are considered premium foods with functional and nutritional benefits. Because of their therapeutic qualities and economic importance, they have gained interest and have been utilised in the production of nutraceutical supplements. Certain fungal taxa, such as *Basidiomycetes* and primarily filamentous endophytic fungi from the *Ascomycete* phylum, produce secondary metabolites (SMs) with bioactive characteristics contributing to their antioxidant and antibacterial activities. Despite having a high protein and significant fat content, these healthy fungi are also an incredible source of several minerals and vitamins, such as mainly Vitamin B, which are crucial for the metabolism of fat and carbohydrates as well as the upkeep of the neurological system (Wadhwa et al., 2024).

Among the documented species (1.5 lakh) only a small number of these species meet the required standards for functional foods. From the beginning of civilisation, many fungal species have been employed to make food and drink items, including cheese, bread, food flavours, and alcohol. The scope has just lately broadened to include further biotechnological applications. Around the world, edible mushrooms from the phyla *Ascomycota* and *Basidiomycota* are used extensively in cooking and considered excellent delicacies. The nutritional properties of the fungi *Penicillium, Fusarium,* and *Aspergillus* sp. are widely established. The fungal mycelium provides therapeutic applications as they are comprised of vitamins, dietary fibre, and lipids. Certain filamentous fungi contain an enormous quantity of protein, whereas some edible fungal species are used for the preparation of probiotics (Tiwari & Park, 2024). To capitalise on fungi's socio-economic advantages, research has focused on creating food ingredients that include medicines, enzymes, nutraceuticals, and functional food items. Species like *Tremella* reduces epidermal bleeding, *Agaricus bisporus* inhibits the formation of cancer cells, *Ganoderma* boosts immunological activities, and *Morchella esculenta* polysaccharides encourage antioxidant enzyme action. Oligosaccharides from *Copropilous* fungi have been shown to have health-promoting properties and are being investigated as a kind of functional food (Tiwari & Park, 2024).

White fungus biotechnology has significantly increased food items' nutritional content and quality, including bread and beverage flavour, single-cell protein (SCP) quality, product production, and shelf life. Mushrooms are regarded as important components of many different cuisines around the world. They are an excellent source of vitamins, amino acids (glutamine, valine, leucine), carbohydrates (60%), protein (27–48%), and lipids (2–8%) (Li et al., 2021). *Agaricus bisporus*, *Pleurotus* spp., *Auricularia auricula*, *Lentinus edodes*, and *Volvariella volvacea* are grown commercially (Boukary et al., 2024). The value of oysters, shiitake, and champignons has increased to over USD 50 billion by 2022 (Shuting, 2024). Sustainable ways are being used to meet the increased demands worldwide owing to a growing understanding of the nutritional components of various fungus species and their increased consumption. One potential strategy to boost the intended product output

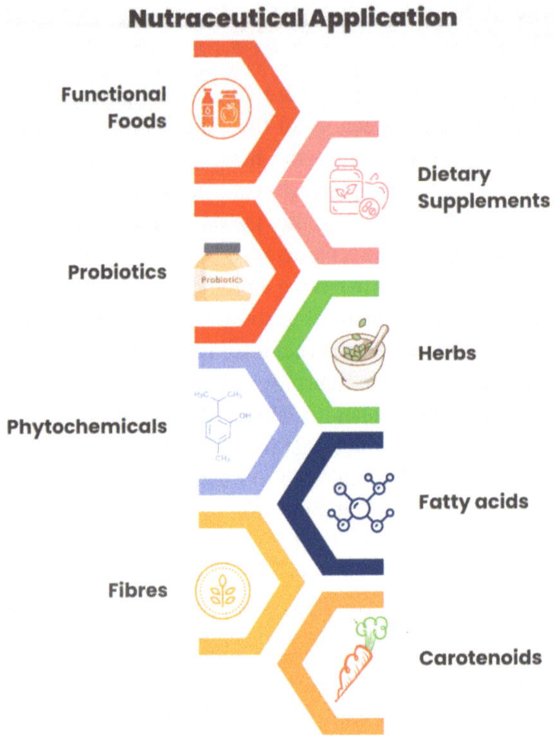

Fig. 3 Categories and applications of nutraceuticals. The diagram presents eight distinct types of nutraceuticals—Functional Foods, Dietary Supplements, Probiotics, Herbs, Phytochemicals, Fatty Acids, Fibres, and Carotenoids—each illustrated with a unique icon and colour-coded arrow. These categories highlight the diverse roles of nutraceuticals in promoting health, from enhancing immunity and digestion to supporting metabolic and cardiovascular functions

and productivity would be the creation of novel strains through genetic engineering research. The applications of fungi in the food industry are illustrated in Fig. 3.

6 Fungal Applications in the Pharmaceutical Industry

Antibiotics and medications are high-value, economically significant products that can be used to treat human illnesses and improve people's health and well-being. Until now, fungi have been used to purify and describe thousands of pharmacologically active metabolites that have shown remarkable therapeutic effects for various illnesses (Rousta et al., 2024).

The ground-breaking discoveries of cephalosporin C and penicillin transformed fungal-mediated drug development. Fungi-derived medications, including established and novel pharmacological classes, demonstrate success in treating autoimmune diseases, hypercholesterolemia, and chronic infections. Among other noteworthy examples, the representative examples are the antibiotics fusidic acid, retapamulin, cephalosporins, penicillin V, enfumafungin, and griseofulvin. This validation of these medications in clinical trials is responsible for their translational success (Prescott et al., 2023). Following this, immunosuppressants produced from fungi (*Tolypocladium inflatum*), including cyclosporin A, have proved helpful in organ transplants because they block the calcineurin pathway, which prevents human T-cell activation (Engwenyu & Anderson, 2021). Mycophenolic acid, another medication that was isolated from *Penicillium brevicompactum*, inhibits the production of guanine and inosine monophosphate dehydrogenase, which limits the growth of lymphocytes (in organ transplants) (Sawant & Sajwan, 2022). Novartis produces the synthetic drug fingolimod, which was inspired by myriocin obtained from fungi. As an immunosuppressant for multiple sclerosis, it was a huge success, earning $1 billion USD in 2012 (Prescott et al., 2023).

A growing medical issue in the current period is the potential of plant-associated endophytes to produce effective antimicrobials and fight drug-resistant bacteria (Tiwari & Park, 2024). The antimicrobials cryptocandin, leucinostatin A, hypericin, munumbicins, colletotric acid, and their derivatives demonstrated clinical efficacy in treating drug-resistant pathogens; however, further research and evaluation are required to ascertain their therapeutic potential and drug development. In obstetric medicine, bromocriptine, a synthetic form of ergocryptine, is used as a dopamine agonist to suppress the pituitary gland's synthesis of prolactin. It is used to treat hyperprolactinemia-related issues. The development of novel therapeutic approaches for managing blood cholesterol levels has been greatly aided by drugs produced by fungi. The discovery of mevastatin, also called compactin, from *Penicillium citrinum* by Japanese scientist Akira Endo marked the beginning of a new age. *Aspergillus terreus* (known as Mevinolin) and *Monascus ruber* (known as Monacolin K) were the initial sources of the statin drug lovastatin, which was shown to be highly effective in lowering blood cholesterol (Prescott et al., 2023; Tiwari & Park, 2024). Statins, one of the most widely promoted drugs in the world, generated around USD 25 billion in sales in 2005 (Lin et al., 2021). Furthermore, several substances originating from fungi and their derivatives are currently being tested in clinical settings for a range of illnesses. These include Wortmannin for prostate cancer, Halimide for cancer, Hypothemycin for solid tumours/melanoma, Cordycepin for lymphoma, and Radicicol for lung cancer (Tiwari & Park, 2024). Additionally, biosynthetic gene clusters (BGCs) found in fungal genomes generate valuable, bioactive compounds (Robey et al., 2021). Furthermore, fungi are employed in the bioremediation of pharmaceutical compounds, which will be discussed in the following section.

6.1 Fungi in the Bioremediation of Emerging Pharmaceutical Pollutants

Humans, wildlife, and the environment are all at risk when excessive amounts of pharmaceutical chemicals (PhCs) are discharged into aquatic habitats. This is a crucial issue requiring immediate attention. New regulations have been implemented to govern the release of medications based on patient safety and efficacy. Nevertheless, research on the environmental impacts of the same medications is still treated with indifference. Because of their extracellular and intracellular enzymes, white rot fungi (WRF), which belong to the Basidiomycota, have gained recognition for their environmental biotechnology use. These enzymes can change the active pharmaceutical compounds (PhACs) and other phenolic and non-phenolic substances in aqueous solutions. Trametes hirsuta demonstrates a multifaceted role in biosolid treatment, including the reduction of biosolid mass, the secretion of diverse extracellular enzymes involved in organic matter degradation, and the efficient removal of pharmaceutical compounds (PhACs) from biosolid slurry, under both natural conditions and when the slurry is supplemented with spiked PhACs. After 35 days of fungal treatment, toxicity was also evaluated. An increase in 25% biosolid content significantly improved the removal of PACs (>50%) but had no discernible effect on biosolid reduction. Additionally, the results demonstrated that both PhACs and biosolids stimulated oxidative enzyme synthesis. These results highlight the potential of fungi to reduce pharmaceutical contamination and provide long-term solutions for environmental clean-up (Torres-Farradá et al., 2024). Different nutraceutical applications of fungi are presented in Table 2. A sustainable and environmentally beneficial method of minimising pharmaceutical pollution, fungal bioremediation lessens the harmful impacts on aquatic ecosystems and human health. However, further research is needed to address issues, including understanding degradation byproducts, scaling up applications, and optimising environmental conditions. By incorporating fungus-based techniques into current wastewater treatment procedures, efficiency might be increased, and a safer and cleaner environment may be achieved. Genetic and metabolic engineering of fungi is important in completely utilising the fungal resources provided in the following section.

7 Role of Genetically Modified Fungal Resources and Recent Advancements

The shift to a more sustainable food system depends on filamentous fungus. Although the sensory appeal, scalability of fungal meals, and nutritional content could be improved through genetic alteration of these organisms, there are currently no genetic tools or proven applications for manufacturing bioengineered foods by edible strains. A modular synthetic biology toolkit was generated for the edible fungus *Aspergillus oryzae*, which is utilised to produce protein, fermented foods,

Table 2 Fungal applications in biorefinery systems

S No	Fungi	Methodology	Results	Advantages	Reference
1.	*Rhizopus oligosporus*	The highest DM of 94.0 ± 0.91%, DP of 86.7 ± 0.1% were achieved and acid-active proteolytic fungus	Chitin extraction was enhanced by successive fermentation by proteolytic fungus and co-LAF	Economical process for removing valuable chemicals from prawn trash	Ruangwicha et al. (2024)
2	*Rhizopus delemar*	To obtain fungal biomass, *Rhizopus delemar* was grown on bread waste in a 1 m^3 airlift bioreactor. Following a gentle enzymatic treatment of fungal biomass with a protease enzyme, the protein hydrolysate was separated as a soluble fraction	The rate of cell migration and wound closure was significantly improved by the monofilaments in the in vitro scratch experiment when compared to viscose fibres, which are frequently used in wound healing dressings. Additionally, at higher concentrations (beyond 500 µg/mL), fungal biomaterials in the form of hydrogel, monofilaments, and microfibers demonstrated improved cell proliferation and great biocompatibility against fibroblast cells	A sustainable method of producing value-added goods for food and medical textile applications from the plentiful food waste	Svensson et al. (2024)
3.	*Ganoderma lucidum*	For basidiocarp formation and laccase isozyme secretion, *G. lucidum* MDU-7 was cultivated in buckets, wood logs, boxes, and an iron mesh-like structure	The highest laccase production (5087.48 U/g ± 81.44 U/g) from mixed agricultural biomass was noted following the ninth day of incubation at pH 6.5. Ethanol with a high yield (89.11 ± 1.35% of the theoretical value) at 6.24 ± 0.14 g/L	The significance of the fungal culture for medicinal and therapeutic purposes	Shankar et al. (2024)

(continued)

Table 2 (continued)

S No	Fungi	Methodology	Results	Advantages	Reference
5.	Aspergillus*oryzae*	Effluent from dairy and ethanol plants to create useful products, such as mycoprotein and bioenergy, using a new three-phase biorefinery based on fungi	The three-phase biorefinery method produced encouraging outcomes, producing 5.0 Nm^3 of biomethane and 17.8 kg of mycoprotein per cubic metre of effluent	The mycoprotein that is produced in this manner has significant amounts of vital minerals, fatty acids, and amino acids	Hashemi et al. (2023)
6.	*Aspergillus niger* CKB	The viability of employing textile waste as a feedstock for solid-state fermentation-based cellulase production was examined in this work. Following 7 days of incubation on pure cotton, Aspergillus niger CKB was chosen as the strain with the highest cellulase activity (0.43 ± 0.01 FPU g^{-1})	According to the results, the maximum cellulase activity (1.18 ± 0.05 FPU g^{-1}) was obtained using autoclaved textile blending cotton/polyester of 80/20, while the CMCase, β-glucosidase, and avicelase activities were 12.19 ± 0.56 U g^{-1}, 1731 ± 4.98 U g^{-1}, and 2.58 ± 0.07 U g^{-1}, respectively. Following its extraction, the fungal cellulase was used to hydrolyse textile waste, yielding a 70.2% sugar recovery	The current work presents a novel circular textile waste-based biorefinery system that recovers polyester and glucose as products with additional value	Hu et al. (2018)

7.	*Aspergillus terreus*	*Aspergillus terreus* strains from mangrove soil produce a yellow colour. In order to forecast extracellular pigment yield and biomass production, response surface modelling in conjunction with the box-Behnken experimental design was used to adjust process parameters such as fermentation period (24–288 h), pH (4–9), and temperature (25 °C–35 °C)	The ideal pigment production (7.59 AU320) and biomass (11.119 g/L) were obtained under the anticipated conditions of a 278-h incubation period, a pH of 6.4, and a temperature of 30 °C	Significant antioxidant activity while remaining stable at different pH and temperature levels. Aspulvinone E is a natural colouring agent that is safe, stable, and potentially antioxidant	Chalakara and Thomas (2025)
8.	*Aspergillus tubingensis*	*Saccharomyces cerevisiae* FAI was rendered inactive by the fungus by co-cultivation and adsorption techniques	With a bioethanol yield of 0.49–0.54 g-ethanol/g-consumed glucose, yeast biocapsules produced 95–98 g/L of bioethanol from glucose	When employing lignocellulosic palm waste enzymatic hydrolysate as an alternative low-cost carbon source, the yeast biocapsules could effectively create bioethanol with a comparable bioethanol yield of 0.49 ± 0.17 g-ethanol/g--consumed glucose	Saeed et al. (2024)

(continued)

Table 2 (continued)

S No	Fungi	Methodology	Results	Advantages	Reference
9.	*Rhodosporidium toruloides* NCIM 3547 and *Saccharomyces cerevisiae*	The agro-waste known as banana inflorescence bract (BIB) is occasionally investigated for the manufacture of second-generation biofuel despite having a significant holocellulosic composition (cellulose 35.56%, w/w; hemicellulose 22.41%, w/w). In order to co-produce microbial lipids and ethanol, this study attempted to use this substrate for the extraction of fermentable sugars (pentose-C5 and hexose-C6)	12.70 ± 0.09 g/L ethanol	A sustainable path for the management of BIB's agro-waste was established	Bharathi and Jacob (2024)

10	*Nemania* sp.KM02 and *Saccharomyces cerevisiae*	Cellulase production on maize cob, eucalyptus, and cypress substrates under solid-state fermentation was assessed for two species of Xylaria (KM01, KM03) and *Nemania* sp. KM02 isolated from decomposing plant biomass	After 96 h of fermentation, maize cobs hydrolysed and fermented simultaneously by *Saccharomyces cerevisiae* and *Nemania* sp. KM02 cellulases produced more bioethanol (28.72 ± 3.82 mg/ml)	Maize cob is a good feedstock for the synthesis of cellulases and bioethanol	Kamande et al. (2024)

and meat substitutes. Neutral loci, programmable promoters, and the CRISPR-Cas9 approach for gene insertion are all part of the toolkit. These instruments increased the edible biomass's intracellular concentrations of the flavour-and-colour molecule heme and the nutraceutical ergothioneine. With minimal processing, the red-coloured strain that overproduces heme can be readily moulded into imitation meat patties. The results demonstrate how synthetic biology can potentially improve fungal foods and offer practical genetic tools for use in food production and other fields (Maini Rekdal et al., 2024). Introducing a gene from a biocontrol fungus increased the disease resistance in transgenic plants. The gene encoding a potent antifungal endochitinase in tobacco and potatoes was transferred from the mycoparasitic fungus *Trichoderma harzianum*. Several plant tissues exhibited high levels of fungal gene expression with no discernible impact on plant growth and development.

Significant variations in endochitinase activity were found between the transformants. Certain transgenic lines exhibited high tolerance or total resistance to the soil-borne pathogen *Rhizoctonia solani*, the foliar pathogens *Alternaria alternata*, *A. solani*, and *Botrytis cinerea*. The limited effectiveness of transgenic expression of bacterial and plant chitinase genes in plants is overcome by an increased level and wide range of resistance achieved with a single *Trichoderma* chitinase gene. These findings show a wealth of genes from biocontrol fungi that can be utilised to manage plant diseases (Khatoon et al., 2024). Because of their exceptional skill in post-translational modifications and their ability to secrete proteins, filamentous fungi have unmatched promise as cell factories for protein production. We investigate the potential of genome engineering, specifically the CRISPR-Cas9 method, to stimulate filamentous fungal protein production. An in-depth analysis of these fungi's protein secretion routes sheds light on their potential industrial uses. Notably, *Aspergillus* and *Trichoderma* species have been the subject of much scientific investigation for the industrial synthesis of proteins and enzymes. Additionally, this paper offers real-world instances of genetic engineering techniques meant to increase filamentous fungal enzyme secretion for various industrial uses. These results highlight the potential of filamentous fungi as adaptable platforms for protein synthesis and point to directions for further study and technical development in this area. *Monascus pilosus* has been utilised for many years to make lipid-lowering medications that are high in Monacolin K (MK). According to genome mining, this fungus still has a large number of potential genes that merit further investigation (Yi et al., 2021). Therefore, effective methods for genetic modification will significantly speed up this development.

8 Challenges and Future Prospects

Although mycoremediation of soil systems has enormous potential for the future, its scope is currently constrained by information gaps and our inability to fully utilise the potential of the fungi involved. Therefore, many studies are needed to enhance the mycoremediation process. Extensive field testing is required to determine how well fungi can be applied for bioremediation in practical settings. Scientists can comprehend the synergistic behaviour of fungi and other soil organisms, such as bacteria and archaea, by using network studies of soil microorganisms, which can shed light on how various microbes interact in the soil (Matchado et al., 2021). This will make it possible to create more comprehensive bioremediation plans that might use a variety of organisms, increasing the process's overall effectiveness.

Future research will focus on genetically altering fungi to improve their capacity to degrade pollutants. This may involve upregulating already-existing enzymes or incorporating enzymes from other organisms (Ahmad et al., 2023). Researchers can understand bioremediation processes better and maybe control the genes and metabolic pathways involved if they have a stronger understanding of metabolomics, proteomics, genomics, and. Hence, research based on omics, artificial intelligence, and other deep machine learning software should be combined with this effort. For the successful use of fungi in large-scale bioremediation initiatives, a multidisciplinary approach involving taxonomists, molecular biologists, and field practitioners is essential. In order to restore soils successfully, researchers must take direct action to address the wide range of contaminants present in soils and their effects on agricultural production, ecosystem health, and soil biodiversity.

9 Conclusion

The vast yet underutilised fungal biodiversity holds enormous potential to mitigate climate change and achieve the SDGs of the UN for a more sustainable future. Given their ability to convert organic materials into a wide variety of high-value products that promote socio-economic sustainability, fungi offer revolutionary opportunities for petroleum-dependent countries to shift toward a bio-based economy. The utilisation of fungi-derived products is a sustainable approach to reducing greenhouse gas emissions while guaranteeing a more reliable and secure food supply for the world's growing population. Alternative fungal-based products include textile biodegradation, biomaterials produced from filamentous fungi, and biorefinery applications for second-generation biofuels. *Crucially, fungi thrive on inexpensive, non-food substrates such as agricultural residues, making them ideal for scalable and decentralised production systems. Their remarkable adaptability to diverse ecological niches enables context-specific innovations, particularly in low- and middle-income countries, where regional challenges demand tailored solutions.*

To unlock the full potential of fungal systems, it is essential to explore their genetic and metabolic diversity through developments in gene editing, metabolic engineering, next-generation sequencing, and artificial intelligence (AI). Multidisciplinary co-operation among the fields of microbiology, genetics, agriculture, and industry is required to address the complexity of fungal biology. *Developing robust fungal strains through AI-driven frameworks is a key to overcoming current limitations in scaling fungal applications for sustainable agriculture and food fermentation. Industry leaders such as Novozymes and Danisco have already demonstrated the potential of fungal biotechnology in producing biofuels from agricultural waste, facilitating carbon sequestration, and generating biofertilisers and biopesticides*. Customised fungal strains can enhance precision agriculture, improve global food security and minimise the environmental impact. As this chapter concludes, fungal resources are essential for promoting sustainable livelihoods and global development—an urgent priority in a world where true sustainability remains an ongoing challenge.

Acknowledgement The authors express sincere gratitude to the editors for their kind invitation and guidance, and to all colleagues and collaborators whose insights and support contributed to the completion of this chapter.

Ethical Statement The authors declare no ethical concerns related to the content presented in this manuscript.

References

Abarca, C., Bidondo, L. F., Bompadre, J., & Velázquez, M. S. (2024). Arbuscular mycorrhizal fungi in tomato tolerance to pathogens and nematodes: A comprehensive review. *Scientia Horticulturae, 329*, 112969.

Aggarwal, S., & Kumari, A. (2024). Role of fungi in biotechnology. In *Entrepreneurship with microorganisms* (pp. 39–67). Academic.

Ahmad, A., Mustafa, G., Rana, A., & Zia, A. R. (2023). Improvements in bioremediation agents and their modified strains in mediating environmental pollution. *Current Microbiology, 80*(6), 208.

Alemu, D., Tafesse, M., & Gudetta Deressa, Y. (2022). Production of mycoblock from the mycelium of the fungus Pleurotus ostreatus for use as sustainable construction materials. *Advances in Materials Science and Engineering, 2022*(1), 2876643.

Bharathi, S. D., & Jacob, S. (2024). Comprehensive treatment strategy for banana inflorescence bract to synthesise biodiesel and bioethanol through fungal biorefinery. *Waste and Biomass Valorization, 15*(1), 417–436.

Boukary, A. A., Olou, A. B., Piepenbring, M., & Yorou, N. S. (2024). Mushroom cultivation in tropical Africa: Successes, challenges, and opportunities. *Journal of Agriculture and Food Research, 101264.*

Carvalho, D., Ferreira, N., França, B., Marques, R., Silva, M., Silva, S., & Oliveira, C. (2024). Advancing sustainability in the automotive industry: Bioprepregs and fully bio-based composites. *Composites Part C: Open Access, 14*, 100459.

Chalakara, A. A., & Thomas, T. (2025). Production kinetics and bioactivity study of Aspulvinone pigment from Aspergillus terreus STCP01 for natural colourant applications. *Bioresource Technology Reports, 29*, 102042.

Chaparro, M. L., Sanabria, P. J., Jiménez, A. M., Gómez, M. I., Bautista, E. J., & Mesa, L. (2021). A circular economy approach for producing a fungal-based biopesticide employing pearl millet as a substrate and its economic evaluation. *Bioresource Technology Reports, 16*, 100869.

Elhalis, H. (2025). Exploring fungal mycelium for sustainable food solutions: From biomass utilization to byproduct innovation. *Food Reviews International*, 1–33.

Engwenyu, L. R., & Anderson, A. S. (2021). A comprehensive review of calcineurin inhibitors used for immunosuppression in cardiac transplantation. *Pharmacology of Immunosuppression*, 27–38.

Espinosa-Páez, E., Hernández-Luna, C. E., Longoria-García, S., Martínez-Silva, P. A., Ortiz-Rodríguez, I., Villarreal-Vera, M. T., & Cantú-Saldaña, C. M. (2021). *Pleurotus ostreatus*: A potential concurrent biotransformation agent/ingredient on development of functional foods (cookies). *LWT, 148*, 111727.

Garg, S., Kim, M., & Romero-Suarez, D. (2024). Current advancements in fungal engineering technologies for sustainable development goals. *Trends in Microbiology, 33*(3), 285–301.

Gharib, M. A. A., Elhassaneen, Y. A. E. E., & Radwan, H. (2022). Nutrients and nutraceuticals content and in vitro biological activities of reishi mushroom (Ganoderma lucidum) fruiting bodies. *Alexandria Science Exchange Journal, 43*(2), 301–316.

Hashemi, S. S., Abbasi-Riyakhuni, M., Denayer, J. F., Tabatabaei, M., Aghbashlo, M., & Karimi, K. (2023). Efficient bioremediation of distillery and dairy wastewaters: A three-stage bio-refinery for high-quality aquaculture feed and bioenergy generation. *Process Safety and Environmental Protection, 180*, 566–574.

Hassan, M. (2025). Fungi as potent bioresource for biofuel production. In *Fungal biotechnology* (pp. 677–694). Academic.

Hu, Y., Du, C., Leu, S. Y., Jing, H., Li, X., & Lin, C. S. K. (2018). Valorisation of textile waste by fungal solid state fermentation: An example of circular waste-based biorefinery. *Resources, Conservation and Recycling, 129*, 27–35.

Kamande, S. M., Omwenga, G. I., & Ngugi, M. P. (2024). Production of cellulases by *Xylaria* sp. and *Nemania* sp. using lignocellulose substrates for bioethanol production from maise cobs. *Heliyon, 10*(17), e36802.

Khan, M. T., Aleinikovienė, J., & Butkevičienė, L. M. (2024). Innovative organic fertilisers and cover crops: Perspectives for sustainable agriculture in the era of climate change and organic agriculture. *Agronomy, 14*(12), 2871.

Khatoon, Z., Orozco-Mosqueda, M. D. C., & Santoyo, G. (2024). Microbial contributions to heavy metal phytoremediation in agricultural soils: A review. *Microorganisms, 12*(10), 1945

Kour, H., Kour, D., Kour, S., Singh, S., Hashmi, S. A. J., Yadav, A. N., & Ahluwalia, A. S. (2022). Bioactive compounds from mushrooms: Emerging bioresources of food and nutraceuticals. *Food Bioscience, 50*, 102124.

Krishnamoorthi, R., Srinivash, M., Mahalingam, P. U., & Malaikozhundan, B. (2022). Dietary nutrients in edible mushroom, *Agaricus bisporus* and their radical scavenging, antibacterial, and antifungal effects. *Process Biochemistry, 121*, 10–17.

Kumar, V., & Dwivedi, S. K. (2020). Multimetal tolerant fungus Aspergillus flavus CR500 with remarkable stress response, simultaneous multiple metal/loid removal ability and bioremediation potential of wastewater. *Environmental Technology & Innovation, 20*, 101075

Li, H., Tian, Y., Menolli, N., Jr., Ye, L., Karunarathna, S. C., Perez-Moreno, J., & Mortimer, P. E. (2021). Reviewing the world's edible mushroom species: A new evidence-based classification system. *Comprehensive Reviews in Food Science and Food Safety, 20*(2), 1982–2014.

Lin, S. Y., Baumann, K., Zhou, C., Zhou, W., Cuellar, A. E., & Xue, H. (2021). Trends in use and expenditures for brand-name statins after introduction of generic statins in the US, 2002-2018. *JAMA Network Open, 4*(11), e2135371.

Maini Rekdal, V., van der Luijt, C. R., Chen, Y., Kakumanu, R., Baidoo, E. E., Petzold, C. J., & Keasling, J. D. (2024). Edible mycelium bioengineered for enhanced nutritional value and sensory appeal using a modular synthetic biology toolkit. *Nature Communications, 15*(1), 2099.

Matchado, M. S., Lauber, M., Reitmeier, S., Kacprowski, T., Baumbach, J., Haller, D., & List, M. (2021). Network analysis methods for studying microbial communities: A mini review. *Computational and Structural Biotechnology Journal, 19*, 2687–2698.

Mayirnao, H. S., Sharma, K., Jangir, P., Kaur, S., & Kapoor, R. (2025). Mushroom-derived nutraceuticals in the 21st century: An appraisal and future perspectives. *Journal of Future Foods, 5*(4), 342–360.

Monclaro, A. V., Gomes, H. A. R., Duarte, G. C., de Souza Moreira, L. R., & Filho, E. X. F. (2024). Unveiling the biomass valorization: The microbial diversity in promoting a sustainable socio-economy. *Bioenergy Research, 17*(3), 1355–1374.

Ojwach, J., Adetunji, A. I., Mutanda, T., & Mukaratirwa, S. (2022). Oligosaccharides production from coprophilous fungi: An emerging functional food with potential health-promoting properties. *Biotechnology Reports, 33*, e00702.

Patel, J. Y., & Sharma, J. (2023). A review of mycelium-based bio-composites and their possible application in architecture. *ShodhKosh: Journal of Visual and Performing Arts, 4*, 213–225.

Periyasamy, S., Isabel, J. B., Kavitha, S., Karthik, V., Mohamed, B. A., Gizaw, D. G., & Aminabhavi, T. M. (2023). Recent advances in consolidated bioprocessing for conversion of lignocellulosic biomass into bioethanol–A review. *Chemical Engineering Journal, 453*, 139783.

Prescott, T. A., Hill, R., Mas-Claret, E., Gaya, E., & Burns, E. (2023). Fungal drug discovery for chronic disease: History, new discoveries and new approaches. *Biomolecules, 13*(6), 986.

Rahamim, V., Nakonechny, F., Azagury, A., & Nisnevitch, M. (2022). Continuous bioethanol production by fungi and yeast working in Tandem. *Energies, 15*(12), 4338.

Rajhans, G., Sen, S. K., Barik, A., & Raut, S. (2021). De-colourisation of textile effluent using immobilised Geotrichum candidum: An insight into mycoremediation. *Letters in Applied Microbiology, 72*(4), 445–457.

Rashad, M., Kenawy, E. R., Hosny, A., Hafez, M., & Elbana, M. (2021). An environmental friendly superabsorbent composite based on rice husk as soil amendment to improve plant growth and water productivity under deficit irrigation conditions. *Journal of Plant Nutrition, 44*(7), 1010–1022.

Robey, M. T., Caesar, L. K., Drott, M. T., Keller, N. P., & Kelleher, N. L. (2021). An interpreted atlas of biosynthetic gene clusters from 1,000 fungal genomes. *Proceedings of the National Academy of Sciences, 118*(19), e2020230118.

Rousta, N., Aslan, M., Yesilcimen Akbas, M., Ozcan, F., Sar, T., & Taherzadeh, M. J. (2024). Effects of fungal based bioactive compounds on human health. *Critical Reviews in Food Science and Nutrition, 64*(20), 7004–7027.

Ruangwicha, J., Cheirsilp, B., & Suyotha, W. (2024). Green biorefinery of shrimp shell waste for α-chitin and high-value co-products through successive fermentation by co-lactic acid bacteria and proteolytic fungus. *Bioresource Technology, 393*, 130106.

Saeed, Z., Cheirsilp, B., Maneechote, W., Kongjan, P., & Jariyaboon, R. (2024). Optimising bio-encapsulation of yeast cells by *Aspergillus tubingensis* TSIP9 and applications in bioethanol production through repeated-batch fermentation. *Biocatalysis and Agricultural Biotechnology, 61*, 103377.

Sansinenea, E. (2021). Application of biofertilisers: Current worldwide status. In *Biofertilizers* (pp. 183–190). Woodhead Publishing.

Sawant, A., & Sajwan, H. (2022). Fungal metabolites: Industrial applications and challenges. *Sustainable Utilization of Fungi in Agriculture and Industry*, 159–201.

Shankar, A., Saini, S., & Sharma, K. K. (2024). Fungal-integrated second-generation lignocellulosic biorefinery: Utilisation of agricultural biomass for co-production of lignocellulolytic enzymes, mushroom, fungal polysaccharides, and bioethanol. *Biomass Conversion and Biorefinery, 14*(1), 1117–1131.

Sharma, R., & Sumbria, R. (2022). Mycelium bricks and composites for sustainable construction industry: A state-of-the-art review. *Innovative Infrastructure Solutions, 7*(5), 298.

Sharma, R., Khatua, S., Acharya, K., & Sharma, Y. P. (2022). Mycochemical composition and antioxidant activity of Flammulina velutipes: A comparative study on hydromethanol, decoction and infusion extracts. *Vegetos, 35*(3), 607–613.

Sharma, E., Bairwa, R., Lal, P., Pattanayak, S., Chakrapani, K., Poorvasandhya, R., & Kumar, R. (2024). Edible mushrooms trending in food: Nutrigenomics, bibliometric, from bench to valuable applications. *Heliyon, 10*(17), e36963.

Shinde, R., Shahi, D. K., Mahapatra, P., Singh, C. S., Naik, S. K., Thombare, N., & Singh, A. K. (2022). Management of crop residues with special reference to the on-farm utilization methods: A review. *Industrial Crops and Products, 181*, 114772.

Shuting, C. (2024). Mushrooms: Fungi for all seasons. *Acta Edulis Fungi, 31*(1).

Silvestri, L. (2022). *Terraforma-material investigation on the possibilities to combine natural growth of mycelium and unfired clay for novel sustainable product design applications*.

Srivastava, S., Mathur, P., Prakash, P., Falletta, E., Katha, U., Pagani, A., & Singh, A. V. (2025). Mushroom-derived innovations: Sustainable biomaterials for biomedical engineering. *Biomedical Materials and Devices, 3*(1), 381–395.

Svensson, S. E., Bucuricova, L., Ferreira, J. A., Souza Filho, P. F., Taherzadeh, M. J., & Zamani, A. (2021). Valorisation of bread waste to a fiber-and protein-rich fungal biomass. *Fermentation, 7*(2), 91.

Svensson, S. E., Abdollahi, M., Moghadam, F. H., Kalita, N. K., Hakkarainen, M., Wijayarathna, E. K. B., & Zamani, A. (2024). Valorisation of bread waste to fungal-based products for medical textile and food applications. *ACS Sustainable Resource Management, 1*(3), 385–394.

Tiwari, P., & Park, K. I. (2024). Advanced fungal biotechnologies in accomplishing sustainable development goals (SDGs): What do we know and what comes next? *Journal of Fungi, 10*(7), 506.

Torres-Farradá, G., Thijs, S., Rineau, F., Guerra, G., & Vangronsveld, J. (2024). White rot fungi as tools for the bioremediation of xenobiotics: A review. *Journal of Fungi, 10*(3), 167.

Usman, M., Murtaza, G., & Ditta, A. (2021). Nutritional, medicinal, and cosmetic value of bioactive compounds in button mushroom (Agaricus bisporus): A review. *Applied Sciences, 11*(13), 5943.

Voutetaki, M. E., & Mpalaskas, A. C. (2024). Natural fiber-reinforced mycelium composite for innovative and sustainable construction materials. *Fibers, 12*(7), 57.

Wadhwa, K., Kapoor, N., Kaur, H., Abu-Seer, E. A., Tariq, M., Siddiqui, S., & Alghamdi, S. (2024). A comprehensive review of the diversity of fungal secondary metabolites and their emerging applications in healthcare and environment. *Mycobiology, 52*, 1–53.

Wang, Y., Liu, Z., Hao, X., Wang, Z., Wang, Z., Liu, S., & Li, R. (2023). Biodiversity of the beneficial soil-borne fungi steered by Trichoderma-amended biofertilisers stimulates plant production. *npj Biofilms and Microbiomes, 9*(1), 46.

Yağmur, A., Demir, S., Canpolat, S., Rezaee Danesh, Y., Farda, B., Djebaili, R., & Pellegrini, M. (2024). Onion Fusarium basal rot disease control by arbuscular mycorrhizal fungi and Trichoderma harzianum. *Plants, 13*(3), 386.

Yang, Y., Zhang, X., Hartley, I. P., Dungait, J. A., Wen, X., Li, D., & Quine, T. A. (2022). Contrasting rhizosphere soil nutrient economy of plants associated with arbuscular mycorrhizal and ectomycorrhizal fungi in karst forests. *Plant and Soil, 470*(1), 81–93.

Yi, D., Bayer, T., Badenhorst, C. P., Wu, S., Doerr, M., Höhne, M., & Bornscheuer, U. T. (2021). Recent trends in biocatalysis. *Chemical Society Reviews, 50*(14), 8003–8049.

Filamentous Fungi: Heterologous Protein Expression

Joshny Joseph, June Susan Abraham, Rini Abraham, Praveen Nagella, and Asha Anand

Abstract Filamentous fungi are versatile microorganisms vital in the industrial production of organic compounds and enzymes. They are crucial for various other ecosystems due to their ability to form symbiotic relationships and thrive in various environments. Their beneficial properties, such as the ability to cultivate in inexpensive media, high secretion, and efficient eukaryotic-like post-translational modification and protein folding, make the fungal-based protein production system superior to the bacterial-based system. The dearth in understanding of the secretory pathways and physiology of fungi in depth has stalled the accelerated modification of strains to produce heterologous proteins. This chapter highlights the various filamentous fungal species in use, including *Aspergillus*, *Trichoderma*, *Penicillium*, and *Myceliophthora,* for the production of heterologous protein. Various techniques used in the enhancement of the fungal strains for efficient heterologous production, such as the regulation of protein folding, codon optimization, carrier protein fusion, glycosylation engineering, selection of promoters, signal peptides, and development of a protease-deficient strain, have also been discussed briefly. Insights on various filamentous fungi used for industrial heterologous protein production, and various examples of proteins produced, are also included. Despite bottlenecks in the expression and production of protein, filamentous fungi are a critical force in the field of biotechnology and molecular biology. Hence, a deeper understanding of secretory pathways will enhance the utilization of filamentous fungi.

Keywords Filamentous fungi · Heterologous protein · Post-translational modifications

---

J. Joseph · J. S. Abraham · P. Nagella · A. Anand (✉)
Department of Life Sciences, School of Sciences, Christ University, Bengaluru, India
e-mail: asha.anand@christuniversity.in

R. Abraham
Department of Biotechnology, Jyoti Nivas Autonomous College, Bengaluru, India

A. Gupta (ed.), *Prospects of Fungal Biotechnologies for Livestock Volume 2*,
Fungal Biology, https://doi.org/10.1007/978-3-032-06478-3_8

1 Introduction

Fungi are an essential part of an ecosystem that balances life on Earth. They are diverse eukaryotic organisms that foster symbiotic relationships with plants for carbon sequestration, are the ultimate decomposers, and can thrive in diverse environmental conditions (Corbu et al., 2023). Apart from being decomposers and symbionts, fungi have been explored for food and beverages for industrial applications like the production of enzymes and proteins (Corbu et al., 2023). Fungi's roles vary from symbionts, saprotrophs, endophytes, or parasites, and they include microscopic and macroscopic forms with complex mycelial structures. Furthermore, when compared to bacterial-based protein production, fungal-based is more advantageous due to the shared gene expression process with other eukaryotic microorganisms (Su et al., 2012). In the last century, filamentous fungi became important in producing various products, such as enzymes, secondary metabolites, and organic acids. These cell factories can be grown on inexpensive, simple media capable of expressing and secreting heterologous proteins (Meshram et al., 2024). Filamentous fungi have become vital microorganisms in biotechnology and molecular biology due to their capabilities for efficient folding and post-translational modifications, easy induction, large quantity protein production and superior secretion pathways (Wang et al., 2020; Madhavan et al., 2022). These capabilities have produced billions of dollars worth of metabolites (primary and secondary) and homologous and heterologous proteins (Cairns et al., 2018; Liu et al., 2023).

2 Heterologous Expression

Heterologous proteins are produced by a different organism than the naturally producing organism, and the biosynthesis of heterologous proteins requires the transformation of foreign DNA with its expression for protein synthesis in the potential host (Meshram et al., 2024). Filamentous fungi have become a potential host due to their complex post-translational modifications, including the production of disulfide links, protein glycosylation or proteolytic cleavage, making them the factory for heterologous protein production (Meshram et al., 2024). Various filamentous fungi are in use as hosts for heterologous protein production; however, the most commonly used species of filamentous fungi include the genera *Aspergillus, Trichoderma, Penicillium*, and *Myceliophthora,* in which few species are considered safe and given the status Generally Recognized As Safe (GRAS) (Liu et al., 2023; Meyer et al., 2020; Nevalainen & Peterson, 2014a). These GRAS species include *Aspergillus niger*, *Aspergillus oryzae*, *Penicillium chrysogenum*, or *Trichoderma reesei*, making them suitable for industrial applications (Meshram et al., 2024; Liu et al., 2023). The following section briefly discusses various filamentous fungal strains generally used for protein expression.

2.1 *Aspergillus*

In the last century, when James Currie discovered that *A. niger* could produce citric acid, *Aspergillus* sp. has been in the limelight in the field of biotechnology (Ntana et al., 2020). Since *Aspergillus* sp. was categorized as GRAS, it is used in the production of enzymes and secondary metabolites, and recently, it has been used for the production of heterologous proteins, too (Liu et al., 2023). There are limitations in the heterologous protein production efficiency due to the lack of a deep understanding of fungal expression and secretion. The capacity of *Aspergillus* sp. to produce proteins in large quantities has facilitated deeper research and alternative methods to increase the efficiency of heterologous protein production (Ntana et al., 2020). *Aspergillus* is a large genus with nearly 400 known species, in which the industrially important species include *Aspergillus niger, Aspergillus oryzae, Aspergillus sojae, Aspergillus tubingensis, Aspergillus terreus, Aspergillus nidulans,* and *Aspergillus unguis* (Liu et al., 2023).

Aspergillus niger is the most highlighted among *Aspergillus* spp. for protein production. It is estimated that *A. niger* together with *A. oryzae* contributes to more than 50% of the industrial enzyme and protein production (Sakekar et al., 2021). Research has shown that *A. niger* can be genetically modified for the heterologous overexpression of non-ribosomal peptide synthetase (NRPS) from *Fusarium* spp. (Cairns et al., 2018). *A. niger* and *A. nidulans* serve as model organisms for fungal research, specifically for protein expression and secretion research (Sakekar et al., 2021).

Aspergillus oryzae is an important filamentous fungus that can secrete enzymes in large quantities and has a long history of producing food products (Alberti et al., 2017). For the high expression of foreign genes, *A. oryzae* has been genetically modified (Pahirulzaman et al., 2012). One example that can be cited here is the replication of biosynthesis of polyketide tenellin (17), which was synthesized originally by *Beauveria bassiana.* Here, *A. oryzae* was modified by inserting four genes with three vectors, where each gene was positioned by a promoter (*PamyB*) and terminator (*TamyB*) from *A. oryzae*'s *Taka amylase A* gene (Jin et al., 2021). This induced the expression of tenellin in the presence of starch and yielded five-fold more than the native host (Hoffmeister & Keller, 2007; Heneghan et al., 2010).

Aspergillus nidulans is used as a model organism for fungal research rather than a host for heterologous protein production due to its inability to produce protein in large quantities (Yan et al., 2023; Kumar, 2020). However, research and study on *A. nidulans* have helped better understand the expression and secretion machinery in filamentous fungi, which has helped modify the other strains like *A. niger* and *A. oryzae* for better protein production (Sun & Su, 2019). A recent study on *A. nidulans* facilitated a higher protein expression of human interleukin-6 (HuIL-6) by reducing the degradation of proteins caused by the endoplasmic reticulum-associated protein degradation (ERAD) and vacuoles in the secretory pathway (Yan et al., 2023). It was also noticed that further disruption of extracellular proteases in *A. nidulans* has increased the yield of protein production (Yan et al., 2023). Apart from this *A. nidulans* has been used for the heterologous production of enzymes like Carbohydrate Active enZymes (CAZymes), endoxylanase and arabinofuranosidase, giving positive yields (Liu et al., 2023; Kumar, 2020).

Aspergillus sojae has commonly been used for soy sauce fermentation for a long time. It has become a potential host for heterologous production in the past few years, it is safe, and its production is void of aflatoxins and cyclopiazonic acid (Chang et al., 2007). As *A. sojae* produces fewer endogenous secondary metabolites, purifying the compound produced under conventional culture conditions is easy (Araki et al., 2022). The production of ilicicolin A in *A. sojae* was increased with the enhanced expression of the rate-determining enzyme AscD using a multicopy integration system (Araki et al., 2022). *Aspergillus vadensis* is a black *Aspergilli* strain that has become a potential host for homologous and heterologous protein production in the last decade due to its particular properties (De Vries et al., 2004). *A. vandensis* differs from other black fungi as this strain produces lesser quantities of extracellular proteases, with media acidification when the nitrogen source of growth media is nitrate (Liu et al., 2023; De Vries et al., 2004; Alberto et al., 2009). Due to these properties, the target protein with enhanced stability is produced in the growth media (Alberto et al., 2009; Gürkök et al., 2010). When the production of feruloyl esterase A (FaeA) was compared in *A. vandensis, A. niger* and *A. tubinfensis*, the production was higher in *A. vandensis* with higher stability and low protease activity, also reported that FaeA production was 12-fold higher than *A. niger* (Alberto et al., 2009; Gürkök et al., 2010). The transformation efficiencies and protoplast transformation are also reported to be higher in *A. vandensis* (De Vries et al., 2004).

Apart from the above-mentioned strains, a few other strains contribute to heterologous protein production, including *A. fumigatus,* which has a high protein secretion capacity and is grown on inexpensive media (Gürkök et al., 2010). Genes from *A. fumigatus* expressed in *A. sojae* have increased the production of α-galactosidase enzyme by threefold. Another strain, *A. awamori*, has been used to express the phytase gene from *A. fumigatus*, resulting in thermostable enzyme production (Martin et al., 2003).

2.2 *Trichoderma*

Trichoderma species of filamentous fungi, usually used as biocontrol agents, is another host widely used for homologous and heterologous protein production due to its high protein production and efficient post-translation modifications (Wei et al., 2021). The saprophytic and mycoparasitic nature of *Trichoderma* species facilitates its growth on diverse nutrients. Certain *Trichoderma* spp. have been categorized in GRAS, recognizing them as safe (Tomico-Cuenca et al., 2021).

T. reesei, recognized as safe, has been the workhorse for heterologous protein production for decades, and the first production of calf chymosin was reported in 1989 (Harkki et al., 1989; Nevalainen & Peterson, 2014b). *T. reesei* is an organism of industrial importance due to its high homologous protein yield (up to 100 g/L) to the culture media (Wei et al., 2021). Though there is a high yield in homologous protein production, there are bottlenecks in heterologous protein production (Liu et al., 2023). As per the reports, *Melanocarpus albomyces* cellulase (Cel5A), a foreign fungal protein, is the highest-produced heterologous protein, yielding 7.7 g/L in *T. reesei* (Wei et al., 2021; Haakana et al., 2004). In an attempt to increase interferon alpha-2b (IFNα-2b) production yield, protease deletion resulted in 4.5 g/L of increased yield (Landowski et al., 2016). Other *Trichoderma* species in focus include *T. virens, T. atroviride,* and *T. harzianum* as hosts for the heterologous protein production (Wei et al., 2021).

T. atroviride promoter *Pccg6* used for the expression of phosphite oxidoreductase (PTXD) of bacterial origin helped *T. atroviride* to outgrow bacterial contamination in culture media (Carreras-Villasenõr et al., 2020). The expression of *T. harzianum* C4 endo-β-1,4-xylanase gene in *Saccharomyces cerevisiae* helped in the secretion of β-xylanase into the growth media (Lee et al., 2009).

2.3 *Penicillium*

Penicillium genus is the largest group of fungi known as one of the sources of antibiotics. There are over 200 species, and they are explored as potential host for heterologous protein production apart from antibiotic potential (Toghueo & Boyom, 2020).

Penicillium oxalicum is one of the species used for heterologous protein production due to the stability of the extracellular protein secreted. Recently, a modified *P. oxalicum* strain, OamyR-QL, was produced through the activation of G protein PGA3. The modification resulted in improved and purified protein expression of cellobiohydrolase I (CBHI) and enhanced cell growth with overexpression of AmyR—a transcription activator, resulting in enhanced Pamy15a promoter efficiency (Xiang et al., 2023).

Penicillium chrysogenum (P. rubens) is a model organism recognized as safe with GRAS status. The first reported heterologous expression in filamentous fungi is the fungal xylanase and cDNA expression of the human tear lipocalin (LCN1) (Graessle et al., 1997). The encoding sequences of cDNA of LCN1 were inserted in *P. chrysogenum* under the control of promoter gene (phoA) responsible for repressible acid phosphatase to achieve its heterologous expression (Graessle et al., 1997). In recent research, when the two expression cassettes were expressed in *P. chrysogenum,* it resulted in the production of purified AFPs (antifungal proteins) with a yield of 80 mg/L (Liu et al., 2023; Sinitsyn et al., 2020).

Penicillium verruculosum is known to be used in biorefinery processes as it is an efficient producer of a cellulase complex (Chulkin et al., 2024). *P. verruculosum* has a rich set of enzymes with specific activity for cellulase complexes, making it a potential host for protein production (Sinitsyn et al., 2020). An example of the heterologous protein production is the expression of endo-xanthanase (EX) in

Penicillium verruculosum 537 (ΔniaD) with the cellobiohydrolase 1 gene promoter, which is originally produced by *Thermogutta terrifontis* strain (Denisenko et al., 2021).

Penicillium funiculosum is a non-pathogenic, non-model hypercellulolytic fungal strain with genes encoded for different extracellular enzymes (Liu et al., 2023). Generally, this strain is used to produce high-quality protein mixtures for the saccharification of lignocellulosic biomass (Ogunmolu et al., 2018). Recently, a random mutagenesis on *Penicillium funiculosum* NCIM 1228 through ethyl methanesulfonate (EMS)-induced mutant D4 resulted in higher enzyme activity and productivity, leading to high saccharification efficiency (Chavan et al., 2024).

Other *Penicillium* spp. that are used for protein production include *P. canescens* due to its capacity to secrete β-galactosidase and xylanase, *P. subrubescens,* due to the enzymes produced related to hemicellulose, pectin, and inulin degradation (Liu et al., 2023).

2.4 *Myceliophthora*

Myceliophthora spp., especially *M. thermophilia*, has become an organism of interest due to its ability to survive in high temperatures, efficient protein secretion capacity and increased growth rate on cellulose media (Liu et al., 2017). This strain was first isolated from the soils of eastern Russia, and it had peculiar characteristics different from other filamentous fungi strains; they were both temperature and acid-base-tolerant (Liu et al., 2023). This strain has the capacity to produce thermostable enzymes for biomass degradation, and the cellulase produced from *M. thermophilia* is given GRAS status (Gu et al., 2018). A study on *M. thermophilia* showed an increase in the production of fumaric acid through metabolic engineering in the strain, and this was the first attempt in filamentous fungi involving multiple genes in metabolic engineering (Gu et al., 2018). In a recent study, *M. thermophilia* C1 variant was improved by adding disulfide bonds, which improved the stability of lytic polysaccharide monooxygenases (LPMO). The β-sheet content of the H1-E34 region was increased, resulting in rigid protein; also, the conversion efficiency compared to wild type was increased to 1.5-fold (Guo et al., 2023).

The other species of *Myceliophthora* that has been investigated as host is *M. heterothallica*, which is similar to *M. thermophilia*, with a difference in the functional mating cycle in *M. heterothallica* (Liu et al., 2023). An example for heterologous production is the expression of xylanase enzyme in *Pichia pastoris*, derived from *Myceliophthora heterothallica*, for enhanced biomass hydrolysis (de Amo et al., 2024). Earlier studies produced heterologous xylanase enzyme from *A. nigerin* in *M. heterothallica*, enhancing xylanase activity (Darmond, 2016).

2.5 Alternative Strains Used for Heterologous Production

The above-mentioned strains are the most widely used host systems apart from these, the strains supporting large-scale protein production include *Fusarium*, *Rhizopus*, or *Mucor* species (Liu et al., 2023). Fusarium is known for mycoprotein production, and it is used in the production of secondary metabolites and as an alternative to protein sources (Gnaim et al., 2025). *Rhizopus oligosporus* can produce high yield of protein by cultivating on food waste-derived substrates (Nitayavardhana et al., 2013). Research on *Mucor circinelloides* shows that it can produce protein, ethanol, oil, glucosamine, and phosphate; it can also ferment polypeptides and amino acids as it facilitates extracellular protease production (Tan et al., 2020). When the fermentation of *M. circinelloides* grown on red grapes pomace substrate produced fungal chitosan, it showed a 61.5% degree of deacetylation, equivalent to commercially available chitosan.

3 Protein Expression and Secretion Pathway in Filamentous Fungi

Filamentous fungi are cell factories of proteins. Their innate properties to secrete a high yield of a variety of proteins, ability to grow on inexpensive media and ease in post-translational modification for greater efficiency make them workhorses of industrial heterologous protein production (Meshram et al., 2024). It is reported that about one-third of the heterologous products produced globally are contributed by fungi and prokaryotic strains. Many fungal strains used for protein production and products produced are recognized in GRAS status (Liu et al., 2023). As filamentous fungi depend on nutrients present in the surroundings from wastes, dead plants, and other living organisms, they are heterotrophic and can produce different amino acids from the inorganic nitrogen present.

The complex protein synthesis pathway in filamentous fungi involves three phases—synthesis, modification, and secretion. Even though the steps seem to be easier, this involves many crucial steps. In the first step (Synthesis), transfer of polypeptide to the endoplasmic reticulum (ER) happens. The translocation of polypeptide to ER can occur through a recognition particle (SRP)-dependent pathway and SRP-independent pathway (Meshram et al., 2024; Wang et al., 2020). In the SRP-dependent pathway, a single peptide part in the N-terminal end (15–36 amino-acid long) is responsible for the extracellular protein genes, where genes get transcribed to mRNA and translated to ribosome-mRNA-nascent peptide complex (Meshram et al., 2024). This complex is the secretory signal sequence associated with the nascent polypeptide chain, is recognized by SRP, and binds to it to prevent translation. This SRP-ribosome-mRNA-nascent peptide complex is directed to the cytosolic part of ER, where it binds to the translocon such as Sec61p complex (peptide translocation complex) (Wang et al., 2020). The polypeptide enters the ER lumen,

T. reesei bip 1 and hac 1 resulted in improved secretion of 1.5–1.8-fold of glucose oxidase from *A. niger* (Wu et al., 2017). Deletion of vital genes in ERAD of protein secretion process has shown non-degradation of misfolded or heterologous proteins in some instances, for example, when doaA ERAD factor was deleted together with overexpression of sttc oligosaccharyl transferase that facilitates glycosylation of secretory proteins in *A. niger,* resulting in increased β-glucuronidase yield (Conesa et al., 2002; Qian et al., 2019). Apart from ERAD, autophagy also causes the degradation of proteins and deletion of the genes related to autophagy, like aoatg15 in *A. oryzae,* which facilitates increased protein secretion like bovine chymosin (Yoon et al., 2013). Even manipulation of certain genes also facilitates positive production of heterologous protein, as in the case of BiP, which is a gene that promotes protein degradation in the ER, translocation and proper folding. Overexpression of BiP1 had a different effect on *T. reesei* where it promoted secretion, while in *A. niger* it had an adverse effect on protein production (Jacobs et al., 2009; Zhang et al., 2014). An alteration in the protein secretion is seen during the regulation of UPR and ERAD, which necessitates a better understanding of the secretory pathway before manipulation.

4.7 Construction of a Protease-Deficient Strain

Among bottlenecks in protein production, degradation of proteins by protease is another factor that limits the yield of production (Qian et al., 2019). The discovery of the role of extracellular proteases in the low yield of heterologous protein production paved the way for the development of protease-deficient strains by the deletion of the protease-encoding gene (Liu et al., 2023). When alkaline serine protease SPW was disrupted in *T. reesei,* there was a reduction in protease activity by half, boosting the production and stability of alkaline endoglucanases EGV of *Humicola insolens* (Zhang et al., 2014). In *T. reesei,* deletion of the subtilisin protease gene slp7 and the metalloprotease gene amp2 has increased the level of human interferon alpha 2b production (Landowski et al., 2016). Apart from deleting the protease gene, disruption of the protease regulatory genes has also given similar results with better yields. For instance, when the transcription factor p53-like Vib1 in *T. reesei* was deleted, there was a drastic decrease in the protease and cellulase activity, and this strain showed an enhanced expression of bglA gene from *A. niger* BGL (Sun & Su, 2019).

Though filamentous fungi are cell factories for heterologous protein production, there are many bottlenecks in the expression and secretion of proteins. The above-mentioned methods can help enhance heterologous production. However, the complexity of many methods is challenging, and it demands a deeper understanding of secretory pathways through various omics studies. The following section will give an insight into the advantages and challenges in filamentous fungi for heterologous protein production.

5 The Advantages and Challenges of Using Filamentous Fungi for Heterologous Protein Production

5.1 *Advantages*

As genetic alterations pave the way for the production of heterologous proteins in filamentous fungi, multiple filamentous fungi have been claimed to be commercially relevant (Liu et al., 2023). Multiple features of filamentous fungi have contributed to their potential to be utilized as molecular factories. Research and drug discovery depend mainly on the development of recombinant proteins. Despite being frequently utilized, due to its ease of use along with rapid growth, *Escherichia coli* fails to undertake crucial post-translational modifications such as glycosylation. Filamentous fungi can address this limitation as they can perform complex post-translational modifications (Vieira Gomes et al., 2018). Despite their capacity for post-translational modifications, yeasts mostly generate high mannose-type glycoproteins. On the other hand, filamentous fungi show less widespread hypermannosylation, which makes it possible to produce glycosylated forms that are better suited for use in medicinal applications in mammalian systems (Deshpande et al., 2008; Punt et al., 2002). Additionally, unlike yeasts, which mainly metabolize glucose and mannose, their extensive metabolic capacity enables the efficient utilization of a variety of monosaccharides, including xylose, arabinose, and galactose (Cavka, 2014). *Trichoderma reesei* was able to secrete 100 g/L of cellulose, while *A. niger* fermentation medium yielded 25–30 g/L of glucoamylase (Ward, 2012).

5.1.1 Metabolic and Genetic Diversity

One of the major advantages of using filamentous fungi is their metabolic and genetic diversity (Madhavan et al., 2022). Varieties of filamentous fungi have been used for the commercial production of heterologous proteins. The differences in metabolic pathways exhibited by the fungi enable them to produce a broad range of heterologous proteins. Recombinant proteins such as *Myceliophthora thermophila* trehalase, *Penicillium citrinum* nuclease P1, and human erythropoietin have been reported to be produced in high titres from *A. niger* (Dong et al., 2020). Apart from the capability to produce high titres of heterologous proteins, both *A. niger* and *A. oryzae* are considered safe to be utilized in food industries (Kamaruddin et al., 2018). Others, notably *A. vadensis*, provide consistent production conditions because of their non-acidifying growing circumstances and minimal protease activity (Alberto et al., 2009). Species of *Penicillium*, including *P. chrysogenum*, *P. oxalicum*, and *P. funiculosum*, also contribute significantly due to their natural enzyme diversity, secretion capacity, and adaptability for genetic manipulation (Liu et al., 2023). *Penicillium* species have successfully produced enzymes like β-glucosidases and cellulases at commercially relevant levels (Ferreira et al., 2020; Crament et al., 2024). Thermophilic fungi like *Myceliophthora thermophila* and *M. heterothallica* expand the host repertoire with their heat and pH tolerance and ability to produce thermostable enzymes, ideal for biomass degradation and industrial applications. This broad metabolic repertoire across species enables filamentous fungi to support a wide range of heterologous proteins, from enzymes and antibodies to human therapeutic proteins, making them invaluable cell factories in biotechnology (Liu et al., 2023). Filamentous fungi live in diverse habitats and require several types of enzymes and proteins that can degrade biomass, anchor to the surface, and defend themselves. Hence, it would have led to the evolution of a multitude of metabolic pathways (Nevalainen & Peterson, 2014a).

5.1.2 Robust Secretory Pathways

Secretory pathways exhibited by filamentous fungi, compared to other cell factories such as bacteria and yeast, are another major advantage, which makes filamentous fungi a better candidate for the large-scale production of recombinant proteins (Fig. 2) (Wang et al., 2020). The adaptability and the efficiency of the secretory pathways promote the synthesis of recombinant proteins in filamentous fungi. Filamentous fungi have a eukaryotic type, an Endoplasmic reticulum-Golgi pathway, which is a more complex pathway compared to other cell production systems and similar to that of eukaryotes (Nevalainen & Peterson, 2014a).

Co-translational translocation of nascent polypeptides into the ER guarantees appropriate signal-peptide recognition. Chaperones and foldases facilitate proper folding, disulfide-bond formation, and N-glycosylation in the ER lumen, which are crucial alterations that improve the stability and functionality of numerous

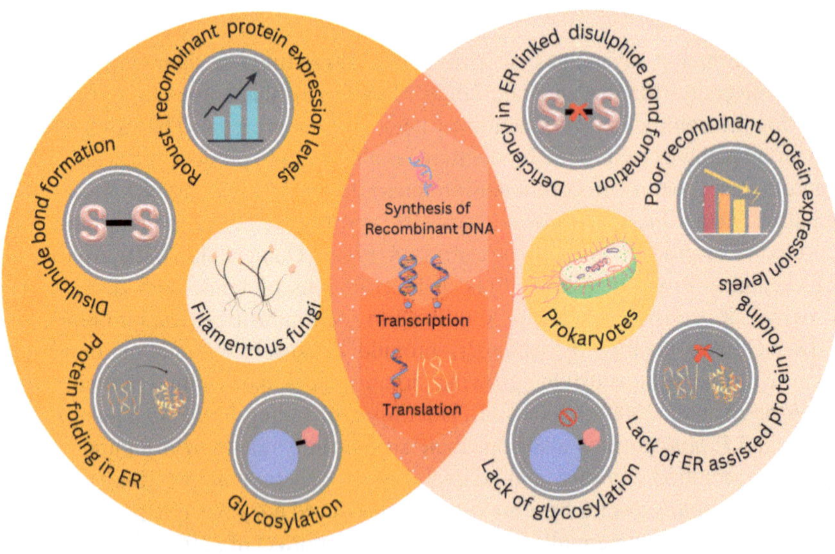

Fig. 2 Comparison between filamentous fungi and prokaryotic secretory pathway

heterologous proteins. Furthermore, processed proteins are organized into secretory transport vesicles in the Golgi and then transported to the plasma membrane for exocytosis into the culture medium (Wang et al., 2020).

5.1.3 Alternative Secretion Pathways

Septal secretion—Reduces the overload in vesicular trafficking at the apex of the hyphae and also protects the proteins from being degraded to an extent, as the secretion would be away from the apex of the septum. Secretory vesicles fuse with the septal plasma membrane. This process involves only microtubules, but secretion via hyphal tips requires both microtubules and F-actin. Septal secretion is significant in industrially beneficial strains such as *Aspergillus* and *Trichoderma*, since both fungi have more septa than hyphal tips (Nevalainen & Peterson, 2014a). Multivesicular bodies help in shunting the transport of the proteins through Golgi-independent vesicles (Alberti et al., 2017).

Glycosylation is an important phase in the secretory pathway, where carbohydrate moieties are added to the newly synthesized protein. It has been reported that glycosylation in filamentous fungi can be engineered to obtain results that can be similar to those of mammals as discussed in the previous Sect. 3 (Maras, 1999). There have been evidences where the glycosylation has been modified in *A. niger* which resulted in enhanced secretion of stable proteins (Liu et al., 2023).

Filamentous fungi have a complete eukaryotic secretory system when compared to other cell factories such as *E. coli*, since the prokaryotic model organism does not have post-translational processing, which is essential for the functioning of many enzymes. It also showcases other challenges, such as differences in the functionality of codons, which will require optimization, and also the inability to produce certain secondary metabolites, such as antibiotics, as it can be harmful to the host (Alberti et al., 2017).

5.1.4 Easy Purification of Desired Products

The ability of filamentous fungi to secrete proteins extracellularly has several benefits. Primarily, the heterologous proteins that are secreted can be easily purified. There would not be a requirement for breaking the cells and separating the desired products from other intracellular components, which would reduce the cost and effort. Secondly, if the recombinant proteins are toxic, still the fungi will not be affected since they will not accumulate within the cells (Sharma et al., 2009). Furthermore, as the filamentous fungi are macroscopic structures, their recovery is comparatively easier when compared to the bacteria as well as yeast, which require more complex techniques such as centrifugation. Due to the filamentous structures, it has more surface area, which is in contact with the substratum, therefore enabling better absorption of nutrients (Ferreira et al., 2020).

5.1.5 Cost-effective

Filamentous fungi have the capability to grow on low-cost substrates and can secrete recombinant proteins (Madhavan et al., 2022). The following evidences suggest the same. Recombinant *A. niger*, when grown on dairy waste, whey was observed to produce high titres of fungal endoglucanases (Crament et al., 2024). The laccase gene of *Pycnoporus sanguineus,* when introduced into *Trichoderma reesei* ZJ-09 strain, the recombinant protein laccase was secreted, utilizing rice straw. Therefore, effective degradation of the lignocellulosic waste, rice straw was possible with the filamentous fungi, *Trichoderma reesei* ZJ-09 in the production of laccase (Xia & Lin, 2022). Agricultural residues such as sugarcane bagasse, corn cob, and rice bran when used as low-cost substrate for production of recombinant enzymes, xylanases in *A. niger* reported high titres of the heterologous proteins (Knob et al., 2014).

5.2 Challenges

Despite various advantages present in filamentous fungi when compared to other microbial systems in heterologous protein production, they also have their own challenges in terms of the yield of heterologous protein production. Below discussed are a few challenges that hinder the abundant production of heterologous proteins (Fig. 3).

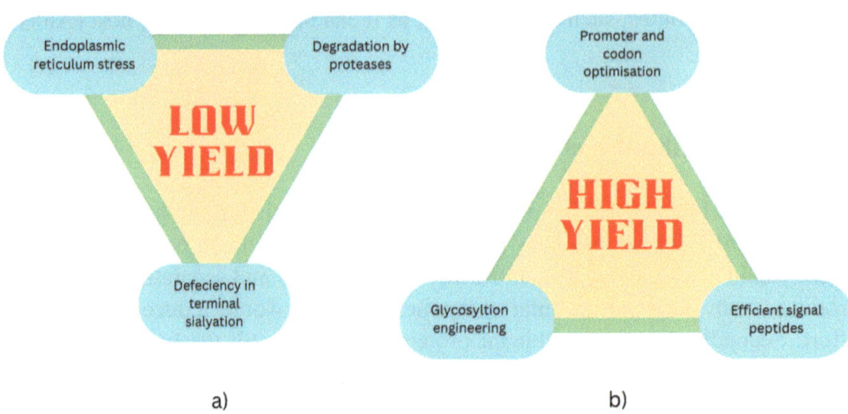

Fig. 3 (**a**) Challenges leading to low yield in heterologous protein production, (**b**) Strategies to increase the yield

5.3 Endoplasmic Reticulum Stress

Introduction of strong promoters is a technique for increasing the secretion of heterologous proteins in filamentous fungi. The functioning of the newly introduced, strong promoters can be taxing for the endoplasmic reticulum (ER) in the course of time, which could result in the unfolding or misfolding of the proteins and eventually result in a reduction of secretion. ER stress is naturally counteracted by two major pathways—Unfolded Protein Response (UPR) and Repression Under Secretory Stress (RESS). Both pathways aim to maintain homeostasis in the ER. UPR is involved in accelerating the production of chaperones, which can help in the proper folding of the nascent proteins as well as the degradation of unfolded or misfolded proteins. Whereas RESS reduces the production of newly released proteins to halt additional workload. ER stress is still a major challenge in industrially important filamentous fungi as there is a void in the knowledge regarding several aspects, such as relation between protein secretion and ER stress, potential role played by antioxidants in relieving oxidative stress in ER and RESS, which cause a reduction in the secretion of industrially significant heterologous proteins (Jadhav et al., 2024). Endoplasmic reticulum-associated degradation (ERAD) is another stress management pathway, which helps in the degradation of misfolded proteins by targeting them to proteosomes. Due to ERAD, foreign proteins that are misfolded are immediately degraded and can lead to a decrease in their yield. According to a study by Carvalho et al., *A. niger* produced a heterologous protein (β-glucuronidase), six times more when the ERAD-related genes derA and hrdC were deleted. This implies that proteins have more time to fold correctly before being degraded when ERAD activity is decreased. But blocking ERAD also causes misfolded proteins to accumulate, which can inhibit fungal development by triggering the unfolded protein response (UPR) (Carvalho et al., 2011).

The unfolded protein response (UPR) and a distinct downregulation of genes encoding secreted proteins, such as glaA (glucoamylase), are both triggered by disturbances in protein folding inside the endoplasmic reticulum (ER) of *Aspergillus niger*. This transcriptional repression is controlled by particular promoter regions and takes place apart from UPR mechanisms. A unique stress-induced regulatory mechanism that can restrict the fungus capacity to generate and secrete essential enzymes is highlighted by the fact that, although ER stress brought on by substances like DTT (dithiothreitol) activates UPR, lowering PDI (Protein Disulfide Isomerase) levels results in glaA downregulation without UPR activation (Al-Sheikh et al., 2004).

5.3.1 Degradation by Proteases

In *Aspergillus* species, titres of secretory proteins fall short of gram per litre production level of native proteins due to degradation of the proteins by proteases. Although several measures have been taken to prevent the degradation of proteins such as maintaining a high pH during fermentation, maintaining a low temperature during downstream processing, separating the product from protease-containing medium, and adding protease inhibitors, proteolysis still occurs reducing the production of industrially significant heterologous proteins (Kamaruddin et al., 2018). Although silencing of protease genes favours higher yield, knocking down *PrtT*, which is a transcription factor responsible for production of proteases, may negatively impact the strain. Because some of the factors involved are linked and may have unanticipated antagonistic or synergistic effects, choosing the right culture conditions to lower protease activity is complicated (Ntana et al., 2020). The integrity of the cell wall of *A. fumigatus* is likewise impacted by *XprG* and *PrtT* knockdown. Additionally, even after *PrtT* was knocked down, the strain continued to produce protease, which continued to degrade the heterologous protein to some extent, indicating that the production of *A. niger*-secreted protease is regulated by other transcription factors (Li et al., 2023a).

5.3.2 Lack of Terminal Sialyation of the Glycans

Sialyation, or the addition of sialic acid, is a process observed during glycosylation in mammalian cells, which is essential for the efficacy and quality of the proteins (Kim et al., 2024). Filamentous fungi are one of the prime producers of heterologous proteins. The data regarding sialyation during the process of glycosylation is scarce (Eneva et al., 2021). There has been evidence of the presence of sialic acid on the surface of *A. fumigatus,* although its genome lacked the genes which are essential for the biosynthesis of sialic acid (Eneva et al., 2021). Several efforts have been made to engineer the glycosylation process of recombinant proteins in *A. nidulans* in order to be acceptable to humans. When the gene ALG3 was deleted in yeast, it resulted in the production of Man5GlcNAc2 which can serve as precursors for further remodelling (Anyaogu et al., 2021). Including this, the homolog of the gene ALG3 in *Aspergilli* was deleted, and more hexose residues were accumulated instead of Man5GlcNAc2. Hence, further optimization is essential to meet the objective of remodelling glycosylation in *A. nidulans* (Anyaogu et al., 2021).

5.3.3 Low Yield

The yield of heterologous protein obtained from filamentous fungi is much lower when compared to the production of homologous proteins. Industrially important *A. niger* produced 400 times more homologous proteins when compared to heterologous proteins (Liu et al., 2023). Filamentous fungus typically lacks the capacity to sustain a self-propagating plasmid. Compared to *E. coli* and *S. cerevisiae*, two model organisms frequently employed for heterologous gene expression, these traits limit the development of effective transformation strategies for filamentous fungi (Su et al., 2012). Compared to the prokaryotic counterparts, as the genome is complex in filamentous fungi, it will be difficult to manipulate the genome for better production of heterologous proteins (Pouresmaeil & Azizi-Dargahlou, 2023).

In spite of these challenges in heterologous expression and secretion in filamentous fungi, there are various examples of commercially available enzymes and proteins that have been produced by filamentous fungi, and the following section discusses the same.

6 Various Examples of Enzymes and Proteins Expressed Using the Filamentous Fungi Expression System

In order to scavenge nutrients from their surroundings, filamentous fungi naturally express and produce a lot of enzymes (Meyer et al., 2016). Filamentous fungus produces around half of all industrial enzymes, and it has increasingly become apparent that filamentous fungi have several distinct advantages over alternative protein expression systems. Filamentous fungal expression platforms are functional despite their complexity. They are generally considered to have safe status, proficiency with fermentation scale-up and the requirement of relatively inexpensive media, the capacity for post-transcriptional and post-translational modifications (particularly helpful in expressing eukaryotic gene products), and the rapidly developing molecular/genetic tools to engineer pertinent strains are the various capabilities exhibited by the filamentous fungi. Generally, filamentous fungi are better at expressing homologous proteins than other forms (Sun & Su, 2019). *Aspergilli*, especially *A. niger*, have much potential to become the future workhorses of protein expression. It is important to achieve the integration of heterologous gene(s) at a transcriptionally active region of the host genome; this facilitates higher transcription and improves mRNA stability (Sakekar et al., 2021). Table 1 lists various examples of enzymes and proteins produced, and the important commercially available enzymes produced by filamentous fungi are discussed below.

6.1 Cellulases

Cellulases are a class of enzymes that includes β-glucosidases, endoglucanases, and exoglucanases, also known as cellobiohydrolases (Jayasekara & Ratnayake, 2019). Cellulase is a key enzyme in biotechnology because of its numerous uses in the food, beverage, textile, paper, and pulp sectors, as well as in agriculture, medicine, and the generation of biofuel. Numerous microorganisms, including bacteria and fungi, can break down cellulose, and some of them generate a sizable quantity of extracellular enzymes (Singh et al., 2019). Fungi are mostly preferred over bacteria due to versatile substrate utilization and penetration ability. Cellulase production has been studied in fungi such as *Trichoderma*, *Aspergillus*, *Cladosporium*,

Table 1 List of commercially available enzymes/proteins produced by fungal proteins expression system through heterologous expression

Enzyme/protein	Expression host	Optimum temperature	Optimum pH	Enzymeactivity	Commercial applications	References
Endoglucanase I	*Trichoderma harzianum* IOC-3844	55 °C	3	75 U/mg	Biofuel production, biomass degradation	Pellegrini et al. (2015)
Xyloglucanase (GH12)	*Aspergillus terreus*	65 °C	5.5	17.4 μmol/ (min mg)	Biomass degradation	Vitcosque et al. (2016)
Cellobiohydrolase II	*Trichoderma reesei*	40 °C	5–6	14.5 U/L	Bioethanol production	Díaz-Rincón et al. (2017)
Polygalacturinase	*Aspergillus niger*	50 °C	5	4617.8 U/ml	Food processing, wastewater treatment, and feed production	Yang (2020)
Aminoacylase	*Aspergillus oryzae*	40 °C	7	7338 U	Industrial production of amino acids	He et al. (2007) and Youshko et al. (2004)
Ferulosyl esterase	*Trichoderma reesei*	Anfae A gene: 50 °C Anfae B gene: 35 °C	5 6	20.69 U/gsd 15.08 U/gsd	Ferulic acid	Long et al. (2018)
Xylanase	*Aspergillus nidulans XynC A773*	–	–	187.9 U/g	Pharmaceutical, food, paper and cellulose production	Fernandes de Souza et al. (2022) and da Silva et al. (2018)
Endoglucanase	*Aspergillus niger*	Cel 9A gene: 70 °C	–	–	Biorefinery, pulp and paper industry	Mandeep and Shukla (2021) and Gladden (2013)
β glucosidase	*Aspergillus niger*	J01 gene: 70 °C	–	–	Food processing, brewery, dairy and pharmaceuticals	Gladden, (2013) and Kannan et al. (2023)

Penicillium, *Scopulariopsis*, *Stachybotrys*, *Verticillium*, and *Chaetomium*. *Trichoderma* and *Aspergillus* are the most researched of these and are employed in both industry and agriculture (Zahra et al., 2020). *A. niger* is a high-protein expression host commonly used for the commercial production of endogenous glucoamylase (Amyloglucosidase (Liquozyme X), xylanase (Ecopulp, Cartazyme HS 10, Cartazyme HT), and protease (Gastrinex)). Most of these enzymes share comparable features, such as being both acidic and mesophilic (Pel et al., 2007). Therefore, native hydrolases (xylanase, glycosylase, and protease) from *A. niger* can be overexpressed simultaneously via gene editing to construct a multi-enzyme high-expression system for β-glucan extraction. The optimization of expression cassettes for recombinant genes is a crucial strategy for enhancing their expression levels, which involves the selection of appropriate expression elements and open reading frames. Among them, promoters such as P*glaA* and P*gpaA* which exhibited the most active transcription efficiency are often used in *A.* n*iger (*Antunes et al., 2016*)*. Screening for novel strong promoters and modifying their core region helps in increasing the efficiency of gene transcription (Li et al., 2023b). Glucoamylase, primarily derived from mesophilic fungus like *Aspergillus* and *Rhizopus* species, is a crucial enzyme for starch saccharification in the food and biofuel industries (Pavezzi et al., 2008). Compared to a mesophile-based fermentation system, enzymes derived from thermophilic fungi can save fermentation energy and expenses. Thermophilic fungus *Myceliophthora thermophila* is an industrially deployed fungus to produce enzymes and biobased chemicals from biomass during optimal growth at 45 °C (Zhu et al., 2023).

6.2 Proteases

Proteases are enzymes that break down protein molecule through peptide bond hydrolysis. The genera *Aspergillus, Penicillium,* and *Rhizopus* are especially useful for producing proteases (Sandhya et al., 2005), as several species of these genera are generally regarded as safe. Fungal proteases offer a distinct advantage over bacterial enzymes in terms of ease of downstream processing. The microbial proteases of *Aspergillus oryzae,* in particular, have been studied in detail since they are known for their capacity to secrete high levels of enzymes in their growth environment (Salihi et al., 2017). Several of these secreted enzymes produced in a large-scale submerged fermentation have been widely used in the food and beverage industry for decades (Mucor rennet, *Aspergillus* proteases (Proctase), and alkaline proteases (oryzin) (de Souza et al., 2015). Proteases are widely used in several industrial sectors such as detergent, food, pharmaceutical, chemical, leather, silk, cleaning contact lenses/denatures, dewooling of animal skin, recovery of soluble proteins and amino acids from chrome leather wastes and waste treatment, etc. (Raghu Ram & Yepuru, 2018). The glucose oxidase and peroxidase were heterologously expressed in an industrially important β-fructofuranosidase-producing fungus *A. niger* ATCC 20611, which was the main commercial fructooligosaccharides (FOS) producer

(Wan et al., 2024). Particularly, the heterologous enzymes were successfully detected in the mycelia because of the fusion with the FopAC domain of the native β-fructofuranosidase. Without requiring additional purifying procedures, the multiple-enzyme-generating mycelia of these strains could be utilized immediately. This is expected to be a cost-effective and efficient fructooligosaccharides production technology with whole-cell catalysis (Wan et al., 2024).

In addition, filamentous fungi can produce a wide range of secondary metabolites, which have been exploited as pharmaceuticals for decades (examples include the antibiotic penicillin [Augmentin amoxicillin/clavulanate, Unasyn (ampicillin/sulbactam)] and the immunosuppressant cyclosporine (Gengraf, Neoral, Sandimmune) (Hoffmeister & Keller, 2007). *Aspergillus nidulans* is a genetic model species among filamentous fungi and has also been used as heterologous host to study gene clusters from other species (Hoffmeister & Keller, 2007). *A. oryzae* is a species that has been used in many cases for heterologous expression (e.g., Adalimumab (Humira), Interleukin-1 Receptor Antagonist (IL-1Ra) (Alberti et al., 2017).

6.3 Ligninolytic Enzymes

The proteins known as ligninolytic enzymes (LEs), also called lignolytic enzymes or lignin-modifying enzymes (LMEs), catalyse the breakdown of lignin into simpler compounds in order to finally create CO_2 (Tien & Kirk, 1983). Laccase (LCC), lignin peroxidase (LiP), manganese peroxidase (MnP), and versatile peroxidase (Vp) are among the complex enzyme systems for lignin breakdown found in many fungal species that also make LEs (Blanchette et al., 1997). These enzymes are distinguished by their wide selectivity in the breakdown and mineralization of several complex substrates, including hydrocarbons, wood, paper, animal feed, insecticides, and biofuels (Asemoloye et al., 2018). It is well acknowledged that LEs are essential for biotechnological, industrial, and environmental applications, particularly in the synthesis of bioenergy or biomaterials (Rodríguez Couto & Toca Herrera, 2006).

6.4 Amylases

α-Amylase has various applications in numerous industries like pharmaceutical, detergent, textile, paper, and of course, the food industry (Bosmans et al., 2013). Because fungal amylases are typically regarded as safe, they are preferred in the food business, particularly when it comes to bread and baking (Fenice et al., 1997). The characterization, purification, expression, and gene cloning of a new α-amylase from the Antarctic mycelial fungus *G. pannorum* were investigated by He et al. (2017). This study successfully cloned the cold-adapted amylase gene AmyA1 from *G. pannorum* R1–2. Industrial applications may benefit from AmyA1's increased stability and longer service life (He et al., 2017).

The α-amylase and a new fungal glucoamylase genes were successfully produced in *Pichia pastoris* after being cloned from *Rhizomucor pusillus*. With optimal pH values of 4.0 (RpGla) and 5.0 (RpAmy) and optimal temperatures as high as 70 °C, respectively, the two recombinant enzymes share similar properties, improving the operating range in terms of pH, temperature, and compatibility with other enzymes (He et al., 2014).

6.5 Chitinase

The hydrolysis of chitin is catalysed by chitinases, which are enzymes with a variety of tasks, including morphogenetic, nutritional, and invasive roles in fungi, defensive roles in plants, developmental roles in insects, and nutritional roles in bacteria and archaea (Hartl et al., 2012). When entomopathogenic fungi penetrate the host's cuticle, they release hydrolytic enzymes such as lipases, chitinases, and proteases that break down the insect's cuticle and can start the infection process (Charnley, 2003). In agroforestry, entomopathogenic fungal chitinases are appealing options for the biological management of insect pests (Zhu et al., 2008).

Several chitinases have been identified from *Metarhizium anisopliae* and *Beauveria bassiana* in vivo (Bhanu Prakash et al., 2012). *M. anisopliae*'s Chit1 chitinase was found to have little to no impact on virulence, however *B. bassiana's* overproduction of the chitinase Bbchit1 increased the fungus's ability to control aphids (Fang et al., 2005). In a bioassay of transgenic isolates with the chitinase gene overexpressed and silenced, it has been demonstrated that *M. anisopliae*'s chitinase chi2 is the cause of virulence (Boldo et al., 2009). Currently, *I. fumosorosea* produces chitinases that effectively hydrolyse and destroy different insects' cuticles (Ali et al., 2010).

When *Aspergillus niveus*, a filamentous fungus, was cultivated under submerged fermentation (SbmF) with crab shells as the carbon source, it produced extracellular antifungal chitinase. The fungus *A. niveus* demonstrated its biotechnological potential by producing a thermostable, denaturation-resistant chitinase that inhibited fungal growth (Alves et al., 2018).

Though the above-mentioned are a few of the examples of filamentous fungi as a potential host for heterologous protein production, they still remain to be explored further as a reservoir for heterologous protein production.

7 Conclusion

Filamentous fungi have become potential workhorses of heterologous protein production due to their efficient secretory pathways with high secretions, efficiency in performing posttranslational modification, and ability to grow in low-cost substrates. The metabolic activities help in diverse heterologous protein production that can pave the way to industrial applications. Every unsuccessful expression and secretion of heterologous protein signals the derisory information on factors impacting the secretion. A deep understanding of secretory pathways of expression and secretion is still in the grey area, which could be vital in increasing heterologous protein production in larger quantities. Through the various experiments carried out to improve the strain for heterologous protein production, the understanding of the complexity of each method has given an opportunity to understand more deeply the secretory pathways. Certain challenges, such as low yield, ER stress, lack of glycan terminalization, and protein degradation by proteases, have been studied and experimented with to find solutions for these problems. In this chapter, we have discussed a few methods experimented with, such as ERAD regulation, protease-deficient strains, codon optimization, native protein fusion, UPR regulation, signal peptide modification, and altered glycosylation pathways that have enhanced the heterologous protein production in some cases. Even with all these challenges, filamentous fungi are still the preferred candidate for heterologous protein production due to their distinct properties. The advancements in omics—proteomics, transcriptomics, and metabolomics—will help expand knowledge on protein production, which will aid in fabricating highly efficient fungal strains by optimizing rate-determining steps.

References

Adney, W. S., Jeoh, T., Beckham, G. T., et al. (2009). Probing the role of N-linked glycans in the stability and activity of fungal cellobiohydrolases by mutational analysis. *Cellulose, 16*, 699–709. https://doi.org/10.1007/S10570-009-9305-1

Alberti, F., Foster, G. D., & Bailey, A. M. (2017). Natural products from filamentous fungi and production by heterologous expression. *Applied Microbiology and Biotechnology, 101*, 493–500. https://doi.org/10.1007/S00253-016-8034-2

Alberto, F., Navarro, D., De Vries, R. P., et al. (2009). Technical advance in fungal biotechnology: Development of a miniaturized culture method and an automated high-throughput screening. *Letters in Applied Microbiology, 49*, 278–282. https://doi.org/10.1111/J.1472-765X.2009.02655.X

Ali, S., Wu, J., Huang, Z., & Ren, S. X. (2010). Production and regulation of extracellular chitinase from the entomopathogenic fungus *Isaria fumosorosea*. *Biocontrol Science and Technology, 20*, 723–738. https://doi.org/10.1080/09583151003714091

Al-Sheikh, H., Watson, A. J., Lacey, G. A., et al. (2004). Endoplasmic reticulum stress leads to the selective transcriptional downregulation of the glucoamylase gene in *Aspergillus niger*. *Molecular Microbiology, 53*, 1731–1742. https://doi.org/10.1111/j.1365-2958.2004.04236.x

Alves, T. B., de Oliveira Ornela, P. H., de Oliveira, A. H. C., et al. (2018). Production and characterization of a thermostable antifungal chitinase secreted by the filamentous fungus

Aspergillus niveus under submerged fermentation, *3 Biotech, 8*, 369. https://doi.org/10.1007/s13205-018-1397-6

Antunes, M. S., Hodges, T. K., & Carpita, N. C. (2016). A benzoate-activated promoter from *Aspergillus niger* and regulation of its activity. *Applied Microbiology and Biotechnology, 100*, 5479–5489. https://doi.org/10.1007/s00253-016-7373-3

Anyaogu, D. C., Hansen, A. H., Hoof, J. B., et al. (2021). Glycoengineering of *Aspergillus nidulans* to produce precursors for humanized N-glycan structures. *Metabolic Engineering, 67*, 153–163. https://doi.org/10.1016/j.ymben.2021.06.001

Arakawa, T., Yphantis, D. A., Lary, J. W., et al. (1991). Glycosylated and unglycosylated recombinant-derived human stem cell factors are dimeric and have extensive regular secondary structure. *Journal of Biological Chemistry, 266*, 18942–18948. https://doi.org/10.1016/S0021-9258(18)55154-1

Araki, Y., Shinohara, Y., Hara, S., et al. (2022). Heterologous production of ascofuranone and ilicicolin a in *Aspergillus sojae*. *Journal of General and Applied Microbiology, 68*, 10–16. https://doi.org/10.2323/jgam.2021.08.001

Asemoloye, M. D., Ahmad, R., & Jonathan, S. G. (2018). Transcriptomic responses of catalase, peroxidase and laccase encoding genes and enzymatic activities of oil spill inhabiting rhizospheric fungal strains. *Environmental Pollution, 235*, 55–64. https://doi.org/10.1016/j.envpol.2017.12.042

Bhanu Prakash, G. V. S., Padmaja, V., Jami, S. K., & Kirti, P. B. (2012). Expression of chitinase genes of *Metarhizium anisopliae* isolates in lepidopteran pests and on synthetic media. *Journal of Basic Microbiology, 52*, 628–635. https://doi.org/10.1002/jobm.201100274

Blanchette, R. A., Krueger, E. W., Haight, J. E., et al. (1997). Cell wall alterations in loblolly pine wood decayed by the white-rot fungus, *Ceriporiopsis subvermispora*. *Journal of Biotechnology, 53*, 203–213. https://doi.org/10.1016/S0168-1656(97)01674-X

Boldo, J. T., Junges, A., do Amaral, K. B., et al. (2009). Endochitinase CHI2 of the biocontrol fungus *Metarhizium anisopliae* affects its virulence toward the cotton stainer bug *Dysdercus peruvianus*. *Current Genetics, 55*, 551–560. https://doi.org/10.1007/s00294-009-0267-5

Bosmans, G. M., Lagrain, B., Fierens, E., & Delcour, J. A. (2013). Impact of amylases on biopolymer dynamics during storage of straight-dough wheat bread. *Journal of Agricultural and Food Chemistry, 61*, 6525–6532. https://doi.org/10.1021/jf402021g

Brandhorst, T., Yang, R., & Kenealy, W. R. (1994). Heterologous expression of the cytotoxin restrictocin in *Aspergillus nidulans* and *Aspergillus niger*. *Protein Expression and Purification, 5*, 486–497. https://doi.org/10.1006/prep.1994.1068

Cairns, T. C., Nai, C., & Meyer, V. (2018). How a fungus shapes biotechnology: 100 years of *Aspergillus niger* research. *Fungal Biology and Biotechnology, 5*, 13. https://doi.org/10.1186/s40694-018-0054-5

Carreras-Villasenõr, N., Rico-Ruiz, J. G., Chávez Montes, R. A., et al. (2020). Assessment of the ptxD gene as a growth and selective marker in *Trichoderma atroviride* using Pccg6, a novel constitutive promoter. *Microbial Cell Factories, 19*, 69. https://doi.org/10.1186/s12934-020-01326-z

Carrez, D., Janssens, W., Degrave, P., et al. (1990). Heterologous gene expression by filamentous fungi: Secretion of human interleukin-6 by *Aspergillus nidulans*. *Gene, 94*, 147–154. https://doi.org/10.1016/0378-1119(90)90381-Z

Carvalho, N. D. S. P., Arentshorst, M., Kooistra, R., et al. (2011). Effects of a defective ERAD pathway on growth and heterologous protein production in *Aspergillus niger*. *Applied Microbiology and Biotechnology, 89*, 357–373. https://doi.org/10.1007/s00253-010-2916-5

Cavka A, Jo¨nsson LJ (2014) Comparison of the growth of filamentous fungi and yeasts in lignocellulose-derived media. Biocatalysis and Agricultural Biotechnology 3:197–204. doi:https://doi.org/10.1016/j.bcab.2014.04.003.

Chang, P. K., Matsushima, K., Takahashi, T., et al. (2007). Understanding nonaflatoxigenicity of *Aspergillus sojae*: A windfall of aflatoxin biosynthesis research. *Applied Microbiology and Biotechnology, 76*, 977–984. https://doi.org/10.1007/s00253-007-1116-4

Charnley, A. K. (2003). *Fungal pathogens of insects: Cuticle degrading enzymes and toxins*, pp. 241–321.
Chavan, S., Shete, A., & Dharne, M. S. (2024). Strain improvement for cellulolytic enzymes for effective saccharification of lignocellulosic biomass by mutant of *Penicillium funiculosum* NCIM 1228. *Systems Microbiology and Biomanufacturing, 4*, 716–730. https://doi.org/10.1007/S43393-024-00247-X
Chulkin, A., Kislitsin, V., Sinelnikov, I., et al. (2024). Transcription factors ClrB and XlnR and their effect on the transcription of cellulase genes in the filamentous fungus *Penicillium verruculosum. International Journal of Molecular Sciences, 25*, 13373. https://doi.org/10.3390/IJMS252413373/S1
Conesa, A., Jeenes, D., Archer, D. B., et al. (2002). Calnexin overexpression increases manganese peroxidase production in *Aspergillus niger. Applied and Environmental Microbiology, 68*, 846–851. https://doi.org/10.1128/AEM.68.2.846-851.2002
Contreras, R., Carrez, D., Kinghorn, J. R., et al. (1991). Efficient KEX2-like processing of a glucoamylase-interleukin-6 fusion protein by *Aspergillus nidulans* and secretion of mature Interleukin-6. *Bio/Technology, 9*, 378–381. https://doi.org/10.1038/NBT0491-378
Corbu, V. M., Gheorghe-Barbu, I., Ştefania, D. A., et al. (2023). Current insights in fungal importance—A comprehensive review. *Microorganisms, 11*, 1384. https://doi.org/10.3390/microorganisms11061384
Crament, T. C., Arendsen, K., Rose, S. H., & Jansen, T. (2024). Cultivation of recombinant *Aspergillus niger* strains on dairy whey as a carbohydrate source. *Journal of Industrial Microbiology & Biotechnology, 51*. https://doi.org/10.1093/jimb/kuae007
da Silva, M. B., Rossi, D. M., Squina, F., & Ayub, M. A. Z. (2018). Comparative production of xylanase and the liberation of xylooligosaccharides from lignocellulosic biomass by *Aspergillus brasiliensis* BLf1 and recombinant *Aspergillus nidulans* XynC A773. *International Journal of Food Science and Technology, 53*, 2110–2118. https://doi.org/10.1111/ijfs.13798
Darmond, C. (2016). *The use of filamentous fungi Myceliophthora heterothallica as a host for heterologous protein production*. Thesis. Concordia University Montreal.
de Amo, G. S., Bezerra-Bussoli, C., Rodrigues da Silva, R., et al. (2024). Heterologous expression of GH11 xylanase from *Myceliophthora heterothallica* F.2.1.4 in Pichia pastoris. Biocatal Agric. *Biotech, 61*, 103343. https://doi.org/10.1016/j.bcab.2024.103343
de Souza, P. M., de Asis Bittencourt, M. L., Caprara, C. C., et al. (2015). A biotechnology perspective of fungal proteases. *Brazilian Journal of Microbiology, 46*, 337–346. https://doi.org/10.1590/S1517-838246220140359
De Vries, R. P., Burgers, K., van de Vondervoort, P. J. I., et al. (2004). A new black *Aspergillus* species, A. vadensis, is a promising host for homologous and heterologous protein production. *Applied and Environmental Microbiology, 70*, 3954–3959. https://doi.org/10.1128/AEM.70.7.3954-3959.2004
Denisenko, Y. A., Korotkova, O. G., Zorov, I. N., et al. (2021). Heterologous expression of Thermogutta terrifontis Endo-Xanthanase in *Penicillium verruculosum*, isolation and primary characterization of the enzyme. *Biochemistry (Moscow), 86*, 489–495. https://doi.org/10.1134/S000629792104009X
Deshpande, N., Wilkins, M. R., Packer, N., & Nevalainen, H. (2008). Protein glycosylation pathways in filamentous fungi. *Glycobiology, 18*, 626–637. https://doi.org/10.1093/glycob/cwn044
Díaz-Rincón, D. J., Duque, I., Osorio, E., et al. (2017). Production of recombinant *Trichoderma reesei* Cellobiohydrolase II in a new expression system based on *Wickerhamomyces anomalus*. *Enzyme Research, 2017*, 1–8. https://doi.org/10.1155/2017/6980565
Dong, L., Lin, X., Yu, D., et al. (2020). High-level expression of highly active and thermostable trehalase from *Myceliophthora thermophila* in *Aspergillus niger* by using the CRISPR/Cas9 tool and its application in ethanol fermentation. *Journal of Industrial Microbiology & Biotechnology, 47*, 133–144. https://doi.org/10.1007/s10295-019-02252-9

Eneva, R., Engibarov, S., Abrashev, R., et al. (2021). Sialic acids, sialoconjugates and enzymes of their metabolism in fungi. *Biotechnology & Biotechnological Equipment, 35*, 346–357. https://doi.org/10.1080/13102818.2021.1879678

Fang, W., Leng, B., Xiao, Y., et al. (2005). Cloning of *Beauveria bassiana* Chitinase gene *Bbchit1* and its application to improve fungal strain virulence. *Applied and Environmental Microbiology, 71*, 363–370. https://doi.org/10.1128/AEM.71.1.363-370.2005

Fenice, M., Selbmann, L., Zucconi, L., & Onofri, S. (1997). Production of extracellular enzymes by Antarctic fungal strains. *Polar Biology, 17*, 275–280. https://doi.org/10.1007/s003000050132

Fernandes de Souza, H., Aguiar Borges, L., Di Próspero, D., Gonçalves, V., et al. (2022). Recent advances in the application of xylanases in the food industry and production by actinobacteria: A review. *Food Research International, 162*, 112103. https://doi.org/10.1016/j.foodres.2022.112103

Ferreira, J. A., Varjani, S., & Taherzadeh, M. J. (2020). A critical review on the ubiquitous role of filamentous fungi in pollution mitigation. *Current Pollution Reports, 6*, 295–309. https://doi.org/10.1007/s40726-020-00156-2

Geysens, S., Pakula, T., Uusitalo, J., et al. (2005). Cloning and characterization of the glucosidase II alpha subunit gene of *Trichoderma reesei*: A frameshift mutation results in the aberrant glycosylation profile of the Hypercellulolytic strain rut-C30. *Applied and Environmental Microbiology, 71*, 2910. https://doi.org/10.1128/AEM.71.6.2910-2924.2005

Gladden, J. (2013). Production of extremophilic bacterial cellulase enzymes in *Aspergillus niger*. Albuquerque/Livermore.

Gnaim, R., Dyer, P. S., & Ledesma-Amaro, R. (2025). *Fusarium*-based mycoprotein: Advancements in the production of sustainable meat substitutes. *Trends in Food Science and Technology, 159*, 104981. https://doi.org/10.1016/j.tifs.2025.104981

Gouka, R. J., Punt, P. J., & Van Den Hondel, C. A. M. J. J. (1997). Glucoamylase gene fusions alleviate limitations for protein production in *Aspergillus awamori* at the transcriptional and (post) translational levels. *Applied and Environmental Microbiology, 63*, 488. https://doi.org/10.1128/AEM.63.2.488-497.1997

Graessle, S., Haas, H., Friedlin, E., et al. (1997). Regulated system for heterologous gene expression in *Penicillium chrysogenum*. *Applied and Environmental Microbiology, 63*, 753–756. https://doi.org/10.1128/AEM.63.2.753-756.1997

Gu, S., Li, J., Chen, B., et al. (2018). Metabolic engineering of the thermophilic filamentous fungus *Myceliophthora thermophila* to produce fumaric acid. *Biotechnology for Biofuels, 11*, 1–10. https://doi.org/10.1186/S13068-018-1319-1

Guo, X., Jiang, L., An, Y., et al. (2023). Construction and characterization of a *Myceliophthora thermophila* lytic polysaccharide monooxygenase mutant S174C/A93C with improved thermostability. *Enzyme and Microbial Technology, 168*, 110255. https://doi.org/10.1016/J.ENZMICTEC.2023.110255

Gürkök, S., Söyler, B., Biely, P., & Ögel, Z. B. (2010). Cloning and heterologous expression of the extracellular alpha-galactosidase from *Aspergillus fumigatus* in *Aspergillus sojae* under the control of gpdA promoter. *Journal of Molecular Catalysis B: Enzymatic, 64*, 146–149. https://doi.org/10.1016/J.MOLCATB.2009.09.012

Haakana, H., Miettinen-Oinonen, A., Joutsjoki, V., et al. (2004). Cloning of cellulase genes from *Melanocarpus albomyces* and their efficient expression in *Trichoderma reesei*. *Enzyme and Microbial Technology, 34*, 159–167. https://doi.org/10.1016/J.ENZMICTEC.2003.10.009

Harkki, A., Uusitalo, J., Bailey, M., et al. (1989). A novel fungal expression system: Secretion of active calf chymosin from the filamentous fungus *trichoderma reesei*. *Bio/Technology, 7*, 596–603. https://doi.org/10.1038/NBT0689-59

Hartl, L., Zach, S., & Seidl-Seiboth, V. (2012). Fungal chitinases: Diversity, mechanistic properties and biotechnological potential. *Applied Microbiology and Biotechnology, 93*, 533–543. https://doi.org/10.1007/s00253-011-3723-3

He, D., Fan, C., Liao, L., et al. (2007). Fermentation conditions of aminoacylase-producing *Aspergillus oryzae* and characteristics of aminoacylase. *Chinese Journal of Applied and Environmental Biology, 13*, 270–273.

He, Z., Zhang, L., Mao, Y., et al. (2014). Cloning of a novel thermostable glucoamylase from thermophilic fungus *Rhizomucor pusillus* and high-level co-expression with α-amylase in *Pichia pastoris*. *BMC Biotechnology, 14*, 114. https://doi.org/10.1186/s12896-014-0114-8

He, L., Mao, Y., Zhang, L., et al. (2017). Functional expression of a novel α-amylase from Antarctic psychrotolerant fungus for baking industry and its magnetic immobilization. *BMC Biotechnology, 17*, 22. https://doi.org/10.1186/s12896-017-0343-8

Heneghan, M. N., Yakasai, A. A., Halo, L. M., et al. (2010). First heterologous reconstruction of a complete functional fungal biosynthetic multigene cluster. *Chembiochem, 11*, 1508–1512. https://doi.org/10.1002/cbic.201000259

Hoffmeister, D., & Keller, N. P. (2007). Natural products of filamentous fungi: Enzymes, genes, and their regulation. *Natural Product Reports, 24*, 393–416. https://doi.org/10.1039/B603084J

Jacobs, D. I., Olsthoorn, M. M. A., Maillet, I., et al. (2009). Effective lead selection for improved protein production in *Aspergillus niger* based on integrated genomics. *Fungal Genetics and Biology, 46 Suppl 1*. https://doi.org/10.1016/J.FGB.2008.08.012

Jadhav, R., Mach, R. L., & Mach-Aigner, A. R. (2024). Protein secretion and associated stress in industrially employed filamentous fungi. *Applied Microbiology and Biotechnology, 108*, 1–14. https://doi.org/10.1007/S00253-023-12985-4.

Jayasekara, S., & Ratnayake, R. (2019). *Microbial Cellulases: An overview and applications*. Cellulose. IntechOpen.

Jeennor, S., Anantayanon, J., Chutrakul, C., et al. (2022). Novel pentose-regulated promoter of *Aspergillus oryzae* with application in controlling heterologous gene expression. *Biotechnology Reports, 33*, e00695. https://doi.org/10.1016/j.btre.2021.e00695

Jin, F.-J., Hu, S., Wang, B.-T., & Jin, L. (2021). Advances in genetic engineering technology and its application in the industrial fungus *Aspergillus oryzae*. *Frontiers in Microbiology, 12*. https://doi.org/10.3389/fmicb.2021.644404

Kamaruddin, N., Storms, R., Mahadi, N. M., et al. (2018). Reduction of extracellular proteases increased activity and stability of heterologous protein in *Aspergillus niger*. *Arabian Journal for Science and Engineering, 43*, 3327–3338. https://doi.org/10.1007/s13369-017-2914-3

Kannan, P., Shafreen, M. M., Achudhan, A. B., et al. (2023). A review on applications of β-glucosidase in food, brewery, pharmaceutical and cosmetic industries. *Carbohydrate Research, 530*, 108855. https://doi.org/10.1016/j.carres.2023.108855

Kim, Y.-G., Raunser, S., Munger, C., et al. (2006). The architecture of the multisubunit TRAPP I complex suggests a model for vesicle tethering. *Cell, 127*, 817–830. https://doi.org/10.1016/j.cell.2006.09.029

Kim, K. H., Ji, E. S., Lee, J. Y., et al. (2024). LC-MS/MS-based site-specific N-glycosylation analysis of VEGFR-IgG fusion protein for sialylation assessment across IEF fractions. *Molecules, 29*, 5393. https://doi.org/10.3390/molecules29225393

Knob, A., Fortkamp, D., Prolo, T., et al. (2014). Agro-residues as alternative for xylanase production by filamentous fungi. *BioResources, 3*, 5738–5773.

Kumar, A. (2020). *Aspergillus nidulans*: A potential resource of the production of the native and heterologous enzymes for industrial applications. *International Journal of Microbiology, 2020*, 1–11. https://doi.org/10.1155/2020/8894215

Landowski, C. P., Mustalahti, E., Wahl, R., et al. (2016). Enabling low cost biopharmaceuticals: High level interferon alpha-2b production in *Trichoderma reesei*. *Microbial Cell Factories, 15*, 1–15. https://doi.org/10.1186/S12934-016-0508-5

Lee, J., Park, J. S., Moon, J. Y., et al. (2003). The influence of glycosylation on secretion, stability, and immunogenicity of recombinant HBV pre-S antigen synthesized in *Saccharomyces cerevisiae*. *Biochemical and Biophysical Research Communications, 303*, 427–432. https://doi.org/10.1016/S0006-291X(03)00351-6

Lee, J. M., Shin, J. W., Nam, J. K., et al. (2009). Molecular cloning and expression of the *Trichoderma harzianum* C4 endo-β-1,4-xylanase gene in *Saccharomyces cerevisiae*. *Journal of Microbiology and Biotechnology, 19*, 823–828. https://doi.org/10.4014/jmb.0810.577

Li, K., Zheng, J., Yu, L., et al. (2023a). Exploration of the strategy for improving the expression of heterologous sweet protein Monellin in *Aspergillus niger*. *Journal of Fungi, 9*, 528. https://doi.org/10.3390/jof9050528

Li, Y., Li, C., Muhammad Aqeel, S., et al. (2023b). Enhanced expression of xylanase in *Aspergillus niger* enabling a two-step enzymatic pathway for extracting β-glucan from oat bran. *Bioresource Technology, 377*, 128962. https://doi.org/10.1016/j.biortech.2023.128962

Liu, Q., Gao, R., Li, J., et al. (2017). Development of a genome-editing CRISPR/Cas9 system in thermophilic fungal *Myceliophthora* species and its application to hyper-cellulase production strain engineering. *Biotechnology for Biofuels, 10*, 1–14. https://doi.org/10.1186/S13068-016-0693-9

Liu, D., Garrigues, S., & de Vries, R. P. (2023). Heterologous protein production in filamentous fungi. *Applied Microbiology and Biotechnology, 107*, 5019–5033. https://doi.org/10.1007/s00253-023-12660-8

Lloyd, A. T., & Sharp, P. M. (1991). Codon usage in *Aspergillus nidulans*. *MGG Molecular & General Genetics, 230*, 288–294. https://doi.org/10.1007/BF00290679

Long, L., Zhao, H., Ding, D., et al. (2018). Heterologous expression of two *Aspergillus niger* feruloyl esterases in *Trichoderma reesei* for the production of ferulic acid from wheat bran. *Bioprocess and Biosystems Engineering, 41*, 593–601. https://doi.org/10.1007/s00449-018-1894-3

Madhavan, A., Arun, K., Sindhu, R., et al. (2022). Engineering interventions in industrial filamentous fungal cell factories for biomass valorization. *Bioresource Technology, 344*, 126209. https://doi.org/10.1016/j.biortech.2021.126209

Mandeep, L. H., & Shukla, P. (2021). Synthetic biology and biocomputational approaches for improving microbial endoglucanases toward their innovative applications. *ACS Omega, 6*, 6055–6063. https://doi.org/10.1021/acsomega.0c05744

Maras, M. (1999). Filamentous fungi as production organisms for glycoproteins of bio-medical interest. *Glycoconjugate Journal, 16*, 99–107. https://doi.org/10.1023/A:1026436424881

Martin, J. A., Murphy, R. A., & Power, R. F. G. (2003). Cloning and expression of fungal phytases in genetically modified strains of *Aspergillus awamori*. *Journal of Industrial Microbiology & Biotechnology, 30*, 568–576. https://doi.org/10.1007/s10295-003-0083-8

Meshram, P., Murmu, M., Barage, S., & Singh, R. (2024). Filamentous fungi as cell factories for heterogeneous protein production. In *Fundamentals of recombinant protein production, purification and characterization* (pp. 143–169). Elsevier.

Meyer, V., Andersen, M. R., Brakhage, A. A., et al. (2016). Current challenges of research on filamentous fungi in relation to human welfare and a sustainable bio-economy: A white paper. *Fungal Biology and Biotechnology, 3*, 6. https://doi.org/10.1186/s40694-016-0024-8

Meyer, V., Basenko, E. Y., Benz, J. P., et al. (2020). Growing a circular economy with fungal biotechnology: A white paper. *Fungal Biology and Biotechnology, 7*, 5. https://doi.org/10.1186/s40694-020-00095-z

Nelson, G., Kozlova-Zwinderman, O., Collis, A. J., et al. (2004). Calcium measurements in living filamentous fungi expressing codon-optimized aequorin. *Molecular Microbiology, 52*, 1437–1450. https://doi.org/10.1111/J.1365-2958.2004.04066.X

Nevalainen, H., & Peterson, R. (2014a). Making recombinant proteins in filamentous fungi – Are we expecting too much? *Frontiers in Microbiology, 5*. https://doi.org/10.3389/FMICB.2014.00075

Nevalainen, H., & Peterson, R. (2014b). Heterologous expression of proteins in *Trichoderma*. *Biotechnology and Biology of Trichoderma*, 89–102. https://doi.org/10.1016/B978-0-444-59576-8.00007-2

Nitayavardhana, S., Issarapayup, K., Pavasant, P., & Khanal, S. K. (2013). Production of protein-rich fungal biomass in an airlift bioreactor using vinasse as substrate. *Bioresource Technology, 133*, 301–306. https://doi.org/10.1016/J.BIORTECH.2013.01.073

Nothwehr, S. F., & Gordon, J. I. (1990). Targeting of proteins into the eukaryotic secretory pathway: Signal peptide structure/function relationships. *BioEssays, 12*, 479–484. https://doi.org/10.1002/BIES.950121005

Ntana, F., Mortensen, U. H., Sarazin, C., & Figge, R. (2020). Aspergillus: A powerful protein production platform. *Catalysts, 10*, 1064. https://doi.org/10.3390/CATAL10091064

Ogunmolu, F. E., Kaur, I., Pasari, N., et al. (2018). Quantitative multiplexed profiling of *Penicillium funiculosum* secretome grown on polymeric cellulase inducers and glucose. *Journal of Proteomics, 179*, 150–160. https://doi.org/10.1016/J.JPROT.2018.03.025

Ohno, A., Maruyama, J. I., Nemoto, T., et al. (2011). A carrier fusion significantly induces unfolded protein response in heterologous protein production by *Aspergillus oryzae*. *Applied Microbiology and Biotechnology, 92*, 1197–1206. https://doi.org/10.1007/S00253-011-3487-9

Okazaki, F., Aoki, J., Tabuchi, S., et al. (2012). Efficient heterologous expression and secretion in *Aspergillus oryzae* of a llama variable heavy-chain antibody fragment VHH against EGFR. *Applied Microbiology and Biotechnology, 96*, 81–88. https://doi.org/10.1007/s00253-012-4158-1

Owji, H., Nezafat, N., Negahdaripour, M., et al. (2018). A comprehensive review of signal peptides: Structure, roles, and applications. *European Journal of Cell Biology, 97*, 422–441. https://doi.org/10.1016/j.ejcb.2018.06.003

Pahirulzaman, K. A. K., Williams, K., & Lazarus, C. M. (2012). A toolkit for heterologous expression of metabolic pathways in *Aspergillus oryzae*. *Methods in Enzymology, 517*, 241–260. https://doi.org/10.1016/B978-0-12-404634-4.00012-7

Pavezzi, F. C., Gomes, E., & da Silva, R. (2008). Production and characterization of glucoamylase from fungus *Aspergillus awamori* expressed in yeast *Saccharomyces cerevisiae* using different carbon sources. *Brazilian Journal of Microbiology, 39*, 108–114. https://doi.org/10.1590/S1517-83822008000100024

Pel, H. J., de Winde, J. H., Archer, D. B., et al. (2007). Genome sequencing and analysis of the versatile cell factory A*spergillus niger* CBS 513.88. *Nature Biotechnology, 25*, 221–231. https://doi.org/10.1038/nbt1282

Pellegrini, V. O. A., Serpa, V. I., Godoy, A. S., et al. (2015). Recombinant *Trichoderma harzianum* endoglucanase I (Cel7B) is a highly acidic and promiscuous carbohydrate-active enzyme. *Applied Microbiology and Biotechnology, 99*, 9591–9604. https://doi.org/10.1007/s00253-015-6772-1

Pouresmaeil, M., & Azizi-Dargahlou, S. (2023). Factors involved in heterologous expression of proteins in E. coli host. *Archives of Microbiology, 205*, 212. https://doi.org/10.1007/s00203-023-03541-9

Punt, P. J., van Biezen, N., Conesa, A., et al. (2002). Filamentous fungi as cell factories for heterologous protein production. *Trends in Biotechnology, 20*, 200–206. https://doi.org/10.1016/S0167-7799(02)01933-9

Qian, Y., Zhong, L., Sun, Y., et al. (2019). Enhancement of Cellulase production in *Trichoderma reesei* via disruption of multiple protease genes identified by comparative Secretomics. *Frontiers in Microbiology, 10*. https://doi.org/10.3389/FMICB.2019.02784

Raghu Ram, M., & Yepuru, S. K. (2018). Production of alkaline protease from *Aspergillus oryzae* isolated from seashore of Bay of Bengal. *Journal of Applied and Natural Science, 10*, 1210–1215. https://doi.org/10.31018/jans.v10i4.1905

Rodríguez Couto, S., & Toca Herrera, J. L. (2006). Industrial and biotechnological applications of laccases: A review. *Biotechnology Advances, 24*, 500–513. https://doi.org/10.1016/j.biotechadv.2006.04.003

Sakekar, A. A., Gaikwad, S. R., & Punekar, N. S. (2021). Protein expression and secretion by filamentous fungi. *Journal of Biosciences, 46*, 1–18. https://doi.org/10.1007/S12038-020-00120-8

Salihi, A., Asoodeh, A., & Aliabadian, M. (2017). Production and biochemical characterization of an alkaline protease from *Aspergillus oryzae* CH93. *International Journal of Biological Macromolecules, 94*, 827–835. https://doi.org/10.1016/j.ijbiomac.2016.06.023

Saloheimo, M., & Pakula, T. M. (2012). The cargo and the transport system: Secreted proteins and protein secretion in *Trichoderma reesei* (*Hypocrea jecorina*). *Microbiology (N Y), 158*, 46–57. https://doi.org/10.1099/mic.0.053132-0
Sandhya, C., Sumantha, A., Szakacs, G., & Pandey, A. (2005). Comparative evaluation of neutral protease production by *Aspergillus oryzae* in submerged and solid-state fermentation. *Process Biochemistry, 40*, 2689–2694. https://doi.org/10.1016/j.procbio.2004.12.001
Sharma, R., Katoch, M., Srivastava, P. S., & Qazi, G. N. (2009). Approaches for refining heterologous protein production in filamentous fungi. *World Journal of Microbiology and Biotechnology, 25*, 2083–2094. https://doi.org/10.1007/s11274-009-0128-x
Singh, N., Devi, A., Bala Bishnoi, M., et al. (2019). Overview of the process of enzymatic transformation of biomass. In *Elements of Bioeconomy*. IntechOpen.
Sinitsyn, A. P., Sinitsyna, O. A., Zorov, I. N., & Rozhkova, A. M. (2020). Exploring the capabilities of the *Penicillium verruculosum* expression system for the development of producers of enzymes for the effective degradation of renewable plant biomass: A review. *Applied Biochemistry and Microbiology, 56*, 638–646. https://doi.org/10.1134/S0003683820060162
Söllner, T., Whiteheart, S. W., Brunner, M., et al. (1993). SNAP receptors implicated in vesicle targeting and fusion. *Nature, 362*, 318–324. https://doi.org/10.1038/362318a0
Su, X., Schmitz, G., Zhang, M., et al. (2012). Heterologous gene expression in filamentous fungi. In *Advances in applied microbiology* (pp. 1–61). Academic.
Sun, X., & Su, X. (2019). Harnessing the knowledge of protein secretion for enhanced protein production in filamentous fungi. *World Journal of Microbiology and Biotechnology, 35*, 54. https://doi.org/10.1007/s11274-019-2630-0
Tan, Y. N., Lee, P. P., & Chen, W. N. (2020). Dual extraction of crustacean and fungal chitosan from a single *Mucor circinelloides* fermentation. *Fermentation, 6*, 84. https://doi.org/10.3390/FERMENTATION6020040
Te'o, V. S. J., Cziferszky, A. E., Bergquist, P. L., & Nevalainen, K. M. H. (2000). Codon optimization of xylanase gene xynB from the thermophilic bacterium *Dictyoglomus thermophilum* for expression in the filamentous fungus *Trichoderma reesei*. *FEMS Microbiology Letters, 190*, 13–19. https://doi.org/10.1111/J.1574-6968.2000.TB09255.X
Tien, M., & Kirk, T. K. (1983). Lignin-degrading enzyme from the Hymenomycete *Phanerochaete chrysosporium* burds. *Science (1979), 221*, 661–663. https://doi.org/10.1126/science.221.4611.661
Toghueo, R. M. K., & Boyom, F. F. (2020). Endophytic *Penicillium* species and their agricultural, biotechnological, and pharmaceutical applications. *3 Biotech, 10*(3), 1–35. https://doi.org/10.1007/S13205-020-2081-1
Tomico-Cuenca, I., Mach, R. L., Mach-Aigner, A. R., & Derntl, C. (2021). An overview on current molecular tools for heterologous gene expression in *Trichoderma*. *Fungal Biology and Biotechnology, 8*, 1–17. https://doi.org/10.1186/S40694-021-00119-2
Upshall, A., Kumar, A. A., Bailey, M. C., et al. (1987). Secretion of active human tissue plasminogen activator from the filamentous fungus *Aspergillus nidulans*. *Nature Biotechnology, 5*, 1301–1304. https://doi.org/10.1038/nbt1287-1301
Valkonen, M., Kalkman, E. R., Saloheimo, M., et al. (2007). Spatially segregated SNARE protein interactions in living fungal cells. *Journal of Biological Chemistry, 282*, 22775–22785. https://doi.org/10.1074/jbc.M700916200
Van Gemeren, I. A., Beijersbergen, A., Musters, W., et al. (1996). The effect of pre- and prosequences and multicopy integration on heterologous expression of the *Fusarium solani* pisi cutinase gene in *Aspergillus awamori*. *Applied Microbiology and Biotechnology, 45*, 755–763. https://doi.org/10.1007/S002530050759
Vieira Gomes, A. M., Souza Carmo, T., Silva Carvalho, L., et al. (2018). Comparison of yeasts as hosts for recombinant protein production. *Microorganisms, 6*, 38. https://doi.org/10.3390/microorganisms6020038
Virag, A., & Harris, S. D. (2006). The Spitzenkörper: A molecular perspective. *Mycological Research, 110*, 4–13. https://doi.org/10.1016/j.mycres.2005.09.005

Vitcosque, G. L., Ribeiro, L. F. C., de Lucas, R. C., et al. (2016). The functional properties of a xyloglucanase (GH12) of *Aspergillus terreus* expressed in *Aspergillus nidulans* may increase performance of biomass degradation. *Applied Microbiology and Biotechnology, 100*, 9133–9144. https://doi.org/10.1007/s00253-016-7589-2

Wakai, S., Yoshie, T., Asai-Nakashima, N., et al. (2014). l-lactic acid production from starch by simultaneous saccharification and fermentation in a genetically engineered *Aspergillus oryzae* pure culture. *Bioresource Technology, 173*, 376–383. https://doi.org/10.1016/j.biortech.2014.09.094

Wan, X., Wang, L., Chang, J., et al. (2024). Effective synthesis of high-content fructooligosaccharides in engineered *Aspergillus niger. Microbial Cell Factories, 23*, 76. https://doi.org/10.1186/s12934-024-02353-w

Wang, J. (2013). Molecular cloning and heterologous expression of an acid-stable endoxylanase gene from *Penicillium oxalicum* in *Trichoderma reesei. Journal of Microbiology and Biotechnology, 23*, 251–259. https://doi.org/10.4014/jmb.1208.08030

Wang, G., Zhang, D., & Chen, S. (2014). Effect of earlier unfolded protein response and efficient protein disposal system on cellulase production in Rut C30. *World Journal of Microbiology and Biotechnology, 30*, 2587–2595. https://doi.org/10.1007/s11274-014-1682-4

Wang, Q., Zhong, C., & Xiao, H. (2020). Genetic engineering of filamentous fungi for efficient protein expression and secretion. *Frontiers in Bioengineering and Biotechnology, 8*. https://doi.org/10.3389/fbioe.2020.00293

Ward, O. P. (2012). Production of recombinant proteins by filamentous fungi. *Biotechnology Advances, 30*, 1119–1139. https://doi.org/10.1016/j.biotechadv.2011.09.012

Ward, M., Wilson, L. J., Kodama, K. H., et al. (1990). Improved production of Chymosin in *Aspergillus* by expression as a glucoamyiase-chymosin fusion. *Bio/Technology, 8*, 435–440. https://doi.org/10.1038/NBT0590-435

Wei, H., Wu, M., Fan, A., & Su, H. (2021). Recombinant protein production in the filamentous fungus *Trichoderma. Chinese Journal of Chemical Engineering, 30*, 74–81. https://doi.org/10.1016/J.CJCHE.2020.11.006

Wohlschlager, L., Csarman, F., Zrilić, M., et al. (2021). Comparative characterization of glyoxal oxidase from Phanerochaete chrysosporium expressed at high levels in *Pichia pastoris* and *Trichoderma reesei. Enzyme and Microbial Technology, 145*, 109748. https://doi.org/10.1016/j.enzmictec.2021.109748

Wu, Y., Sun, X., Xue, X., et al. (2017). Overexpressing key component genes of the secretion pathway for enhanced secretion of an *Aspergillus niger* glucose oxidase in *Trichoderma reesei. Enzyme and Microbial Technology, 106*, 83–87. https://doi.org/10.1016/J.ENZMICTEC.2017.07.007

Xia, Y., & Lin, X. (2022). Efficient biodegradation of straw and persistent organic pollutants by a novel strategy using recombinant *Trichoderma reesei. Bioresources and Bioprocessing, 9*, 91. https://doi.org/10.1186/s40643-022-00581-9

Xiang, B., Zhao, S., Chen, J., et al. (2023). Engineering the filamentous fungus *Penicillium oxalicum* for rapid, low-background and efficient protein expression. *Enzyme and Microbial Technology, 162*, 110150. https://doi.org/10.1016/J.ENZMICTEC.2022.110150

Xu, Y., Wang, Y., Liu, T., et al. (2018). The GlaA signal peptide substantially increases the expression and secretion of α-galactosidase in *Aspergillus niger. Biotechnology Letters, 40*, 949–955. https://doi.org/10.1007/s10529-018-2540-5

Yan, Q., Han, L., Liu, Z., et al. (2023). Stepwise genetic modification for efficient expression of heterologous proteins in *Aspergillus nidulans. Applied Microbiology and Biotechnology, 107*, 6923–6935. https://doi.org/10.1007/s00253-023-12755-2

Yang, S. (2020). Over expression and its enzymatic properties of Polygalacturonase in *Aspergillus niger. Science and Technology of Food Industry, 41*, 80–85.

Yoon, J., Kikuma, T., Maruyama, J., & Kitamoto, K. (2013). Enhanced production of bovine chymosin by autophagy deficiency in the filamentous fungus *Aspergillus oryzae. PLoS One, 8*, e62512. https://doi.org/10.1371/journal.pone.0062512

Youshko, M. I., van Langen, L. M., Sheldon, R. A., & Švedas, V. K. (2004). Application of aminoacylase I to the enantioselective resolution of α-amino acid esters and amides. *Tetrahedron: Asymmetry, 15*, 1933–1936. https://doi.org/10.1016/j.tetasy.2004.05.018

Zahra, T., Irfan, M., Nadeem, M., et al. (2020). Cellulase production by *Trichoderma viride* in submerged fermentation using response surface methodology. Punjab University. *Journal of Zoology, 35*. https://doi.org/10.17582/journal.pujz/2020.35.2.223.228

Zhang, G., Zhu, Y., Wei, D., & Wang, W. (2014). Enhanced production of heterologous proteins by the filamentous fungus *Trichoderma reesei* via disruption of the alkaline serine protease SPW combined with a pH control strategy. *Plasmid, 71*, 16–22. https://doi.org/10.1016/J.PLASMID.2014.01.001

Zhu, Y., Pan, J., Qiu, J., & Guan, X. (2008). Isolation and characterization of a chitinase gene from entomopathogenic fungus *Verticillium lecanii*. *Brazilian Journal of Microbiology, 39*, 314–320. https://doi.org/10.1590/S1517-83822008000200022

Zhu, Z., Zhang, M., Liu, D., et al. (2023). Development of the thermophilic fungus *Myceliophthora thermophila* into glucoamylase hyperproduction system via the metabolic engineering using improved AsCas12a variants. *Microbial Cell Factories, 22*, 150. https://doi.org/10.1186/s12934-023-02149-4

Engineered Fungal Cell Factories: Pioneering Sustainable Protein and Peptide Production

Charu Gupta and Shikha Baghel Chauhan

Abstract The growing need for sustainable protein and peptide sources has spurred advancements in biomanufacturing, with engineered fungal cell factories emerging as a viable solution. Fungi offer unique advantages, including metabolic versatility, rapid growth, and scalability, making them ideal platforms for producing biopharmaceuticals, nutraceuticals, and alternative proteins. Recent progress in genetic engineering—such as CRISPR-Cas9, RNA interference, and omics-driven strategies—has enhanced the ability to modify fungi for high-yield production of therapeutic proteins and bioactive peptides.

Engineered fungi like *Aspergillus*, *Trichoderma*, and *Saccharomyces* are increasingly used to produce valuable biomolecules such as enzymes, insulin, and antimicrobial peptides. These organisms naturally secrete large quantities of proteins, making them particularly attractive for industrial-scale applications.

This article reviews technological innovations transforming fungi into efficient production platforms. The integration of systems biology, synthetic biology, and advanced fermentation has enabled optimization of metabolic pathways, reduced byproduct formation, and improved protein expression. Fungal cell factories are also environmentally sustainable, requiring fewer resources and offering a lower carbon footprint than conventional protein production systems, such as animal-based or bacterial processes.

Despite these advantages, challenges persist in scale-up, cost-effectiveness, and regulatory compliance. The article discusses current limitations and presents strategies to address them, including adaptive laboratory evolution and bioprocess optimization. It also highlights successful case studies where engineered fungi have been used in the production of biopharmaceuticals and sustainable proteins, underlining their transformative potential in biomedical and industrial applications.

C. Gupta (✉)
Amity Institute of Herbal Research and Studies, Amity University Uttar Pradesh, Noida, Uttar Pradesh, India

S. B. Chauhan
Amity Institute of Pharmacy, Amity University Uttar Pradesh, Noida, India

A. Gupta (ed.), *Prospects of Fungal Biotechnologies for Livestock Volume 2*, Fungal Biology, https://doi.org/10.1007/978-3-032-06478-3_9

Keywords Bioactive peptides · CRISPR-Cas9 · Engineered fungal cell factories · Fungal biotechnology · Sustainable protein production

1 Introduction

In the quest for sustainable food production, the search for alternative protein sources has intensified, driven by the growing global population and the environmental impact of traditional livestock farming. One of the most promising avenues in this field involves the use of engineered fungal cell factories, which are poised to revolutionize the production of proteins and peptides. Fungi, diverse organisms that include yeasts, molds, and mushrooms, possess unique metabolic pathways and growth characteristics that can be harnessed for biotechnological applications. By genetically modifying these organisms, researchers aim to create cell factories capable of producing high-quality proteins and bioactive peptides sustainably.

Recent studies have highlighted the potential of fungi as bioreactors to produce various proteins, including those derived from plants and animals. For example, a study by Wang et al. (2020) demonstrated that genetically engineered filamentous fungi could efficiently produce plant-based proteins with enhanced nutritional profiles, addressing both food security and sustainability challenges (Garg et al., 2024). This approach not only alleviates the pressure on traditional agricultural systems but also reduces greenhouse gas emissions associated with animal farming.

The appeal of using fungal cell factories extends beyond mere protein production; they can also synthesize bioactive peptides with health benefits. Bioactive peptides, short chains of amino acids, have garnered attention for their potential therapeutic effects, including anti-inflammatory, antioxidant, and antihypertensive properties. A recent investigation by Liu et al. (2023) explored the use of engineered yeast strains to produce bioactive peptides from food waste, showcasing how fungal cell factories can contribute to a circular economy by converting waste into valuable products (Wang et al., 2020). This dual benefit of protein production and waste valorization underscores the versatility of engineered fungi in addressing current global challenges.

Moreover, the scalability and rapid growth rates of certain fungi make them ideal candidates for industrial applications. Unlike traditional crops, which require extensive land and resources, fungi can be cultivated in controlled environments, utilizing minimal space and water. According to a review by Lübeck and Lübeck (2022), the efficiency of fungal fermentation processes can lead to higher yields of protein and peptides in shorter timeframes compared to conventional agricultural practices (Lübeck & Lübeck, 2022). This efficiency is critical as the world seeks to meet the protein demands of an ever-increasing population while minimizing environmental degradation.

However, the development of engineered fungal cell factories is not without challenges. The optimization of metabolic pathways, the identification of suitable host organisms, and the regulatory hurdles associated with the use of genetically modified organisms (GMOs) are significant barriers that researchers must navigate. Nevertheless, advancements in synthetic biology and genomics are paving the way for more efficient and precise engineering of fungal strains. Techniques such as CRISPR-Cas9 gene editing have enabled researchers to make targeted modifications that enhance the productivity and functionality of these organisms (Hassan et al., 2021). Such innovations are crucial for the sustainable production of proteins and peptides tailored to specific dietary and nutritional needs.

Furthermore, the market for alternative proteins is rapidly expanding, driven by consumer demand for more sustainable and ethical food sources. According to a report by the Good Food Institute, 2022), the alternative protein market is projected to reach $290 billion by 2035, reflecting a significant shift in consumer preferences and dietary habits. This growing interest in plant-based and fungal-derived proteins presents an opportunity for the development of engineered fungal cell factories that can meet market demands while aligning with sustainability goals.

2 Fungal Cell Factories: A Versatile Biomanufacturing Platform

Fungal cell factories are emerging as a groundbreaking biomanufacturing platform, harnessing the unique metabolic capabilities of fungi to produce a wide array of valuable products. Unlike traditional production methods that rely heavily on bacterial systems, fungal organisms offer distinct advantages, including their ability to secrete complex proteins, synthesize secondary metabolites, and perform post-translational modifications. This versatility allows for the production of pharmaceuticals, enzymes, biofuels, and biopolymers, making fungal cell factories an attractive option for sustainable manufacturing processes. The adaptability of fungal species, such as yeast and filamentous fungi, enables them to thrive in diverse environments and utilize various substrates, including waste products, for growth and production. As a result, these organisms not only contribute to the circular economy by reducing waste but also support the development of eco-friendly bioprocessing methods. The advances in genetic engineering and synthetic biology further enhance the potential of fungal cell factories, allowing researchers to optimize pathways and improve yields.

Key fungal species used in protein and peptide production:

2.1 **Aspergillus** *spp.—Enzymes, Biopharmaceuticals*

Aspergillus species, particularly *Aspergillus niger* and *Aspergillus oryzae*, play a crucial role in industrial biotechnology due to their ability to produce a variety of enzymes and biopharmaceuticals. These fungi are widely used for the production of enzymes such as amylases, proteases, and cellulases, which are essential in food processing, biofuel production, and pharmaceuticals. Recent studies have highlighted their potential in biopharmaceutical applications, including the production of therapeutic proteins and vaccines, thus expanding their utility in the healthcare sector (Salazar-Cerezo et al., 2023; Henrique & Nirlane, 2016).

2.2 Trichoderma *spp.—Industrial Enzymes, Bioactive Compounds*

Trichoderma species are renowned for their ability to produce industrially significant enzymes and bioactive compounds. They are particularly effective in producing cellulases and hemicellulases, which are instrumental in biomass conversion and biofuel production. Additionally, *Trichoderma* spp. is studied for its potential in agriculture as a biocontrol agent and for its ability to produce secondary metabolites with antimicrobial properties. Recent research has focused on the genetic manipulation of Trichoderma to enhance enzyme production and optimize bioactive compound synthesis (Tyśkiewicz et al., 2022; Druzhinina & Kubicek, 2017).

2.3 Saccharomyces cerevisiae*—Recombinant Proteins, Therapeutic Peptides*

Saccharomyces cerevisiae, commonly known as baker's yeast, is one of the most extensively studied organisms in biotechnology. It is widely used for the production of recombinant proteins and therapeutic peptides due to its well-characterized genetics, ease of manipulation, and ability to perform post-translational modifications. *S. cerevisiae* has been utilized to produce a variety of biopharmaceuticals, including insulin and vaccines. Recent advancements in synthetic biology have further improved its efficiency in producing complex proteins, making it a vital tool in modern biomanufacturing (Zhao et al., 2024; Milne et al., 2023).

2.4 Pichia pastoris*—High-Yield Heterologous Protein Production*

Pichia pastoris has gained prominence as a preferred host for high-yield heterologous protein production. This methylotrophic yeast can utilize methanol as a carbon source and is capable of producing complex proteins with high levels of glycosylation. Its robust growth characteristics and strong promoter systems enable the efficient expression of recombinant proteins, making *P. pastoris* an attractive platform for pharmaceutical and industrial enzyme production. Recent studies have focused on optimizing fermentation conditions and metabolic pathways to enhance protein yields and reduce production costs (Karbalaei et al., 2020; Feng et al., 2023).

3 Natural Ability of Fungi to Secrete Proteins Efficiently

Fungi possess a remarkable natural ability to secrete proteins efficiently, which is a critical advantage in biomanufacturing. This secretion system allows fungi to release enzymes and other proteins directly into the culture medium, simplifying downstream processing and purification. The ability to secrete large quantities of proteins has made fungi, especially filamentous species like *Aspergillus* and *Trichoderma*, ideal candidates for industrial applications. Recent advances in genomics and proteomics have provided insights into the secretion mechanisms of fungi, paving the way for engineered strains with enhanced secretion capabilities (Reilly et al., 2016; Lübeck & Lübeck, 2022).

4 Advances in Genetic Engineering of Fungal Cell Factories

Advances in genetic engineering of fungal cell factories are revolutionizing biomanufacturing by enhancing the productivity and efficiency of fungi in producing valuable bioproducts. Techniques such as CRISPR-Cas9, synthetic biology, and metabolic pathway optimization have enabled researchers to precisely modify fungal genomes, improving traits such as substrate utilization, protein secretion, and metabolite production. These advancements allow for the development of fungal strains with tailored characteristics, capable of producing complex proteins, enzymes, and bioactive compounds at higher yields. Recent studies have demonstrated successful applications in both industrial enzyme production and biopharmaceutical development, highlighting the potential of engineered fungal cell factories to meet growing global demands sustainably. The diagrammatic view of advances in genetic engineering of fungal cell factories is depicted in Fig. 1.

4.1 CRISPR-Cas9 and Genome Editing

CRISPR-Cas9 technology has emerged as a powerful tool for precise genetic modifications in fungal cell factories, significantly optimizing protein expression.

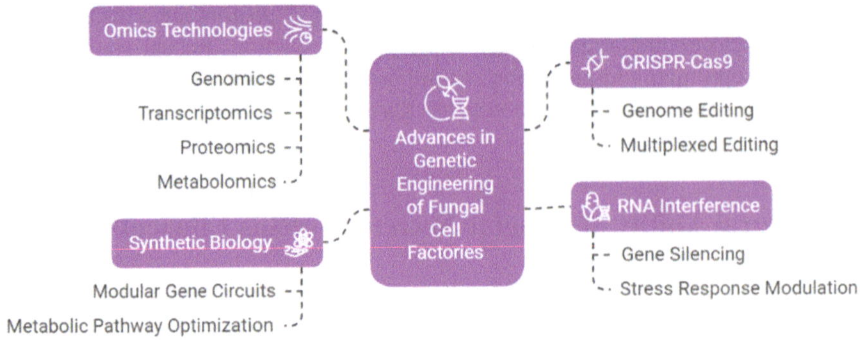

Fig. 1 Advances in genetic engineering of fungal cell factories

4.1.1 Precise Genetic Modifications for Optimized Protein Expression

By enabling targeted edits in the genome, CRISPR-Cas9 facilitates the knockout or enhancement of specific genes involved in metabolic pathways, ultimately leading to improved yields of desired proteins. For instance, researchers have utilized CRISPR-Cas9 to knock out genes that compete for precursors or resources, redirecting metabolic flux toward the desired product (Zhu et al., 2024). Additionally, by inserting or modifying promoter regions, scientists can regulate gene expression levels, allowing for fine-tuning of protein production rates (Leal et al., 2024). Recent studies have demonstrated the successful application of CRISPR-Cas9 in filamentous fungi, such as *Aspergillus niger* and *Trichoderma reesei*, where targeted modifications resulted in enhanced secretion of industrial enzymes and recombinant proteins (Adiego-Pérez et al., 2019). Moreover, this technology allows for multiplexed editing, meaning multiple genes can be modified simultaneously, further accelerating strain development for biotechnological applications (Tavakoli et al., 2021). The ability to create precise genetic alterations not only enhances the productivity of fungal cell factories but also reduces the time and resources required for strain optimization. As research continues to evolve, CRISPR-Cas9 holds immense promise in revolutionizing protein expression in fungal systems, paving the way for sustainable biomanufacturing of high-value products.

4.1.2 Overcoming Traditional Strain Improvement Limitations

Traditional strain improvement methods, such as random mutagenesis and classical breeding, often face significant limitations, including lengthy development cycles, low efficiency, and the potential for unintended mutations. However, the advent of CRISPR-Cas9 and genome editing techniques has transformed this landscape, allowing researchers to overcome these hurdles with unprecedented precision and speed. By utilizing CRISPR-Cas9, scientists can target specific genes for knockout, activation, or modification, enabling the optimization of metabolic pathways crucial for bioproduction. For example, recent studies have demonstrated that targeting genes involved in competing metabolic pathways can redirect flux toward desired products, thereby enhancing yield and productivity significantly (Leal et al., 2024). Furthermore, CRISPR technology allows for the simultaneous editing of multiple genes through multiplexing, which greatly accelerates the strain development process compared to traditional methods (Liu et al., 2015). This capability not only shortens the timeline for developing optimized strains but also increases the likelihood of achieving desirable phenotypic traits.

Recent applications of CRISPR-Cas9 in filamentous fungi, such as *Aspergillus* and *Trichoderma* species, have shown promising results, with enhanced production of enzymes and bioactive compounds reported (Song et al., 2019).

Additionally, the precision of genome editing minimizes off-target effects, a common drawback in traditional methods, allowing for the creation of stable, high-performing strains tailored for specific biotechnological applications (Lee & Lee, 2021). As the field continues to evolve, the integration of CRISPR-Cas9 with other synthetic biology tools holds immense potential for further enhancing fungal strains, leading to more efficient and sustainable biotechnological processes. By overcoming the limitations of traditional strain improvement methods, CRISPR technology paves the way for innovative solutions to meet the growing demands for bioproducts in various industries.

4.2 RNA Interference (RNAi) and Gene Silencing

RNA interference (RNAi) and gene silencing have emerged as powerful tools in the genetic engineering of fungal cell factories, offering innovative approaches to regulate gene expression and enhance bioproduction. RNAi employs small interfering RNAs (siRNAs) or short hairpin RNAs (shRNAs) to specifically target and degrade mRNA transcripts of undesirable genes, effectively silencing their expression. This technique allows for the precise downregulation of genes that may be competing for resources or negatively impacting the yield of target proteins. Recent studies have successfully applied RNAi in filamentous fungi, such as Aspergillus and Trichoderma species, to silence genes associated with metabolic pathways that hinder enzyme production, resulting in significantly increased yields of industrial enzymes (Dang et al., 2011).

Furthermore, RNAi can be employed to knock down genes responsible for stress responses, thereby improving the robustness of fungal strains in industrial settings. By fine-tuning gene expression, researchers can optimize metabolic pathways and improve the secretion efficiency of desired proteins (Garg et al., 2024).

Additionally, combining RNAi with CRISPR technology can provide a more comprehensive approach to strain optimization, allowing for both the knockout of target genes and the silencing of non-essential genes simultaneously (Rajput et al., 2021). As the field of fungal biotechnology continues to advance, RNAi and gene silencing techniques are poised to play a critical role in the development of high-performing fungal cell factories, enhancing the efficiency and sustainability of bioproduction processes.

4.3 Fine-Tuning Metabolic Pathways

Fine-tuning metabolic pathways using RNA interference (RNAi) and gene silencing techniques is revolutionizing the field of metabolic engineering in various organisms, including fungal cell factories. By selectively silencing genes involved in specific metabolic pathways, researchers can redirect the flow of metabolites toward the synthesis of desired products. For example, RNAi has been employed to downregulate competing pathways that detract from the production of high-value metabolites, such as pharmaceuticals and industrial enzymes. Recent studies have demonstrated the successful application of RNAi in filamentous fungi, such as *Aspergillus niger*, to silence genes associated with negative regulatory pathways, thereby enhancing the production of citric acid and other organic acids (Upton et al., 2020).

Moreover, gene silencing techniques allow for the modulation of transcription factors that control entire metabolic networks, providing a broader strategy for optimizing complex pathways (Yin et al., 2024). This approach not only improves product yields but also enhances the overall efficiency of the metabolic processes involved. Additionally, combining RNAi with other genetic engineering tools, such as CRISPR-Cas9, allows for a comprehensive strategy that can target multiple genes simultaneously, further enhancing pathway optimization (Liu et al., 2017). As these techniques continue to evolve, they offer promising avenues for developing fungal strains with tailored metabolic capabilities, ultimately leading to more sustainable production methods for a wide array of bioproducts.

4.4 Omics Technologies (Genomics, Transcriptomics, Proteomics, Metabolomics)

Omics technologies, encompassing genomics, transcriptomics, proteomics, and metabolomics, are increasingly pivotal in advancing our understanding of biological systems and enhancing biotechnological applications. Genomics focuses on the comprehensive analysis of an organism's entire genome, enabling insights into genetic variations and their implications for traits and functions. Transcriptomics examines the complete set of RNA transcripts, shedding light on gene expression patterns under various conditions. Proteomics delves into the structure and function of proteins, providing insights into their roles in cellular processes and metabolic pathways. Metabolomics offers a snapshot of metabolites present in biological samples, revealing the metabolic state of an organism and its response to environmental changes. Together, these integrated omics approaches facilitate a systems biology perspective, enabling researchers to unveil complex interactions within biological systems, optimize microbial strains, and develop more efficient bioprocesses for the sustainable production of valuable bioproducts (Wijayawardene et al., 2023; Dai & Shen, 2022).

4.4.1 Enhancing Fungal Metabolic Engineering Strategies

Enhancing fungal metabolic engineering strategies through omics technologies has significantly advanced the field of biotechnology, enabling more efficient production of valuable metabolites. By integrating genomics, transcriptomics, proteomics, and metabolomics, researchers can gain comprehensive insights into the metabolic pathways and regulatory networks within fungi. Genomic analyses facilitate the identification of key genes involved in metabolic processes, while transcriptomic studies reveal gene expression patterns under various growth conditions, allowing for targeted modifications to optimize pathways (Tamano & Yoshimi, 2021). Proteomics provides information on protein abundance and post-translational modifications, offering clues to enzyme activity and functionality, which can be crucial for metabolic flux analysis (Jo et al., 2023). Moreover, metabolomics allows for the direct measurement of metabolites, helping to identify bottlenecks in production and evaluate the impact of genetic modifications on metabolic outputs (Woodward et al., 2021). The integration of these omics approaches enables a systems biology perspective, allowing for iterative cycles of design, build, test, and learning, ultimately improving the efficiency and sustainability of fungal cell factories for biotechnological applications (Culibrk et al., 2016).

4.5 Synthetic Biology Approaches

Synthetic biology approaches are transforming the genetic engineering of fungal cell factories by enabling the design, construction, and manipulation of biological systems with precision and efficiency. These methodologies allow researchers to engineer complex genetic circuits that can regulate gene expression in response to specific stimuli, optimizing metabolic pathways for enhanced production of valuable compounds (Yan et al., 2023). By employing modular components, such as synthetic promoters, ribosome binding sites, and transcription factors, synthetic biology facilitates the fine-tuning of metabolic networks in fungi, leading to improved yields and product specificity (Hyde et al., 2024). Furthermore, recent advancements in genome editing technologies, such as CRISPR-based tools, combined with synthetic biology, enhance the ability to create tailored strains that exhibit desired phenotypic traits, such as increased tolerance to stress or improved substrate utilization (Jeong et al., 2023). As these approaches continue to evolve, they hold significant potential for developing next-generation fungal cell factories that can sustainably produce high-value bioproducts.

4.5.1 Modular Gene Circuits for Controlled Protein Production

Modular gene circuits represent a significant advancement in synthetic biology, allowing for the controlled production of proteins in various organisms, including fungi. These circuits consist of interchangeable genetic parts, such as promoters, ribosome binding sites, and terminators, that can be assembled to create specific regulatory functions tailored to desired outcomes. By designing modular circuits that respond to environmental stimuli or specific metabolites, researchers can finely tune protein expression levels, timing, and duration (Baldin et al., 2022). This precision enables the optimization of metabolic pathways to enhance yield and improve the quality of bioproducts.

Recent studies have demonstrated successful applications of modular gene circuits in fungal species, resulting in the enhanced production of enzymes and therapeutic proteins (Pullen et al., 2025). Furthermore, these circuits can be multiplexed, allowing for the simultaneous regulation of multiple genes, thus providing a robust platform for complex biotechnological applications (Yang et al., 2024).

5 Bioprocess Optimization for Enhanced Protein and Peptide Yields

Bioprocess optimization is a crucial aspect of biotechnology aimed at enhancing protein and peptide yields in microbial systems, particularly in fungal cell factories. This optimization involves fine-tuning various parameters, including nutrient composition, pH, temperature, and cultivation time, to create an ideal environment for the production of target molecules. Recent advancements in automated high-throughput screening and process monitoring technologies have enabled researchers to systematically evaluate the effects of these parameters on protein production (Sarnaik et al., 2020). Additionally, the implementation of advanced fermentation strategies, such as fed-batch and continuous cultivation, has shown promising results in improving yield and productivity by ensuring a steady supply of nutrients and maintaining optimal growth conditions (do Nascimento Muniz et al., 2024). Metabolic engineering approaches, combined with bioprocess optimization, allow for the modification of metabolic pathways to enhance the flux toward desired products, further increasing yields (Jiang et al., 2022). The integration of omics technologies, such as metabolomics and proteomics, provides insights into metabolic bottlenecks, enabling targeted interventions and iterative optimization of the bioprocess (Kaur et al., 2023). These strategies collectively contribute to the sustainable and efficient production of high-value proteins and peptides in fungal biomanufacturing. The diagrammatic representations of optimizing processes for sustainable protein production are depicted in Fig. 2.

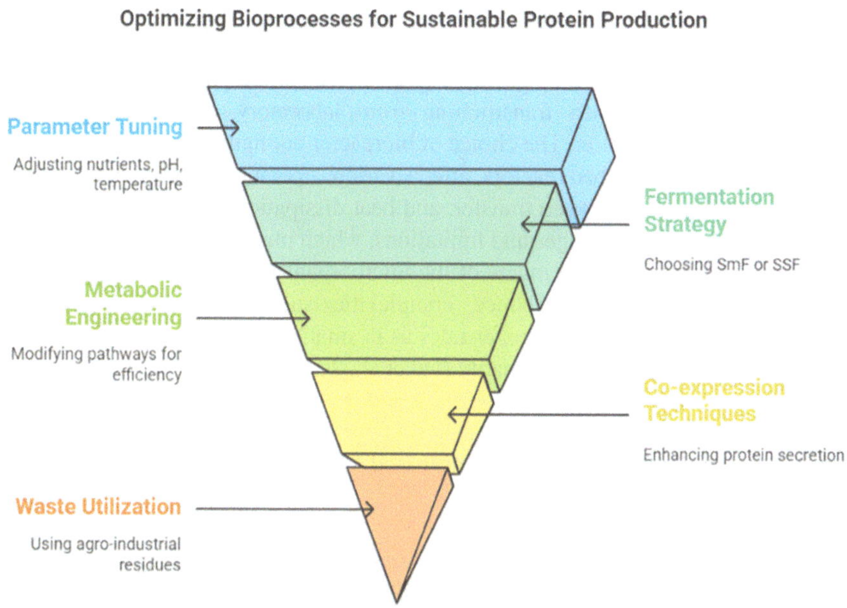

Fig. 2 Optimizing process for sustainable protein production

5.1 Fermentation Strategies

5.1.1 Submerged vs. Solid-State Fermentation

Submerged fermentation (SmF) and solid-state fermentation (SSF) are two predominant fermentation strategies employed in biotechnological processes. In submerged fermentation, microorganisms are cultivated in liquid media, allowing for better control of environmental parameters such as temperature, pH, and oxygen levels. This method is particularly effective for producing high yields of enzymes and proteins, as it facilitates nutrient accessibility and efficient mixing (Mesbah, 2022). However, SmF often requires extensive downstream processing to separate and purify the desired products from the liquid medium.

On the other hand, solid-state fermentation involves cultivating microorganisms on solid substrates, such as agricultural wastes or lignocellulosic materials. SSF offers advantages such as lower water usage, reduced risk of contamination, and the potential for enhanced production of secondary metabolites, including bioactive compounds (Yafetto, 2022). The choice between SmF and SSF largely depends on the specific bioproduct being targeted, the characteristics of the microorganism being used, and the economic considerations related to substrate availability and processing costs. Both methods have unique advantages and challenges that researchers must consider when developing efficient fermentation processes.

5.1.2 Bioreactor Designs and Scale-Up Challenges

Bioreactor design and scale-up present significant challenges in the field of biotechnology, particularly when transitioning from laboratory-scale experiments to industrial-scale production. The choice of bioreactor configuration, such as stirred-tank, airlifts, or packed-bed reactors, affects various operational parameters, including mixing efficiency, oxygen transfer, and heat dissipation (Mattedi et al., 2023). Each design has its advantages and limitations, which must be carefully considered based on the specific requirements of the microorganism being cultivated and the desired product. Scale-up introduces complexities such as maintaining the same growth conditions and mass transfer rates as in smaller reactors, which can lead to variations in product yield and quality (Du et al., 2022).

5.1.3 Metabolic Engineering for Increased Yield and Reduced Byproduct Formation

Metabolic engineering has emerged as a pivotal strategy for enhancing the yield and reducing byproduct formation in engineered fungal cell factories, particularly in the context of sustainable protein and peptide production. By strategically modifying metabolic pathways, researchers can optimize the allocation of cellular resources toward the desired product while minimizing the diversion of carbon and energy into unwanted byproducts (Du et al., 2022). For instance, recent advancements have demonstrated that the overexpression of key enzymes involved in amino acid biosynthesis can significantly boost protein yields in fungi like *Aspergillus niger* and *Pichia pastoris*. Additionally, the deletion or downregulation of competing pathways, such as those responsible for ethanol or organic acid production, further refines metabolic flux toward target compounds (Jian Zha et al., 2023). These strategies are complemented by systems biology tools, including transcriptomics and metabolomics, which provide insights into global metabolic shifts and guide iterative optimization (Knychala et al., 2024). Such approaches align with the goals of sustainable bioproduction, as they enhance efficiency and reduce waste. Furthermore, integrating machine learning models to predict pathway performance has revolutionized strain design, enabling faster development cycles (Guillaume Gricourt et al., 2024). Collectively, these innovations underscore the transformative potential of metabolic engineering in creating robust fungal platforms for eco-friendly protein and peptide synthesis.

5.1.4 Co-expression Strategies to Enhance Secretion Efficiency

Co-expression strategies have become a cornerstone in enhancing the secretion efficiency of recombinant proteins and peptides, particularly in fungal cell factories. By co-expressing chaperones, foldases, or other auxiliary proteins alongside the target product, cellular machinery can be optimized to improve protein folding, trafficking, and secretion. For instance, the co-expression of endoplasmic reticulum (ER)-resident chaperones such as BiP (binding immunoglobulin protein) or PDI (protein disulfide isomerase) has been shown to alleviate ER stress and enhance secretion yields (Wang et al., 2022).

A successful case study involves the production of human lysozyme in *Pichia pastoris*, where co-expression of the chaperone HAC1 significantly improved secretion efficiency by activating the unfolded protein response (UPR) pathway, leading to a threefold increase in yield (Liu et al., 2020).

Another notable example is the co-expression of the signal sequence receptor (SSR) complex in *Aspergillus niger* during the production of glucoamylase. This strategy enhanced the translocation of the enzyme into the secretory pathway, resulting in a 50% higher secretion rate compared to conventional systems (Wallis et al., 1999).

These advancements are complemented by recent studies leveraging synthetic biology tools to design multi-gene constructs for the simultaneous expression of multiple auxiliary proteins, further boosting secretion efficiency (Li et al., 2024). Such approaches not only address bottlenecks in protein secretion but also align with sustainable bioproduction goals by reducing metabolic burden and improving productivity. The integration of these strategies into fungal cell factories underscores their potential to revolutionize industrial-scale protein production.

5.1.5 Media Optimization: Use of Agro-Industrial Waste as Substrates for Sustainability

Media optimization using agro-industrial waste as substrates has gained significant attention for fostering sustainability in bioproduction. By repurposing low-cost, renewable resources like sugarcane bagasse, wheat bran, and corn-steep liquor, industries can reduce dependency on expensive synthetic media while addressing waste management challenges. A successful case study involves the use of sugarcane bagasse hydrolysate for the production of cellulases by *Trichoderma reesei*. Researchers demonstrated that this waste substrate not only supported fungal growth but also enhanced enzyme yields by 40% compared to conventional media, highlighting its economic and environmental benefits (Rocha et al., 2013).

Another example is the utilization of wheat bran as a substrate for the production of bioactive peptides by *Aspergillus oryzae*. This approach achieved a 30% reduction in production costs while maintaining high peptide titers, showcasing the potential of agro-industrial residues in sustainable bioprocessing (Hyde et al., 2024). These strategies align with circular economy principles, transforming waste into value-added products.

Recent studies emphasize the integration of pretreatment technologies, such as enzymatic hydrolysis and fermentation optimization, to maximize substrate utilization efficiency (Milad Hadidi et al., 2024). Such innovations underscore the dual role of agro-industrial waste in reducing environmental footprints and enhancing the economic viability of bioprocesses.

6 Sustainable and Economic Aspects of Fungal-Based Protein Production

Fungal-based protein production has emerged as a sustainable and economically viable alternative to traditional protein sources, addressing global challenges such as food security, resource scarcity, and environmental degradation. Fungi, with their rapid growth rates, ability to utilize diverse substrates, and low ecological footprint, offer a promising platform for producing proteins like enzymes, bioactive peptides, and even food-grade proteins. A successful case study is the production of mycoprotein using *Fusarium venenatum*, commercialized under the brand Quorn™. This fungal biomass, grown on glucose derived from starch hydrolysates, has a carbon footprint 90% lower than beef and requires significantly less land and water, making it a sustainable protein source (Souza Filho et al., 2019). Another notable example is the use of *Aspergillus oryzae* for producing koji-based fermented foods, which has been scaled up in Asia. By utilizing agro-industrial residues like rice bran and soybean meal, this process reduces costs while promoting circular economy principles (Milad Hadidi et al., 2024). Economically, fungal systems benefit from low-cost substrates, high yields, and minimal processing requirements, enabling competitive pricing compared to animal-derived proteins. Furthermore, advancements in genetic engineering have enhanced strain performance, reduced production costs, and improved scalability (Tiwari & Park, 2024). These innovations highlight the dual advantages of sustainability and profitability in fungal-based systems. By integrating renewable resources and waste streams into bioprocessing, fungal platforms exemplify a transformative approach to meeting global protein demands while minimizing environmental impacts.

6.1 Low-Resource Requirements of Fungi Compared to Traditional Protein Production

Fungi exhibit significantly lower resource requirements compared to traditional protein production systems, such as livestock farming, making them a sustainable alternative for meeting global protein demands. Fungal cultivation requires minimal land, water, and energy while producing high yields of biomass or specific proteins. For instance, *Fusarium venenatum*, the fungus used to produce mycoprotein (Quorn™), requires only 0.5 m^2 of land per kilogram of protein, compared to 125 m^2 for beef production, and its water usage is approximately 20 times lower (Cellura et al., 2022). This efficiency stems from fungi's ability to grow on diverse substrates, including agricultural waste, and their rapid growth rates under controlled conditions.

A successful case study is the use of *Aspergillus oryzae* to produce koji-based fermented foods in Asia. By utilizing low-cost substrates like rice bran and soybean meal, this process achieves high protein yields while minimizing resource inputs (Cerimi et al., 2019).

Another example is the production of single-cell protein from *Candida utilis* using food industry byproducts like molasses. This approach not only reduces waste but also generates nutritious protein-rich biomass with minimal environmental impact (Tiwari & Park, 2024). Fungi's ability to thrive on renewable resources and industrial residues further enhances their sustainability. Additionally, advancements in bioreactor design and fermentation optimization have reduced energy consumption during fungal cultivation (Leal et al., 2024). These innovations highlight fungi's potential to revolutionize protein production by addressing resource scarcity and environmental concerns.

6.2 Reduction in Carbon Footprint and Waste Generation

The reduction in carbon footprint and waste generation is a critical advantage of fungal-based protein production systems compared to traditional agriculture and livestock farming. Fungi inherently produce fewer greenhouse gas emissions due to their efficient metabolic processes and ability to grow on low-impact substrates. For example, mycoprotein production using *Fusarium venenatum* has been shown to generate 90% less CO_2 per kilogram of protein compared to beef, with significantly lower methane emissions (Cellura et al., 2022). This reduction is attributed to the fungi's ability to convert renewable resources like glucose into biomass with minimal energy loss. A successful case study is the integration of *Aspergillus niger* in biorefineries for enzyme production, where agro-industrial residues such as sugarcane bagasse are utilized as substrates. This approach not only reduces waste but also lowers the overall carbon footprint by repurposing waste streams into valuable products (Rocha et al., 2013). Another notable example is the use of *Trichoderma reesei* for cellulase production, where lignocellulosic waste is converted into biofuels and biochemicals, achieving a circular economy model that minimizes waste generation (Dai & Shen, 2022). Additionally, advancements in precision fermentation technologies have further reduced energy consumption and emissions during fungal cultivation (Liu et al., 2023). These innovations demonstrate how fungi can transform industrial waste into sustainable outputs while mitigating environmental impacts. By adopting fungal platforms, industries can significantly contribute to global sustainability goals by reducing reliance on resource-intensive systems.

6.3 Cost-Effectiveness and Economic Feasibility of Large-Scale Production

Engineered fungal cell factories offer a transformative approach to sustainable protein and peptide production, combining high yields with low-input bioprocessing. Fungi such as *Aspergillus niger* and *Trichoderma reesei* thrive on inexpensive agro-industrial byproducts (e.g., lignocellulosic waste), reducing substrate costs by up to 60% compared to traditional systems. Their scalability in submerged fermentation and compatibility with metabolic engineering tools further enhance cost efficiency. For instance, CRISPR-Cas9-mediated strain optimization in *Aspergillus* has boosted protein titers by 3–5 times, slashing per-unit production costs. A 2022 techno-economic analysis revealed that fungal platforms can achieve production costs below $2/kg for recombinant proteins, rivaling bacterial systems while offering superior post-translational modifications (Liu et al., 2023).

A landmark case is Novozymes' use of *Trichoderma reesei* for industrial enzyme production, where strain engineering and fermentation optimization reduced manufacturing costs by 40% between 2019 and 2023, enabling market expansion into biofuels and food processing. Similarly, Mycorena's mycoprotein platform employs *Fusarium venenatum* to produce meat alternatives at $3.50/kg, leveraging fungal biomass's rapid growth on starch waste. This model reduced capital expenditure by 30% compared to plant-based protein facilities, as highlighted in a 2023 lifecycle assessment (Cerimi et al., 2019).

Recent studies underscore the role of metabolic engineering in enhancing feasibility. For example, Cerimi et al., 2019 demonstrated that glycoengineered *Komagataella phaffii* (formerly *Pichia pastoris*) reduced downstream processing costs by 25% through improved secretion efficiency. With advancements in bioreactor design and AI-driven bioprocess optimization, fungal systems are poised to dominate sustainable protein markets, aligning economic viability with circular bioeconomy goals.

6.4 Potential for Integrating Fungal Systems into Circular Bioeconomy Models

Engineered fungal cell factories are uniquely positioned to advance circular bioeconomy frameworks by valorizing waste streams into high-value proteins and peptides. Fungi such as *Aspergillus oryzae* and *Fusarium* species thrive on low-cost, non-food substrates like agricultural residues (e.g., wheat bran, sugarcane bagasse) and food-processing byproducts, aligning with "zero-waste" objectives. Their ability to hydrolyze complex polymers (e.g., cellulose, chitin) into nutrients enables upcycling of organic waste into biomass rich in proteins, enzymes, or bioactive peptides, closing resource loops. For example, Nature's Fynd employs a patented *Fusarium* strain to convert plant starch waste from tofu production into nutritional proteins, diverting 10,000+ tons of annual waste from landfills while generating meat alternatives. This process, validated in a 2022 Science of the Total Environment study, reduced water and land use by 60% compared to soy protein.

Another success is the EU-funded FUNGATERIAN project (2021–2024), which engineered *Aspergillus niger* to produce food-grade enzymes from brewery-spent grain. The initiative achieved 85% substrate-to-biomass efficiency, with residual fungal biomass repurposed as animal feed, creating a closed-loop system. A 2023 Nature Sustainability analysis highlighted that such fungal platforms could reduce agro-industrial carbon footprints by 30%–50% by 2040.

Recent advances in synthetic biology further enhance circularity. For instance, Wang et al. (2024) designed *Rhizopus oligosporus* strains to co-produce proteases and single-cell protein from dairy effluent, cutting downstream purification costs by 35%. By coupling waste remediation with high-value output, fungal systems exemplify the circular bioeconomy's dual economic and environmental promise.

7 Challenges and Bottlenecks in Commercialization

The commercialization of fungal-derived biopharmaceuticals faces several challenges, including low production yields, complex downstream processing, and high purification costs. Fungal systems often produce heterogeneous glycosylation patterns, which can affect drug efficacy and safety, necessitating extensive glycoengineering (Rajput et al., 2021). Additionally, regulatory hurdles and the need for scalable fermentation technologies hinder rapid market entry. Contamination risks during cultivation and genetic instability of engineered strains further complicate production. Overcoming these bottlenecks requires advancements in strain optimization, process standardization, and cost-effective purification methods to ensure consistent quality and affordability of fungal-derived therapeutics.

7.1 Regulatory Hurdles for Fungal-Derived Biopharmaceuticals and Alternative Proteins

The development and commercialization of fungal-derived biopharmaceuticals and alternative proteins face significant regulatory challenges. A primary concern is ensuring product safety, particularly regarding potential toxic metabolites and contaminants. Fungi can produce mycotoxins, which pose health risks if present in consumable products. Therefore, rigorous purification processes and comprehensive safety assessments are essential to confirm the absence of such harmful substances in the final products (Arnau et al., 2019).

Another regulatory hurdle involves the potential allergenicity of novel fungal proteins. Regulatory bodies require thorough evaluations to determine if these proteins might trigger allergic reactions in consumers. This necessitates extensive in silico and in vitro testing to assess allergenic potential before market approval (Hadi & Brightwell, 2021).

The use of genetically modified (GM) fungal strains for enhanced production introduces additional complexities. Regulatory frameworks often mandate detailed assessments of GM organisms to evaluate any environmental risks, gene transfer possibilities, and long-term health implications. Navigating these regulations can be time-consuming and requires substantial documentation to demonstrate safety and compliance (Food Policy, 2022).

Moreover, the novelty of fungal-derived products means they may fall under "novel food" regulations in various jurisdictions. This classification demands comprehensive dossiers detailing the production process, nutritional profile, toxicological data, and historical consumption evidence, if available. The absence of a history of human consumption for certain fungal species can complicate and prolong the approval process (Alternative Proteins for Human Consumption, 2022).

7.2 Scaling Up Production While Maintaining Product Consistency

Scaling up the production of fungal-derived biopharmaceuticals and alternative proteins while maintaining product consistency presents several challenges. One primary concern is ensuring that the quality and characteristics of the product remain uniform across different production scales. Variations in parameters such as nutrient availability, oxygen transfer, and mixing efficiency can lead to inconsistencies in protein expression and post-translational modifications, directly impacting product efficacy and safety. To address these issues, implementing Process Analytical Technology (PAT) is essential. PAT enables real-time monitoring and control of critical process parameters, facilitating immediate adjustments to maintain the desired product quality. By utilizing PAT, manufacturers can detect deviations promptly and apply corrective measures, ensuring consistency throughout the production process (Challener, 2023).

Another significant aspect is the selection of appropriate bioreactor systems and fermentation strategies. Transitioning from small-scale to large-scale production requires careful consideration of bioreactor design to maintain optimal growth conditions and product yield. Employing single-use technologies (SUTs) can offer flexibility and reduce contamination risks during scale-up. SUTs simplify the fermentation process, making it more adaptable to different production scales and potentially allowing for on-site production facilities, which can be particularly beneficial for personalized medicine applications.

7.3 Post-translational Modifications (PTMs) and Their Impact on Protein Functionality

Post-translational modifications (PTMs) are chemical alterations made to proteins after their synthesis, profoundly influencing their functionality. These modifications can regulate protein activity, stability, localization, and interactions, thereby expanding the functional diversity of the proteome. Common PTMs include phosphorylation, glycosylation, ubiquitination, methylation, and acetylation.

Phosphorylation, the addition of a phosphate group, is pivotal in modulating protein activity and signaling pathways. It can activate or deactivate enzymes and receptors, affecting various cellular processes. Glycosylation, the attachment of sugar moieties, influences protein folding, stability, and cell-cell communication. Ubiquitination involves adding ubiquitin molecules, marking proteins for degradation, or altering their cellular location. Methylation and acetylation modify histones and other proteins, impacting gene expression and chromatin structure.

The dynamic nature of PTMs allows cells to respond swiftly to environmental changes. However, aberrant PTMs can lead to diseases. For instance, abnormal phosphorylation patterns are linked to cancer and neurodegenerative disorders. Similarly, defective glycosylation can result in congenital disorders. Understanding PTMs is crucial for developing therapeutic strategies, as targeting specific modifications can modulate protein function and ameliorate disease conditions.

Recent research emphasizes the importance of PTMs in regulating protein stability and function. For example, a study highlighted that PTMs can affect all aspects of protein function, including proteolytic stability. Abnormal PTMs may cause changes in protein properties and loss of function, underscoring their significance in health and disease (Lee et al., 2023; Zhong et al., 2023).

8 Adaptive Laboratory Evolution (ALE)

ALE is a powerful strategy that enhances fungal strains for higher yield, robustness, and efficiency under industrial conditions. It involves subjecting fungal cells to selective pressure over multiple generations, allowing them to evolve beneficial traits naturally. ALE has been used to improve stress tolerance (e.g., resistance to low pH, high osmolarity) and optimize metabolic pathways for better protein secretion. A study on *Aspergillus niger* demonstrated improved enzyme secretion and metabolic efficiency through ALE (Zhong et al., 2023).

8.1 AI-Driven Bioprocess Modeling

AI and machine learning are transforming fungal bioprocess optimization by enabling predictive modeling and real-time adjustments. AI can predict the best fermentation conditions by analyzing vast datasets, reducing trial-and-error experiments. Deep learning models have successfully predicted fungal growth patterns and optimized substrate utilization (Lee et al., 2023). AI-based sensors continuously monitor fermentation variables, adjusting parameters for consistent product quality.

8.2 Smart Fermentation Technologies

Next-generation fermentation systems leverage automation, IoT (Internet of Things), and real-time monitoring to enhance scalability. Smart bioreactors equipped with AI-driven control systems dynamically adjust temperature, pH, and oxygen levels to maintain optimal conditions. Biosensors detect metabolic shifts, allowing precise adjustments to prevent bottlenecks. A novel bioreactor system integrating AI and automation led to a 30% increase in recombinant protein yield in engineered *Trichoderma reesei* (Ghita Guessous et al., 2024).

By integrating *ALE, AI-driven modeling, and smart fermentation*, fungal cell factories can achieve higher efficiency, consistency, and scalability. These cutting-edge strategies pave the way for sustainable and cost-effective fungal-based protein production.

9 Real-World Applications and Case Studies

Successful industrial production examples:

9.1 Fungal-Derived Insulin and Therapeutic Enzymes

Fungal-derived insulin and therapeutic enzymes have revolutionized biopharmaceuticals, with real-world applications spanning diabetes management and enzyme replacement therapies. Insulin production using *Pichia pastoris* has been a landmark success, as this fungal system enables high-yield expression and accurate folding of human insulin. A case study by Novo Nordisk highlights the scalability of fungal platforms, producing over 50% of global recombinant insulin using engineered fungi (Wang et al., 2022). Similarly, therapeutic enzymes like lipase and lactase, used in treating metabolic disorders, are efficiently produced by *Aspergillus niger*. For instance, Biocon's fungal-derived lipase has been widely adopted for pancreatic enzyme replacement therapy, demonstrating cost-effectiveness and high activity (Zhu et al., 2024). Another example is the industrial-scale production of chymosin by *Aspergillus oryzae*, which is used in cheese production but also holds potential for therapeutic applications. These successes underscore fungi's versatility in producing complex biomolecules. Advances in genetic engineering and fermentation optimization have further enhanced yields and reduced costs (Liu et al., 2023). These examples highlight fungi's pivotal role in meeting global healthcare demands sustainably.

9.2 Antimicrobial Peptides from Engineered Fungal Strains

Antimicrobial peptides (AMPs) derived from engineered fungal strains have gained significant attention for their applications in medicine, agriculture, and food preservation. Fungi like *Aspergillus niger* and *Pichia pastoris* have been genetically optimized to produce AMPs such as defensins and plectasin, which exhibit potent activity against multidrug-resistant pathogens. A notable case study is the production of plectasin, a peptide active against Gram-positive bacteria, using *Pichia pastoris*. This system achieved high yields and demonstrated efficacy in treating methicillin-resistant *Staphylococcus aureus* (MRSA) infections (Jian Zha et al., 2023). Another example is the use of *Trichoderma reesei* to produce antifungal peptides for crop protection, reducing reliance on chemical pesticides while enhancing agricultural sustainability (Feng et al., 2023). In the food industry, AMPs from *Fusarium venenatum* have been successfully applied to extend shelf life by inhibiting spoilage microorganisms (Liu et al., 2023). These industrial successes are supported by advancements in synthetic biology, enabling precise control over peptide expression and secretion. Additionally, fungal platforms offer scalability and cost-effectiveness, making them ideal for large-scale AMP production. These real-world applications highlight fungi's potential to address global challenges like antibiotic resistance and food security.

9.3 Mycoproteins and Alternative Fungal-Based Proteins for Food Applications

Mycoproteins, derived from filamentous fungi, have emerged as a sustainable and nutritious alternative to traditional animal-based proteins in food applications. Produced through the fermentation of fungi such as *Fusarium venenatum*, mycoprotein offers a protein-rich, high-fiber food source with a meat-like texture, making it an appealing substitute for meat products (Derbyshire et al., 2023).

Beyond *Fusarium venenatum*, other fungi are being explored for their potential in alternative protein production. For instance, *Rhizomucor pusillus* has been studied for its safety and application in food, with research indicating its potential as a novel mycoprotein source (Derbyshire et al., 2023).

The production of mycoprotein involves fermenting fungi in controlled environments, resulting in a product that is not only rich in protein but also offers additional health benefits. Studies have shown that mycoprotein consumption can aid in muscle protein synthesis and improve lipid profiles, contributing to better cardiometabolic health (Majumder et al., 2024).

Several companies are pioneering the development of mycoprotein-based products. For example, Nature's Fynd utilizes a unique fungus strain discovered in geothermal springs to produce a versatile protein suitable for various food applications (Majumder et al., 2024).

Similarly, Mycorena, a Swedish food tech company, has developed "Promyc," a mycoprotein ingredient designed for use in meat and fish analogues, as well as other food products (Asgar et al., 2010).

The environmental benefits of mycoprotein are noteworthy. Its production requires less land and water compared to traditional livestock farming, and it generates a lower carbon footprint, making it a more sustainable protein source (Majumder et al., 2024).

In summary, mycoproteins and alternative fungal-based proteins present a promising solution to meet the growing demand for sustainable and health-promoting protein sources in the food industry.

10 Contributions of Leading Biotech Companies and Research Institutions

Leading biotech companies and research institutions have significantly advanced the development of mycoproteins and alternative fungal-based proteins for food applications. These organizations are pioneering sustainable protein solutions to address global food security and environmental challenges.

Mycorena, a Swedish food tech company, specializes in fungi fermentation to produce mycoprotein-based products. Their flagship product, *Promyc*, serves as a versatile ingredient in various vegan food applications, including meat and fish analogues. In 2022, Mycorena expanded its Innovation and Development Center (MIND), making it Europe's largest mycoprotein facility. This expansion underscores their commitment to scaling sustainable protein production.

Nature's Fynd, based in Chicago, utilizes a unique fungus strain discovered in Yellowstone National Park's geothermal springs. They have developed a fermentation process to produce "Fy," a protein-rich ingredient applicable in various food products. In 2020, Nature's Fynd secured an $80 million Series B funding round, facilitating the establishment of a 35,000-square-foot production facility aimed at bringing their innovative protein products to market.

Perfect Day, a California-based startup, employs precision fermentation to produce dairy-identical proteins without animals. By bioengineering microflora to produce whey protein, they offer sustainable alternatives to traditional dairy products. As of 2020, Perfect Day raised $300 million in Series C funding, reflecting strong investor confidence in their approach to revolutionizing the dairy industry.

Research institutions are also contributing to this field. For instance, studies have explored the potential of *Rhizomucor pusillus* in producing novel mycoproteins through fermentation. These investigations focus on the safety and nutritional aspects of fungal biomass, aiming to diversify the sources of alternative proteins for human consumption.

Collectively, these efforts by biotech companies and research institutions are propelling the development of sustainable, nutritious, and scalable fungal-based protein sources, offering promising solutions to meet the growing global demand for alternative proteins.

11 Future Directions and Innovations in Fungal Biotechnology

Future directions in fungal biotechnology focus on enhanced strain engineering, precision fermentation, and AI-driven bioprocess optimization. Advanced genetic tools such as CRISPR-Cas9 and synthetic biology will enable the development of high-yield fungal strains for protein and bioactive peptide production. AI and machine learning will optimize fermentation conditions, ensuring scalability and consistency. Additionally, smart bioreactors and biosensors will revolutionize process monitoring, reducing costs and improving efficiency. The integration of novel substrates, metabolic pathway engineering, and sustainable feedstocks will further enhance fungal-based protein production, making it a key player in the future of sustainable food and biopharmaceutical industries.

11.1 Emerging Technologies (AI-Guided Metabolic Engineering, Synthetic Biology Advancements)

AI-guided metabolic engineering and synthetic biology are revolutionizing biomanufacturing by enabling precise design and optimization of fungal cell factories. AI algorithms predict metabolic fluxes and identify optimal genetic edits, accelerating strain development (Guillaume et al., 2024). Synthetic biology tools like CRISPR-Cas9 and modular gene circuits allow targeted genome editing and dynamic pathway regulation, enhancing the production of biofuels, enzymes, and therapeutics.

For example, AI-driven models optimized *Aspergillus niger* for 40% higher enzyme yields by rerouting carbon flux (Wang et al., 2022), while CRISPR-engineered *Pichia pastoris* achieved high-titer insulin production (Wallis et al., 1999). These innovations bridge computational design with biological execution, enabling scalable, sustainable bioproduction.

11.2 Engineering Fungi for Novel Biomaterials and Next-Generation Biopharmaceuticals

Engineering fungi for novel biomaterials and biopharmaceuticals is revolutionizing industries by enabling the production of sustainable and high-value products. Fungi like *Aspergillus* and *Pichia* are engineered to produce biodegradable polymers, such as fungal-derived chitin and polyesters, for eco-friendly packaging and biomedical applications (Dai & Shen, 2022). In biopharmaceuticals, fungi are optimized to synthesize complex molecules like glycoproteins and antibody fragments with precise post-translational modifications (Woodward et al., 2021). Advances in synthetic biology and CRISPR-Cas9 have enhanced strain performance, enabling scalable production of these materials. For instance, *Trichoderma reesei* has been engineered to produce spider silk proteins for durable textiles (Feng et al., 2023). These innovations highlight fungi's versatility in addressing global challenges.

11.3 Integrating Fungal Cell Factories with Precision Fermentation and AI-Driven Optimization

Integrating fungal cell factories with precision fermentation and AI-driven optimization enables unprecedented control over bioproduction. Precision fermentation ensures optimal growth conditions via real-time monitoring and adaptive control systems, while AI algorithms predict metabolic bottlenecks and optimize strain performance (Guillaume et al., 2024). For example, machine learning models have enhanced *Pichia pastoris* productivity by 30% by fine-tuning oxygen and nutrient delivery (Leal et al., 2024). Similarly, AI-guided metabolic engineering of *Aspergillus niger* improved enzyme yields by predicting optimal gene-editing targets (Hassan et al., 2021). These synergistic approaches reduce costs, minimize waste, and accelerate the development of high-value products like biotherapeutics and industrial enzymes, positioning fungi as versatile platforms for next-generation biomanufacturing.

12 Conclusions and Prospects

Engineered fungal cell factories have emerged as powerful tools in the sustainable production of proteins and bioactive peptides. Their intrinsic attributes—such as rapid growth, metabolic versatility, and a natural ability to secrete complex proteins—make fungi highly suitable for industrial biomanufacturing. This review highlights how recent advancements in genetic engineering (CRISPR-Cas9, RNAi), synthetic biology, and omics technologies have significantly expanded the utility of fungal platforms in producing a wide range of value-added bioproducts for the food, pharmaceutical, and agricultural sectors.

Species such as *Aspergillus niger*, *Trichoderma reesei*, *Pichia pastoris*, and *Saccharomyces cerevisiae* are now routinely engineered for the production of enzymes, antimicrobial peptides, recombinant therapeutics, and mycoproteins. These organisms offer distinct advantages over traditional systems, including post-translational modification capabilities and the ability to thrive on low-cost substrates, including agro-industrial residues. Consequently, fungal platforms are increasingly being integrated into circular bioeconomy models, where they contribute to waste valorization and environmental sustainability.

However, several challenges remain. Key among these are scale-up limitations, production consistency, product heterogeneity, and regulatory hurdles—especially for genetically modified fungal strains used in food or therapeutic applications. Addressing these challenges requires multidisciplinary solutions, including process analytical technologies, bioreactor innovation, and the development of stable, high-performing chassis strains.

Looking ahead, integration of AI-driven tools in strain design, metabolic modeling, and fermentation control is expected to revolutionize fungal biotechnology. Machine learning models can predict optimal gene edits, fermentation parameters, and pathway optimizations, drastically reducing development cycles. Adaptive laboratory evolution (ALE) strategies will also be pivotal in enhancing strain robustness, stress tolerance, and yield performance under industrial conditions.

The advancement of modular genetic circuits and programmable expression systems through synthetic biology will allow for dynamic control over fungal metabolism, enabling efficient on-demand production of target molecules. Moreover, emerging technologies such as smart fermentation systems, biosensor-integrated bioreactors, and continuous production models will improve scalability and reduce energy and water footprints.

Real-world industrial examples, such as fungal-derived insulin, enzymes, and mycoproteins (e.g., Quorn™, Fy™, and Promyc™), demonstrate the commercial viability of these systems. These successes highlight the ability of fungal platforms to meet growing global demands for protein while aligning with sustainability goals.

In conclusion, engineered fungal cell factories are poised to play a central role in the transition toward sustainable and circular bio-economies. Their potential to transform underutilized biomass into high-value proteins, peptides, and bioproducts offers a compelling solution to current global challenges in food security, environmental sustainability, and bio-based manufacturing. With continued innovation in strain engineering, regulatory frameworks, and process optimization, fungal biotechnology is well-positioned to drive the next generation of sustainable industrial solutions.

Acknowledgements The author is grateful to Dr. Ashok K. Chauhan, Founder President, Dr. Atul Chauhan, Chancellor, and Prof. Dr. Balvinder Shukla, Vice Chancellor, Amity University-UP, Noida, India for the motivation and research facilities.

Funding The authors thankfully acknowledge the research project funded by DSIR, New Delhi, India [TDUPW-11011/1/2024-IRD-DSIR] and are grateful to the funding agency for financial support.

References

Adiego-Pérez, B., Randazzo, P., Daran, J. M., Verwaal, R., Roubos, J. A., Daran-Lapujade, P., & van der Oost, J. (2019). Multiplex genome editing of microorganisms using CRISPR-Cas. *FEMS Microbiology Letters, 366*(8). https://doi.org/10.1093/femsle/fnz086

Alternative Proteins for Human Consumption FS900199/FS900200. (2022, March). https://doi.org/10.46756/sci.fsa.wdu243.

Arnau, J., Yaver, D., & Hjort, C. M. (2019). Strategies and challenges for the development of industrial enzymes using fungal cell factories. *Grand Challenges Fungal Biotechnol., 179–210.* https://doi.org/10.1007/978-3-030-29541-7_7

Asgar, M. A., Fazilah, A., Huda, N., Bhat, R., & Karim, A. A. (2010). Nonmeat protein alternatives as meat extenders and meat analogs. *Comprehensive Reviews in Food Science and Food Safety, 9*(5), 513–529. https://doi.org/10.1111/j.1541-4337.2010.00124.x

Baldin, C., Kühbacher, A., Merschak, P., Wagener, J., & Gsaller, F. (2022). Modular inducible multigene expression system for filamentous fungi. *Microbiology Spectrum, 10*(6). https://doi.org/10.1128/spectrum.03670-22

Cellura, M., Cusenza, M. A., Longo, S., Luu, L. Q., & Skurk, T. (2022). Life cycle environmental impacts and health effects of protein-rich food as meat alternatives: A review. *Sustainability, 14*(2), 979. https://doi.org/10.3390/su14020979

Cerimi, K., Akkaya, K. C., Pohl, C., Schmidt, B., & Neubauer, P. (2019). Fungi as source for new bio-based materials: A patent review. *Fungal Biology and Biotechnology, 6*(1). https://doi.org/10.1186/s40694-019-0080-y

Challener, C. (2023). Optimizing fermentation processes for biologics manufacturing. *BioPharm International, 36*(10), 15–18. https://www.biopharminternational.com/view/optimizing-fermentation-processes-for-biologics-manufacturing

Culibrk, L., Croft, C. A., & Tebbutt, S. J. (2016). Systems biology approaches for host–fungal interactions: An expanding multi-omics frontier. *OMICS: A Journal of Integrative Biology, 20*(3), 127–138. https://doi.org/10.1089/omi.2015.0185

Dai, X., & Shen, L. (2022). Advances and trends in omics technology development. *Frontiers in Medicine, 9*, 911861. https://doi.org/10.3389/fmed.2022.911861

Dang, Y., Yang, Q., Xue, Z., & Liu, Y. (2011). RNA interference in fungi: Pathways, functions, and applications. *Eukaryotic Cell, 10*(9), 1148–1155. https://doi.org/10.1128/ec.05109-11

Derbyshire, E. J., Theobald, H., Wall, B. T., & Stephens, F. (2023). Food for our future: The nutritional science behind the sustainable fungal protein—Mycoprotein. A symposium review. *Journal of Nutritional Science, 12*. https://doi.org/10.1017/jns.2023.29

do Nascimento Muniz, E., de Queiroz Montenegro, R. T., da Silva, D. N., D'Almeida, A. P., Gonçalves, L. R. B., & de Albuquerque, T. L. (2024). Advances in biotechnological strategies for sustainable production of non-animal proteins: Challenges, innovations, and applications. *Fermentation, 10*(12), 638–638. https://doi.org/10.3390/fermentation10120638

Druzhinina, I. S., & Kubicek, C. P. (2017). Genetic engineering of Trichoderma reesei cellulases and their production. *Microbial Biotechnology, 10*(6), 1485–1499. https://doi.org/10.1111/1751-7915.12726

Du, Y., Wang, M., Yang, L.-H., Tong, L., Guo, D.-S., & Ji, X.-J. (2022). Optimization and scale-up of fermentation processes driven by models. *Bioengineering (Basel), 9*(9), 473–473. https://doi.org/10.3390/bioengineering9090473

Feng, L., Xu, J., Ye, C., Gao, J., Huang, L., Xu, Z., & Lian, J. (2023). Metabolic engineering of *Pichia pastoris* for the production of triacetic acid lactone. *Journal of Fungi, 9*(4), 494–494. https://doi.org/10.3390/jof9040494

Food Policy. (2022). https://gfi.org/blog/a-rare-glimpse-into-the-global-regulatory-landscape-for-alternative-proteins/?utm

Garg, S., Kim, M., & Romero-Suarez, D. (2024). Current advancements in fungal engineering technologies for sustainable development goals. *Trends in Microbiology, 33*(3), 285–301. https://doi.org/10.1016/j.tim.2024.11.001

Good Food Institute. (2022). *The state of the alternative protein industry*. Retrieved from Good Food Institute website https://gfi.org/industry/

Gricourt, G., Meyer, P., Duigou, T., & Faulon, J.-L. (2024). Artificial intelligence methods and models for retro-biosynthesis: A scoping review. *ACS Synthetic Biology, 13*(8), 2276–2294. https://doi.org/10.1021/acssynbio.4c00091

Guessous, G., Blake, L., Bui, A., Woo, Y., & Manzanarez, G. (2024). Disentangling the web: An interdisciplinary review on the potential and feasibility of spider silk bioproduction. *ACS Biomaterilas Science and Engineering Journal, 10*(9), 5412–5438. https://doi.org/10.1021/acsbiomaterials.4c00145

Gricourt, G., Meyer, P., Duigou, T., & Faulon, J-L. (2024). Artificial Intelligence Methods and Models for Retro-Biosynthesis: A Scoping Review. *ACS Synthetic Biol, 13*(8), 2276–2294. https://doi.org/10.1021/acssynbio.4c00091

Hadi, J., & Brightwell, G. (2021). Safety of alternative proteins: Technological, environmental and regulatory aspects of cultured meat, plant-based meat, insect protein and single-cell protein. *Foods, 10*(6), 1226. https://doi.org/10.3390/foods10061226

Hadidi, M., Aghababaei, F., Gonzalez-Serrano, D. J., Goksen, G., Trif, M., McClements, D. J., & Moreno, A. (2024). Plant-based proteins from agro-industrial waste and by-products: Towards a more circular economy. *International Journal of Biological Macromolecules, 261*, 129576–129576. https://doi.org/10.1016/j.ijbiomac.2024.129576

Hassan, M. M., Zhang, Y., Yuan, G., De, K., Chen, J.-G., Muchero, W., Tuskan, G. A., Qi, Y., & Yang, X. (2021). Construct design for CRISPR/Cas-based genome editing in plants. *Trends in Plant Science, 26*(11), 1133–1152. https://doi.org/10.1016/j.tplants.2021.06.015

Henrique, L., & Nirlane, P. (2016). *Aspergillus* biotechnology: An overview on the production of hydrolases and secondary metabolites. *Curr. Biotechnology, 6*(4). https://doi.org/10.2174/2211550105666160607071134

Hyde, K. D., Baldrian, P., Chen, Y., Thilini Chethana, K. W., De Hoog, S., Doilom, M., Gomes, A. R., Gonçalves, M. F. M., Gonkhom, D., Gui, H., Hilário, S., Hu, Y., Jayawardena, R. S., Khyaju, S., Kirk, P. M., Kohout, P., Luangharn, T., Maharachchikumbura, S. S. N., Manawasinghe, I. S., & Mortimer, P. E. (2024). Current trends, limitations and future research in the fungi? *Fungal Diversity, 125*. https://doi.org/10.1007/s13225-023-00532-5

Jeong, S. H., Lee, H. J., & Lee, S. J. (2023). Recent advances in CRISPR-Cas technologies for synthetic biology. *Journal of Microbiology, 61*(1), 13–36. https://doi.org/10.1007/s12275-022-00005-5

Jiang, W., Li, C., Li, Y., & Peng, H. (2022). Metabolic engineering strategies for improved lipid production and cellular physiological responses in yeast *Saccharomyces cerevisiae*. *Journal of Fungi, 8*(5), 427. https://doi.org/10.3390/jof8050427

Jo, C., Zhang, J., Tam, J. M., Church, G. M., Khalil, A. S., Segrè, D., & Tang, T.-C. (2023). Unlocking the magic in mycelium: Using synthetic biology to optimize filamentous fungi for biomanufacturing and sustainability. *Materials Today Bio, 19*, 100560. https://doi.org/10.1016/j.mtbio.2023.100560

Karbalaei, M., Rezaee, S. A., & Farsiani, H. (2020). *Pichia pastoris*: A highly successful expression system for optimal synthesis of heterologous proteins. *Journal of Cellular Physiology, 235*(9), 5867–5881. https://doi.org/10.1002/jcp.29583

Kaur, H., Kaur, G., Gupta, T., Mittal, D., & Syed, A. A. (2023). Integrating omics technologies for a comprehensive understanding of the microbiome and its impact on cattle product. *Biology (Basel), 12*(9), 1200–1200. https://doi.org/10.3390/biology12091200

Knychala, M. M., Boing, L. A., Ienczak, J. L., Trichez, D., & Stambuk, B. U. (2024). Precision fermentation as an alternative to animal protein, a review. *Fermentation, 10*(6), 315–315. https://doi.org/10.3390/fermentation10060315

Leal, K., Rojas, E., Madariaga, D., Contreras, M. J., Nuñez-Montero, K., Barrientos, L., Goméz-Espinoza, O., & Iturrieta-González, I. (2024). Unlocking fungal potential: The CRISPR-Cas system as a strategy for secondary metabolite discovery. *Journal of Fungi, 10*(11), 748–748. https://doi.org/10.3390/jof10110748

Lee, H. J., & Lee, S. J. (2021). Advances in accurate microbial genome-editing CRISPR technologies. *Journal of Microbiology and Biotechnology, 31*(7), 903–911. https://doi.org/10.4014/jmb.2106.06056

Lee, J. M., Hammarén, H. M., Savitski, M. M., & Baek, S. H. (2023). Control of protein stability by post-translational modifications. *Nature Communications, 14*(1), 201. https://doi.org/10.1038/s41467-023-35795-8

Li, W., Zou, G., Bao, D., & Wu, Y. (2024). Current advances in the functional genes of edible and medicinal fungi: Research techniques, functional analysis, and prospects. *Journal of Fungi, 10*(5), 311–311. https://doi.org/10.3390/jof10050311

Liu, R., Chen, L., Jiang, Y., Zhou, Z., & Zou, G. (2015). Efficient genome editing in filamentous fungus *Trichoderma reesei* using the CRISPR/Cas9 system. *Cell Discovery, 1*(1), 1–11. https://doi.org/10.1038/celldisc.2015.7

Liu, Q., Gao, R., Li, J., Lin, L., Zhao, J., Sun, W., & Tian, C. (2017). Development of a genome-editing CRISPR/Cas9 system in thermophilic fungal Myceliophthora species and its application to hyper-cellulase production strain engineering. *Biotechnology for Biofuels, 10*(1). https://doi.org/10.1186/s13068-016-0693-9

Liu, J., Han, Q., Cheng, Q., Chen, Y., Wang, R., Li, X., Liu, Y., & Yan, D. (2020). Efficient expression of human lysozyme through the increased gene dosage and co-expression of transcription factor Hac1p in *Pichia pastoris*. *Current Microbiology, 77*(5), 846–854. https://doi.org/10.1007/s00284-019-01872-9

Liu, Z., de Souza, T. S. P., Holland, B., Dunshea, F., Barrow, C., & Suleria, H. A. R. (2023). Valorization of food waste to produce value-added products based on its bioactive compounds. *PRO, 11*(3), 840. https://doi.org/10.3390/pr11030840

Lübeck, M., & Lübeck, P. S. (2022). Fungal cell factories for efficient and sustainable production of proteins and peptides. *Microorganisms, 10*(4), 753. https://doi.org/10.3390/microorganisms10040753

Majumder, R. K., Miatur, S., Saha, A., & Hossain, S. (2024). Mycoprotein: Production and nutritional aspects: A review. *Sustainable Food Technology, 2*, 81–91. https://doi.org/10.1039/d3fb00169e

Mattedi, A., Sabbi, E., Farda, B., Djebaili, R., Mitra, D., Ercole, C., Cacchio, P., Del Gallo, M., & Pellegrini, M. (2023). Solid-state fermentation: Applications and future perspectives for biostimulant and biopesticides production. *Microorganisms, 11*(6), 1408. https://doi.org/10.3390/microorganisms11061408

Mesbah, N. M. (2022). Industrial biotechnology based on enzymes from extreme environments. *Frontiers in Bioengineering and Biotechnology, 10*, 870083. https://doi.org/10.3389/fbioe.2022.870083

Milne, N., Sáez-Sáez, J., Nielsen, A. M., Dyekjær, J. D., Rago, D., Kristensen, M., Wulff, T., & Borodina, I. (2023). Engineering *Saccharomyces cerevisiae* for the de novo production of halogenated tryptophan and tryptamine derivatives. *ChemistryOpen, 12*(4), e202200266. https://doi.org/10.1002/open.202200266

Pullen, R. M., Decker, S. R., Subramanian, V., Adler, M. J., Tobias, A. V., Perisin, M., Sund, C. J., Servinsky, M. D., & Kozlowski, M. T. (2025). Considerations for domestication of novel strains of filamentous fungi. *ACS Synthetic Biology, 14*(2), 343–362. https://doi.org/10.1021/acssynbio.4c00672

Rajput, M., Choudhary, K., Kumar, M., Vivekanand, V., Chawade, A., Ortiz, R., & Pareek, N. (2021). RNA interference and CRISPR/Cas gene editing for crop improvement: Paradigm shift towards sustainable agriculture. *Plants, 10*(9), 1914. https://doi.org/10.3390/plants10091914

Reilly, M. C., Magnuson, J. K., & Baker, S. E. (2016). Approaches to understanding protein hypersecretion in fungi. *Fungal Biology Reviews, 30*(4), 145–151. https://doi.org/10.1016/j.fbr.2016.06.002

Rocha, V. A. L., Maeda, R. N., Santa Anna, L. M. M., & Pereira, N., Jr. (2013). Sugarcane bagasse as feedstock for cellulase production by *Trichoderma harzianum* in optimized culture medium. *Electronic Journal of Biotechnology, 16*(5). https://doi.org/10.2225/vol16-issue5-fulltext-1

Salazar-Cerezo, S., de Vries, R. P., & Garrigues, S. (2023). Strategies for the development of industrial fungal producing strains. *Journal of Fungi, 9*(8), 834–834. https://doi.org/10.3390/jof9080834

Sarnaik, A., Liu, A., Nielsen, D., & Varman, A. M. (2020). High-throughput screening for efficient microbial biotechnology. *Current Opinion in Biotechnology, 64*, 141–150. https://doi.org/10.1016/j.copbio.2020.02.019

Song, R., Zhai, Q., Sun, L., Huang, E., Zhang, Y., Zhu, Y., Guo, Q., Tian, Y., Zhao, B., & Lu, H. (2019). CRISPR/Cas9 genome editing technology in filamentous fungi: Progress and perspective. *Applied Microbiology and Biotechnology, 103*(17), 6919–6932. https://doi.org/10.1007/s00253-019-10007-w

Souza Filho, P. F., Andersson, D., Ferreira, J. A., & Taherzadeh, M. J. (2019). Mycoprotein: Environmental impact and health aspects. *World Journal of Microbiology and Biotechnology, 35*(10), 147. https://doi.org/10.1007/s11274-019-2723-9

Tamano, K., & Yoshimi, A. (2021). Metabolic engineering techniques to increase the productivity of primary and secondary metabolites within filamentous fungi. *Frontiers in Fungal Biology, 2*, 743070. https://doi.org/10.3389/ffunb.2021.743070

Tavakoli, K., Pour-Aboughadareh, A., Kianersi, F., Poczai, P., Etminan, A., & Shooshtari, L. (2021). Applications of CRISPR-Cas9 as an advanced genome editing system in life sciences. *Biotech, 10*(3), 14. https://doi.org/10.3390/biotech10030014

Tiwari, P., & Park, K.-I. (2024). Advanced fungal biotechnologies in accomplishing sustainable development goals (SDGs): What do we know and what comes next? *Journal of Fungi, 10*(7), 506–506. https://doi.org/10.3390/jof10070506

Tyśkiewicz, R., Nowak, A., Ozimek, E., & Jaroszuk-Ściseł, J. (2022). Trichoderma: The current status of its application in agriculture for the biocontrol of fungal Phytopathogens and stimulation of plant growth. *International Journal of Molecular Sciences, 23*(4), 2329. https://doi.org/10.3390/ijms23042329

Upton, D. J., McQueen-Mason, S. J., & Wood, A. J. (2020). In silico evolution of *Aspergillus niger* organic acid production suggests strategies for switching acid output. *Biotechnology for Biofuels, 13*(1), 27. https://doi.org/10.1186/s13068-020-01678-z

Wallis, G. L., Swift, R. J., Hemming, F. W., Trinci, A. P., & Peberdy, J. F. (1999). Glucoamylase overexpression and secretion in *Aspergillus niger*: Analysis of glycosylation. *Biochimica et Biophysica Acta, 1472*(3), 576–586. https://doi.org/10.1016/s0304-4165(99)00188-9. PMID: 10564772.

Wang, Q., Zhong, C., Zhong, C., & Xiao, H. (2020). Genetic engineering of filamentous fungi for efficient protein expression and secretion. *Frontiers in Bioengineering and Biotechnology, 8*, 293. https://doi.org/10.3389/FBIOE.2020.00293

Wang, Y., Chen, H., Ma, L., Gong, M., Wu, Y., Bao, D., & Zou, G. (2022). Use of CRISPR-Cas tools to engineer *Trichoderma* species. *Microbial Biotechnology, 15*, 2521–2532. https://doi.org/10.1111/1751-7915.14126

Wang, X., Dong, X., Zhang, Z., & Wang, Y. (2024). Transportation carbon reduction technologies: A review of fundamentals, application, and performance. *Journal of Traffic and Transportation Engineering (English Edition), 11(6*, 1340–1377. https://doi.org/10.1016/j.jtte.2024.11.001

Wijayawardene, N. N., Boonyuen, N., Ranaweera, C. B., De Zoysa, H. K. S., Padmathilake, R. E., Nifla, F., Dai, D.-Q., Liu, Y., Suwannarach, N., Kumla, J., Bamunuarachchige, T. C., & Chen, H. (2023). OMICS and other advanced technologies in mycological applications. *Journal of Fungi, 9*(6), 688–688. https://doi.org/10.3390/jof9060688

Woodward, A., Pandele, A., Abdelrazig, S., Ortori, C. A., Khan, I., Uribe, M. C., May, S., Barrett, D. A., Grundy, R. G., Kim, D.-H., & Rahman, R. (2021). Integrated metabolomics and transcriptomics using an optimised dual extraction process to study human brain cancer cells and tissues. *Metabolites, 11*(4), 240–240. https://doi.org/10.3390/metabo11040240

Yafetto, L. (2022). Application of solid-state fermentation by microbial biotechnology for bioprocessing of agro-industrial wastes from 1970 to 2020: A review and bibliometric analysis. *Heliyon, 8*(3), e09173. https://doi.org/10.1016/j.heliyon.2022.e09173

Yan, X., Liu, X., Zhao, C., & Chen, G.-Q. (2023). Applications of synthetic biology in medical and pharmaceutical fields. *Signal Transduction Targeted Therapy, 8*(1), 1–33. https://doi.org/10.1038/s41392-023-01440-5

Yang, P., Condrich, A., Lu, L., Scranton, S., Hebner, C., Sheykhhasan, M., & Ali, M. A. (2024). Genetic engineering in bacteria, fungi, and oomycetes, taking advantage of CRISPR. *DNA, 4*(4), 427–454. https://doi.org/10.3390/dna4040030

Yin, J., Li, X., Dong, L., Zhu, X., Chen, Y., Zhao, W., Liu, Y., Shan, J., Liu, W., Lin, C., & Miao, W. (2024). Transformation-based gene silencing and functional characterization of an ISC effector reveal how a powdery mildew fungus disturbs salicylic acid biosynthesis and immune response in the plant. *Molecular Plant Pathology, 25*(11), e70030. https://doi.org/10.1111/mpp.70030

Zha, J., Liu, D., Ren, J., Liu, Z., & Wu, X. (2023). Advances in metabolic engineering of *Pichia pastoris* strains as powerful cell factories. *Journal of Fungi, 9*(10), 1027–1027. https://doi.org/10.3390/jof9101027

Zhao, M., Ma, J., Zhang, L., & Qi, H. (2024). Engineering strategies for enhanced heterologous protein production by *Saccharomyces cerevisiae. Microbial Cell Factories, 23*(1), 32. https://doi.org/10.1186/s12934-024-02299-z

Zhong, Q., Xiao, X., Qiu, Y., Xu, Z., Chen, C., Chong, B., Zhao, X., Hai, S., Li, S., An, Z., & Dai, L. (2023). Protein posttranslational modifications in health and diseases: Functions, regulatory mechanisms, and therapeutic implications. *Med Comm, 4*(3), e261. https://doi.org/10.1002/mco2.261

Zhu, H., Wang, H., Wang, L., & Zheng, Z. (2024). CRISPR/Cas9-based genome engineering in the filamentous fungus *Rhizopus oryzae* and its application to L-lactic acid production. *Biotechnology Journal, 19*(9), 2400309. https://doi.org/10.1002/biot.202400309

Fungi as Bioconverters: Advancing Circular Economy Through Waste Recycling and Resource Optimization

Sweta Gaikwad

Abstract Fungal biotechnology is rapidly gaining recognition for its potential to drive sustainable innovations across industries. Harnessing the unique biological properties of fungi, such as their potent enzymatic activity and resource-efficient growth, enables transformative applications in bioremediation, renewable energy production, food innovation, and biomaterials development. Fungi play a critical role in promoting the circular economy (CE) principles by effectively recycling trash, turning biomass into useful products, and regenerating resources. This chapter examines the integration of fungal biotechnology within the CE framework, showcasing how fungi enable organic waste recycling, bioconversion of agricultural and industrial residues, and sustainable product development. Real-world examples, such as oil spill remediation using *Pleurotus* species, heavy metal detoxification by *Aspergillus* and *Trametes*, and innovations like mycelium-based packaging and fungal proteins for alternative food sources, highlight fungi's practical impact. The chapter is organized to first introduce fungal biotechnology and CE fundamentals, followed by sections detailing fungal contributions to waste upcycling, regenerative resource use, and material loop closure. It further presents success stories, discusses existing challenges in scaling and regulation, and proposes future research priorities including advances in genetic engineering, artificial intelligence applications, and policy frameworks to foster large-scale fungal innovation within a sustainable circular economy.

Keywords Bioremediation · Circular economy · Fungal biotechnology · Sustainable resource utilization · Waste upcycling · Waste valorization

S. Gaikwad (✉)
Department of Life-Science, Shri Shankaracharya Professional University, Bhilai, Chhattisgarh, India
e-mail: drswetagaikwad@sspubhilai.com

A. Gupta (ed.), *Prospects of Fungal Biotechnologies for Livestock Volume 2*, Fungal Biology, https://doi.org/10.1007/978-3-032-06478-3_10

1 Introduction

Humanity has benefited from fungi throughout history, and they have a lot of potential to improve it even more (Roth et al., 2023). These are unique organisms that differ from plants, animals, and bacteria due to their heterotrophic mode of nutrition and filamentous growth. As a vast kingdom, fungi contain a variety of species having a wide range of properties and useful enzymes. Bread, cheese, beer, wine, and many other items have been made using them for centuries in the food and beverage business. However, the biology of fungi has enormous promise for solving some of the most important problems facing the globe today, including environmental sustainability, human medicine, food security, and energy insecurity. Their biotechnological significance is due to several features like their fast growth and adaptability, as they can thrive in diverse environmental conditions, including extreme temperatures, low pH, and nutrient-deficient soils. They produce extracellular enzymes like cellulases, ligninases, proteases, and lipases, which are useful for industrial applications specially in breaking down complex polymers present in plant cell walls. It is undisputed that fungal-based biotechnology is of pivotal importance to our daily lives. Despite decades of research on fungi as plant diseases, saprotrophs, and decomposers, fungal biotechnology applications are only now beginning to gain traction.

Fungal biotechnology involves the manipulation and enhancement of fungal processes to produce bioactive compounds, enzymes, pharmaceuticals, biofuels, and sustainable materials. It has revolutionized several industries by providing sustainable alternatives to chemical processes. Fungi produce industrially significant enzymes used in food, detergents, textiles, and pharmaceuticals. Yeasts, mushrooms, and filamentous fungi are examples of this complex community, which has been used to create a wide range of food products for human and animal use. The yeast *Saccharomyces cerevisiae* is used extensively in winemaking, brewing, and baking. *Fusarium venenatum* is used to produce Quorn, a high-protein, meat substitute. Some fungi produce natural preservatives like natamycin, which prevent spoilage in dairy and bakery products. Fungal fermentation is crucial for producing fermented foods. Citric acid is predominantly formed by fungus like *Aspergillus niger*, and is commonly used in food, pharmaceuticals, and beverages. Some of the very widely consumed fungal-based fermented foods like Tempeh are prepared with the help of *Rhizopus oligosporus* (Massa et al., 2024), soy sauce by *Aspergillus oryzae* (Daba et al., 2021), and cheese is ripened by the fermentation activity of *Penicillium roqueforti* a best best-known species worldwide for its technological use for blue-veined cheese production. Our food source was sustained and expanded with the help of our understanding of fungal variety and how to identify beneficial fungi for the biotransformation of food items. The commercial usage of fungus has been greatly facilitated by their amenability to several species and subsequent simplicity of upgrading to large-scale cultures. These elements have helped to bring together several research teams and businesses that have taken part in successive

"Eurofung" initiatives that have been sponsored by the European Union in an effort to speed up the industry's growth. This has led to a remarkable increase in the average yearly production of mushrooms to 6161 thousand tons and a thriving intercontinental trade.

Due to the detrimental effects of nonrenewable fossil fuels on the environment, sustainable and renewable energy sources must be developed. Enzymatic capabilities and metabolic pathways of fungi make them valuable in sustainable energy production. That is why fungal biotechnology plays a crucial role in sustainable fuel production due to its ability to break down complex biomass into fermentable sugars and produce bioethanol, biodiesel, and other biofuels. *Saccharomyces cerevisiae* are commonly used in this process, especially for ethanol production. Some oleaginous fungi, such as *Mortierella* and *Rhodosporidium*, accumulate high amounts of lipids that can be converted into biodiesel (Chang et al., 2021). Fungal lipases aid in the transesterification process, improving biodiesel production efficiency. Besides, fungi contribute to anaerobic digestion processes by breaking down organic waste, increasing methane and hydrogen gas production (Saye et al., 2021). White-rot fungi, like *Phanerochaete chrysosporium*, help degrade lignin-rich biomass, making it more accessible for microbial digestion in biogas plants. Advances in fungal metabolic engineering are intended to improve the efficiency of converting lignocellulose into biofuels.

Bioremediation is the process of employing biological organisms to break down, purify, or eliminate environmental contaminants. The unique enzymatic capabilities of fungi allow them to break down complex pollutants, including heavy metals, hydrocarbons, and synthetic chemicals, making them valuable in environmental restoration. White-rot fungi like *Phanerochaete chrysosporium* and *Pleurotus ostreatus* and *Trametes versicolor* (Beltrán-Flores et al., 2021) produce ligninolytic enzymes (laccases, peroxidases) that degrade toxic pollutants such as polycyclic aromatic hydrocarbons (PAHs) and pesticides. These fungi can break down dyes from textile industries and persistent organic pollutants (POPs) like dioxins and PCBs. Fungi such as *Aspergillus niger* and *Penicillium* species can absorb and bioaccumulate heavy metals like lead (Pb), cadmium (Cd), and mercury (Hg) from contaminated soil and water (El-Mahdy et al., 2021). At the same time, mycorrhizal fungi enhance phytoremediation by increasing plant uptake of metals. Fungal biofilms help absorb and degrade hydrocarbons in marine environments. Hydrocarbon-degrading fungi, including *Amorphotheca resinae* and *Aspergillus* species, can break down petroleum-based pollutants in oil spills. Some fungi, such as *Aspergillus tubingensis*, have shown potential in breaking down plastics like polyurethane, offering a sustainable approach to plastic waste management (Ullah et al., 2025). Ongoing research in fungal bioremediation holds promise for addressing global pollution challenges sustainably.

A lot of emphasis has been paid to the shift to a circular bioeconomy in light of growing environmental concerns and the need for sustainable resource management. Although it was first proposed in the 1970s to reduce the use of industrial production inputs, the Circular Economy (CE) concept has the potential to be applied to any resource. By using a "reduce, reuse, recycle"strategy to establish a closed-loop system, CE is an economic system that aims to eradicate waste and encourage the continuous use of resources. With the potential to increase human activity's resilience through the use of the natural cycle model, CE suggests a shift from the "extraction-production-disposal" paradigm of linear economy (LE), which is currently widely used in the industrial setting. In order to achieve the objective of a circular bioeconomy, fungi are crucial because of their exceptional capacity to transform complex lignocellulosic materials into beneficial products.

Saprophytic fungi play major role in material recovery by breaking down organic and even some inorganic waste. Fungus like *Pleurotus ostreatus* can break down toxic substances, including plastics, petroleum, and heavy metals, making polluted sites safe again (Di Piazza et al., 2021). *Trametes versicolor* also called "Turkey Tail" is used in the process of bioremediation to degrade harmful pollutants and recover valuable elements from waste (Tišma et al., 2021). Fungi play crucial role in waste upcycling because they have the ability to convert agricultural, industrial, and food waste into valuable products, reducing landfill burden. For example, fungal mycelium can be used to create biodegradable packaging, and even construction materials are sustainable alternatives to plastics and synthetic leather (e.g., *Ecovative's MycoComposite*). Some fungi which can grow on straw, coffee grounds, and sawdust can convert low-value agricultural waste into edible mushrooms. *Pleurotus ostreatus* and its usefulness in bioremediation of petroleum and aromatic polycyclic hydrocarbons are highlighted by several researchers. Food waste valorization is another important aspect related to fungal biotechnology where food waste is transformed into fungal biomass or valuable sources of protein. Utilizing waste materials as soil amendments can be a cost-effective and environmentally friendly way to clean up polluted soils. This is because it turns waste into useful resources, which improves the properties of the soil (Paniagua-López et al., 2024).

This chapter examines the transformative role of fungal biotechnology in advancing circular economy principles. By addressing critical CE goals—such as minimizing waste, regenerating resources, and establishing closed-loop systems—fungi emerge as a cornerstone for sustainable innovation. The chapter explores diverse fungal applications, including the recycling of organic waste, bioconversion of industrial by-products, and the development of bio-based solutions like bioplastics, biofuels, and alternative textiles. Real-world examples illustrate fungi's potential to remediate pollution, enhance industrial efficiency, and pioneer eco-friendly materials. Concluding with an analysis of challenges and future prospects, the chapter highlights cutting-edge advancements like genetic engineering, artificial intelligence, and policy integration as key drivers for scaling fungal biotechnology. Together, these insights underscore the immense potential of fungi in fostering a sustainable, regenerative economy.

2 How Fungal Biotechnology Aligns with Circular Economy Principles?

Fungal biotechnology aligns with the circular economy by transforming waste into valuable bio-based products, reducing environmental impact, and fostering sustainable industrial practices. Its applications in biodegradable materials, biofuels, waste management, and sustainable agriculture make it a key player in building a regenerative and eco-friendly economy (Fig. 1). Since filamentous fungi and yeasts are true cellular factories, they play a significant role in biotechnology and offer beneficial outlets for environmental and health protection by decomposing a variety of industrial pollutants (Niego et al., 2023). Since the 1990s, there has been an increasing interest in employing fungi for other biotechnological uses; in recent years, the terms "fungal biotechnology"and "circular economy"have increased tenfold in publications (Roth et al., 2023). Fungal biotechnology plays a transformative role in advancing the principles of the circular economy (CE) by leveraging the natural capabilities of fungi to recycle, regenerate, and optimize resources.

2.1 *Waste Reduction and Upcycling Along with the Role of Fungi in Breaking Down Waste Materials*

Fungi play a crucial role in waste reduction and upcycling. Their enzymatic capabilities allow them to decompose complex waste substances efficiently, making them an essential component of sustainable waste management strategies. Below is a broad exploration of how fungi contribute to waste reduction across various waste streams (Fig. 2). Their ability to break down lignocellulosic materials, proteins, and carbohydrates makes them highly effective in reducing different types of waste.

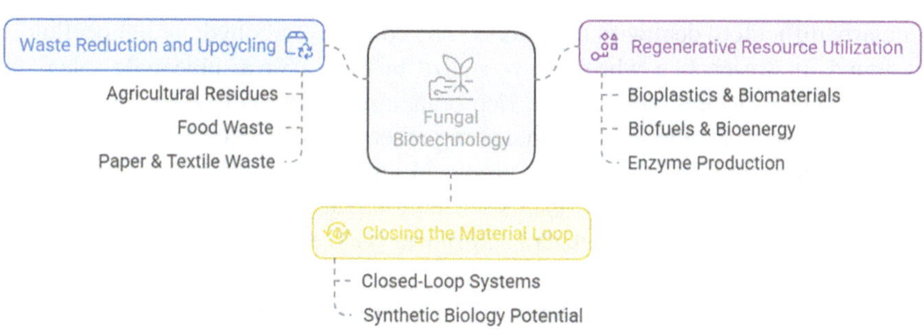

Fig. 1 Integration of fungal biotechnology in the circular economy framework

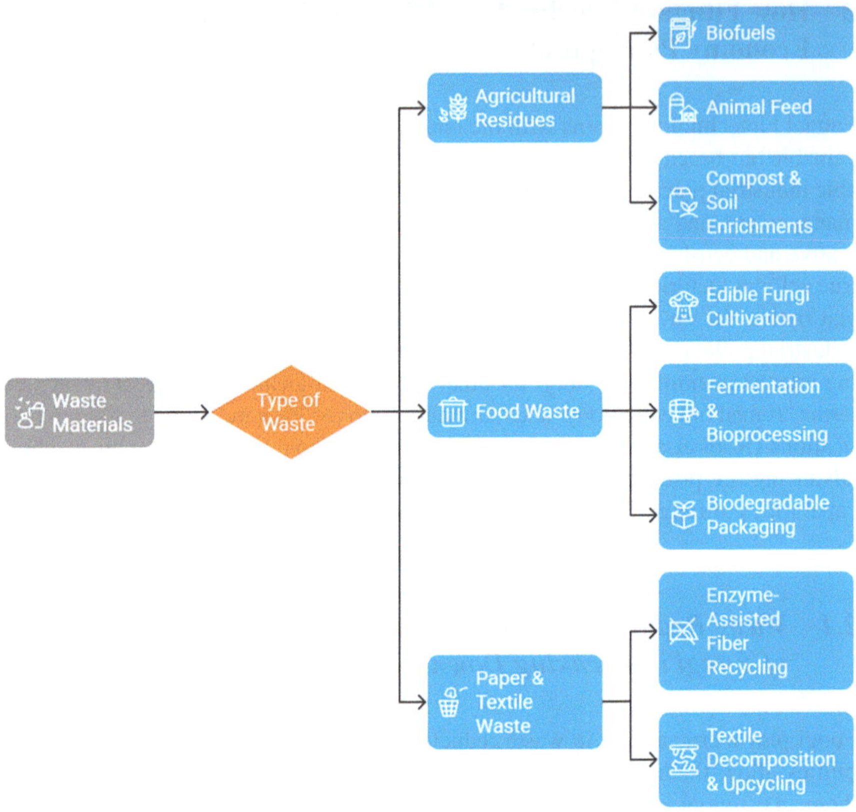

Fig. 2 Fungal applications in waste valorization and circular resource flows

2.1.1 Agricultural Residues

There are many agricultural wastes in our environment every day, and they are particularly difficult to deal with these days. The complex carbohydrate lignocellulose is found in wastes like wheat straw, maize husks, cassava, pineapple, plantain, banana, and yam peels, rice husks, rice bran, corn husks, corn cobs, palm kernel cake, soybean meal, and wheat bran, among others. An estimated 140 billion metric tons of lignocellulose biomass are produced annually by agriculture worldwide. Burning these trashes typically results in environmental degradation. Certain fungi, particularly species from the genera *Alternaria* (Gao et al., 2021), *Phanerochaete*, *Trametes*, *Pleurotus*, *Fusarium*, *Aspergillus*, *Amycolatopsis* (Tan et al., 2021), and *Ascocoryne* produce enzymes like peroxidase, laccase, protease, lipase, lignin peroxidase, cutinase, and cellulase that break down these materials into simpler compounds (Table 1). These broken-down residues can be further processed into different forms.

Table 1 Role of fungi in breaking down agricultural waste

Waste type	Fungi involved	Type of residue	Enzymes produced	End products reused as/in	Reference
Used for biofuel production	*Irpex lacteus*	Corn stover, barley straw, corn cob, wheat straw	Laccase, manganese peroxidase, Endo and exoglucanase, β-glucosidase, Cellulase, Xylanase	Bioethanol	Metreveli et al. (2022), Li et al. (2023), and Cao et al. (2024)
	Aspergillus candidus, A. niger	Banana peel, sugarcane bagasse, soyabean hulls	Endo and exoglucanase, β-glucosidase, Xylanase, Lipase Arabinofuranosidase, Pectinase, Galactosidase	Biodiesel	Rosas-vega et al. (2025)
	Neurospora sp.	Corn stover, wheat and rice straw, cotton stalks, fruit peel, sawdust	Endoglucanases, Cellobiohydrolases, β-glucosidase, Xylanases, β-xylosidase, Arabinofuranosidase, Polygalacturonase	Bioethanol	Miteva-Staleva et al. (2022)
	Trichoderma viride, T. reesei	Soyabean hulls	Xylanase, Endoglucanase exoglucanase	Biodiesel	Markt et al. (2024)
	Myceliophthora thermophila	Wheat bran	Xylanase, Glucosidase	Bioethanol	da Rosa-Garzon et al., (2022)
	Streptomyces sp.	Rice bran and straw	Endo and exoglucanase, Cellobiase, Xylanase	Bioethanol	Wang et al. (2024)
	Trametes versicolor Pleurotus eryngii	Rice husk, corn stalk, Vegetable peel (potato)	Laccase, Arabinofuranosidases, Xylanase, β-Glucosidase, β-Xylosidase, Manganese peroxidase	Bioethanol	Tišma et al. (2021)
	Mortierella isabellina	Wheat straw	Laccase, Peroxidase, Xylanase, Endo and exoglucanase, Mannanase,β-Glucosidase	Biodiesel	Sallet et al. (2025)
	Rhodosporidiumtoruloides	Sugarcane molasses and bagasse, wheat straw, orange peel	ATP-citrate lyase, Acetyl-CoA carboxylase, Fatty acid synthase, Malic enzyme, Hexokinase		Ngamsirisomsakul et al. (2021) and Karim et al. (2021)
	Yarrowia lipolytica,	Fruit and vegetable waste	Endo and exoglucanase, Lignin peroxidase, Versatile peroxidase	Biodiesel	Pereira et al. (2022)

(continued)

Table 1 (continued)

Waste type	Fungi involved	Type of residue	Enzymes produced	End products reused as/in	Reference
For improved animal feed					
	Trichoderma viridi, T. harzianum, and *T. reesei*	Sugarcane bagasse, rice straw, sugar beet pulp, tomato leaves, and sugar beet leaves	Pectin lyase, Polygalacturonase, Acid and alkaline protease, Exo and Endoglucanase, β-Glucosidase, Xylanase, Mannanase, Arabinofuranosidase	Cattle feed, poultry, and aquaculture	Gebeyew et al. (2023)
	Saccharomyces crevisiae	Irish potato peel, Yam peels, cassava peels	Invertase, Maltase, Alcohol dehydrogenase, Phytase	Cattle, poultry, pigs, fish, shrimp feed	Shi et al. (2025)
	Aspergillus niger	Rapeseed cake, Groundnut and soybean hulls, Rice straw, husks	Pectin lyase, Polygalacturonase, Acid protease, Xylanase, Mannanase, Cellobiohydrolase, Exo and Endoglucanase, Tannase	Monogastric, ruminant animals feed	Aydın Altop et al. (2024)
	R. oligosporus, R. oryzae	Brewery-spent grain, rice bran, Soybean hulls, oilseed cakes	α-Amylase, Glucoamylase, Phytase, Mannanase, Exo and Endoglucanase, Lipase	Poultry, cattle, and small ruminants feed	Ndego et al. (2022)
	Pleurotus sajor-caju	Oil palm waste, Rice bran	Laccase, Lignin and Manganese peroxidase, Endo and exoglucanase,	Cattle and ruminants feed Poultry, pigs, fish feed	Bansal and Singh (2021)
For compost and soil enrichment	*Trichoderma harzianum, Pleurotus ostreatus, Polyporus ostriformis, Phanerochaete chrysosporium*	Paddy straw, leaves, trash	Polygalacturonase, Pectin lyase, Xylanase, Mannanase, Laccase,Endoglucanase, Exoglucanase, β-Glucosidase, Ureases, and Phytases	Agricultural composting units, Organic farms, Vermicomposting beds, Forest litter decomposition projects, Bioremediation of contaminated soils	Li et al. (2022), Sharma et al. (2023) and Saha et al. (2024)
	Streptomyces sp.	Paddy straw	α-Amylase, Glucoamylase, Lipase, Laccases, Peroxidases, Chitinases	Bioremediation, Spent mushroom substrate, Manure composting, Humus formation in degraded soils	Ning et al. (2024)
	Rhizoctonia solani and Fusarium oxysporum	Sugar beet and potato waste	Serine and metalloprotease, Polygalacturonase, Pectin lyase, Xylanase, Mannanase, Polygalacturonase, Pectin lyase	Sterilized substrate treatment, Thermophilic composting, Enzyme supplementation to soils	Russ et al. (2024)
	Mucor circinelloides, Fusarium verticillioides	Areca husk, coffee husk, maize waste, paddy straw	α-Amylase, Glucoamylase, Xylanase, Mannanase, Endo and exoglucanase, β-glucosidase,	Agro-waste composting, Soil amendment production, Biofertilizer consortia	de Melo Cavalcante et al. (2023) and Elsayed et al. (2023)
	Aspergillus sp., *Penicillium* sp.	Wheat straw, vegetable, and fruit wastes	Acidic, neutral, alkaline proteases, Polygalacturonase, Pectin lyase, α-Amylase, Glucoamylase	Soil inoculants for nutrient cycling, Biocontrol of plant pathogens, Phosphorus bioavailability improvement	Mironov et al. (2021)

2.1.1.1 Biofuels Production

Fungi are a profitable option that have drawn a lot of interest in the production of biofuels such as bioethanol, biodiesel, and biogas. By using fungal-assisted enzymatic pretreatment to saccharify lignocellulosic biomass (agricultural waste) into a variety of sugars, bioethanol is produced (Ramachandraiah et al., 2021). Meanwhile, a variety of processes, including fast pyrolysis, gasification, liquefaction, distillation, combustion, and thermochemical fermentation, are employed to create biofuels from agricultural wastes including maize stover, wheat straw, and rice straw, which are promising raw materials. Agricultural residues are a cheap and plentiful supply of lignocellulosic resources in the European Union that can be utilized to produce important products like bioethanol. It is based on the unique skills of white-rot fungi (WRF), which are members of the phylum basidiomycota. These fungi have evolved the ability to break down plant cell wall lignin using a set of extracellular ligninolytic enzymes.

2.1.1.2 Animal Feed

Agro-residues, which are utilized as a less expensive feed source for ruminants, contain three main polymers: cellulose, hemicellulose, and lignin. For animals, the degradation of fiber enhances digestibility. In solid-state fermentation, a specific set of agricultural residues was successfully biotransformed by an endophytic fungal strain derived from *Celastrus paniculatus*. Solid-state fermentation is a microbial fermentation process that reduces anti-nutritional factors that impede the digestibility and bioavailability of nutrients in agro-industrial wastes, converting them into a variety of valuable bioproducts with added value, such as protein-rich animal feed for the livestock industry. In Solid-state fermentation (SSF), filamentous fungi such as *Aspergillus niger*, various species of *Trichoderma, Rhizopus, Mucor, Penicillium, Fusarium, Candida*, and yeasts are used. They have demonstrated the biotechnological potential to valorize agro-industrial wastes like the peels of Irish potatoes, yams, cassavas, brewery-spent grain, rice bran, rapeseed cake, etc. (Yafetto et al., 2023). Consequently, their utilization enhances the nutrient composition and caliber of agro-industrial wastes that have been valorized for use as animal feed. It is less expensive, safer, and improves the general health and growth performance of animals. Research on treating rice straw and maize stalks with *Trichoderma viride* demonstrated that it improved their feeding values without having a negative impact on animal performance. There is compelling evidence that the fungus *Ceriporiopsis subvermispora* exhibits diverse lignocellulose breakdown patterns with varying substrate types.

2.1.1.3 Compost and Soil Enrichment

The compost produced by using microorganisms to bioconvert agroresidues has a number of advantages, including improved soil fertility and health, which can raise agricultural output, higher soil biodiversity, lower ecological risks, and a healthier environment. Composting is the best way to handle the massive amounts of agrowaste produced worldwide because of these benefits. Fungi are essential for the breakdown of soil residues, immobilization, nutrient retention, and stabilization of soil organic matter. Their development habits, including the external digestion of food resources by their digestive enzymes and secondary metabolites, have a significant impact on biogeochemical cycles. Additionally, they are crucial for the breakdown of residues and the stabilization of soil organic matter. Both lignin and cellulose can be broken down by white rot fungi. Nonetheless, *Trichoderma* and *Phanerochaete* are the lignocellulolytic fungi that have been investigated the most. A recent and extremely promising development in plant productivity, which could be referred to as "the Second Green Revolution," has been the potential use of cultivating soil fungal biodiversity to enhance soil quality and boost agricultural ecosystem productivity. Using such methods could provide a more advanced way to manipulate plant productivity as an alternative to the current overuse of fertilizers.

2.1.2 Food Waste

One of the biggest causes of waste in the world is food waste. Food waste accumulates globally as a result of a number of factors, including inadequate planning, excessive food preparation during weddings, parties, hotels, and restaurants, as well as noncompliance with safety regulations. Utilizing microbes to break down garbage can be a significant way to address the expanding issues with food waste (Suresh et al., 2021).

2.1.2.1 Edible Fungi Cultivation

Fungi offer a promising solution by converting food scraps into new, valuable products. For example, when it comes to edible fungi cultivation, they not only break down food waste but also transform it into a valuable substrate that supports their growth. The microbes found in these settings have a significant impact on the development and fructification of farmed mushrooms, and in certain situations, they are necessary. Some fungi, such as morels and truffles, associate with their host plants in an ectomycorrhizal manner. This is ultimately represented in the structure of the mushroom holobiont, which is defined as the fungal host plus related microorganisms. The characteristics of this holobiont can significantly affect the production of mushrooms.

2.1.2.2 Fermentation and Bioprocessing

Fungi play a crucial role in the breakdown of waste materials, making them valuable agents in fermentation and bioprocessing. Extracellular enzymatic activity shown by them convert bulky, insoluble biomass into simpler sugars and compounds that microbes can ferment. This enhances the efficiency of microbial fermentation in bioethanol or biogas production. Some fungi, such as *Aspergillus oryzae*, *Trichoderma* spp., *Aspergillus* spp., *Penicillium* spp. are widely used for the purpose. Many fungal species are used as mycoremediators, such as *Rhizopus arrhizus, Phanerochaete chrysosporium, Phanerochaete sordida, Trametes hirsuta, Trametes versicolor, Pleurotus ostreatus, Lentinus tigrinus,* and *Lentinus edodes* (Hussain et al., 2022). Fungi have garnered interest in contemporary food science and technology in addition to their traditional uses because of their adaptability and potential for new uses.

2.1.2.3 Biodegradable Packaging

Growing concern about plastic waste in the environment has accelerated the development of biodegradable materials for food packaging. Fungi are emerging as powerful tools in the development of biodegradable packaging by not only breaking down waste materials but also transforming them into sustainable, eco-friendly alternatives to plastics. Fungal mycelium can transform food waste into compostable packaging materials. Fungi, especially basidiomycetes develop mycelium—a root-like network—which can be grown into molds or forms using agricultural waste as a base. Popular fungi used for this process are *Ganoderma lucidum, Pleurotus ostreatus, Trametes versicolor,* and *Fomes fomentarius.* Applications for fungal functionalized nanostructures are promising at many food processing, packaging, and safety stages. A viable and efficient strategy to prevent microbial contamination is the incorporation of fungal-derived nanostructures into food packaging materials (Rami et al., 2024). Because of their tolerance to metals and ability to bioaccumulate them, fungi are being investigated for the biological synthesis of metallic nanomaterials. Fungi have the potential to generate metal nanoparticles in a variety of shapes, including meso- and nanostructures, by biomimetic mineralization. In this process, Ecovative Design (USA) uses mycelium to make packaging for electronics, furniture, wine bottles, etc. Similarly, Grown.bio (EU) produces custom biodegradable packaging using fungal technology.

2.1.3 Paper and Textile Waste

Paper and textile waste, primarily composed of cellulose and synthetic fibers, contribute significantly to environmental pollution. Textile waste is challenging to decompose and occupies substantial landfill space. On the other hand, the textile industry has one of the biggest effects on the environment worldwide, both in terms of pre- and post-consumption waste and the logistics that make up the supply chain. Because post-consumer textile waste is difficult to digest, too much of it can disseminate scents throughout the environment, attract pests, and cause infectious diseases. Paper waste alone constitutes a significant portion of landfill content. In the European Union, paper and cardboard accounted for 40.3% of packaging waste in 2021 (Eurostat, 2023). Their decomposition releases methane, contributing to climate change. Inks and other chemicals in paper products can leach into soil and water, posing risks to ecosystems and human health. Fungi play a crucial role in addressing the environmental challenges posed by textile and paper waste. Through their natural decomposition abilities and potential for upcycling, fungi offer sustainable solutions that align with circular economy principles.

2.1.3.1 Enzyme-Assisted Fiber Recycling

Cellulose majorly contributes to the textile waste. Degradation of highly crystalline structure of cellulose present in textile and paper wastes requires synergy of endoglucanases, exoglucanases, and β-glucosidases in a complete cellulase system. *Aspergillus niger* and *Trichoderma reesei*, two microorganisms that produce fungal cellulase on cotton-based textile waste, can be used as a substrate and source of carbon for the manufacture of cellulase. On semi-synthetic textiles, the white-rot fungus *Hypholoma fasciculare* has been demonstrated to thrive (Harper & Moody, 2023). One technique that contributes to good waste management and the appropriate use of natural resources is recycling used paper. Sixty percent of recycled paper is used by European nations (Mondal et al., 2022). Fungi can degrade natural fibers like cotton, linen, and hemp. Advanced biotechnological applications also explore fungal treatment for synthetic fibers to reduce microplastic pollution. By hydrolyzing the starch in coated waste papers, fungi make it easier to remove ink from the fiber surface. Different enzymes for the deinking of waste papers are endo 1, 4β-D-glucanase, amylase, lipase, and cutinase produced by fungi. Certain fungi produce cellulase and hemicellulase enzymes that break down old paper fibers, allowing them to be repurposed into new paper products. *Aspergillus niger* and *Phanerochaete chrysosporium* are used as a biopretreatment agent to improve degradation of lignocellulosic paper by Ofon et al. (2025). Three isolates among the collection of 20 fungal strains isolated from composting soil, three (*Aspergillus* spp. AMA, *A. terreus* AN1, and *Myceliopthora fergusii*, T4I) were identified showing significant deinking of composite paper waste.

2.1.3.2 Textile Decomposition and Upcycling

Traditional disposal methods of textile wastes, such as landfilling and incineration, exacerbate pollution and resource depletion. In this context, fungi have emerged as promising agents for sustainable textile waste management, offering capabilities for both decomposition and upcycling into valuable materials. Fletcher (2025) studied a sustainable method for upcycling such wastes using one of the edible mushrooms, *Pleurotus ostreatus*. Fungi, especially white-rot species like *Pleurotus ostreatus* and *Trametes versicolor*, possess enzymatic systems capable of breaking down complex polymers found in textiles. These fungi secrete ligninolytic enzymes that degrade cellulose, hemicellulose, and lignin, facilitating the decomposition of natural fibers. Recent studies have demonstrated the efficacy of *P. ostreatus* in colonizing various textile substrates, including cotton and polyester blends. The fungus not only decomposes the fibers but also transforms the waste into mycelium-based composites with potential applications in packaging and construction materials (Saini et al., 2024). Moreover, *T. versicolor* has shown potential in degrading synthetic fibers like polyethylene terephthalate (PET) when co-cultured with bacteria such as *Bacillus subtilis*. This synergistic approach enhances the breakdown of recalcitrant polymers, paving the way for efficient bioremediation strategies (Sangosanya et al., 2024). Innovative research has also explored the cultivation of edible fungi on textile waste, indicating the potential for producing food resources while managing waste (Hazelgrove & Moody, 2024).

2.2 *Regenerative Resource Utilization by Fungi*

Fungi have emerged as powerful agents in the sustainable transformation of waste materials into valuable resources. In this sequence comes the production of bioplastics and biomaterials which is the mycelium-based packaging. Mycelium, the vegetative part of fungi, has been successfully used to produce biodegradable packaging materials that serve as a sustainable alternative to petroleum-based Styrofoam. These mycelium-based composites are created by growing fungal mycelia on agricultural residues such as corn stalks, husks, or sawdust. The final product is lightweight, biodegradable, and can be molded into various shapes for packaging. This highlights that mycelium-based packaging is a biomaterial that offers a green, biodegradable alternative to Styrofoam, which is widely used but environmentally damaging. Similarly, white-rot and brown-rot species are capable of degrading lignocellulosic biomass, making them key players in biofuel production. They secrete a range of lignin-degrading enzymes that allow for more efficient conversion of plant residues into fermentable sugars, which are then used to produce ethanol, methane, or biodiesel. Fungi produce a wide range of extracellular enzymes such as cellulases, ligninases, xylanases, and proteases, which are pivotal in breaking down complex organic matter. These enzymes facilitate the conversion of waste into simpler molecules that can be used in agriculture (e.g., composting), bioremediation, and even food industries. Thus, fungi serve as foundational components in regenerative systems, leveraging waste as a resource to produce environmentally friendly materials and energy (Fig. 3). Their integration into industrial processes offers a sustainable pathway to waste valorization and environmental conservation.

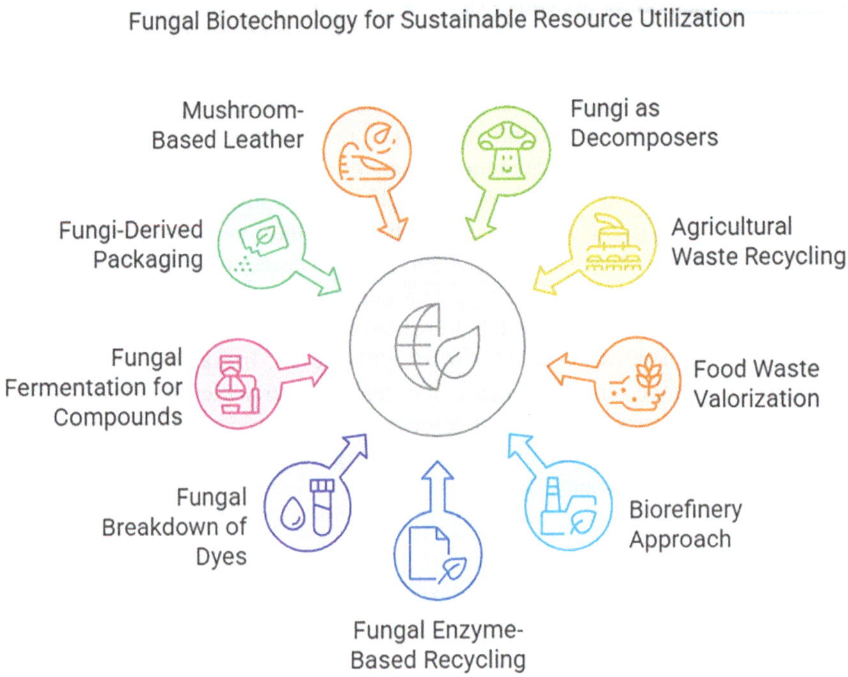

Fig. 3 Fungal innovations in sustainable biomass conversion and resource utilization

2.3 *Closing the Material Loop with Fungal-Based Circular Systems*

In the face of mounting environmental challenges—ranging from excessive waste generation to resource depletion, the key goal of the circular economy is to close the material loop, meaning that waste products are continuously regenerated into valuable resources, minimizing the need for virgin materials and reducing environmental impact. Among the most promising tools in this regenerative approach are fungi. Fungal-based systems fit naturally into the biological cycle of the circular economy, where materials can be safely returned to the environment. Through biotransformation, fermentation, and composting, fungi enable the valorization of organic and even semi-organic waste, offering an environmentally friendly and low-energy solution to one of humanity's most pressing problems: waste management. Recent innovations and real-world applications of fungal biotechnology are not only theoretically sound but also economically viable—transforming industries such as construction, packaging, textiles, and agriculture.

Example 1: Mycocycle It is a biotechnology startup, utilizes fungi to convert construction waste, such as carpet fibers and tires—into reusable materials like "MycoFIBER", "MycoFILL", and "MycoFOAM". This process not only diverts waste from landfills but also produces low-carbon raw materials for the built environment (Mycocycle, n.d.; BioSourced, 2022). This white paper provides a comprehensive overview of how fungal biotechnology, particularly the use of mycelium, can contribute to a circular economy by transforming waste materials into valuable products.

Example 2: Ecovative Design for Sustainable Packaging Solutions Such designs employ mycelium to create biodegradable packaging materials. Their process involves growing mycelium on agricultural waste, forming products like protective packaging and insulation panels. This approach offers a sustainable alternative to polystyrene, with companies like IKEA adopting it for eco-friendly packaging. Some of the studies evaluated the use of fungal mycelium combined with various cotton byproducts to produce biodegradable molded packaging materials. It highlights the potential of mycelium-based composites as sustainable alternatives to traditional polystyrene packaging.

Example 3: Valorization of Textile Waste Through Fungal Fermentation Research has demonstrated that fungal solid-state fermentation can convert textile waste into valuable products. For instance, fungi can break down cotton-polyester blends, enabling the recovery of cotton fibers and the production of glucose syrup, contributing to a circular textile economy.

Example 4: Spent Mushroom Substrate (SMS) in Circular Applications After mushroom cultivation, the remaining substrate, known as spent mushroom substrate (SMS), can be repurposed in various ways. Applications include use as compost, animal feed, biofuels, and even as a base for new packaging and construction materials, thereby closing the loop in mushroom production.

Example 5: Mycelium-Based Composites for Waste Management Mycelium-based composites (MBCs) offer a sustainable solution for waste management. By combining mycelium with agricultural residues, these composites can be used in construction, packaging, and textiles, reducing reliance on synthetic materials and promoting biodegradability (Barta et al., 2024).

Synthetic biology holds immense promise in revolutionizing the field of sustainable bioconversion, especially through the genetic optimization of fungal strains. Fungi are nature's powerful decomposers, by leveraging synthetic biology; researchers can enhance these natural capabilities to create efficient, eco-friendly bioprocesses. Key advancements include Metabolic Pathway Engineering where synthetic biology enables the modification or introduction of specific metabolic pathways in fungi to improve the yield and rate of desired bioconversion products like biofuels, bioplastics, and organic acids. Secondly, tolerance and robustness of engineered fungal strains can be made more resilient to environmental stressors such as pH, temperature fluctuations, or toxic compounds commonly found in industrial waste (Kun et al., 2021). Substrate specificity gained through gene editing tools like CRISPR/Cas9, fungi can be tailored to degrade specific waste types ranging from agricultural residues to synthetic polymers thereby improving selectivity and process efficiency (Ding et al., 2021). Synthetic circuits can be embedded into fungal genomes to tightly control their growth, product formation, or environmental containment, enhancing the safety of industrial applications.

In essence, synthetic biology serves as a precision toolkit to unlock the full potential of fungi as biofactories. Optimized fungal strains can not only accelerate the transition to a circular bioeconomy but also contribute to cleaner industrial processes with reduced dependency on fossil fuels and toxic chemicals.

3 Success Stories Where Fungi Contribute to Closing Material Loops

In recent years, both academic research and industrial initiatives have increasingly harnessed fungal biotechnology to develop scalable, eco-friendly solutions that contribute to waste reduction, resource recovery, and environmental restoration (Fig. 4). The following section highlights a series of real-world success stories where fungi have been effectively deployed to close material loops and drive sustainable innovation across multiple sectors (Fig. 5).

3.1 *Mycoremediation of Pollutants and Wastewater*

Mycoremediation is a form of bioremediation that uses fungi to degrade or remove environmental pollutants, including hydrocarbons, heavy metals, pesticides, dyes, and various industrial wastes. This eco-friendly technique leverages the natural

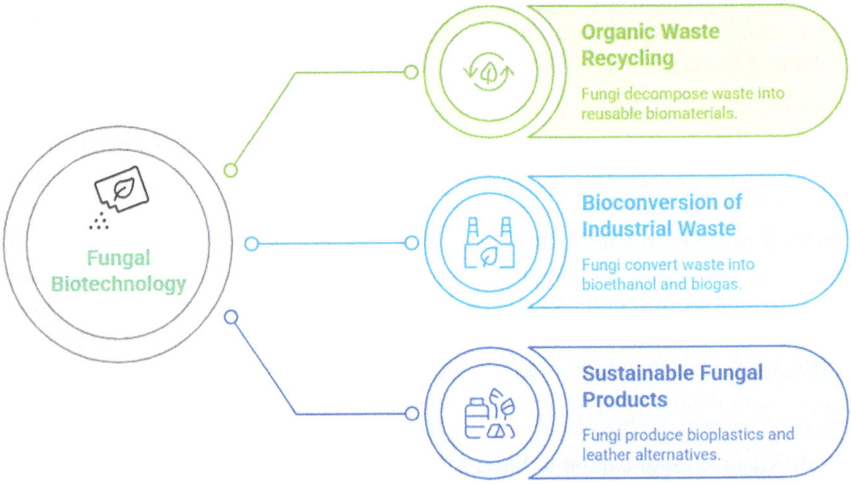

Fig. 4 Regenerative fungal systems for circular resource use

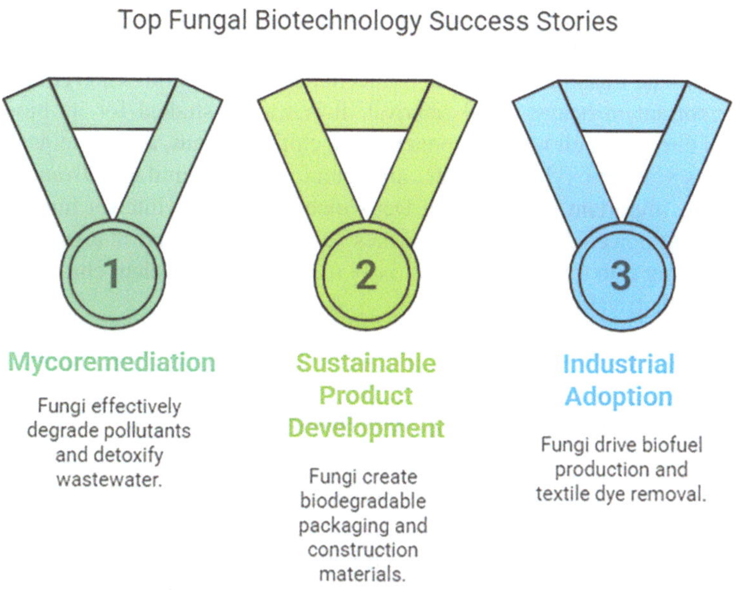

Fig. 5 Landmark applications of fungal biotechnology in industry and sustainability

enzymatic systems of fungi, especially white-rot fungi, to break down complex and toxic compounds into simpler, less harmful substances.

Example 1: Oil Spill Cleanup with *Pleurotus* species *Pleurotus* species, commonly known as oyster mushrooms, have shown significant potential in degrading hydrocarbons, such as those found in crude oil and petroleum products, making them promising agents in mycoremediation. *Pleurotus* spp. secrete a variety of extracellular enzymes such as laccases, manganese peroxidases (MnPs), and lignin peroxidases (LiPs), which are capable of breaking down complex organic pollutants including Polycyclic Aromatic Hydrocarbons (PAHs) and Alkanes and aromatic compounds in petroleum. *Pleurotus* belongs to white-rot fungi, which are well-known for their ability to degrade lignin in wood. This lignin-degrading capacity is closely linked to their ability to decompose hydrocarbons, since lignin and hydrocarbons share similar structural complexity and chemical stability. This group of fungi also contributes to environmental restoration as they can grow on a variety of substrates including straw, sawdust, and agricultural waste, allowing them to colonize and remediate oil-polluted soils effectively. During colonization, fungal mycelia absorb and transform oil residues, and the improved soil conditions can support the re-growth of plants and other organisms. It plays a powerful role in closing material loops by transforming hazardous oil-based pollutants into non-toxic substances, thereby restoring environmental health.

Example 2: Heavy Metal Removal in Industrial Wastewater Heavy metals such as lead (Pb), cadmium (Cd), chromium (Cr), and mercury (Hg) are persistent pollutants found in industrial effluents. Fungi, particularly *Aspergillus* and *Trametes* spp., offer a sustainable and effective approach to remove these metals through biosorption, enzymatic transformation, and intracellular accumulation. *Aspergillus niger* is very important in heavy metal removal. It is widely studied for its biosorption potential due to its cell wall components like chitin, glucans, and proteins that bind metal ions via carboxyl, phosphate, and amino groups. Similarly, *Trametes versicolor* plays important role in Metal Detoxification. It is a white-rot fungus known for secreting powerful oxidative enzymes such as laccase and manganese peroxidase, enabling it to transform and detoxify metals like hexavalent chromium (Cr^{6+}) into less toxic trivalent forms (Cr^{3+}).

3.2 Fungal Biofactories for Sustainable Product Development

Mycelium of fungi has emerged as a biofabrication material that can replace synthetic plastics, foams, and even wood or leather. Due to its biodegradable, lightweight, fire-resistant, and insulating properties, mycelium-based composites are being developed for use in packaging, interior design, construction, and fashion industries. These materials not only reduce plastic waste and carbon emissions but also utilize agricultural residues as growth substrates, contributing to a circular bioeconomy. Leading companies like Ecovative, Mogu, and MycoWorks are pioneering this space with real-world applications ranging from eco-packaging to interior design panels and even leather alternatives.

Example 1: Mycelium-Based Packaging and Construction Materials Ecovative Design (USA) has developed mushroom packaging, a compostable alternative to polystyrene foam, using *Ganoderma* and *Pleurotus* species which involve growing mycelium on hemp or agricultural waste in custom molds to form protective packaging. The physical and mechanical properties of mycelium-based materials were explored, showing that *Ganoderma lucidum* grown on hemp substrate produced structurally robust and fully compostable packaging alternatives. Similarly, Mogu (Italy) specializes in acoustic panels, tiles, and flooring from *Trametes* and *Pleurotus* species. These are made by binding fungal mycelium with low-value lignocellulosic residues and growing them into finished products under sterile conditions. It was demonstrated that mycelium grown on lignocellulosic waste can be processed into durable and aesthetically pleasing construction materials suitable for commercial use. MycoWorks has developed "Reishi", a high-end, leather-like biomaterial made from fungal mycelium, offering an animal-free, plastic-free alternative to traditional leather. Mycelium sheets derived from *Ganoderma lucidum* are studied and characterized for flexibility, strength, and biodegradability, validating their use in fashion and upholstery. Thus, mycelium-based materials are a cutting-edge example of fungal biofactories closing the material loop by transforming low-value agricultural waste into high-value, biodegradable products showing how fungal biotechnology can lead to scalable solutions for sustainable product development.

Example 2: Food & Feed Industry Innovations The global demand for sustainable protein sources is rising in response to the environmental and ethical concerns associated with animal agriculture. Fungal-based proteins, or mycoproteins, offer a promising alternative due to their low environmental footprint, high protein content, and fibrous meat-like texture. Mycoproteins are produced through controlled fermentation processes, using carbohydrate substrates like glucose or wheat-derived starch. These proteins are rich in essential amino acids, dietary fiber, and are cholesterol-free, making them not only sustainable but also nutritionally beneficial. One of the most successful commercial applications is Quorn, a meat substitute produced by Marlow Foods in the UK derived from the filamentous fungus *Fusarium venenatum* and has been on the market since the 1980s. It uses a continuous fermentation process to grow the fungus, which is then heat-treated to reduce RNA content, harvested, and shaped into various meat-like products. It promotes muscle protein synthesis and shows benefits in appetite regulation and cholesterol reduction. The mycoprotein consumption lowers LDL cholesterol and may improve blood glucose control. Derbyshire and Delange (2021) highlighted the environmental advantages of mycoprotein production—requiring 90% less land and water and generating lower GHG emissions than beef production. Its low ecological impact and positive health outcomes make mycoprotein a key player in transforming the future of food and feed industries.

3.3 Large-Scale Industrial Adoption of Fungal Biotechnology

As global energy demands rise, sustainable biofuels like bioethanol and biogas have emerged as viable alternatives to fossil fuels. The use of fungi particularly filamentous fungi like *Trichoderma reesei* and *Aspergillus* spp. in industrial-scale biofuel production enables the conversion of biomass into fermentable sugars, supporting a circular bioeconomy and closing material loops.

Example 1: Fungal Enzyme-Driven Biofuel Production in Brazil and India Brazil, the world's second-largest producer of bioethanol, has integrated fungal enzymes into its second-generation (2G) ethanol plants. *Trichoderma reesei* is utilized for its production of cellulases and hemicellulases, critical for the hydrolysis of pretreated sugarcane bagasse into fermentable sugars. The potential of fungal enzymatic cocktails was demonstrated to enhance saccharification efficiency of sugarcane residues, improving ethanol yields. Companies like GranBio and Raízen have adopted fungal enzyme-based hydrolysis in their commercial 2G ethanol plants.

India's push toward biogas production from agricultural residues also heavily relies on fungal enzymes. *Aspergillus niger* and *Penicillium* spp. are employed for pretreatment and saccharification, improving methane yields by making cellulose and hemicellulose more accessible to methanogenic bacteria. Fungal pretreatments improve biogas production from paddy straw by enhancing lignin degradation and cellulose accessibility. Initiatives like the SATAT scheme (Sustainable Alternative Towards Affordable Transportation) promote fungal biotechnology adoption in decentralized biogas plants across rural India. The deployment of fungal biotechnology for biofuel production in countries like Brazil and India exemplifies how industrial-scale fungal enzyme systems can close material loops. By turning agricultural waste into clean energy, fungi help drive the transition toward a sustainable, low-carbon future.

Example 2: Fungal-Based Textile Dye Removal The textile industry is one of the largest contributors to water pollution globally, discharging massive amounts of synthetic dyes into aquatic systems. Many synthetic dyes are toxic, carcinogenic, and resistant to degradation by conventional treatment methods. Fungal enzymes or myco-enzymes are used to break down and remove synthetic dyes from polluted textile wastewater. Fungi, particularly white-rot fungi, produce powerful extracellular enzymes (e.g., laccases, manganese peroxidases, lignin peroxidases) capable of breaking down these complex dye molecules into non-toxic forms. They oxidize, depolymerize, or mineralize complex dye molecules, turning them into non-toxic, simpler compounds like carbon dioxide, water, and harmless salts. This myco-enzymatic degradation presents a cost-effective, eco-friendly solution for industrial-scale wastewater treatment. Fungal-based enzyme systems for dye degradation offer a sustainable and scalable solution for textile wastewater management. Mycoremediation technology not only treats harmful effluents but also aligns with industrial goals for green manufacturing and closed-loop systems, marking a critical advancement in environmental biotechnology.

4 Challenges and Future Directions

Fungal biotechnology holds immense promise for advancing the circular economy (CE) by offering sustainable solutions for waste management, biofabrication, renewable materials, and clean energy production. However, despite its success in pilot studies and niche markets, large-scale integration into mainstream industries faces significant challenges that need to be addressed to fully unlock its potential.

4.1 Technical and Economic Challenges

One major technical hurdle is the scalability of fungal bioprocessing. Fungi grow under specific environmental conditions—temperature, humidity, and aeration which can be difficult and costly to maintain consistently on an industrial scale. Unlike traditional chemical synthesis or bacterial fermentation, fungal cultivation often requires longer processing times and larger bioreactor designs, limiting its competitiveness. Another important concern is the cost-effectiveness of fungal-based materials. Although products like mycelium packaging, bio-leathers, and fungal bioplastics are biodegradable and eco-friendly, their production costs remain higher than those of conventional plastics and synthetic materials. There are challenges with cost competitiveness as factors such as substrate preparation, slow growth rates, and downstream processing add to the expense, making it harder for fungal materials to replace petroleum-based products economically. Additionally, regulatory hurdles pose a significant barrier to fungal bio-manufacturing. Benz and Schipper (2021) cover regulation-related issues. Approval processes for novel fungal strains, genetically modified organisms (GMOs), and new biomaterials can be lengthy, expensive, and vary across regions. There is a pressing need for clear, harmonized regulations to ensure safety while encouraging innovation in the fungal biotechnology sector.

4.2 Innovations Driving the Future of Fungal Biotechnology

Despite these challenges, groundbreaking innovations are reshaping the future of fungal applications. CRISPR and genetic engineering are being employed to optimize fungal strains for enhanced enzyme production, faster growth rates, improved substrate utilization, and stress tolerance. By precisely editing fungal genomes, researchers can design superior fungal "biofactories" tailored for specific industrial needs. Moreover, artificial intelligence (AI) is playing a transformative role in fungal metabolic pathway engineering. Machine learning algorithms can predict optimal genetic modifications, simulate metabolic fluxes, and accelerate strain development processes, greatly reducing the trial-and-error phase traditionally involved in fungal research. To support the integration of fungal biotechnology into the circular economy, policy recommendations are essential. Governments and international organizations should incentivize research and development, support startup ecosystems, standardize regulatory frameworks, and promote public-private partnerships. Embedding fungal technologies into CE frameworks will require collaborative efforts across scientific, industrial, and regulatory landscapes to foster a bio-based, low-carbon economy.

5 Conclusions and Prospects

Fungal biotechnology stands as a transformative force in advancing the principles of the circular economy (CE), offering sustainable solutions for waste management, material regeneration, bioproduct development, and environmental remediation. Through diverse applications—from organic waste recycling to bioplastics and biofuels—fungi demonstrate remarkable versatility in closing material loops and creating high-value products from low-value residues. The success stories outlined, such as mycoremediation of pollutants, mycelium-based packaging, fungal protein for sustainable food systems, and bioethanol production, highlight not only the technical feasibility but also the enormous commercial potential of fungal technologies.

These innovations align closely with CE objectives: reducing resource extraction, reusing waste streams, recycling materials into new applications, and regenerating ecosystems.

To fully unlock the potential of fungal biotechnology in advancing a circular economy, future research must address several strategic areas. First, it is essential to develop resilient fungal strains that can withstand industrial-scale stresses such as high temperatures, variable pH, and contaminant exposure, ensuring stable and efficient bioprocessing under real-world conditions. Second, there is a growing need to create modular and decentralized fungal bioprocessing units that can enable local communities, industries, and farms to valorize their waste streams on-site, reducing logistical costs and carbon footprints associated with centralized waste management systems. Third, researchers should investigate novel fungal species sourced from extreme environments such as deep-sea habitats, arid deserts, or polluted ecosystems, as these fungi may possess unique metabolic pathways with immense potential for innovative biotechnological applications. Fourth, expanding interdisciplinary collaborations will be critical. Integrating expertise from mycology, biotechnology, environmental science, materials science, engineering, and policy studies can accelerate technological breakthroughs and ensure that fungal solutions are both scientifically robust and socially acceptable. Finally, it is imperative to formulate supportive policies and incentives that actively embed fungal biotechnology into national and global circular economy (CE) strategies. Clear regulatory frameworks, funding mechanisms, and public-private partnerships will catalyze the scaling of fungal innovations, bridging the gap between laboratory research and industrial reality.

By focusing on these future directions, fungal biotechnology can evolve from a promising alternative to a mainstream pillar of sustainable development and circular economy practices worldwide. Investing in this field today will not only transform industries but also contribute meaningfully to global efforts toward environmental restoration, climate change mitigation, and responsible resource stewardship.

Acknowledgements The author would like to thank Dr. Vijita Diwan from Shri Shankaracharya Professional University, Bhilai, Chhattisgarh, for insightful guidance during the preparation of this chapter. I also extend my gratitude to the Dept. of Life-Science, S.S.P.U., Bhilai, for providing the necessary resources and infrastructure. I am also thankful to my colleagues at Shri Shankaracharya Professional University, for their valuable feedback and suggestions. Finally, I express my sincere appreciation to my family for their continuous encouragement and support throughout this research endeavor.

References

Aydın Altop, E., Güngör, E., & Erener, G. (2024). Solid-state fermentation of wheat bran by Aspergillus niger strains: Effect on nutritional composition and in vitro digestibility. *Anadolu Journal of Agricultural Sciences, 39*(2), 239–246. https://doi.org/10.7161/omuanajas.1314081

Bansal, V., & Singh, S. (2021). The role of fungal biomass in animal feed and its economic value: A case study on Pleurotus sajor-caju. *International Journal of Animal Sciences, 16*(2), 45. https://doi.org/10.3923/ijasci.2021.45.52

Barta, D. G., Simion, I., Tiuc, A. E., & Vasile, O. (2024). Mycelium-based composites as a sustainable solution for waste management and circular economy. *Materials, 17*(2), 404. https://doi.org/10.3390/ma17020404

Beltrán-Flores, E., Sarrà, M., & Blánquez, P. (2021). Pesticide bioremediation by Trametes versicolor: Application in a fixed-bed reactor, sorption contribution and bioregeneration. *Science of the Total Environment, 794*, 148386. https://doi.org/10.1016/j.scitotenv.2021.148386

Benz, J. P., & Schipper, K. (2021). Fungal biotechnology: Future perspectives and regulatory issues. *Applied Microbiology and Biotechnology, 105*(13), 5259–5267. https://doi.org/10.1007/s00253-021-11387-1

BioSourced. (2022, September 21). How Mycocycle turns waste into value using mycelium. *Chemistry for Sustainability*.

Cao, X., Cai, R., Zuo, S., Niu, D., Yang, F., & Xu, C. (2024). Enhanced lignin degradation by Irpex lacteus through expanded sterilization further improved the fermentation quality and microbial community during the silage preservation process. *Bioresources and Bioprocessing, 11*(1), 14. https://doi.org/10.1186/s40643-024-00730-2

Chang, L., Chen, H., Tang, X., Zhao, J., Zhang, H., Chen, Y. Q., & Chen, W. (2021). Advances in improving the biotechnological application of oleaginous fungus Mortierella alpina. *Applied Microbiology and Biotechnology, 105*(16), 6275–6289. https://doi.org/10.1007/s00253-021-11440-2

da Rosa-Garzon, N. G., Laure, H. J., Rosa, J. C., & Cabral, H. (2022). Valorization of agricultural residues using Myceliophthora thermophila as a platform for production of lignocellulolytic enzymes for cellulose saccharification. *Biomass and Bioenergy, 161*, 106452. https://doi.org/10.1016/j.biombioe.2022.106452

Daba, G. M., Mostafa, F. A., & Elkhateeb, W. A. (2021). The ancient koji mold (*Aspergillus oryzae*) as a modern biotechnological tool. *Bioresources and Bioprocessing, 8*(1), 52. https://doi.org/10.1186/s40643-021-00408-z

de Melo Cavalcante, R. R., da Silva Rabelo, R., Ribeiro De Oliveira, J. E., de Sousa Lima, D., & Teodoro, M. L. T. (2023). Eco-friendly bioemulsifier production by Mucor circinelloides UCP0001 isolated from mangrove sediments using renewable substrates. *Biomolecules, 10*(3), 365. https://doi.org/10.3390/biomolecules10030365

Derbyshire, E., & Delange, J. (2021). Replacing meat with mycoprotein: A healthy and sustainable substitute rich in protein and fiber. *Food, 10*(2), 200. https://doi.org/10.3390/foods10020200

Di Piazza, S., Benvenuti, M., Damonte, G., Cecchi, G., Mariotti, M. G., & Zotti, M. (2021). Fungi and circular economy: Pleurotus ostreatus grown on a substrate with agricultural waste of lavender, and its promising biochemical profile. *Recycling, 6*(2), 40.

Ding, M. Z., Zhang, X., Qi, F. B., Zheng, P., & Sun, J. (2021). Synthetic biology: A powerful tool for the improved production of fungal natural products. *Frontiers in Microbiology, 12*, 708929. https://doi.org/10.3389/fmicb.2021.708929

El-Mahdy, O. M., Mohamed, H. I., & Mogazy, A. M. (2021). Biosorption effect of *Aspergillus niger* and *Penicillium chrysosporium* for Cd-and Pb-contaminated soil and their physiological effects on Vicia faba L. *Environmental Science and Pollution Research, 28*(47), 67608–67631. https://doi.org/10.1007/s11356-021-15382-4

Elsayed, Z. M., Sehsah, M. S. A., Oueslati, S., Ibrahim, E., Hamden, S., Seddek, R., Abo-Elmagd, F., Alkhalifah, D. H. M., Sheteiwy, M. S., AbdElgawad, H., El-Saadony, M. T., & El-Tahan, S. M. (2023). The effect of using fresh vs. stored farmyard manure on Fusarium verticillioides infection, lodging, and borer incidence in maize (Zea mays). *Frontiers in Plant Science, 13*, 998440. https://doi.org/10.3389/fpls.2022.998440

Eurostat. (2023, October 19). *EU packaging waste generation with record increase*. Recycling Magazine.

Fletcher, I. (2025). Evaluating Pleurotus ostreatus growth for upcycling textile waste from the fashion industry. *Frontiers in Sustainability, 6*, 1522527. https://doi.org/10.3389/frsus.2025.1522527

Gao, R., Pan, H., & Lian, J. (2021). Recent advances in the discovery, characterization, and engineering of poly (ethylene terephthalate) (PET) hydrolases. *Enzyme and Microbial Technology, 150*, 109868. https://doi.org/10.1016/j.enzmictec.2021.109868

Gebeyew, K., Ayele, S., Mohammed, A., & Tefera, B. (2023). Intake, digestibility, growth performance and blood profile of rams fed sugarcane bagasse or rice husk treated with Trichoderma viride and effective microorganisms. *Journal of Animal Science and Technology, 65*(10). https://doi.org/10.5187/JAST.2023.e65i10.128

Harper, R., & Moody, S. C. (2023). Filamentous fungi are potential bioremediation agents of semi-synthetic textile waste. *Journal of Fungi, 9*(6), 661. https://doi.org/10.3390/jof9060661

Hazelgrove, L., & Moody, S. C. (2024). Successful cultivation of edible fungi on textile waste offers a new avenue for bioremediation and potential food production. *Scientific Reports, 14*(1), 11510. https://doi.org/10.1038/s41598-024-61680-5

Hussain, A., Rehman, F., Rafeeq, H., Waqas, M., Asghar, A., Afsheen, N., et al. (2022). In-situ, Ex-situ, and nano-remediation strategies to treat polluted soil, water, and air–A review. *Chemosphere, 289*, 133252. https://doi.org/10.1016/j.chemosphere.2021.133252

Karim, A., Islam, M. A., Khalid, Z. B., Yousuf, A., Khan, M. M. R., & Faizal, C. K. M. (2021). Microbial lipid accumulation through bioremediation of palm oil mill effluent using a yeast-bacteria co-culture. *Renewable Energy, 176*, 106–114. https://doi.org/10.1016/j.renene.2021.04.043

Kun, R. S., Wehrs, M., Bommareddy, R. R., Moon, T. S., & Gladden, J. M. (2021). Engineering fungal hosts for synthetic biology applications. *Fungal Biology and Biotechnology, 8*(1), 6. https://doi.org/10.1186/s40694-021-00110-2

Li, Y., Guo, Q., Wang, Q., Hu, B., Tian, S., Yang, Q., Cheng, Z., Chen, Q., & Zhang, G. (2022). Impacts of composting duration on physicochemical properties and microbial communities during short-term composting for oyster mushrooms. *Science of the Total Environment, 847*, 157673. https://doi.org/10.1016/j.scitotenv.2022.157673

Li, S., Chen, Y., Liang, X., Ni, Z., & Chen, H. (2023). Expanding the application of lignin peroxidases from Irpex lacteus in lignin degradation by exploring the effects of small molecules on enzyme activity. *E3S Web of Conferences, 393*, 02036. https://doi.org/10.1051/e3sconf/202339302036

Markt, R., Gehringer, D., Thalhammer, J., & Braun, R. (2024). Pre-treatment with Trichoderma viride: Towards a better understanding of its consequences for anaerobic digestion. *Environmental Microbiology Reports, 16*(1), e13572. https://doi.org/10.1111/1758-2229.13572

Massa, A., Baiget, M., Rothschild, L. J., Axpe, E., & Carrero-Carralero, C. (2024). New food ingredient via acid-tolerant Rhizopus oligosporus growth. *Applied Food Research, 4*(2), 100583. https://doi.org/10.1016/j.afres.2024.100583

Metreveli, E., Khardziani, T., Jokharidze, T., & Elisashvili, V. (2022). Production and properties of Irpex lacteus cellulase and xylanase. *Journal of Biological Studies, 5*(1), 239–239. https://doi.org/10.62400/jbs.v5il.6635

Mironov, V., Vanteeva, A., & Merkel, A. (2021). Microbiological activity during co-composting of food and agricultural waste for soil amendment. *Agronomy, 11*(5), 928.

Miteva-Staleva, J., Eneva, R., & Hubenov, V. (2022). Fungi and their potential for biofuels production. *Bulgarian Journal Of Agricultural Science, 28*(2), 265–270. https://doi.org/10.5281/zenodo.1234567

Mondal, S., Biswal, D., Pal, K., Rakshit, S., Halder, S. K., Mandavgane, S. A., et al. (2022). Biodeinking of waste papers using combinatorial fungal enzymes and subsequent production of butanol from effluent. *Bioresource Technology, 353*, 127078. https://doi.org/10.1016/j.biortech.2022.127078

Mycocycle. (n.d.). *Mycocycle—Waste to value using fungi.*

Ndego, A., Ezedom, T., Egbune, E. O., & Tonukari, N. J. (2022). Biochemical characterization of solid-state fermented maize cob (Zea mays) using Rhizopus oligosporus and its application in poultry feed production. *International Journal of Recycling of Organic Waste in Agriculture, 12*(2). https://doi.org/10.30486/ijrowa.2022.1939020.1323

Ngamsirisomsakul, M., Reungsang, A., & Kongkeitkajorn, M. B. (2021). Assessing oleaginous yeasts for their potentials on microbial lipid production from sugarcane bagasse and the effects of physical changes on lipid production. *Bioresource Technology Reports, 14*, 100650. https://doi.org/10.1016/j.biteb.2021.100650

Niego, A. G. T., Lambert, C., Mortimer, P., Thongklang, N., Rapior, S., Grosse, M., Schery, H., Charria-Giron, E., Walker, A., Hyde, K. D., & Stadler, M. (2023). The contribution of fungi to the global economy. *Fungal Diversity, 121*(1), 95–137. https://doi.org/10.1007/s13225-023-00529-5

Ning, Y., Liu, Y., Guo, H., Wang, X., Yang, Y., Zhou, D. (2024). Enhancing C and N turnover, microbial community succession, and compost maturation using Streptomyces–Bacillus inoculants during cow manure and corn straw co-composting. Environmental Science and Pollution Research, 31(24), 66346–66358. doi:https://doi.org/10.1007/s11356-023-27092-0.

Ofon, U. A., Ndubuisi-Nnaji, U. U., Adegoke, A. A., Offiong, N. A. O., Ewenike, O. P., & Shaibu, S. E. (2025). Bioenergy potential of paper waste: Fungal pretreatment and kinetics modelling. *Next Sustainability, 5*, 100086. https://doi.org/10.1016/j.nsus.2024.100086

Paniagua-López, M., Silva-Castro, G. A., Romero-Freire, A., Martín-Peinado, F. J., Sierra-Aragón, M., & García-Romera, I. (2024). Integrating waste valorization and symbiotic microorganisms for sustainable bioremediation of metal (loid)-polluted soils. *Science of the Total Environment, 945*, 174030. https://doi.org/10.1016/j.scitotenv.2024.174030

Pereira, A. S., Belo, I., & Lopes, M. (2022). Enhancing microbial lipids synthesis for biodiesel production by Yarrowia lipolytica W29 from volatile fatty acids: Two-stage batch strategies. *Applied Sciences, 12*(17), 8614. https://doi.org/10.3390/app12178614

Ramachandraiah, A., Kaushik, J., Gowda, L., Chigadannavar, P. S., Krishnappa, R., Venkatesh, S., Sivadas, S., Judith, S., Ramesh, U. M., Quadri, Z., Dey, P., Chinthala, M., Gola, D., Chauhan, N., & Bharti, R. K. (2021). Prospect of biofuel production by fungus. In *Current developments in biotechnology and bioengineering. Fungal biotechnology* (pp. 285–313). Elsevier. https://doi.org/10.1016/B978-0-12-821394-0.00011-1

Rami, M. R., Forouzandehdel, S., & Aalizadeh, F. (2024). Enhancing biodegradable smart food packaging: Fungal-synthesized nanoparticles for stabilizing biopolymers. *Heliyon, 10*(18), e37692. https://doi.org/10.1016/j.heliyon.2024.e37692

Rosas-Vega, F. E., Pozzan, R., Martínez-Burgos, W. J., Letti, L. A. J., de Mattos, P. B. G., Ramos-Neyra, L. C., et al. (2025). Enzymes produced by the genus aspergillus integrated into the biofuels industry using sustainable raw materials. *Fermentation, 11*(2), 62. https://doi.org/10.3390/fermentation11020062

Roth, M. G., Westrick, N. M., & Baldwin, T. T. (2023). Fungal biotechnology: *From yesterday to tomorrow. Frontiers in Fungal Biology, 4,* 1135263. https://doi.org/10.3389/ffunb.2023.1135263

Russ, L., Andreo-Jimenez, B., Nijhuis, E., & Postma, J. (2024). Rhizoctonia solani disease suppression: Addition of keratin-rich soil amendment leads to functional shifts in soil microbial communities. *FEMS Microbiology Ecology, 100*(4), fiae024. https://doi.org/10.1093/femsec/fiae024

Saha, K. C., Uddin, M. K., Shaha, P. K., Chowdhury, M. A. H., Hassan, L., & Saha, B. K. (2024). Application of Trichoderma harzianum enhances salt tolerance and yield of Indian mustard through increased antioxidant enzyme activity. *Heliyon, 11*(1), e41114. https://doi.org/10.1016/j.heliyon.2024.e41114

Saini, R., Kaur, G., & Brar, S. K. (2024). Textile residue-based mycelium biocomposites from Pleurotus ostreatus. *Mycology, 15*(4), 683–689. https://doi.org/10.1080/21501203.2023.2278308

Sallet, D., Ugalde, G. A., Tres, M. V., Mazutti, M. A., Zabot, G. L., & Kuhn, R. C. (2025). Oil and biodiesel production from Mortierella isabellina biomass by a direct near-critical fluid extraction and transesterification method. *Biomass, 5*(1), 6. https://doi.org/10.3390/biomass5010006

Sangosanya, A. O., Vandelook, S., Elsacker, E., & Peeters, E. (2024, September). Bioremediation and upcycling of textile fibers using fungal-bacterial co-cultivation. In *2st annual meeting of the ELMs pathfinder portfolio: For all project consortia within the EIC "Pathfinder Challenge programme" engineered living materials (ELMs).*

Saye, L. M., Navaratna, T. A., Chong, J. P., O'Malley, M. A., Theodorou, M. K., & Reilly, M. (2021). The anaerobic fungi: Challenges and opportunities for industrial lignocellulosic biofuel production. *Microorganisms, 9*(4), 694. https://doi.org/10.3390/microorganisms9040694

Sharma, N., Singh, J., & Malik, V. (2023). Improving the agronomic value of paddy straw using Trichoderma harzianum, Eisenia fetida, and cow dung. *Fermentation, 9*(7), 671. https://doi.org/10.3390/fermentation9070671

Shi, C.-X., Yang, S.-R., Li, Y.-Q., Wang, H.-L., Min, S.-N., Zhang, S., Zhang, H.-L., Luo, Y.-W., Zhao, W.-X., He, Y., Cao, B.-H., & Su, H.-W. (2025). Saccharomyces cerevisiae supplementation improves growth performance and heat stress tolerance in Angus steers. *Agriculture, 15*(4), 439. https://doi.org/10.3390/agriculture15040439

Suresh, A., Alex, R. A., & Abraham, J. (2021). Fungal mediated conversion of food waste to compost. *Current Trends in Biotechnology and Pharmacy, 15*(6), 4–9. https://doi.org/10.5530/ctbp.2021.6.2

Tan, X., Zhang, S., & Liu, G. (2021). Amycolatopsis species: Prospects for biodegradation of plastic waste. *Bioresource Technology, 322*, 124589. https://doi.org/10.1016/j.biortech.2020.124589

Tišma, M., Žnidaršič-Plazl, P., Šelo, G., Tolj, I., Šperanda, M., Bucić-Kojić, A., & Planinić, M. (2021). Trametes versicolor in lignocellulose-based bioeconomy: State of the art, challenges and opportunities. *Bioresource Technology, 330*, 124997.

Ullah, H., Hayat, M., Raza, B., Alam, K., Ziad, M., Afridi, J., & Khan, F. (2025). Biodegradation of microplastic: The role of Aspergillus species in sustainable plastic waste management. *The Research of Medical Science Review, 3*(1), 314–329.

Wang, Y., Zhang, J., Chen, X., & Li, H. (2024). Optimized production and characterization of a thermostable cellulase from Streptomyces thermodiastaticus strain. *AMB Express, 14*(1), 1787. https://doi.org/10.1186/s13568-024-01787-0

Yafetto, L., Odamtten, G. T., & Wiafe-Kwagyan, M. (2023). Valorization of agro-industrial wastes into animal feed through microbial fermentation: A review of the global and Ghanaian case. *Heliyon, 9*(4), e14814. https://doi.org/10.1016/j.heliyon.2023.e14814

The Fungal Edible Proteins' Industrial Biotechnology: Where Are We?

Amro Abd Al Fattah Amara and Arti Gupta

Abstract Different kinds of fungi are widely involved in full or partial biotechnological applications. Fungal cells can stand alone, each, as a cell factory unit to produce one or more than one of the respective numbers of biotechnological products. Such products can be edible protein, enzymes, pharmaceuticals' products, secondary metabolites, and the like. Applying a powerful hydrocarbon degrading machine and safe microbe such as the fungi will be a perfect choice for any biotechnological process aiming to produce inexpensive edible protein. This chapter is concerned with the edible fungal protein. The protein produced from the fungal biotechnological production processes could add privilege to the other protein resources. Such protein can be consumed directly by humans or indirectly by fortifying the animal feeds.

Keywords Edible protein · Mycoprotein · Mushrooms · Yeast · Filamentous fungi · Single-cell protein

1 Introduction

All the creatures need food to supply their bodies with the energy that is essential for any external or internal reaction, to build tissues, and to repair degraded ones. In one simple definition, food is defined as any substance that can be metabolized by an animal to give energy and build tissue. For each creature, there are some kinds of

A. A. A. F. Amara (✉)
Protein Research Department, Genetic Engineering and Biotechnology Research Institute [GEBRI], City of Scientific Research and Technological Applications [SRTA-City], Universities and Research Center District, Alex, Egypt
e-mail: aamara@srtacity.sci.eg

A. Gupta
Department of Zoology, Shri Awadh Raj Singh Degree College, Ram Manohar Lohia University, Ayodhya, India

A. Gupta (ed.), *Prospects of Fungal Biotechnologies for Livestock Volume 2*, Fungal Biology, https://doi.org/10.1007/978-3-032-06478-3_11

food that are preferable to others. The mammalian including human needed protein that can be obtained from different resources. For humans, essential amino acids are not produced by our bodies and are needed to be supplemented through foods. There are different kinds of food resources like plants, animals, fish, birds, and the like. Recently, human added new food varieties that some are not commercially known before. Insects become currently a respective source of edible protein for humans and animals (Adámková et al., 2017; Alejandro Ruiz et al., 2025; Coppola et al., 2025; de Gier & Verhoeckx, 2018; Gałęcki et al., 2024; Kępińska-Pacelik & Biel, 2022). Microbes as well become a part of human and animal foods (Bartholomai et al., 2022). After the huge amount of research done on single-cell microbes, other horizons are becoming evaluable; it is the single-cell protein (SCP) biotechnological production (Ghaly et al., 1993; Sun et al., 2024) which followed with the mycoprotein (MyPr) production (Linder, 2024). In fact, using fungi as a food is not new but mushrooms were used 100 years ago. The new is in the ability to ferment the unicellular mushroom's unit in a fermentation process like that used for processing filamentous fungi (Hadar & Cohen-Arazi, 1986; Taskin et al., 2012). Other fungal structures that proved to have high safe edible protein content are not away from all those chances; they already were used historically in food fermentation (Guo et al., 2025), bread, bear, or used as a single source of wild type protein in form of mushrooms (Bakratsas et al., 2023; Li et al., 2024), that produced in the wild ecosystem and adapted early by humans to be produced in-house. The generic name for food derived from fungi is 'Mycoprotein'. MyPr is in its simple definition is the protein derived from fungi, used to make food stuff, especially for meat replacements. Some yeast could produce color that gives the produced product the texture of the meat (Somboonchai et al., 2022). MyPr prepared to be a meat alternative is primarily produced from the fungus *Fusarium venenatum* (Tong et al., 2024a; Tong et al., 2024b). However, other fungal species like *Monascus purpureus* (Souza Filho et al., 2018), *Aspergillus oryzae* (Pratama et al., 2025), *Neurospora intermedia*, and *Rhizopus oryzae* (Maini Rekdal et al., 2024; Bátori et al., 2015; Toghiani et al., 2025) can also be used. These fungi are cultivated through fermentation, where they convert different wastes particularly hydrocarbons from plant biomass into protein. Other fungi that have high protein content include *Saccharomyces* cerevisiae (Doaa et al., 2018; Pereira et al., 2021; Pobiega et al., 2024), *Kluyveromyces marxianus* (Pobiega et al., 2024; Carrasco de Mendoza et al., 1991; Yadav et al., 2014), *Candida* utilis (Pobiega et al., 2024; Buitrago Mora et al., 2019), and *Yarrowia lipolytica* (Pobiega et al., 2024; Vong et al., 2016). Considering mushrooms as a fungus, their proteins can also be given the name, 'MyPr' (Bakratsas et al., 2023). Mushrooms are dried and grained to powder, packed and sealed as food for soup preparation or mixed with other food products (Fu et al., 2022; Gong et al., 2022; Ma et al., 2020; Ming et al., 2015; Safari & Sarkheil, 2018; Sivamaruthi et al., 2024),

Fusarium venenatum MyPr is the most common and well-established source, known for its nutritional value and ability to mimic meat texture (Tong et al., 2024a; Tong et al., 2024b). *Monascus purpureus* MyPr is used in traditional Chinese medicine and can be utilized for MyPr production. *Aspergillus oryzae* MyPr is with various industrial applications, including fermentation. *Neurospora intermedia* MyPr is with potential for MyPr production. *Rhizopus oryzae* MyPr is in fermentation processes. *Fusarium yellowstonensis* (also known as *Fusarium* strain *flavolapis* or *Fusarium oxysporum* MK7) (Kubíková, 1968): An extremophile discovered in Yellowstone National Park and used by Nature's Fynd to produce Fy Protein™. *Neurospora crassa* is used by Meati Inc. to produce MushroomRoot™ (Linder, 2024).

One might ask why we are conducting research on fungal protein? And why produce food from fungi or from a fungal origin? One could answer this question in different ways that include different bases and purposes like:

1. Food is not available in every part of the earth. And if it is available, it might be so expensive for many.
2. Farming animals consume several resources that interfere with the cultivated area for producing human food. Finding new resources that can be used as an alternative to animal food will decrease the dependence on the food produced from the agriculture sector.
3. In particular cases food is needed with certain qualities to help patients in some illnesses like diabetes, cancer, and the like.
4. Fungi including mushrooms could be produced as food from wastes like agriculture wastes (Kuforiji & Fasidi, 2009; Singh & Singh, 2012; Singh & Singh, 2014) which adding other valuables and help in getting rid from the accumulated wastes in the nature.
5. Some foods like bread are fortified by the fungal protein to increase their nutritional values (Doaa et al., 2018; Laurujisawat et al., 2025).
6. According to their high nutritional properties, economic processing, safety, and digestibility MyPrs in some successful cases succeeded to replace meat products.
7. For some ethnic groups that prohibited animal eating, MyPr might be the best alternative for the meat.

2 Edible Mushrooms

Various edible mushrooms were used traditionally by many as a food (Rauf et al., 2023; Bano & Rajarathnam, 1988; Das et al., 2021a; Debnath & Sen, 2022; Glamočlija et al., 2015; González et al., 2020; Losoya-Sifuentes et al., 2025; O'Brien et al., 2019) and medicine (Jiang et al., 2025; Shamim et al., 2023; Leonti et al., 2020; Lallawmsanga et al., 2016; Toledo et al., 2016). The term mushroom was coined to edible species only. The term toadstool (Yamada et al., 2012) is applied to other species that considered fetal. Species that result in death or serious illness when eaten are less in numbers. Mushrooms must be well inspected as in some cases even two closely related species; one of them might be fetal while the second is edible. The majority of mushrooms are tough, woody, bitter, and tasteless. The best way to utilize mushrooms is to seal them from the market or to grow edible known variants.

They are large fleshy or woody fungi. Mushrooms are the higher fungi, belonging to the classes Ascomycetes (Morchella, Tuber, and the like), and Basidiomycetes (Agaricus, Auricularia, Tremella, and the like). They are characterized by having a heterotrophic mode of nutrition (Riviere, 1977). They are rich in protein and constitute a valuable source of supplementary food. The value in promoting cultivating of mushrooms are able to grow on inexpensive substrates, industrial and agricultural waste (rich in nitrogen and carbon) (Chang & Hayes, 1978). Today there are hundreds of mushroom-based dishes particularly for their protein content (Bakratsas et al., 2023; Maseko et al., 2025; Scholtmeijer et al., 2023). The practices of collecting wild-type mushrooms from forests and other wild habitats (Lallawmsanga et al., 2016; Toledo et al., 2016; Khaund & Joshi, 2016) have become a rising business (Singhal et al., 2019; Zhang et al., 2025; Zheng et al., 2023). Global mushroom cultivation was estimated at about 11.9 million tons per year in 2019 (FAOSTAT, 2020). China (Li et al., 2021; Liu et al., 2024; Liu et al., 2012; Mi et al., 2016; Wang et al., 2014) alone produced 8.9 million tons of them, followed by Japan (Aoki et al., 2020; Shimizu et al., 2018) (0.47 million tons) and the USA (Li et al., 2019) (0.38 million tons). It is a gift for the human. From a source for survive to a source of business that highlight the huge progress that happened in the last years. Some of the edible mushrooms are popular than the other ones and are currently cultivated commercially (Das et al., 2021b; Ho et al., 2020).

The mushroom provides a daily dish for millions, it is acceptable to economize the family incomes. Putting the spores of a mushroom on an agricultural waste and keep it in a dark room for weeks and then harvest the mushroom bodies is not a hard work. But this practice which costs nearly nothing could satisfy a family food demand for weeks. The mushrooms taste is like umami taste (a savory taste sensation produced by glutamates) and for their nutritional properties; they are low in fat, high in protein (Stoffel et al., 2021; Trujillo et al., 2021), and high in dietary fibers (Chen & Cheung, 2014). They contain vitamins (including vitamin B), minerals, antioxidants, and nutraceuticals (Abo Nahas et al., 2021; Dikkala et al., 2021; Jagadish et al., 2021). They are traditionally eaten fresh but are more and more used as dried powder (flour) (Fu et al., 2022; Gong et al., 2022; Ma et al., 2020; Ming et al., 2015; Safari & Sarkheil, 2018) to fortify various food types with nutrients and especially proteins as they contain up to 20–25% protein (Ho et al., 2020). Such dried fungal products are called "MyPr" (Bakratsas et al., 2023).

The mushroom species are usually grown commercially (Gunawardena et al., 2014; O'Connor et al., 2019) to attain a size of 5–10 cm tall and have a fleshy cap from about 2–10 cm across. Historically, mushrooms are grown for profit purposes in caves, dark cellars, or specially constructed mushroom houses, in which the proper humidity and temperature are maintained. The mini-commercial production of edible mushrooms occurs in buildings in which temperature and humidity are strictly regulated. A special bedding culture is prepared and inoculated with a pure culture of the fungus mycelium. Several harvests of mushrooms are produced from each inoculation.

2.1 Examples About the Most Famous Edible Mushrooms

2.1.1 White Bottom Mushroom (*Agaricus bisporus*)

The most important cultivated mushrooms are *Agaricus bisporus* (Cavusoglu et al., 2021) (common mushroom). *A. bisporus* was domesticated in 1650s, and described scientifically in 1707. Native to North American and Eurasian grasslands it can be found in fields and grasslands following rain (Spencer, 1985; Galán et al., 1999; mushroom. mushroom, 2013). It is reported to have antiviral, cytotoxic, and antioxidant activities (Elhusseiny et al., 2021; Amara & El-Baky, 2023), and antioxidant enzyme activity (Amara & El-Baky, 2023; Liu et al., 2013; Chen et al., 2017; Ragucci et al., 2023). It has antioxidant potential and protein hydrolysates (Kimatu et al., 2017). It can be harvested at different stages of maturity to obtain more antioxidant enzyme activity (Liu et al., 2013). It has antibacterial, antibiofilm, and antihemolytic activity of the antimicrobial peptides from *A. bisporus* against oropathogenic bacteria (Gangwar & Maurya, 2022). *A. bisporus* lectin (ABL) has activity against human liver cancer cell line Hep G2 and mouse lymphocytic leukemia cell line L1210 (Zhang et al., 2019a; Drzewiecka et al., 2024; mushroom. mushroom, 2013).

2.1.2 The Chanterelle

Several species in the genera include *Cantharellus*, *Craterellus*, *Gomphus,* and *Polyozellus* as in Table 1. They have nut-like flavor, popular in Europe. Abundant in midsummer in coniferous and hardwood *Cantharellus cibarius* has anti-angiogenic activity (Marathe et al., 2022). *C. cibarius* Fr reported to have neuroprotection of low-molecular-weight galactan against Alzheimer's disease. *C. rutilus* regulates bone metabolism to prevent periodontal bone loss during orthodontic tooth movement in osteoporotic rats (Zhou et al., 2023). It has inhibitory effects of polyozellin from Polyozellus multiplex on HMGB1-mediated septic responses (Yang et al., 2015; mushroom. mushroom, 2013).

2.1.3 The Edible Porcini Mushroom (*Boletus edulis*)

B. edulis is growing during summer and early autumn in open deciduous woods as in Table 1. The king mushroom has a stem of 5–15 cm tall and a brown cap of 10–15 cm across. These mushrooms are most tender when the veins that cover their cap are pale yellow (Amara & El-Baky, 2023; Nuhn et al., 2013; Ennamany et al., 1994; Landi et al., 2021; Zhang et al., 2021a; Zhang et al., 2021b). Mostly ectomycorrhizal fungi. Has lectin with antitumoral properties (Bovi et al., 2011). It has interrelationships among paraptosis, apoptosis, and autophagy in lung cancer A549 cells induced by BEAP, an antitumor protein (Zhang et al., 2021a). A purification

Table 1 The most common edible mushrooms

Mushroom commercial name	Mushroom scientific name	Importance
White bottom mushroom	*Agaricus bisporus*	Edible (Spencer, 1985; Galán et al., 1999), has antiviral, cytotoxic, and antioxidant activities (Elhusseiny et al., 2021; Amara & El-Baky, 2023), antioxidant enzyme activity (Amara & El-Baky, 2023; Liu et al., 2013; Chen et al., 2017; Ragucci et al., 2023), antibacterial, antibiofilm, and anti-hemolytic activity (Gangwar & Maurya, 2022) and lectin (ABL) (Zhang et al., 2019a; Drzewiecka et al., 2024)
Shiitake mushrooms	*Lentinula edodes*	Edible (Li et al., 2021; Amara & El-Baky, 2023; Efsa Panel on Nutrition NF et al., 2022; Song et al., 2019; Wang et al., 2016), laccase (Zhao & Kwan, 1999), hydrolytic enzymes (Mata et al., 2016), endoglucanase, xylanase, manganese peroxidase (Kobayashi et al., 2023), and borosin peptides (Ford et al., 2022)
The chanterelle	Several species in the genera *Cantharellus*, *Craterellus*, *Gomphus*, and *Polyozellus*	Edible (Das et al., 2021b; Amara & El-Baky, 2023; Ghosh et al., 2023; Hong et al., 2012), anti-angiogenic activity (Marathe et al., 2022). Low-molecular-weight galactan against Alzheimer's disease, regulates bone metabolism (Zhou et al., 2023) and has inhibitory effects of polyozellin HMGB1-mediated septic responses (Yang et al., 2015)
The edible pore mushrooms Porcini mushroom	*Boletus edulis-*	Edible (Amara & El-Baky, 2023; Nuhn et al., 2013; Ennamany et al., 1994; Landi et al., 2021; Zhang et al., 2021a, b; Agueda et al., 2006), lectin with antitumoral properties (Bovi et al., 2011), and anticancer (Zhang et al., 2021a, b; Bovi et al., 2011)
Oyster mushroom	*Pleurotus ostreatus*	Edible (Amara & El-Baky, 2023; Hassan et al., 2020; Lee et al., 2018; Mihai et al., 2022; Uriarte-Frías et al., 2021; Zhou et al., 2022) (Ota et al., 2013), MyPr (Bakratsas et al., 2023), anti-human immunodeficiency virus (Wang & Ng, 2000), toxic and lethal effects of ostreolysin (Zuzek et al., 2006), and has an antifungal peptide called pleurostrin (Wang & Ng, 2006)
Enoki mushrooms	*Flammulina velutipes*	Edible (Ko et al., 1995) (Amara & El-Baky, 2023; Ge et al., 2015; Lin et al., 1975; Tomita et al., 1998; Wang & Ng, 2001), and produce Flammin and velin: new ribosome inactivating polypeptides (Ng & Wang, 2004)
Jelly Ears	Auricularia auricula-judae	Edible (Amara & El-Baky, 2023; Han et al., 2023; Looney, 2013), and used in bread enrichment (Yuan et al., 2017)
The sulfur mushroom	*Laetiporus sulphureus*	Edible (Amara & El-Baky, 2023; Dong et al., 2022; Mancheño et al., 2005; Toumi et al., 2021)
The morel mushroom	*Morchella esculenta*	Edible (Amara & El-Baky, 2023; Gao et al., 2025; Ower, 1982; Zhang et al., 2019b)
The shaggy-mane	*Coprinus comatus*	Edible (Amara & El-Baky, 2023; Luo et al., 2007; Xiao et al., 2020; Yuan et al., 2022), and has laccase with potent antiproliferative and HIV-1 (Zhao et al., 2014; Gu et al., 2014)
The giant puffballs	*Calvatia gigantea*	Edible (Amara & El-Baky, 2023; Cicha-Jeleń et al., 2024) with many medicinal properties (Cicha-Jeleń et al., 2024), unsaturated fatty acids, free amino acids, polysaccharides, and bioactive metabolites (Cicha-Jeleń et al., 2024)

and characterization of a novel protein with activity against non-small-cell lung cancer in vitro and in vivo from the edible mushroom *B. edulis* was reported (Zhang et al., 2021b). Described as a bit nutty and meaty. It has porcini smelling such as freshly baked (Amara & El-Baky, 2023; Landi et al., 2021; Zhang et al., 2021a; Zhang et al., 2021b; Agueda et al., 2006). Active protein against non-small-cell lung cancer (Zhang et al., 2021b). It has lectin with antitumoral properties (Bovi et al., 2011).

2.1.4 Oyster Mushroom

Pleurotus ostreatus (Bakratsas et al., 2023; Bautista Justo et al., 1999) (oyster mushroom) has a pleasant, oyster-like flavor. Often dipped in egg and fried slowly as in Table 1. Grows on decaying tree trunks in bracket-like clusters. It is almost stemless abundant from June to November (Amara & El-Baky, 2023; Hassan et al., 2020; Lee et al., 2018; Mihai et al., 2022; Uriarte-Frías et al., 2021; Zhou et al., 2022; mushroom. mushroom, 2013). It is used for MyPr production by submerged fermentation in a batch stirred tank bioreactor using agro-industrial hydrolysate (Bakratsas et al., 2023). Ostreopexin is a hemopexin fold protein from the oyster mushroom, *P. ostreatus* (Ota et al., 2013). It has a protein with anti-human immunodeficiency virus, translation-inhibitory, and ribonuclease activities (Wang & Ng, 2000). Toxic and lethal effects of ostreolysin, a cytolytic protein from edible oyster mushroom (*P. ostreatus*), have been studied against rodents (Zuzek et al., 2006). It has an antifungal peptide called pleurostrin (Wang & Ng, 2006). *Pleurotus albidus* MyPr flour was used to replace wheat flour to produce cookies (Stoffel et al., 2021). Used for MyPr production (Bakratsas et al., 2023). Fresh commercially grown mushrooms can always be eaten safely. *Pleurotus citrinopileatus* has an alkaline protease from fresh fruiting bodies and is considered as an edible mushroom (mushroom. mushroom, 2013; Cui et al., 2007).

2.1.5 Enoki Mushrooms (*Flammulina velutipes*)

Flammulina velutipes (Ko et al., 1995) is very popular in several Asian dishes (mushroom. mushroom, 2013) as in Table 1. Have a mild, earthy flavor. Have crispy texture when served in soups. Immunomodulatory protein, FIP-fve isolated from the edible mushroom, *F. velutipes* (Amara & El-Baky, 2023; Ko et al., 1995; Ge et al., 2015; Lin et al., 1975; Tomita et al., 1998; Wang & Ng, 2001). Flammin and velin are new ribosome-inactivating polypeptides isolated *F. velutipes* (Ng & Wang, 2004). Bioactive sesquiterpenoids were produced from the solid culture of the *F. velutipes* when grown on cooked rice (Wang et al., 2012). Active immunomodulatory protein from *F. velutipes* was expressed in *Pichia pastoris* GS115 (Lin et al., 2013).

2.1.6 The Shaggy-Mane (*Coprinus comatus*)

The shaggy-mane is a common and widespread mushroom species appearing from spring until fall in lawns, gardens, and other open spaces as in Table 1. The shaggy-mane is considered one of the most sought-after edible species. Blackened portions should be discarded before the mushroom is eaten. Some related species cause poisoning when alcohol is consumed within 5 days after ingestion of the mushroom (Amara & El-Baky, 2023; Luo et al., 2007; Xiao et al., 2020; Yuan et al., 2022; mushroom. mushroom, 2013). A novel laccase was identified with potent antiproliferative and HIV-1 reverse transcriptase inhibitory activities from mycelia of mushroom *C. comatus* (Zhao et al., 2014). Two novel laccase isoenzymes from *C. comatus* were engineered in *P. pastoris* by fusing an additional ten amino acids tag at N-terminus (Gu et al., 2014).

2.1.7 The Giant Puffballs (*Calvatia gigantean*)

They are very large and globose and have no gills or pores. Their spores are born internally as in Table 1. Their fruiting body is creamy white in the edible stage. Grow during late summer in grassy places and at the edge of woods. They significantly differ from fetal or offensive fungi. They are edible as long as the tissues within are not discolored or larva-infested. Species that are brown to purple within should be avoided (Amara & El-Baky, 2023; Cicha-Jeleń et al., 2024). It has many medicinal properties (Cicha-Jeleń et al., 2024). The mushroom is recognized for its nutritional content, including easily digestible protein, carbohydrates, fiber, phenolic compounds, vitamins, and minerals, while being low in calories, cholesterol, and sodium. Furthermore, *C. gigantea* exhibits a range of biological effects, such as antioxidants, anticancer, antimicrobial, antidiabetic, and wound-healing properties, attributed to its diverse chemical composition that includes unsaturated fatty acids, free amino acids, polysaccharides, and bioactive metabolites (Cicha-Jeleń et al., 2024).

2.1.8 Other Species

Other species include *Volvariella volvacea* (Lin et al., 1973) (paddy straw mushroom) (Das et al., 2021b; Ho et al., 2020). Tremella such as *Tremella fuciformis* (Hung et al., 2014) (silver ear mushroom), Agrocybe such as *Agrocybe cylindracea* (Berne et al., 2002), nameko (*Pholiota nameko*) (Zhang et al., 2014), monkey head mushroom (*Hypsizygus marmoreus*) (Tsai et al., 2013), as well as two famous medicinal mushrooms, namely Maitake (*Grifola frondosa*) and the Reishi (*Ganoderma lucidum* (An et al., 2010)) (Rai, 2004; Suwannarach et al., 2022). Jelly Ears (*Auricularia auricula-judae*) Auricularia such as *Auricularia* ssp. (Zhao et al., 2023) (wood ear or black ear mushrooms) is very common in Asian countries. In China, it is also used in medicine. It produces bioactive peptides (Amara & El-Baky,

2023; Han et al., 2023; Looney, 2013). It is used for the enrichment of bread (Yuan et al., 2017).

Shiitake mushrooms (*Lentinula edodes*) (Zhao et al., 2024) (shiitake) grow predominantly in China and Japan, and are popular in Asia and have a woody taste. It is used to prepare several local dishes (Li et al., 2021; Amara & El-Baky, 2023; Efsa Panel on Nutrition NF et al., 2022; Song et al., 2019; Wang et al., 2016). Some important proteins can be isolated from it like laccase (Zhao & Kwan, 1999), hydrolytic enzymes (Mata et al., 2016), lignocellulose degradation enzymes, such as endoglucanase, xylanase, and manganese peroxidase (Kobayashi et al., 2023). Borosin peptides are also reported (Ford et al., 2022).

The sulfur mushroom (*Laetiporus sulphureus*) develops on rotten logs, stumps, and standing trees, producing a brown wood rot. They may reach a breadth of several meters and a weight of several kilograms. Spores are produced in enormous numbers in minute pores on the lower surface. The fungus is edible if picked in the young, growing stage, yet rapidly becomes dry, tough, and honeycombed by insect larvae. It is considered as medicinal edible mushroom (Amara & El-Baky, 2023; Dong et al., 2022; Mancheño et al., 2005; Toumi et al., 2021). The morel mushroom (*Morchella esculenta*) has a unique irregular honeycombed appearance and excellent flavor. Morels are found during the spring in old orchards especially in areas that were burned out. Most highly prized of all mushrooms. Show preventive effect on nonalcoholic fatty liver disease in mice (Zhao & Kwan, 1999; mushroom. mushroom, 2013; Kimatu et al., 2017; Bautista Justo et al., 1999).

Commercial mushrooms are safer than those that grow in the nature. Additionally, wild mushrooms are growing seasonally, in particular times and conditions. Meanwhile, some species are preferred to be collected from the wild. Currently, mushrooms add flavor, texture, and nutritional value to several dishes. Mushrooms are widely consumed throughout the world and are produced in a multi-stage process. China, Japan, and the USA are largest producers (FAOSTAT, 2020). In different geographical areas, there is a particular kind of mushroom that is preferable to other types which might be more consumed in another area. In western Europe, the most commonly consumed mushroom is *A. bisporus*, but in the Far East, species of *Pleurotus*, and *Auricularia* besides *Lentinus edodes* are the dominant cultivated mushrooms (Kavanagh, 2005). The presence of a suitable inexpensive substrate may favor producing one type over the others. *A. bisporus* is grown in the UK, the USA, and France on wheat straw, *V. volvacea* is grown in South-East Asia on damp rice straw and in Hong Kong on cotton waste, and *Lentinus edodes* is cultivated on fresh oak logs in Japan. In the industry, media for fungal fermentation need to be optimized regarding the specific application and production process (Gupta et al., 2019; Koivurinta et al., 1979).

Wild and cultivated edible mushrooms as well as their mycelium and hyphae grown in a fermentation process are considered as a source of different beneficial products from different grades like pharmaceutics, nutraceuticals, edible protein, fibrinolytic/thrombolytic enzymes, etc. (Kwok et al., 2005; Guillamon et al., 2010). Fibrinolytic enzymes from medicinal and edible mushrooms and their application as therapeutic agents have been known since the first publication in 1982 (Morozova et al., 1982). *Flammulina velutipes* proved to produce a fibrinolytic that comprised two metalloproteases with pI values of 6.1 and 7.1, which have fibrinolytic, thrombolytic, and endopeptidase activities. Other and similar fibrinolytic or thrombolytic effects have been found within the fruit body of *Cordyceps militaris*, *Armillaria mellea*, *Pleurotus ostreatus*, and *Tricholoma saponaceum* (Kim & Kim, 2001; Lee et al., 2005; Choi et al., 2011a; Liu et al., 2014; Morozova et al., 1982; Nonaka et al., 1995; Choi & Shin, 1998; Shen et al., 2007; Kim & Kim, 1999; Choi & Sa, 2000; Kim et al., 2006; Cui et al., 2008; Lee et al., 2006; Li et al., 2007; Kim et al., 2008; Cha et al., 2010; Lu et al., 2010; Park et al., 2010; Kim et al., 2011; Choi et al., 2017).

3 Edible Yeasts

The edible yeast accounts for hundreds. The most famous is *Saccharomyces cerevisiae*. Traditionally, it was used to produce bread, beer, and wine. The yeast could ferment its weight of glucose per hour. *Saccharomyces* yeasts have high protein content. It is considered as SCP. *Saccharomyces bayanus* (used for making wine) and *Saccharomyces boulardii* are used as a probiotic (Jin et al., 2021; Rengel Dos Passos et al., 2021; van der Aa Kühle et al., 2005; Vieira et al., 2012; Johnston et al., 2011; Dinleyici et al., 2012; Shan et al., 2013; Szajewska & Mrukowicz, 2005; Choi et al., 2011b; Guslandi et al., 2000; Dugoua et al., 2009). *Saccharomyces* yeast also forms symbiotic matrices with bacteria to form kefir (Witthuhn et al., 2005). Kefir is an effervescent liquor like kumiss, made from fermented milk, used as a food and as a medicine in the northern Caucasus. They are components of kombucha (Marsh et al., 2014). Kombucha is a drink of fermented sweetened tea. Yeast and their metabolic products were used in food processing (Bigey et al., 2024; Boyaci Gunduz et al., 2022; Hittinger et al., 2018; Li et al., 2023; Menezes et al., 2019; Tofalo et al., 2020; Valdez Castillo et al., 2023; Watanabe & Takagi, 2019) and for preservation (Tamang et al., 2015; Anupma & Tamang, 2016).

Pan bread was progressing and fortified by raising its protein content with high levels of live yeast cells *S. cerevisiae* (Doaa et al., 2018) that enhance bread protein content, vitamin B complexes, and minerals such as calcium, phosphorus, manganese, magnesium, zinc, and copper and has high biological value of essential and nonessential amino acids so it has several health benefits (Doaa et al., 2018). *S. cerevisiae* as a SCP is a rich source of proteins which are necessary for replacing worn-out tissues or recovery after infections and contains 18 amino acids and is considered to be 55% high-quality protein. *S. cerevisiae* contains gluthanione, an antioxidant and beta-glucan which stimulates the immune system (Hong et al., 2007). *S. cerevisiae* cells contain chromium, a trace mineral which is currently identified as glucose tolerance factor which is vital in preventing diabetes, lowering blood pressure and fluctuating blood sugar (Zetić et al., 2001; Yamada & Sgarbieri, 2005). *S. cerevisiae* species secretes peptides (Pereira et al., 2021; Albergaria et al., 2010; Alcaide-Hidalgo et al., 2007; Branco et al., 2014; Branco et al., 2017; Branco et al., 2015;

Rizk et al., 2016; Rizk et al., 2018). *S, cerevisiae* transcriptome during bread dough fermentation was inspected (Aslankoohi et al., 2013).

3.1 Single-Cell Protein

A wide range of fungi were considered for use as SCP (Ritala et al., 2017; Nasseri et al., 2011). Products from *Fusarium* are commercially available. Fungal SCP generally contains protein content of 30–45% (Ritala et al., 2017; Nasseri et al., 2011; Anupama, 2000). The fungal SCP content of threonine and lysine is typically high (Ritala et al., 2017). SCP was given for some fungal proteins before the rising of the new name 'Mycoprotein'. However, one should differentiate between the use of the whole cell (SCP) as a protein source and the protein that produced through the fungi. In fact, there is some mix between the two terms SCP and MyPr.

Different companies are involved in the production of the fermented fungal protein. Each has designed it with other food gradient to reach certain texture and taste that could be accepted by the consumers. Many lessons were learned from the history of the SCP. Other facts, that the agriculture net production from soybeans and other plant-based protein is insufficient to feed the world. And many limitations were removed to fill in the gap. Moreover, the biotechnology, the fermentation, the recombinant gene technology, etc. become available everywhere. The chances have been shared by different countries in the world. While wonderful strains, fermented food tactics, and the like were transferred from the old world to the new one, other beneficial technological discoveries were transferred vice versa. In another angle, the old world market is huge and accepts many of the newly produced products.

Vegemite is an Australian yeast extract manufactured by Bega Cheese (Bega, New South Wales, Australia). It was sold since 1922. Marmite is a British yeast extract produced by Unilever company (London, UK), *VITAM-R* is a yeast extract produced by VITAM Hefe-Produkt GmbH company, Hameln, Germany. Cenovis is a yeast extract from Switzerland, produced by the company Cenovis SA, Berstelstrasse, Switzerland (Amara & El-Baky, 2023). Provesta Corporation (1980s) used *Torula* in their product Provesteen®T. *Pichia* and *Kluyveromyces* yeast are used in other similar products (Hitzman, 1986). Methylotrophic yeasts as *Pichia pastoris* can produce biomass and protein from the fermentation of the methanol. Phillips Petroleum Company produced *Pichia pastoris* at industrial scale (Johnson, 2012). Perfect Day's animal-free ice cream uses recombinant yeast to produce cow casein and whey. The size of the yeast market for animal feed is above 1.5 billion $ in 2022 (Inc, G.M.I, 2022).

The market value of yeast for poultry feed segment is expected to reach above 1.7 billion $ by 2032. The value of animal feed yeast market in North America in 2022 is 375 million $ at CAGR (2023–2032). Other companies in the animal feed include Cangzhou Tianyu Feed Additive Co., Ltd. (Chinese company for yeast powder production as additives for feed), and Tangshan Top Biotechnology Co., Ltd. (Chinese manufacturer of brewer's yeast, autolyzed yeast, yeast extract, and natural, non-GM pure yeast powder as animal feed additive) (Inc, G.M.I, 2022). Market size of yeast extract for human consumption is anticipated to reach more than 2.5 billion $ by 2030 (Yeast-Based Spreads Market Key Vendors, Regional Analysis, Key Players and Competitive Landscape Forecasts by 2027, 2022). LeSaffre (French company that produces yeast (*S. cerevisiae*) and yeast-derived products including SCP as well as yeast-based flavor ingredients), and Unilever (producers of Marmite yeast-based spreads) (Lähteenmäki-Uutela et al., 2021). Another SCP product was produced by the PEKILO process from the filamentous fungus *Paecilomyces varioti* grown on sugars in the sulphite waste liquor or wood hydrolysates. They had two factories, but as the cellulose mills ceased operations, the factories were closed in 1991. Even though the product was sold primarily as animal feed. It was also studied as a supplement in meat products (Koivurinta et al., 1979; Ritala et al., 2017). The term SCP refers to any protein from microbial sources in the form of biomass or extracted protein (Leger et al., 2021). Using them in SCP production is a modern concept (Ritala et al., 2017; Mahasneh, 1997; Omar & Sabry, 1991; Trinci, 1992; Bourdichon et al., 2012a; Bourdichon et al., 2012b). Yeast (such as *S. cerevisiae*, *Pichia pastoris*, *Candida utilis*, *Torulopsis glabrata*, and *Geotrichum candidum*), and molds (such as *Aspergillus oryzae*, *Fusarium venenatum*, *Trichoderma*, and *Rhizopus*) (Ritala et al., 2017). Filamentous fungi are easy to harvest from the SCP fermentation medium, and fungi including yeasts can also provide vitamins (group-B). They have cell walls rich in glucans. The size of SCP market in 2021 was above 8 billion $ and is expected to expand at a compound annual growth rate (CAGR) of more than 9% from 2022 to 2030. The market size will exceed 18.5 billion $ by 2030 (Inc, G.M.I, 2022). It is also anticipated that the SCP market size will archive 4.5 billion $ by 2030 in EU countries. North America and western Europe are anticipated to remarkable contribute to SCP market growth. Asia Pacific and the Middle East are predicted to offer significant opportunities for market growth. The SCP market in the ASEAN region is led by Malaysia show great raise (Inc, G.M.I, 2022).

4 Mycoprotein

Edible fungi have traditionally been important food, especially mushrooms, where the fruiting bodies are used as food, or through food fermentation, including their widespread use in cheese production, e.g., blue and white cheeses (Dai et al., 2021). Fungal fermented non-animal foods are an integrated part of the diet in several countries, especially in East and South Asia and certain African regions. The fermented foods are divided into high-protein meat alternatives from legumes or cereals, such as Tempeh and Oncom, and into salty amino acid sauce and paste, e.g., shoyu and miso (Dai et al., 2021; Steinkraus, 1997).

The MyPr flour significantly increased the nutritional value of the cookies because of the contents of protein, dietary fiber, and phenolic compounds. Furthermore, the MyPr increased the hardness and altered the color of the cookies (Stoffel et al., 2021) as in Table 2.

Table 2 Different mycoprotein products and companies

Company	Product	Web address
Marlow Foods Ltd (Quorn) (Stokesley, UK) (Linder, 2024)	A well-established company that pioneered MyPr production and offers a wide range of meat alternatives under the Quorn brand.	https://marlowfood.co.uk/index.html
3FBIO Ltd (ENOUGH)	Produces ABUNDA MyPr, a fermented food ingredient, and is scaling up production to meet growing demand	https://www.enough-food.com
Meati Foods (Boulder, CO, USA) (Linder, 2024)	Specializes in producing MushroomRoot MyPr, focusing on creating meat-like products with a focus on texture and taste	https://www.meati.com
Enifer Espoo (Finland)	A Finnish startup focused on MyPr production using a unique fermentation process that can utilize various feedstocks, including agricultural byproducts	https://enifer.com
MycoTechnology	Involved in MyPr production, though specific details about their operations and products may vary.	https://www.mycoiq.com
MyForest Foods	Uses solid-state fermentation to grow mycelium in indoor vertical farms, creating products like MyBacon	https://myforestfoods.com
Aqua Cultured Foods	Develops seafood alternatives using microbial fermentation	https://www.aquaculturedfoods.com
Mush Foods (New York, NY, USA) (Linder, 2024)	Produce as MyPr ingredient called 50CUT	https://eatmush.com
Libre Foods	Also involved in MyPr production. These companies are driving innovation in the alternative protein space, with a focus on sustainability and providing consumers with appealing and nutritious food options	https://www.librefoods.co
The Better Meat Co. (West Sacramento, CA, USA) (Linder, 2024)	Produce as MyPr ingredient called Rhiza	https://www.bettermeat.co
ENOUGH (Glasgow UK) (Linder, 2024)	Produce as MyPr ingredient called ABUNDA	https://www.enough-food.com
Eternal (San Francisco CA, US) (Linder, 2024)	Produce as MyPr ingredient called Mycofood	https://eternal.bio
Hyfé Foods (Chicago, IL, USA) (Linder, 2024)	Produce as MyPr ingredient	https://hyfe.tech

Besides Basidiomycetes mushrooms, certain species of Ascomycetes and Zygomycetes microfungi have traditionally been used for food fermentation into meat-like products such as Tempeh (*Rhizopus oligosporus*) (Nout & Rombouts, 1990; Nout & Kiers, 2005) and Oncom (*Neurospora* spp.) (Moo-Young et al., 1993; Hellwig et al., 2020; Adebo et al., 2022) in Asia. These microfungi were found to contain higher protein in their mycelium than most Basidiomycetes mushrooms. For example, *Neurospora sitophila* mycelium contains 39–45% protein

(Moo-Young et al., 1993). Traditionally, soybeans were used. But other substrates such as chickpeas, lupines, and various cereals are used as well (Nout & Rombouts, 1990). Several novel Western companies have started production of "Tempeh" style food kinds using local products, such as lupines and peas, especially using *R. oligosporus*.

Fungal cell factories are used for efficient and sustainable production of proteins and peptides (Lübeck & Lübeck, 2022; Madhavan et al., 2022; Meyer, 2021; Wei et al., 2021). Filamentous fungi are excellent organisms as cell factories for producing a variety of products (Lübeck & Lübeck, 2022). Filamentous fungus lifestyle grows on organic carbon (Lübeck & Lübeck, 2022; Papagianni, 2004). They play an important role in the environment as decomposers (Spatafora et al., 2017; Sakekar et al., 2021; Sørensen et al., 2013; Zoglowek et al., 2015; Chen et al., 2019). Fungi were used for the benefit of human life (Barzee et al., 2021a; Hyde et al., 2019; Meyer et al., 2020; Meyer et al., 2021; Paul & Joshi, 2022). Hyde et al. (Hyde et al., 2019) detail 50 ways in which fungi potentially can be exploited (Barzee et al., 2021a; Tubb & Seba, 2019). A shift in the dietary pattern toward more fungal protein-based products is remarkable (Tubb & Seba, 2019; Enzing et al., 2014).

The harvested MyPr is often cooked, textured, and sometimes flavored to enhance its taste and mouthfeel, making it suitable for various culinary applications. MyPr is rich in protein, typically containing around 45–50% protein by dry weight. It provides all essential amino acids, making it a complete protein source. MyPr is low in saturated fat and contains no cholesterol, making it a heart-healthy option. It is also a good source of dietary fiber, which can aid in digestion and promote a feeling of fullness. MyPr contains various vitamins (such as B vitamins) and minerals (like iron and zinc), contributing to its nutritional profile. MyPr is commonly used in various food products, including meat substitutes, ready-to-eat meals, and snacks. MyPr can be used in a variety of dishes, including stir-fries, curries, burgers, and sausages, because of its meat-like texture and ability to absorb flavors. The fermentation process allows for quick production of MyPr, making it an efficient protein source. MyPr can be flavored and textured to mimic various meat products. Some individuals may have allergies or sensitivities to fungi, which can limit consuming MyPr (Dębińska & Sozańska, 2022).

Producing MyPr using solid-state fermentation (Lin et al., 2024; Zhou et al., 2024) has recently been expanded to other substrates than the already eatable substrates. Low-value side streams from the food industry such as brewers' spent grain were used (Hellwig et al., 2020; Cooray & Chen, 2018; Ibarruri et al., 2019). Other structures include pulp from sugar beets, potato starch production, coffee production or bran from flour production (Ibarruri et al., 2021). They can provide new food properties, in terms of texture, flavor, and increased solubility (Dai et al., 2021; Hellwig et al., 2020).

Any new or used strains should possess a GRAS (generally regarded as safe) or Qualified Presumed of Safety (QPS) status (Bourdichon et al., 2012a; Hazards et al., 2020). There is a rising interest in growing fungal mycelium in bioreactors. Submerged fermentation in sugar-rich substrates was used (Barzee et al., 2021a; Meyer et al., 2021; Derbyshire & Delange, 2021).

The most well-currently MyPr product on the market is called Quorn™ (Table 2), which has been produced since 1985 (Derbyshire & Delange, 2021). Prior to commercialization, there was 20 years of research and progress, including a major screening process involving over 3000 fungal species (Finnigan et al., 2019). *Fusarium venenatum* is approved for sale as a food protein source by the United Kingdom Ministry of Agriculture, Fisheries and Food (Wiebe & Quorn, 2004), and is currently sold in 17 countries (Finnigan et al., 2019). MyPr from *F. venenatum* is used as an ingredient in various products such as alternative chicken patties, sausages, and burgers with more than 60% protein (Barzee et al., 2021b) and amino acids (Derbyshire & Delange, 2021), fibrous chitin–glucan matrix, (Derbyshire & Delange, 2021), reducing overall cholesterol levels and short-term energy intake (Derbyshire & Delange, 2021) vitamin B9 (folate), vitamin B12, calcium, phosphorous, magnesium, and zinc. Their fermentation processing will lower the dependence on the greenhouse which a source of gas emission (Sharif et al., 2021).

5 Conclusion

Fungi can be a source of protein and SCP. But only edible fungi can be used as SCP. They have some unique properties mainly their high protein content and their ability to degrade different kinds of wastes. They also have low colterol. Many companies are involved in the production of fermented fungal protein which has gained a new name 'mycoprotein' MyPr and is produced in different countries as a sustainable source of protein that needs only a fermentation plant and a source of waste. The progress that happened in the previous years is coined to the scientists, companies, government, consumers, and everyone sharing effort to success such source of food. From Asia and the Middle as well as other places fermented food to a submerged fermentation (and the like), useful fungal strains continue to serve human. Some of them are well known and used historically like the yeast *Saccharomyces cerevisiae*, the filamentous fungi Aspergillus oryzae, and of course the different kinds of edible mushrooms. Many and many other species are included. The fermented food that has been used safely through thousands of years could be a treasure for isolating new wonderful strains. Besides being a protein and food producer, they are safe for the nature, reduce the dependence on the agriculture landscapes, are consumed by humans and animals, have medicinal properties, and high economic value.

References

Abo Nahas, H. H., Darwish, A. M. G., Abo Nahas, Y. H., Elsayed, M. A., Abdel-Azeem, M. A., & Abdel-Azeem, A. M. (2021). Fungi as a gold mine of antioxidants. *Fungal Biology*, 73–113. https://doi.org/10.1007/978-3-030-85603-8_2

Adámková, A., Mlček, J., Kouřimská, L., Borkovcová, M., Bušina, T., Adámek, M., Bednářová, M., & Krajsa, J. (2017). Nutritional potential of selected insect species reared on the Island of Sumatra. *International Journal of Environmental Research and Public Health, 14*. https://doi.org/10.3390/ijerph14050521

Adebo, J. A., Njobeh, P. B., Gbashi, S., Oyedeji, A. B., Ogundele, O. M., Oyeyinka, S. A., & Adebo, O. A. (2022). Fermentation of cereals and legumes: Impact on nutritional constituents and nutrient bioavailability. *Fermentation, 8*, 63. https://doi.org/10.3390/fermentation8020063

Agueda, B., Parlade, J., de Miguel, A. M., & Martinez-Pena, F. (2006). Characterization and identification of field ectomycorrhizae of Boletus edulis and Cistus ladanifer. *Mycologia, 98*, 23–30. https://doi.org/10.3852/mycologia.98.1.23

Albergaria, H., Francisco, D., Gori, K., Arneborg, N., & Gírio, F. (2010). Saccharomyces cerevisiae CCMI 885 secretes peptides that inhibit the growth of some non-Saccharomyces wine-related strains. *Applied Microbiology and Biotechnology, 86*, 965–972. https://doi.org/10.1007/s00253-009-2409-6

Alcaide-Hidalgo, J. M., Pueyo, E., Polo, M. C., & Martínez-Rodríguez, A. J. (2007). Bioactive peptides released from Saccharomyces cerevisiae under accelerated autolysis in a wine model system. *Journal of Food Science, 72*, M276–M279. https://doi.org/10.1111/j.1750-3841.2007.00450.x

Alejandro Ruiz, F. E., Ortega Jácome, J. F., Tejera, E., & Alvarez-Suarez, J. M. (2025). Edible insects as functional foods: Bioactive compounds, health benefits, safety concerns, allergenicity, and regulatory considerations. *Frontiers in Nutrition, 12*, 1571084. https://doi.org/10.3389/fnut.2025.1571084

Amara, A. A., & El-Baky, N. A. (2023). Fungi as a source of edible proteins and animal feed. *Journal of Fungi (Basel), 9*. https://doi.org/10.3390/jof9010073

An, M., Gao, G. F., Qi, J., Li, F., & Liu, X. (2010). Expression and crystallographic studies of a fungal immunomodulatory protein LZ-8 from a medicinal fungus Ganoderma lucidum. *Sheng Wu Gong Cheng Xue Bao, 26*, 1563–1568.

Anupama, R. P. (2000). Value-added food: Single cell protein. *Biotechnology Advances, 18*, 459–479.

Anupma, A., & Tamang, J. P. (2016). Diversity of filamentous fungi isolated from some amylase and alcohol-producing starters of India. *Frontiers in Microbiology, 11*. https://doi.org/10.3389/fmicb.2020.00905

Aoki, W., Watanabe, M., Watanabe, M., Kobayashi, N., Terajima, J., Sugita-Konishi, Y., Kondo, K., & Hara-Kudo, Y. (2020). Discrimination between edible and poisonous mushrooms among Japanese Entoloma sarcopum and related species based on phylogenetic analysis and insertion/deletion patterns of nucleotide sequences of the cytochrome oxidase 1 gene. *Genes & Genetic Systems, 95*, 133–139. https://doi.org/10.1266/ggs.19-00032

Aslankoohi, E., Zhu, B., Rezaei, M. N., Voordeckers, K., De Maeyer, D., Marchal, K., Dornez, E., Courtin, C. M., & Verstrepen, K. J. (2013). Dynamics of the Saccharomyces cerevisiae transcriptome during bread dough fermentation. *Applied and Environmental Microbiology, 79*, 7325–7333. https://doi.org/10.1128/aem.02649-13

Bakratsas, G., Polydera, A., Nilson, O., Chatzikonstantinou, A. V., Xiros, C., Katapodis, P., & Stamatis, H. (2023). Mycoprotein production by submerged fermentation of the edible mushroom *Pleurotus ostreatus* in a batch stirred tank bioreactor using agro-industrial hydrolysate. *Food, 12*. https://doi.org/10.3390/foods12122295

Bano, Z., & Rajarathnam, S. (1988). Pleurotus mushrooms. Part II. Chemical composition, nutritional value, post-harvest physiology, preservation, and role as human food. *Critical Reviews in Food Science and Nutrition, 27*, 87–158. https://doi.org/10.1080/10408398809527480

Bartholomai, B. M., Ruwe, K. M., Thurston, J., Jha, P., Scaife, K., Simon, R., Abdelmoteleb, M., Goodman, R. E., & Farhi, M. (2022). Safety evaluation of Neurospora crassa mycoprotein for use as a novel meat alternative and enhancer. *Food and Chemical Toxicology, 168*, 113342. https://doi.org/10.1016/j.fct.2022.113342

Barzee, T. J., Cao, L., Pan, Z., & Zhang, R. (2021a). Fungi for future foods. *Journal of Future Foods, 1*, 25–37. https://doi.org/10.1016/j.jfutfo.2021.09.002

Barzee, T. J., Cao, L., Pan, Z., & Zhang, R. (2021b). Fungi for future foods. *Journal of Future Foods, 1*, 25–37.

Bátori, V., Ferreira, J. A., Taherzadeh, M. J., & Lennartsson, P. R. (2015). Ethanol and protein from ethanol plant by-products using edible fungi Neurospora intermedia and Aspergillus oryzae. *BioMed Research International, 2015*, 176371. https://doi.org/10.1155/2015/176371

Bautista Justo, M., Alanís Guzmán, M. G., González de Mejía, E., García Díaz, C. L., Martínez, G., & Barboza Corona, E. (1999). Protein quality of three strains of Mexican mushrooms (Pleurotus ostreatus). *Archivos Latinoamericanos de Nutrición, 49*, 81–85.

Berne, S., Krizaj, I., Pohleven, F., Turk, T., Macek, P., & Sepcić, K. (2002). Pleurotus and Agrocybe hemolysins, new proteins hypothetically involved in fungal fruiting. *Biochimica et Biophysica Acta, 1570*, 153–159. https://doi.org/10.1016/s0304-4165(02)00190-3

Bigey, F., Menatong Tene, X., Wessner, M., Pradal, M., Aury, J. M., Cruaud, C., & Neuvéglise, C. (2024). Differential adaptation of the yeast Candida anglica to fermented food. *Food Microbiology, 123*, 104584. https://doi.org/10.1016/j.fm.2024.104584

Bourdichon, F., Casaregola, S., Farrokh, C., Frisvad, J. C., Gerds, M. L., Hammes, W. P., Harnett, J., Huys, G., Laulund, S., Ouwehand, A., et al. (2012a). Food fermentations: Microorganisms with technological beneficial use. *International Journal of Food Microbiology, 154*, 87–97. https://doi.org/10.1016/j.ijfoodmicro.2011.12.030

Bourdichon, F. O., Casaregola, S., Farrokh, C., Frisvad, J. C., Gerds, M. L., Hammes, W. P., Harnett, J., Huys, G., Laulund, S., Ouwehand, A., et al. (2012b). Food fermentations: Microorganisms with technological beneficial use. *International Journal of Food Microbiology, 154*, 87–97. https://doi.org/10.1016/j.ijfoodmicro.2011.12.030

Bovi, M., Carrizo, M. E., Capaldi, S., Perduca, M., Chiarelli, L. R., Galliano, M., & Monaco, H. L. (2011). Structure of a lectin with antitumoral properties in king bolete (Boletus edulis) mushrooms. *Glycobiology, 21*, 1000–1009. https://doi.org/10.1093/glycob/cwr012

Boyaci Gunduz, C. P., Agirman, B., & Erten, H. (2022). Identification of yeasts in fermented foods and beverages using MALDI-TOF MS. *FEMS Yeast Research, 22*. https://doi.org/10.1093/femsyr/foac056

Branco, P., Francisco, D., Chambon, C., Hébraud, M., Arneborg, N., Almeida, M. G., Caldeira, J., & Albergaria, H. (2014). Identification of novel GAPDH-derived antimicrobial peptides secreted by Saccharomyces cerevisiae and involved in wine microbial interactions. *Applied Microbiology and Biotechnology, 98*, 843–853. https://doi.org/10.1007/s00253-013-5411-y

Branco, P., Viana, T., Albergaria, H., & Arneborg, N. (2015). Antimicrobial peptides (AMPs) produced by Saccharomyces cerevisiae induce alterations in the intracellular pH, membrane permeability and culturability of Hanseniaspora guilliermondii cells. *International Journal of Food Microbiology, 205*, 112–118. https://doi.org/10.1016/j.ijfoodmicro.2015.04.015

Branco, P., Kemsawasd, V., Santos, L., Diniz, M., Caldeira, J., Almeida, M. G., Arneborg, N., & Albergaria, H. (2017). Saccharomyces cerevisiae accumulates GAPDH-derived peptides on its cell surface that induce death of non-saccharomyces yeasts by cell-to-cell contact. *FEMS Microbiology Ecology, 93*. https://doi.org/10.1093/femsec/fix055

Buitrago Mora, H. M., Piñeros, M. A., Espinosa Moreno, D., Restrepo Restrepo, S., Cardona Jaramillo, J. E. C., Álvarez Solano, Ó. A., Fernandez-Niño, M., & González Barrios, A. F. (2019). Multiscale design of a dairy beverage model composed of Candida utilis single cell protein supplemented with oleic acid. *Journal of Dairy Science, 102*, 9749–9762. https://doi.org/10.3168/jds.2019-16729

Carrasco de Mendoza, M. S., Basílico, J. C., Umansky, G., & Scarinci, H. E. (1991). Biological value of the unicellular protein of Kluyveromyces marxianus var. lactis. *Archivos Latinoamericanos de Nutrición, 41*, 72–78.

Cavusoglu, S., Uzun, Y., Yilmaz, N., Ercisli, S., Eren, E., Ekiert, H., Elansary, H. O., & Szopa, A. (2021). Maintaining the quality and storage life of button mushrooms (Agaricus bisporus) with gum, agar, sodium alginate, egg white protein, and lecithin coating. *Journal of Fungi (Basel)*, 7. https://doi.org/10.3390/jof7080614

Cha, W. S., Park, S. S., Kim, S. J., & Choi, D. (2010). Biochemical and enzymatic properties of a fibrinolytic enzyme from *Pleorotus eryngii* cultivated under solid-state conditions using corn cob. *Bioresource Technology, 101*, 6475–6481.

Chang, S. T., & Hayes, W. A. (1978). *Preface. Biology and cultivation edible mushrooms; academic press* (pp. xix–xxi).
Chen, L., & Cheung, P. C. (2014). Mushroom dietary fiber from the fruiting body of Pleurotus tuber-regium: Fractionation and structural elucidation of nondigestible cell wall components. *Journal of Agricultural and Food Chemistry, 62*, 2891–2899. https://doi.org/10.1021/jf500112j
Chen, M., Liao, J., Li, H., Cai, Z., Guo, Z., Wach, M. P., & Wang, Z. (2017). iTRAQ-MS/MS proteomic analysis reveals differentially expressed proteins during post-harvest maturation of the white button mushroom Agaricus bisporus. *Current Microbiology, 74*, 641–649. https://doi.org/10.1007/s00284-017-1225-y
Chen, X., Wang, B., & Pan, L. (2019). Heterologous expression and characterization of Penicillium citrinum nuclease P1 in Aspergillus niger and its application in the production of nucleotides. *Protein Expression and Purification, 156*, 36–43. https://doi.org/10.1016/j.pep.2018.12.004
Choi, H. S., & Sa, Y. S. (2000). Fibrinolytic and antithrombotic protease from *Ganoderma lucidum. Mycologia, 92*, 545–552.
Choi, H. S., & Shin, P. H. (1998). Purification and partial characterization of a fibrinolytic protease in *Pleurotus ostreatus. Mycologia, 90*, 674–679.
Choi, D., Cha, W. S., Park, N., Kim, H. W., Lee, J. H., Park, J. S., & Park, S. S. (2011a). Purification and characterization of a novel fibrinolytic enzyme from fruiting bodies of Korean *Cordyceps militaris. Bioresource Technology, 102*, 3279–3285.
Choi, C. H., Jo, S. Y., Park, H. J., Chang, S. K., Byeon, J.-S., & Myung, S.-J. (2011b). A randomized, double-blind, placebo-controlled multicenter trial of saccharomyces boulardii in irritable bowel syndrome. *Journal of Clinical Gastroenterology, 45*, 679–683. https://doi.org/10.1097/mcg.0b013e318204593e
Choi, J. H., Kim, D. W., Kim, S., Kim, S., & J. (2017). Purification and partial characterization of a fibrinolytic enzyme from the fruiting body of the medicinal and edible mushroom *Pleurotus ferulae. Preparative Biochemistry & Biotechnology, 47*, 539–546.
Cicha-Jeleń, M., Muszynska, B., Kala, K., & Sulkowska-Ziaja, K. (2024). Medicinal potential of the giant puffball mushroom Calvatia gigantea (Agaricomycetes): A review. *International Journal of Medicinal Mushrooms, 26*, 13–25. https://doi.org/10.1615/IntJMedMushrooms.2024054161
Cooray, S. T., & Chen, W. N. (2018). Valorization of brewer's spent grain using fungi solid-state fermentation to enhance nutritional value. *Journal of Functional Foods, 42*, 85–94. https://doi.org/10.1016/j.jff.2017.12.027
Coppola, F., Lombardi, S. J., & Tremonte, P. (2025). Edible insect meals as bioactive ingredients in sustainable snack bars. *Food, 14*. https://doi.org/10.3390/foods14040702
Cui, L., Liu, Q. H., Wang, H. X., & Ng, T. B. (2007). An alkaline protease from fresh fruiting bodies of the edible mushroom Pleurotus citrinopileatus. *Applied Microbiology and Biotechnology, 75*, 81–85. https://doi.org/10.1007/s00253-006-0801-z
Cui, L., Dong, M. S., Chen, X. H., Jiang, M., Lv, X., & Yan, G. J. (2008). A novel fibrinolytic enzyme from *Cordyceps militaris*, a Chinese traditional medicinal mushroom. *World Journal of Microbiology and Biotechnology, 24*, 483–489.
Dai, X., Sharma, M., & Chen, J. (2021). Fungi in sustainable food production. *Fungal Biology*. https://doi.org/10.1007/978-3-030-64406-2
Das, A. K., Nanda, P. K., Dandapat, P., Bandyopadhyay, S., Gullón, P., Sivaraman, G. K., McClements, D. J., Gullón, B., & Lorenzo, J. M. (2021a). Edible mushrooms as functional ingredients for development of healthier and more sustainable muscle foods: A flexitarian approach. *Molecules, 26*. https://doi.org/10.3390/molecules26092463
Das, A. K., Nanda, P. K., Dandapat, P., Bandyopadhyay, S., Gullón, P., Sivaraman, G. K., McClements, D. J., Gullón, B., & Lorenzo, J. M. (2021b). Edible mushrooms as functional ingredients for development of healthier and more sustainable muscle foods: A flexitarian approach. *Molecules (Basel, Switzerland), 26*, 2463. https://doi.org/10.3390/molecules26092463
de Gier, S., & Verhoeckx, K. (2018). Insect (food) allergy and allergens. *Molecular Immunology, 100*, 82–106. https://doi.org/10.1016/j.molimm.2018.03.015

Dębińska, A., & Sozańska, B. (2022). Fermented Food in asthma and respiratory allergies-chance or failure? *Nutrients, 14*. https://doi.org/10.3390/nu14071420

Debnath, S., & Sen, D. (2022). Mushrooms are potential foods against cancer: Identified by molecular docking and molecular dynamics simulation. *Natural Product Research, 36*, 2604–2609. https://doi.org/10.1080/14786419.2021.1912041

Derbyshire, E. J., & Delange, J. (2021). Fungal Protein – What is it and what is the health evidence? A systematic review focusing on Mycoprotein. *Frontiers in Sustainable Food Systems, 5*. https://doi.org/10.3389/fsufs.2021.581682

Dikkala, P. K., Usmani, Z., Kumar, S., Gupta, V. K., Bhargava, A., & Sharma, M. (2021). Fungal production of vitamins and their Food industrial applications. *Fungal Biology*, 563–596. https://doi.org/10.1007/978-3-030-85603-8_16

Dinleyici, E. C., Eren, M., Ozen, M., Yargic, Z. A., & Vandenplas, Y. (2012). Effectiveness and safety of *Saccharomyces boulardiifor* acute infectious diarrhea. *Expert Opinion on Biological Therapy, 12*, 395–410. https://doi.org/10.1517/14712598.2012.664129

Doaa, A. S., Omyma, E. S. S., Nader, R. A., Mona, M. S., & Amara, A. A. (2018). The development fortified Pan bread by increasing its protein content with high levels of live yeast cells *Saccharomyces cerevisiae. Corpus Journal of Dairy and Veterinary Science, 1*(2). https://doi.org/10.32474/CDVS.MS.ID.000109

Dong, W. G., Wang, Z. X., Feng, X. L., Zhang, R. Q., Shen, D. Y., Du, S., Gao, J. M., & Qi, J. (2022). Chromosome-level genome sequences, comparative genomic analyses, and secondary-metabolite biosynthesis evaluation of the medicinal edible mushroom Laetiporus sulphureus. *Microbiol Spectr, 10*, e0243922. https://doi.org/10.1128/spectrum.02439-22

Drzewiecka, B., Wessely-Szponder, J., Swieca, M., Espinal, P., Fusté, E., & Fernández-De La Cruz, E. (2024). Bioactive peptides and other Immunomodulators of mushroom origin. *Biomedicine, 12*, 1483.

Dugoua, J.-J., Machado, M., Zhu, X., Chen, X., Koren, G., & Einarson, T. R. (2009). Probiotic safety in pregnancy: A systematic review and meta-analysis of randomized controlled trials of lactobacillus, Bifidobacterium, and saccharomyces spp. *Journal of Obstetrics and Gynaecology Canada, 31*, 542–552. https://doi.org/10.1016/s1701-2163(16)34218-9

Efsa Panel on Nutrition NF, Food, A., Turck, D., Bohn, T., Castenmiller, J., De Henauw, S., Hirsch-Ernst, K. I., Maciuk, A., Mangelsdorf, I., McArdle, H. J., et al. (2022). Safety of pea and rice protein fermented by shiitake (Lentinula edodes) mycelia as a novel food pursuant to regulation (EU) 2015/2283. *EFSA Journal, 20*, e07205–e07205. https://doi.org/10.2903/j.efsa.2022.7205

Elhusseiny, S. M., El-Mahdy, T. S., Awad, M. F., Elleboudy, N. S., Farag, M. M. S., Aboshanab, K. M., & Yassien, M. A. (2021). Antiviral, cytotoxic, and antioxidant activities of three edible agaricomycetes mushrooms: Pleurotus columbinus, Pleurotus sajor-caju, and Agaricus bisporus. *Journal of Fungi (Basel)*, 7. https://doi.org/10.3390/jof7080645

Ennamany, R., Kretz, O., Badoc, A., Deffieux, G., & Creppy, E. E. (1994). Effect of bolesatine, a glycoprotein from Boletus satanas, on rat thymus in vivo. *Toxicology, 89*, 113–118. https://doi.org/10.1016/0300-483x(94)90219-4

Enzing, C., Ploeg, M., Barbosa, M., & Sijtsma, L. (2014). Microalgae-based products for the food and feed sector: An outlook for Europe. In JRC Publications Repository, M. Vigani, C. Parisi, E. R. Cerezo (Eds.), *EUR 26255* (pp. 1–82). Luxembourg: Publications Office of the European Union. Available online: https://ec.europa.eu/jrc/en/publication/eur-scientific-and-technical-researchreports/microalgae-based-products-food-and-feed-sector-outlook-europe. Accessed on 5 July 2022.

FAOSTAT. (2020). *Food and Agriculture Data*. Available online: http://www.fao.org/faostat/en/#home. Accessed on 20 May 2020.

Finnigan, T. J. A., Wall, B. T., Wilde, P. J., Stephens, F. B., Taylor, S. L., & Freedman, M. R. (2019). Mycoprotein: The future of nutritious nonmeat protein, a symposium review. *Current Developments in Nutrition, 3*, nzz021. https://doi.org/10.1093/cdn/nzz021

Ford, R. E., Foster, G. D., & Bailey, A. M. (2022). Exploring fungal RiPPs from the perspective of chemical ecology. *Fungal Biology and Biotechnology, 9*, 12.

Fu, Q., Shi, H., Hu, D., Cheng, J., Chen, S., & Ben, A. (2022). Pork longissimus dorsi marinated with edible mushroom powders: Evaluation of quality traits, microstructure, and protein degradation. *Food Research International, 158*, 111503. https://doi.org/10.1016/j.foodres.2022.111503

Galán, B., Mendoza, C. G., Calonje, M., & Novaes-Ledieu, M. (1999). News & notes: Production, purification, and properties of an Endo-1,3-β-Glucanase from the basidiomycete Agaricus bisporus. *Current Microbiology, 38*, 190–193. https://doi.org/10.1007/pl00006785

Gałęcki, R., Hanuszewska-Dominiak, M., & Kaczmar, E. (2024). Edible insects as a source of dietary protein for companion animals with food responsive enteropathies - perspectives and possibilities. *Polish Journal of Veterinary Sciences, 27*, 309–318. https://doi.org/10.24425/pjvs.2024.149361

Gangwar, R., & Maurya, V. K. (2022). Antibacterial, antibiofilm, and anti-hemolytic activity of the anti-microbial peptides from Agaricus bisporus against Oropathogenic bacteria. *Journal of Environment and Bio-Sciences, 36*, 183–191.

Gao, C., Liu, P., Li, W., Chen, W., Zhang, Z., Wu, D., Huang, J., Dong, G., & Yang, Y. (2025). Preparation of Morchella esculenta protein and its preventive effect on nonalcoholic fatty liver disease in mice. *Food & Function, 16*, 1086–1099. https://doi.org/10.1039/d4fo04489d

Ge, Z. W., Liu, X. B., Zhao, K., & Yang, Z. L. (2015). Species diversity of Flammulina in China: New varieties and a new record. *Mycosystema, 34*, 589–603.

Ghaly, A. E., Ben-Hassan, R. M., Mansour, M. H., & Nassar, M. A. (1993). Modeling batch production of single cell protein from cheese whey. *Applied Biochemistry and Biotechnology, 43*, 15–24. https://doi.org/10.1007/bf02916426

Ghosh, S., Basak, G., Saha, R., & Acharya, K. (2023). Bioactivity profiling of infusion, decoction, and hydroalcoholic fraction obtained from wild Golden Chanterelle mushroom, Cantharellus cibarius (Agaricomycetes). *International Journal of Medicinal Mushrooms, 25*, 23–33. https://doi.org/10.1615/IntJMedMushrooms.2022046751

Glamočlija, J., Stojković, D., Nikolić, M., Ćirić, A., Reis, F. S., Barros, L., Ferreira, I. C., & Soković, M. (2015). A comparative study on edible Agaricus mushrooms as functional foods. *Food & Function, 6*, 1900–1910. https://doi.org/10.1039/c4fo01135j

Gong, P., Huang, Z., Guo, Y., Wang, X., Yue, S., Yang, W., Chen, F., Chang, X., & Chen, L. (2022). The effect of superfine grinding on physicochemical properties of three kinds of mushroom powder. *Journal of Food Science, 87*, 3528–3541. https://doi.org/10.1111/1750-3841.16239

González, A., Cruz, M., Losoya, C., Nobre, C., Loredo, A., Rodríguez, R., Contreras, J., & Belmares, R. (2020). Edible mushrooms as a novel protein source for functional foods. *Food & Function, 11*, 7400–7414. https://doi.org/10.1039/d0fo01746a

Gu, C., Zheng, F., Long, L., Wang, J., & Ding, S. (2014). Engineering the expression and characterization of two novel laccase isoenzymes from Coprinus comatus in Pichia pastoris by fusing an additional ten amino acids tag at N-terminus. *PLoS One, 9*, e93912. https://doi.org/10.1371/journal.pone.0093912

Guillamon, E., Garcia-Lafuente, A., Lozano, M., D'arrigo, M., Rostagno, M. A., Villares, A., & Alfredo Martinez, J. (2010). Edible mushrooms: Role in the prevention of cardiovascular diseases. *Fitoterapia, 81*, 715–723.

Gunawardena, D., Bennett, L., Shanmugam, K., King, K., Williams, R., Zabaras, D., Head, R., Ooi, L., Gyengesi, E., & Münch, G. (2014). Anti-inflammatory effects of five commercially available mushroom species determined in lipopolysaccharide and interferon-γ activated murine macrophages. *Food Chemistry, 148*, 92–96. https://doi.org/10.1016/j.foodchem.2013.10.015

Guo, A., Hui, C., Ma, Y., Zhang, X., Zhang, L., Xu, S., & Li, C. (2025). Upcycling chitin waste and aged Rice into fungi protein through fermentation with Cordyceps militaris. *J Fungi (Basel), 11*. https://doi.org/10.3390/jof11040315

Gupta, S., Summuna, B., Gupta, M., & Annepu, S. K. (2019). Edible mushrooms: Cultivation, bioactive molecules, and health benefits. *Reference Series in Phytochemistry*, 1815–1847. https://doi.org/10.1007/978-3-319-78030-6_86

Guslandi, M., Mezzi, G., Sorghi, M., & Testoni, P. A. (2000). Saccharomyces boulardii in maintenance treatment of Crohn's disease. *Digestive Diseases and Sciences, 45*, 1462–1464. https://doi.org/10.1023/a:1005588911207

Hadar, Y., & Cohen-Arazi, E. (1986). Chemical composition of the edible mushroom Pleurotus ostreatus produced by fermentation. *Applied and Environmental Microbiology, 51*, 1352–1354. https://doi.org/10.1128/aem.51.6.1352-1354.1986

Han, Q., Li, H., Zhao, F., Gao, J., Liu, X., & Ma, B. (2023). Auricularia auricula peptides nutritional supplementation delays H(2)O(2)-induced senescence of HepG2 cells by modulation of MAPK/NF-κB signaling pathways. *Nutrients, 15*. https://doi.org/10.3390/nu15173731

Hassan, R. A., Shafi, M. E., Attia, K. M., & Assar, M. H. (2020). Influence of oyster mushroom waste on growth performance, immunity and intestinal morphology compared with antibiotics in broiler chickens. *Frontiers in Veterinary Science, 7*, 333–333. https://doi.org/10.3389/fvets.2020.00333

Hazards, E., Panel, O. B., Koutsoumanis, K., Allende, A., Alvarez-Ordóñez, A., Bolton, D., Bover-Cid, S., Chemaly, M., Davies, R., De Cesare, A., Hilbert, F., et al. (2020). Scientific opinion on the update of the list of QPS-recommended biological agents intentionally added to food or feed as notified to EFSA (2017–2019). *EFSA Journal, 18*, e05966. https://doi.org/10.2903/j.efsa.2020.5966

Hellwig, C., Gmoser, R., Lundin, M., Taherzadeh, M. J., & Rousta, K. (2020). Fungi burger from stale bread? A case study on perceptions of a novel protein-rich Food product made from an edible fungus. *Foods (Basel, Switzerland), 9*, 1112. https://doi.org/10.3390/foods9081112

Hittinger, C. T., Steele, J. L., & Ryder, D. S. (2018). Diverse yeasts for diverse fermented beverages and foods. *Current Opinion in Biotechnology, 49*, 199–206. https://doi.org/10.1016/j.copbio.2017.10.004

Hitzman, D. O. (1986). The Provesteen process—An ultra-high densityfermentation. In M. Moo-Young, K. F. Gregory, (Eds.), *Microbial biomass proteins* (pp. 27–32). Elsevier.

Ho, L.-H., Asyikeen Zulkifli, N., & Tan, T.-C. (2020). Edible mushroom: Nutritional properties, potential nutraceutical values, and its utilisation in food product development. *An Introduction to Mushroom*. https://doi.org/10.5772/intechopen.91827

Hong, L., Xun, M., & Wutong, W. (2007). Anti-diabetic effect of an A-glucan from fruit body of Maitake (Grifola Frondosa) on KK-Ay mice. *The Journal of Pharmacy and Pharmacology, 59*, 575–582.

Hong, S. S., Lee, J. H., Jeong, W., Kim, N., Jin, H. Z., Hwang, B. Y., Lee, H. J., Lee, S. J., Jang, D. S., & Lee, D. (2012). Acetylenic acid analogues from the edible mushroom Chanterelle (Cantharellus cibarius) and their effects on the gene expression of peroxisome proliferator-activated receptor-gamma target genes. *Bioorganic & Medicinal Chemistry Letters, 22*, 2347–2349. https://doi.org/10.1016/j.bmcl.2012.01.070

Hung, C. L., Chang, A. J., Kuo, X. K., & Sheu, F. (2014). Molecular cloning and function characterization of a new macrophage-activating protein from Tremella fuciformis. *Journal of Agricultural and Food Chemistry, 62*, 1526–1535. https://doi.org/10.1021/jf403835c

Hyde, K. D., Xu, J., Rapior, S., Jeewon, R., Lumyong, S., Niego, A. G. T., Abeywickrama, P. D., Aluthmuhandiram, J. V. S., Brahamanage, R. S., Brooks, S., et al. (2019). The amazing potential of fungi: 50 ways we can exploit fungi industrially. *Fungal Diversity, 97*, 1–136. https://doi.org/10.1007/s13225-019-00430-9

Ibarruri, J., Cebrián, M., & Hernández, I. (2019). Solid state fermentation of brewer's spent grain using Rhizopus sp. to enhance nutritional value. *Waste and Biomass Valorization, 10*, 3687–3700. https://doi.org/10.1007/s12649-019-00654-5

Ibarruri, J., Cebrián, M., & Hernández, I. (2021). Valorisation of fruit and vegetable discards by fungal submerged and solid-state fermentation for alternative feed ingredients production. *Journal of Environmental Management, 281*, 111901. https://doi.org/10.1016/j.jenvman.2020.111901

Inc, G.M.I. (2022). *Single cell protein market share industry statistics—2030*. Available online: https://www. gminsights.com/industry-analysis/single-cell-proteinmarket. Accessed on 12 December 2022.

Jagadish, B. R., Sridhar, K. R., Dattaraj, H. R., Chandramohana, N., & Mahadevakumar, S. (2021). Nutraceutical potential of wild edible mushroom Hygrocybe alwisii. *Fungal Biology*, 597–615. https://doi.org/10.1007/978-3-030-85603-8_17

Jiang, N., Huang, Y. L., Zhang, Z., Qi, Q., & Li, C. (2025). Evaluation of natural polysaccharides from edible mushrooms for the treatment of male testicular injury caused by X-ray. *Journal of Food Science, 90*, e17679. https://doi.org/10.1111/1750-3841.17679

Jin, Y., Yu, S., Liu, J. J., Yun, E. J., Lee, J. W., Jin, Y. S., & Kim, K. H. (2021). Production of neoagarooligosaccharides by probiotic yeast Saccharomyces cerevisiae var. boulardii engineered as a microbial cell factory. *Microbial Cell Factories, 20*, 160. https://doi.org/10.1186/s12934-021-01644-w

Johnson, E. A. (2012). Biotechnolog0y of non-saccharomyces yeasts—The ascomycetes. *Applied Microbiology and Biotechnology, 97*, 503–517. https://doi.org/10.1007/s00253-012-4497-y

Johnston, B. C., Goldenberg, J. Z., Vandvik, P. O., Sun, X., & Guyatt, G. H. (2011). Probiotics for the prevention of pediatric antibiotic-associated diarrhea. In *Cochrane database of systematic reviews*. Wiley.

Kavanagh, K. (2005). Fungal fermentation systems and products. *Fungi*, 89–112. https://doi.org/10.1002/0470015330.ch4

Kępińska-Pacelik, J., & Biel, W. (2022). Insects in pet food industry-hope or threat? *Animals (Basel), 12*. https://doi.org/10.3390/ani12121515

Khaund, P., & Joshi, S. R. (2016). Lentinula edodes based GIS mapping, biometabolites and antiinflamatory activity of wild edible mushrooms from tropical 'sacred grove' forests of Meghalaya, India. *Revista de Biologia Tropical, 64*, 247–257. https://doi.org/10.15517/rbt.v64i1.20349

Kim, J. H., & Kim, Y. S. (1999). A fibrinolytic metalloprotease from the fruiting bodies of an edible mushroom, Armillariella mellea. *Bioscience, Biotechnology and Biochemistry, 63*, 2130–2136.

Kim, J. H., & Kim, Y. S. (2001). Characterization of a metalloenzyme from a wild mushroom, *Tricholoma saponaceum. Bioscience, Biotechnology and Biochemistry, 65*, 356–362.

Kim, J.-S., Sapkota, K., Park, S.-E., Choi, B.-S., Kim, S., Nguyen, T. H., Kim, C.-S., Choi, H.-S., Kim, M.-K., Chun, H.-S., et al. (2006). A fibrinolytic enzyme from the medicinal mushroom *Cordyceps militaris. Journal of Microbiology, 44*, 622–631.

Kim, J.-S., Kim, J.-E., Choi, B.-S., Park, S.-E., Sapkota, K., Kim, S., Lee, H.-H., Kim, C.-S., Park, Y., Kim, M.-K., et al. (2008). Purification and characterization of fibrinolytic metalloprotease from *Perenniporia fraxinea* mycelia. *Mycological Research, 112*, 990–998.

Kim, H. C., Choi, B. S., Sapkota, K., Kim, S., Lee, H. J., Yoo, J. C., & Kim, S.-J. (2011). Purification and characterization of a novel, highly potent fibrinolytic enzyme from *Paecilomyces tenuipes. Process Biochemistry, 46*, 1545–1553.

Kimatu, B. M., Zhao, L., Biao, Y., Ma, G., Yang, W., Pei, F., & Hu, Q. (2017). Antioxidant potential of edible mushroom (Agaricus bisporus) protein hydrolysates and their ultrafiltration fractions. *Food Chemistry, 230*, 58–67. https://doi.org/10.1016/j.foodchem.2017.03.030

Ko, J. L., Hsu, C. I., Lin, R. H., Kao, C. L., & Lin, J. Y. (1995). A new fungal immunomodulatory protein, FIP-fve isolated from the edible mushroom, Flammulina velutipes and its complete amino acid sequence. *European Journal of Biochemistry, 228*, 244–249.

Kobayashi, N., Wada, N., Yokoyama, H., Tanaka, Y., Suzuki, T., Habu, N., & Konno, N. (2023). Extracellular enzymes secreted in the mycelial block of Lentinula edodes during hyphal growth. *AMB Express, 13*, 36. https://doi.org/10.1186/s13568-023-01547-6

Koivurinta, J., Kurkela, R., & Koivistoinen, P. (1979). Uses of Pekilo, a microfungus biomass from *Paecilomyces varioti* in sausage and meat balls. *International Journal of Food Science & Technology, 14*, 561–570. https://doi.org/10.1111/j.1365-2621.1979.tb00902.x

Kubíková, J. (1968). Fusarium oxysporum (Schlecht.) Snyd. et Hans. – A dominant fungus species on the root-surface of woody plant seedlings. *Plant and Soil, 28*, 306–312. https://doi.org/10.1007/bf01880248

Kuforiji, O. O., & Fasidi, I. O. (2009). Biodegradation of agro-industrial wastes by a edible mushroom Pleurotus tuber-regium (Fr.). *Journal of Environmental Biology, 30*, 355–358.

Kwok, Y., Ng, K. F., Li, C. C., Lam, C. C., & Man, R. Y. (2005). A prospective, randomized, double-blind, placebo-controlled study of the platelet and global hemostatic effects of *Ganoderma lucidum* (Ling-Zhi) in healthy volunteers. *Anesthesia and Analgesia, 101*, 423–426.

Lähteenmäki-Uutela, A., Rahikainen, M., Lonkila, A., & Yang, B. (2021). Alternative proteins and EU food law. *Food Control, 130*, 108336. https://doi.org/10.1016/j.foodcont.2021.108336

Lallawmsanga, Passari, A. K., Mishra, V. K., Leo, V. V., Singh, B. P., Valliammai Meyyappan, G., Gupta, V. K., Uthandi, S., & Upadhyay, R. C. (2016). Antimicrobial potential, identification and phylogenetic affiliation of wild mushrooms from two sub-tropical semi-evergreen Indian Forest ecosystems. *PLoS One, 11*, e0166368. https://doi.org/10.1371/journal.pone.0166368

Landi, N., Ragucci, S., Culurciello, R., Russo, R., Valletta, M., Pedone, P. V., Pizzo, E., & Di Maro, A. (2021). Ribotoxin-like proteins from Boletus edulis: Structural properties, cytotoxicity and in vitro digestibility. *Food Chemistry, 359*, 129931. https://doi.org/10.1016/j.foodchem.2021.129931

Laurujisawat, P., Dumrongchai, T., Rodklongtan, A., & Chitprasert, P. (2025). Spray-dried microencapsulation of probiotics with Genipin-crosslinked whey protein isolate for enhanced stability in fortified instant cereal drinks. *Journal of Food Science, 90*, e70257. https://doi.org/10.1111/1750-3841.70257

Lee, S. Y., Kim, J. S., Kim, J. E., Sapkota, K., Shen, M. H., Kim, S., Chun, H. S., Yoo, J. C., Choi, H. S., Kim, M. K., et al. (2005). Purification and characterization of fibrinolytic enzyme from cultured mycelia of Armillaria mellea. *Protein Expression and Purification, 43*, 10–17.

Lee, J. S., Baik, H.-S., & Park, S.-S. (2006). Purification and characterization of two novel fibrinolytic proteases from mushroom. *Fomitella fraxinea. J. Microbiol. Biotechnol., 16*, 264–271.

Lee, T.-K., Nguyen, T. T. H., Park, N., Kwak, S.-H., Kim, J., Jin, S., Son, G.-M., Hur, J., Choi, J.-I., & Kim, D. (2018). The use of fermented buckwheat to produce L-carnitine enriched oyster mushroom. *AMB Express, 8*, 138–138. https://doi.org/10.1186/s13568-018-0664-6

Leger, D., Matassa, S., Noor, E., Shepon, A., Milo, R., & Bar-Even, A. (2021). Photovoltaic-driven microbial protein production can use land and sunlight more efficiently than conventional crops. *Proceedings of the National Academy of Sciences of the United States of America, 118*, e2015025118. https://doi.org/10.1073/pnas.2015025118

Leonti, M., Bellot, S., Zucca, P., & Rescigno, A. (2020). Astringent drugs for bleedings and diarrhoea: The history of Cynomorium coccineum (Maltese mushroom). *Journal of Ethnopharmacology, 249*, 112368. https://doi.org/10.1016/j.jep.2019.112368

Li, H. P., Hu, Z., Yuan, J. L., Fan, H. D., Chen, W., Wang, S. J., Zheng, S. S., Zheng, Z. L., & Zou, G. L. (2007). A novel extracellular protease with fibrinolytic activity from the culture supernatant of *Cordyceps sinensis*: Purification and characterization. *Phytotherapy Research, 21*, 1234–1241.

Li, D., Sossah, F. L., Sun, L., Fu, Y., & Li, Y. (2019). Genome analysis of Hypomyces perniciosus, the causal agent of wet bubble disease of button mushroom (Agaricus bisporus). *Genes (Basel), 10*. https://doi.org/10.3390/genes10060417

Li, Q., Liu, J., Shang, X., Li, Y., Zhang, L., Li, Z., Jiang, N., Tan, Q., Yu, H., & Song, C. (2021). Characterizing diversity based on the chemical and nutritional composition of shiitake culinary-medicinal mushroom Lentinula edodes (Agaricomycetes) commonly cultivated in China. *International Journal of Medicinal Mushrooms, 23*, 51–64. https://doi.org/10.1615/IntJMedMushrooms.2021039240

Li, Q., Yi, P., Zhang, J., Shan, Y., Lin, Y., Wu, M., Wang, K., Tian, G., Li, J., & Zhu, T. (2023). Bioconversion of food waste to crayfish feed using solid-state fermentation with yeast. *Environmental Science and Pollution Research International, 30*, 15325–15334. https://doi.org/10.1007/s11356-022-23100-x

Li, X., Tao, Q., Hu, Q., Ma, N., & Ma, G. (2024). In vitro gastrointestinal digestion and fecal fermentation of Pleurotus eryngii proteins extracted using different methods: Insights for the utilization of edible mushroom-based proteins as novel nutritional and functional components. *Food & Function, 15*, 8865–8877. https://doi.org/10.1039/d4fo02604g

Lin, J. Y., Jeng, T. W., Chen, C. C., Shi, G. Y., & Tung, T. C. (1973). Isolation of a new cardiotoxic protein from the edible mushroom, Volvariella volvacea. *Nature, 246*, 524–525. https://doi.org/10.1038/246524a0

Lin, J. Y., Wu, H. L., & Shi, G. Y. (1975). Toxicity of the cardiotoxic protein flammutoxin, isolate from edible mushroom Flammulina velutipes. *Toxicon, 13*, 323–331. https://doi.org/10.1016/0041-0101(75)90191-9

Lin, J. W., Jia, J., Shen, Y. H., Zhong, M., Chen, L. J., Li, H. G., Ma, H., Guo, Z. F., Qi, M. F., Liu, L. X., et al. (2013). Functional expression of FIPfve, a fungal immunomodulatory protein from the edible mushroom *Flammulina velutipes* in *Pichia pastoris* GS115. *Journal of Biotechnology, 168*, 527–533.

Lin, T., Li, Z., Fan, G., & Xie, C. (2024). Enhancing the nutritional value and antioxidant properties of foxtail millet by solid-state fermentation with edible fungi. *Food Science & Nutrition, 12*, 6660–6672. https://doi.org/10.1002/fsn3.4203

Linder, T. (2024). What next for mycoprotein? *Current Opinion in Food Science, 58*, 101199.

Liu, Y. T., Sun, J., Luo, Z. Y., Rao, S. Q., Su, Y. J., Xu, R. R., & Yang, Y. J. (2012). Chemical composition of five wild edible mushrooms collected from Southwest China and their antihyperglycemic and antioxidant activity. *Food and Chemical Toxicology, 50*, 1238–1244. https://doi.org/10.1016/j.fct.2012.01.023

Liu, J., Wu, Y. C., Kan, J., Wang, Y., & Jin, C. H. (2013). Changes in reactive oxygen species production and antioxidant enzyme activity of Agaricus bisporus harvested at different stages of maturity. *Journal of the Science of Food and Agriculture, 93*, 2201–2206. https://doi.org/10.1002/jsfa.6027

Liu, X., Zheng, X., Qian, P., Kopparapu, N., Deng, Y., Nonaka, M., & Harada, N. (2014). Purification and characterization of a novel fibrinolytic enzyme from culture supernatant of *Pleurotus ostreatus*. *Journal of Microbiology and Biotechnology, 24*, 245–253.

Liu, J., Sun, J., He, R., Xia, J., & He, P. (2024). The situation of counterfeited and mislabeled commercialized edible mushrooms in China and the development of possible controls. *Food, 13*. https://doi.org/10.3390/foods13193097

Looney, B. (2013). Systematics of the genus Auricularia with an emphasis on species from the southeastern United States, North American. *Fungi*. https://doi.org/10.2509/naf2013.008.006

Losoya-Sifuentes, C., Cruz, M., Del Refugio Rocha-Pizaña, M., Loredo-Treviño, A., & Belmares, R. (2025). Edible mushrooms: A nutrient-rich ingredient for healthier Food products – A review. *Current Nutrition Reports, 14*, 9. https://doi.org/10.1007/s13668-024-00605-0

Lu, C. L., Chen, S., & Chen, S. N. (2010). Purification and characterization of a novel fibrinolytic protease from *Schizophyllum commune*. *Journal of Food and Drug Analysis, 18*, 69–76.

Lübeck, M., & Lübeck, P. S. (2022). Fungal cell factories for efficient and sustainable production of proteins and peptides. *Microorganisms, 10*. https://doi.org/10.3390/microorganisms10040753

Luo, H., Liu, Y., Fang, L., Li, X., Tang, N., & Zhang, K. (2007). Coprinus comatus damages nematode cuticles mechanically with spiny balls and produces potent toxins to immobilize nematodes. *Applied and Environmental Microbiology, 73*, 3916–3923. https://doi.org/10.1128/AEM.02770-06

Ma, N., Tao, H., Du, H., Zhao, L., Hu, Q., & Xiao, H. (2020). Antifatigue effect of functional cookies fortified with mushroom powder (Tricholoma Matsutake) in mice. *Journal of Food Science, 85*, 4389–4395. https://doi.org/10.1111/1750-3841.15510

Madhavan, A., Arun, K. B., Sindhu, R., Alphonsa Jose, A., Pugazhendhi, A., Binod, P., Sirohi, R., Reshmy, R., & Kumar Awasthi, M. (2022). Engineering interventions in industrial filamentous fungal cell factories for biomass valorization. *Bioresource Technology, 344*, 126209. https://doi.org/10.1016/j.biortech.2021.126209

Mahasneh, I. A. (1997). Production of single cell protein from five strains of the microalga *Chlorella* sp. (Chlorophyta). *Cytobiosciences, 90*, 153–161.

Maini Rekdal, V., Villalobos-Escobedo, J. M., Rodriguez-Valeron, N., Olaizola Garcia, M., Prado Vásquez, D., Rosales, A., Sörensen, P. M., Baidoo, E. E. K., Calheiros de Carvalho, A., Riley, R., et al. (2024). Neurospora intermedia from a traditional fermented food enables waste-to-food conversion. *Nature Microbiology, 9*, 2666–2683. https://doi.org/10.1038/s41564-024-01799-3

Mancheño, J. M., Tateno, H., Goldstein, I. J., Martínez-Ripoll, M., & Hermoso, J. A. (2005). Structural analysis of the Laetiporus sulphureus hemolytic pore-forming lectin in complex with sugars. *Journal of Biological Chemistry, 280*, 17251–17259. https://doi.org/10.1074/jbc.m413933200

Marathe, S. J., Hamzi, W., Bashein, A. M., Deska, J., Seppänen-Laakso, T., Singhal, R. S., & Shamekh, S. (2022). Anti-Angiogenic effect of Cantharellus cibarius extracts, its correlation with lipoxygenase inhibition, and role of the bioactives therein. *Nutrition and Cancer, 74*, 724–734. https://doi.org/10.1080/01635581.2021.1909739

Marsh, A. J., O'Sullivan, O., Hill, C., Ross, R. P., & Cotter, P. D. (2014). Sequence-based analysis of the bacterial and fungal compositions of multiple kombucha (tea fungus) samples. *Food Microbiology, 38*, 171–178. https://doi.org/10.1016/j.fm.2013.09.003

Maseko, K. H., Regnier, T., Bartels, P., & Meiring, B. (2025). Mushroom mycelia as sustainable alternative proteins for the production of hybrid cell-cultured meat: A review. *Journal of Food Science, 90*, e70060. https://doi.org/10.1111/1750-3841.70060

Mata, G., Salmones, D., & Pérez-Merlo, R. (2016). Hydrolytic enzyme activities in shiitake mushroom (Lentinula edodes) strains cultivated on coffee pulp. *Revista Argentina de Microbiología, 48*, 191–195. https://doi.org/10.1016/j.ram.2016.05.008

Menezes, A. G. T., Ramos, C. L., Cenzi, G., Melo, D. S., Dias, D. R., & Schwan, R. F. (2019). Probiotic potential, antioxidant activity, and phytase production of indigenous yeasts isolated from indigenous fermented foods. *Probiotics and Antimicrobial Proteins, 12*, 280–288. https://doi.org/10.1007/s12602-019-9518-z

Meyer, V. (2021). Metabolic engineering of filamentous fungi. *Metabolic Engineering*, 765–801. https://doi.org/10.1002/9783527823468.ch20

Meyer, V., Basenko, E. Y., Benz, J. P., Braus, G. H., Caddick, M. X., Csukai, M., de Vries, R. P., Endy, D., Frisvad, J. C., Gunde-Cimerman, N., et al. (2020). Growing a circular economy with fungal biotechnology: A white paper. *Fungal Biology and Biotechnology, 7*, 5–5. https://doi.org/10.1186/s40694-020-00095-z

Meyer, V., Cairns, T., Barthel, L., King, R., Kunz, P., Schmideder, S., Müller, H., Briesen, H., Dinius, A., & Krull, R. (2021). Understanding and controlling filamentous growth of fungal cell factories: Novel tools and opportunities for targeted morphology engineering. *Fungal Biology and Biotechnology, 8*, 8–8. https://doi.org/10.1186/s40694-021-00115-6

Mi, F., Zhang, Y., Yang, D., Tang, X., Wang, P., He, X., Zhang, Y., Dong, J., Cao, Y., Liu, C., et al. (2016). Evidence for inbreeding and genetic differentiation among geographic populations of the saprophytic mushroom Trogia venenata from southwestern China. *PLoS One, 11*, e0149507. https://doi.org/10.1371/journal.pone.0149507

Mihai, R. A., Melo Heras, E. J., Florescu, L. I., & Catana, R. D. (2022). The edible gray oyster fungi pleurotus ostreatus (Jacq. ex Fr.) P. Kumm a potent waste consumer, a biofriendly species with antioxidant activity depending on the growth substrate. *Journal of fungi (Basel, Switzerland), 8*, 274. https://doi.org/10.3390/jof8030274.

Ming, J., Chen, L., Hong, H., & Li, J. (2015). Effect of superfine grinding on the physico-chemical, morphological and thermogravimetric properties of Lentinus edodes mushroom powders. *Journal of the Science of Food and Agriculture, 95*, 2431–2437. https://doi.org/10.1002/jsfa.6967

Moo-Young, M., Chisti, Y., & Vlach, D. (1993). Fermentation of cellulosic materials to mycoprotein foods. *Biotechnology Advances, 11*, 469–479. https://doi.org/10.1016/0734-9750(93)90015-f

Morozova, E., Falina, N. N., Denisova, N. P., Barkova, L. V., & Psurtseva, N. V. (1982). Analiz komponentnogo sostava i substratnoÄ spetsifichnosti fibrinoliticheskogo preparata iz griba *Flammulina vetutipes* [Analysis of the component constitution and substrate specificity of a fibrinolytic preparation from the fungus *Flammulina velutipes*]. *Biokhimiia, 47*, 1181–1185.

mushroom. mushroom. (2013). *Encyclopædia Britannica student and Home Edition*. Encyclopædia Britannica.

Nasseri, A. T., Rasoul-Ami, S., Morowvat, M. H., & Ghasemi, Y. (2011). Single cell protein: Production and process. *American Journal of Food Technology, 6*, 103–116. https://doi.org/10.3923/ajft.2011.103.116

Ng, T. B., & Wang, H. X. (2004). Flammin and velin: New ribosome inactivating polypeptides from the mushroom Flammulina velutipes. *Peptides, 25*, 929–933. https://doi.org/10.1016/j.peptides.2004.03.007

Nonaka, T., Ishikawa, H., Tsumuraya, Y., Hashimoto, Y., & Dohmae, N. (1995). Characterization of a thermostable lysine-specific metalloendopeptidase from the fruiting bodies of a basidiomycete, Grifola frondosa. *The Journal of Biochemistry, 118*, 1014–1020.

Nout, M. J. R., & Kiers, J. L. (2005). Tempe fermentation, innovation and functionality: Update into the third millenium. *Journal of Applied Microbiology, 98*, 789–805. https://doi.org/10.1111/j.1365-2672.2004.02471.x

Nout, M. J. R., & Rombouts, F. M. (1990). Recent developments in tempe research. *Journal of Applied Bacteriology, 69*, 609–633. https://doi.org/10.1111/j.1365-2672.1990.tb01555.x

Nuhn, M. E., Binder, M., Taylor, A. F. S., Halling, R. E., & Hibbett, D. S. (2013). Phylogenetic overview of the Boletineae. *Fungal Biology, 117*, 479–511. https://doi.org/10.1016/j.funbio.2013.04.008

O'Connor, E., McGowan, J., McCarthy, C. G. P., Amini, A., Grogan, H., & Fitzpatrick, D. A. (2019). Whole genome sequence of the commercially relevant mushroom strain Agaricus bisporus var. bisporus ARP23. *G3 (Bethesda), 9*, 3057–3066. https://doi.org/10.1534/g3.119.400563

O'Brien, B. J., Milligan, E., Carver, J., & Roy, E. D. (2019). Integrating anaerobic co-digestion of dairy manure and food waste with cultivation of edible mushrooms for nutrient recovery. *Bioresource Technology, 285*, 121312. https://doi.org/10.1016/j.biortech.2019.121312

Omar, S., & Sabry, S. (1991). Microbial biomass and protein production from whey. *J. Islam. World Acad. Sci., 4*, 170–172.

Ota, K., Mikelj, M., Papler, T., Leonardi, A., Križaj, I., & Maček, P. (2013). Ostreopexin: A hemopexin fold protein from the oyster mushroom, Pleurotus ostreatus. *Biochimica et Biophysica Acta, 1834*, 1468–1473. https://doi.org/10.1016/j.bbapap.2013.03.027

Ower, R. (1982). Notes on the development of the Morel Ascocarp: Morchella Esculenta. *Mycologia, 74*, 142–144. https://doi.org/10.1080/00275514.1982.12021480

Papagianni, M. (2004). Fungal morphology and metabolite production in submerged mycelial processes. *Biotechnology Advances, 22*, 189–259. https://doi.org/10.1016/j.biotechadv.2003.09.005

Park, I. S., Park, J. U., Seo, M. J., Kim, M. J., Lee, H. H., Kim, S. R., Kang, B. W., Choi, Y. H., Joo, W. H., & Jeong, Y. K. (2010). Purification and biochemical characterization of a 17 kDa fibrinolytic enzyme from *Schizophyllum commune*. *Journal of Microbiology, 48*, 836–841.

Paul, S., & Joshi, S. R. (2022). Industrial perspectives of fungi. *Industrial Microbiology and Biotechnology*, 81–105. https://doi.org/10.1007/978-981-16-5214-1_3

Pereira, P. R., Freitas, C. S., & Paschoalin, V. M. F. (2021). *Saccharomyces cerevisiae* biomass as a source of next-generation food preservatives: Evaluating potential proteins as a source of antimicrobial peptides. *Comprehensive Reviews in Food Science and Food Safety, 20*, 4450–4479. https://doi.org/10.1111/1541-4337.12798

Pobiega, K., Sękul, J., Pakulska, A., Latoszewska, M., Michońska, A., Korzeniowska, Z., Macherzyńska, Z., Pląder, M., Duda, W., Szafraniuk, J., et al. (2024). Fungal proteins: Sources, production and purification methods, industrial applications, and future perspectives. *Applied Sciences, 14*, 6259.

Pratama, F., Rahardja, R. T., Rachmadi, A. R., Salam, W. Q., Kho, K., Adelie, A., & Devanthi, P. V. P. (2025). Optimizing Mycoprotein production by aspergillus oryzae using soy whey as a substrate. *Journal of Fungi (Basel), 11*. https://doi.org/10.3390/jof11050349

Ragucci, S., Hussain, H. Z. F., Bosso, A., Landi, N., Clemente, A., Pedone, P. V., Pizzo, E., & Di Maro, A. (2023). Isolation, characterization, and biocompatibility of bisporitin, a ribotoxin-like protein from white button mushroom (Agaricus bisporus). *Biomolecules, 13*. https://doi.org/10.3390/biom13020237

Rai, R. D. (2004). Chapter 21: Production of edible fungi. In D. K. Arora (Ed.), *Fungi biotechnology in agricultural, food, and environmental applications* (pp. 233–246). Marcel Dekker.

Rauf, A., Joshi, P. B., Ahmad, Z., Hemeg, H. A., Olatunde, A., Naz, S., Hafeez, N., & Simal-Gandara, J. (2023). Edible mushrooms as potential functional foods in amelioration of hypertension. *Phytotherapy Research, 37*, 2644–2660. https://doi.org/10.1002/ptr.7865

Rengel Dos Passos, F., Lopes Maestre, K., Florêncio da Silva, B., Rodrigues, A. C., Contini Triques, C., Alves Garcia, H., Fagundes-Klen, M. R., Antonio da Silva, E., & Fiorese, M. L. (2021). Production of a synbiotic composed of galacto-oligosaccharides and Saccharomyces boulardii using enzymatic-fermentative method. *Food Chemistry, 353*, 129486. https://doi.org/10.1016/j.foodchem.2021.129486

Ritala, A., Häkkinen, S. T., Toivari, M., & Wiebe, M. G. (2017). Single cell protein-state-of-the-art, industrial landscape and patents 2001-2016. *Frontiers in Microbiology, 8*, 2009–2009. https://doi.org/10.3389/fmicb.2017.02009

Riviere, J. (1977). Microbial Proteins. In M. O. Moss & J. E. Smith (Eds.), *Industrial applications of microbiology*. Surrey University Press.

Rizk, Z., El Rayess, Y., Ghanem, C., Mathieu, F., Taillandier, P., & Nehme, N. (2016). Impact of inhibitory peptides released by Saccharomyces cerevisiae BDX on the malolactic fermentation performed by Oenococcus oeni Vitilactic F. *International Journal of Food Microbiology, 233*, 90–96. https://doi.org/10.1016/j.ijfoodmicro.2016.06.018

Rizk, Z., Rayess, Y. E., Ghanem, C., Mathieu, F., Taillandier, P., & Nehme, N. (2018). Identification of multiple-derived peptides produced by Saccharomyces cerevisiae involved in malolactic fermentation inhibition. *FEMS Yeast Research, 18*. https://doi.org/10.1093/femsyr/foy080

Safari, O., & Sarkheil, M. (2018). Dietary administration of eryngii mushroom (Pleurotus eryngii) powder on haemato-immunological responses, bactericidal activity of skin mucus and growth performance of koi carp fingerlings (Cyprinus carpio koi). *Fish & Shellfish Immunology, 80*, 505–513. https://doi.org/10.1016/j.fsi.2018.06.046

Sakekar, A. A., Gaikwad, S. R., & Punekar, N. S. (2021). Protein expression and secretion by filamentous fungi. *Journal of Biosciences, 46*. https://doi.org/10.1007/s12038-020-00120-8

Scholtmeijer, K., van den Broek, L. A. M., Fischer, A. R. H., & van Peer, A. (2023). Potential protein production from Lignocellulosic materials using edible mushroom forming fungi. *Journal of Agricultural and Food Chemistry, 71*, 4450–4457. https://doi.org/10.1021/acs.jafc.2c08828

Shamim, M. Z., Mishra, A. K., Kausar, T., Mahanta, S., Sarma, B., Kumar, V., Mishra, P. K., Panda, J., Baek, K. H., & Mohanta, Y. K. (2023). Exploring edible mushrooms for diabetes: Unveiling their role in prevention and treatment. *Molecules, 28*. https://doi.org/10.3390/molecules28062837

Shan, L. S., Hou, P., Wang, Z. J., Liu, F. R., Chen, N., Shu, L. H., Zhang, H., Han, X. H., Han, X. X., Cai, X. X., et al. (2013). Prevention and treatment of diarrhoea with saccharomyces boulardii in children with acute lower respiratory tract infections. *Beneficial Microbes, 4*, 329–334. https://doi.org/10.3920/bm2013.0008

Sharif, M., Zafar, M. H., Aqib, A. I., Saeed, M., Farag, M. R., & Alagawany, M. (2021). Single cell protein: Sources, mechanism of production, nutritional value and its uses in aquaculture nutrition. *Aquaculture, 531*, 735885. https://doi.org/10.1016/j.aquaculture.2020.735885

Shen, M. H., Kim, J. S., Sapkota, K., Park, S. E., Choi, B. S., Kim, S., Lee, H. H., Kim, C. S., Chun, H. S., Ryoo, C. I., et al. (2007). Purification, characterization, and cloning of fibrinolytic metalloprotease from *Pleurotus ostreatus* mycelia. *Journal of Microbiology and Biotechnology, 17*, 1271–1283.

Shimizu, T., Mori, K., Ouchi, K., Kushida, M., & Tsuduki, T. (2018). Effects of dietary intake of Japanese mushrooms on visceral fat accumulation and gut microbiota in mice. *Nutrients, 10*. https://doi.org/10.3390/nu10050610

Singh, M. P., & Singh, V. K. (2012). Biodegradation of vegetable and agrowastes by Pleurotus sapidus: A novel strategy to produce mushroom with enhanced yield and nutrition. *Cellular and Molecular Biology (Noisy-le-Grand, France), 58*, 1–7.

Singh, V. K., & Singh, M. P. (2014). Bioremediation of vegetable and agrowastes by Pleurotus ostreatus: A novel strategy to produce edible mushroom with enhanced yield and nutrition. *Cellular and Molecular Biology (Noisy-le-Grand, France), 60*, 2–6.

Singhal, S., Rasane, P., Kaur, S., Garba, U., Singh, J., Raj, N., & Gupta, N. (2019). Mushroom cultivation, processing and value-added products: A patent based review. *Recent Patents on Food, Nutrition & Agriculture, 10*, 3–19. https://doi.org/10.2174/2212798410666180604101353

Sivamaruthi, B. S., Sisubalan, N., Kesika, P., Sureka, I., & Chaiyasut, C. (2024). A concise review of the nutritional profiles, microbial dynamics, and health impacts of fermented mushrooms. *Journal of Food Science, 89*, 3973–3994. https://doi.org/10.1111/1750-3841.17172

Somboonchai, T., Foiklang, S., Panatuk, J., Cherdthong, A., Laorodphan, N., Wanapat, M., Yammuen-Art, S., & Kang, S. (2022). Replacement of soybean meal by red yeast fermented tofu waste on feed intake, growth performance, carcass characteristics, and meat quality in Thai Brahman crossbred beef cattle. *Tropical Animal Health and Production, 54*, 133. https://doi.org/10.1007/s11250-022-03127-5

Song, X., Zhao, Y., Song, C., Chen, M., Huang, J., Bao, D., Tan, Q., & Yang, R. (2019). Mitogenome types of two Lentinula edodes sensu lato populations in China. *Scientific Reports, 9*, 9421. https://doi.org/10.1038/s41598-019-45922-5

Sørensen, A., Lübeck, M., Lübeck, P. S., & Ahring, B. K. (2013). Fungal Beta-glucosidases: A bottleneck in industrial use of lignocellulosic materials. *Biomolecules, 3*, 612–631. https://doi.org/10.3390/biom3030612

Souza Filho, P. F., Nair, R. B., Andersson, D., Lennartsson, P. R., & Taherzadeh, M. J. (2018). Vegan-mycoprotein concentrate from pea-processing industry byproduct using edible filamentous fungi. *Fungal Biology and Biotechnolgoy, 5*, 5. https://doi.org/10.1186/s40694-018-0050-9

Spatafora, J. W., Aime, M. C., Grigoriev, I. V., Martin, F., Stajich, J. E., & Blackwell, M. (2017). The fungal tree of life: From molecular systematics to genome-scale phylogenies. *The Fungal Kingdom*, 1–34. https://doi.org/10.1128/9781555819583.ch1

Spencer, D. M. (1985). The mushroom–its history and importance. In P. B. Flegg, D. M. Spencer, & D. A. Wood (Eds.), *The biology and technology of the cultivated mushroom*. Wiley.

Steinkraus, K. H. (1997). Classification of fermented foods: Worldwide review of household fermentation techniques. *Food Control, 8*, 311–317. https://doi.org/10.1016/s0956-7135(97)00050-9

Stoffel, F., de Oliveira Santana, W., Fontana, R. C., & Camassola, M. (2021). Use of Pleurotus albidus mycoprotein flour to produce cookies: Evaluation of nutritional enrichment and biological activity. *Innovative Food Science & Emerging Technologies, 68*, 102642. https://doi.org/10.1016/j.ifset.2021.102642

Sun, W., Zhang, Z., Li, X., Lu, X., Liu, G., Qin, Y., Zhao, J., & Qu, Y. (2024). Production of single cell protein from brewer's spent grain through enzymatic saccharification and fermentation enhanced by ammoniation pretreatment. *Bioresource Technology, 394*, 130242. https://doi.org/10.1016/j.biortech.2023.130242

Suwannarach, N., Kumla, J., Zhao, Y., & Kakumyan, P. (2022). Impact of cultivation substrate and microbial community on improving mushroom productivity: A review. *Biology (Basel), 11*, 569. https://doi.org/10.3390/biology11040569

Szajewska, H., & Mrukowicz, J. (2005). Meta-analysis: Non-pathogenic yeast *Saccharomyces boulardii* in the prevention of antibiotic-associated diarrhoea. *Alimentary Pharmacology and Therapeutics, 22*, 365–372. https://doi.org/10.1111/j.1365-2036.2005.02624.x

Tamang, J., Thapa, N., Tamang, B., Rai, A., & Chettri, R. (2015). Microorganisms in fermented foods and beverages. *Health Benefits of Fermented Foods and Beverages*, 1–110. https://doi.org/10.1201/b18279-2

Taskin, M., Ozkan, B., Atici, O., & Aydogan, M. N. (2012). Utilization of chicken feather hydrolysate as a novel fermentation substrate for production of exopolysaccharide and mycelial biomass from edible mushroom Morchella esculenta. *International Journal of Food Sciences and Nutrition, 63*, 597–602. https://doi.org/10.3109/09637486.2011.640309

Tofalo, R., Fusco, V., Böhnlein, C., Kabisch, J., Logrieco, A. F., Habermann, D., Cho, G. S., Benomar, N., Abriouel, H., Schmidt-Heydt, M., et al. (2020). The life and times of yeasts in traditional food fermentations. *Critical Reviews in Food Science and Nutrition, 60*, 3103–3132. https://doi.org/10.1080/10408398.2019.1677553

Toghiani, J., Fallah, N., Nasernejad, B., Mahboubi, A., Taherzadeh, M. J., & Afsham, N. (2025). Production of protein-rich fungal biomass from pistachio dehulling waste using edible Neurospora intermedia. *Scientific Reports, 15*, 5873. https://doi.org/10.1038/s41598-024-81941-7

Toledo, C. V., Barroetaveña, C., Fernandes, Â., Barros, L., & Ferreira, I. C. (2016). Chemical and antioxidant properties of wild edible mushrooms from native Nothofagus spp. Forest, Argentina. *Molecules*, 21. https://doi.org/10.3390/molecules21091201

Tomita, T., Ishikawa, D., Noguchi, T., Katayama, E., & Hashimoto, Y. (1998). Assembly of flammutoxin, a cytolytic protein from the edible mushroom Flammulina velutipes, into a pore-forming ring-shaped oligomer on the target cell. *The Biochemical Journal, 333*(Pt 1), 129–137. https://doi.org/10.1042/bj3330129

Tong, S., Chen, W., Hong, R., Chai, M., Sun, Y., Wang, Q., & Li, D. (2024a). Efficient Mycoprotein production with low CO(2) emissions through metabolic engineer and fermentation optimization of Fusarium venenatum. *Journal of Agricultural and Food Chemistry, 72*, 604–612. https://doi.org/10.1021/acs.jafc.3c08509

Tong, S., Hong, R., Chen, W., Chai, M., Zhang, Y., Sun, Y., Wang, Q., & Li, D. (2024b). Synchronous bioproduction of Betanin and Mycoprotein in the engineered edible fungus *Fusarium venenatum*. *Journal of Agricultural and Food Chemistry, 72*, 19462–19469. https://doi.org/10.1021/acs.jafc.4c06071

Toumi, M. E., Perduca, M., Necib, Y., Kebaili, F. F., Toumi, M., Torche, I., Gaglio, S. C., Rebai, R., Lauriola, A., Ayadi, M., et al. (2021). Characterization of cytotoxic lactose binding lectin from sulphur polypore, Laetiporus sulphureus (Agaricomycetes), from Algeria. *International Journal of Medicinal Mushrooms, 23*, 45–57. https://doi.org/10.1615/IntJMedMushrooms.2021040303

Trinci, A. P. J. (1992). Myco-protein: A twenty-year overnight success story. *Mycological Research, 96*, 1–13. https://doi.org/10.1016/s0953-7562(09)80989-1

Trujillo, K. G., Sánchez, J. E., Guillén, K., de la Rosa, A. P. B., & Liedo, P. (2021). Edible mushroom protein content and antioxidant capacity increase the longevity of Anastrepha ludens fruit flies. *International Journal of Medicinal Mushrooms, 23*, 33–43. https://doi.org/10.1615/IntJMedMushrooms.2021038305

Tsai, P. F., Ma, C. Y., & Wu, J. S. (2013). A novel glycoprotein from mushroom Hypsizygus marmoreus (Peck) Bigelow with growth inhibitory effect against human leukemic U937 cells. *Food Chemistry, 141*, 1252–1258. https://doi.org/10.1016/j.foodchem.2013.04.024

Tubb, C., & Seba, T. (2019). Rethinking food and agriculture 2020-2030: The second domestication of plants and animals, the disruption of the cow, and the collapse of industrial livestock farming. *RethinkX, 2019.*

Uriarte-Frías, G., Hernández-Ortega, M. M., Gutiérrez-Salmeán, G., Santiago-Ortiz, M. M., Morris-Quevedo, H. J., & Meneses-Mayo, M. (2021). Pre-hispanic foods oyster mushroom (Pleurotus ostreatus), Nopal (Opuntia ficus-indica) and Amaranth (Amaranthus sp.) as new alternative ingredients for developing functional cookies. *Journal of fungi (Basel, Switzerland), 7*, 911. https://doi.org/10.3390/jof7110911.

Valdez Castillo, M., Brar, S. K., Arriaga, S., Blais, J. F., Heitz, M., & Avalos Ramirez, A. (2023). Co-fermentation of agri-food residues using a co-culture of yeasts as a new bioprocess to produce 2-phenylethanol. *Molecules, 28*. https://doi.org/10.3390/molecules28145536

van der Aa Kühle, A., Skovgaard, K., & Jespersen, L. (2005). In vitro screening of probiotic properties of Saccharomyces cerevisiae var. boulardii and food-borne Saccharomyces cerevisiae strains. *International Journal of Food Microbiology, 101*, 29–39. https://doi.org/10.1016/j.ijfoodmicro.2004.10.039

Vieira, A. T., Teixeira, M. M., & Martins, F. S. (2012). The role of probiotics and prebiotics in inducing gut immunity. *Frontiers in Immunology, 4*. https://doi.org/10.3389/fimmu.2013.00445

Vong, W. C., Au Yang, K. L., & Liu, S. Q. (2016). Okara (soybean residue) biotransformation by yeast Yarrowia lipolytica. *International Journal of Food Microbiology, 235*, 1–9. https://doi.org/10.1016/j.ijfoodmicro.2016.06.039

Wang, H. X., & Ng, T. B. (2000). Isolation of a novel ubiquitin-like protein from *Pleurotus ostreatus* mushroom with anti-human immunodeficiency virus, translation-inhibitory and ribonuclease activities. *Biochemical and Biophysical Research Communications, 276*, 587–593.

Wang, H. X., & Ng, T. B. (2001). Isolation and characterization of velutin, a novel low-molecular-weight ribosome-inactivating protein from winter mushroom (*Flammulina velutipes*) fruiting bodies. *Life Sciences, 68*, 2151–2158.

Wang, H., & Ng, T. B. (2006). Ganodermin, an antifungal protein from fruiting bodies of the medicinal mushroom Ganoderma lucidum. *Peptides, 27*, 27–30.

Wang, Y., Bao, L., Yang, X., Li, L., Li, S., Gao, H., Yao, X.-S., Wen, H., & Liu, H.-W. (2012). Bioactive Sesquiterpenoids from the solid culture of the edible mushroom Flammulina velutipes growing on cooked rice. *Food Chemistry, 132*, 1346–1353.

Wang, X. M., Zhang, J., Wu, L. H., Zhao, Y. L., Li, T., Li, J. Q., Wang, Y. Z., & Liu, H. G. (2014). A mini-review of chemical composition and nutritional value of edible wild-grown mushroom from China. *Food Chemistry, 151*, 279–285. https://doi.org/10.1016/j.foodchem.2013.11.062

Wang, G., Cao, X., Ma, X., Guo, M., Liu, C., Yan, L., & Bian, Y. (2016). Diversity and effect of Trichoderma spp. associated with green mold disease on Lentinula edodes in China. *Microbiology, 5*, 709–718. https://doi.org/10.1002/mbo3.364

Watanabe, D., & Takagi, H. (2019). Yeast prion-based metabolic reprogramming induced by bacteria in fermented foods. *FEMS Yeast Research, 19*. https://doi.org/10.1093/femsyr/foz061

Wei, H., Wu, M., Fan, A., & Su, H. (2021). Recombinant protein production in the filamentous fungus Trichoderma. *Chinese Journal of Chemical Engineering, 30*, 74–81. https://doi.org/10.1016/j.cjche.2020.11.006

Wiebe, M. G., & Quorn, T. M. (2004). Mycoprotein—Overview of a successful fungal product. *Mycologist, 18*, 17–20.

Witthuhn, R. C., Schoeman, T., & Britz, T. J. (2005). Characterisation of the microbial population at different stages of Kefir production and Kefir grain mass cultivation. *International Dairy Journal, 15*, 383–389. https://doi.org/10.1016/j.idairyj.2004.07.016

Xiao, H., Bian, Y., Huang, H., Zhang, Z., Wu, L., & Wu, L. (2020). Inhibitory effect of protein Y3 from Coprinus comatus on tobacco mosaic virus. *Pesticide Biochemistry and Physiology, 168*, 104474. https://doi.org/10.1016/j.pestbp.2019.09.012

Yadav, J. S., Bezawada, J., Ajila, C. M., Yan, S., Tyagi, R. D., & Surampalli, R. Y. (2014). Mixed culture of Kluyveromyces marxianus and Candida krusei for single-cell protein production and organic load removal from whey. *Bioresource Technology, 164*, 119–127. https://doi.org/10.1016/j.biortech.2014.04.069

Yamada, E. A., & Sgarbieri, V. C. (2005). Yeast (*Saccharomyces cerevisiae*) protein concentrate: Preparation, chemical composition, and nutritional and functional properties. *Journal of Agricultural and Food Chemistry., 53*, 3931–3936.

Yamada, M., Tokumitsu, N., Saikawa, Y., Nakata, M., Asano, J., Miyairi, K., Okuno, T., Konno, K., & Hashimoto, K. (2012). Molybdophyllysin, a toxic metalloendopeptidase from the tropical toadstool, Chlorophyllum molybdites. *Bioorganic and Medicinal Chemistry, 20*, 6583–6588. https://doi.org/10.1016/j.bmc.2012.09.036

Yang, E. J., Ku, S. K., Lee, W., Song, K. S., & Bae, J. S. (2015). Inhibitory effects of polyozellin from Polyozellus multiplex on HMGB1-mediated septic responses. *Inflammation Research, 64*, 733–746. https://doi.org/10.1007/s00011-015-0856-3

Yeast-Based Spreads Market Key Vendors, Regional Analysis, Key Players and Competitive Landscape Forecasts by 2027. (2022). *Fairfield Market Research*. Available online: https://www.openpr.com/news/2674900/yeast-based-spreads-market-key-vendorsregional-analysis-key. Accessed on 14 Dec 2022.

Yuan, B., Zhao, L., Yang, W., McClements, D. J., & Hu, Q. (2017). Enrichment of bread with nutraceutical-rich mushrooms: Impact of Auricularia auricula (mushroom) flour upon quality attributes of wheat dough and bread. *Journal of Food Science, 82*, 2041–2050. https://doi.org/10.1111/1750-3841.13812

Yuan, X., Zhu, X., Sun, R., Jiang, W., Zhang, D., Liu, H., & Sun, B. (2022). Sensory attributes and characterization of aroma profiles of fermented sausages based on fibrous-like meat substitute from soybean protein and Coprinus comatus. *Food Chemistry, 373*, 131537. https://doi.org/10.1016/j.foodchem.2021.131537

Zetić, V. G., Stehlik-Tomas, V., Grba, S., Lutilsky, L., & Kozlek, D. (2001). Chromium uptake by *Saccharomyces cerevisiae* and isolation of glucose tolerance factor from yeast biomass. *Journal of Bioscience and Bioengineering, 26*, 217–223.

Zhang, Y., Liu, Z., Ng, T. B., Chen, Z., Qiao, W., & Liu, F. (2014). Purification and characterization of a novel antitumor protein with antioxidant and deoxyribonuclease activity from edible mushroom Pholiota nameko. *Biochimie, 99*, 28–37. https://doi.org/10.1016/j.biochi.2013.10.016

Zhang, G. Q., Chen, Q. J., Hua, J., Liu, Z. L., Sun, Y., Xu, X., Han, P., & Wang, H. X. (2019a). An inulin-specific lectin with anti-HIV-1 reverse transcriptase, antiproliferative, and mitogenic activities from the edible mushroom Agaricus bitorquis. *BioMed Research International, 2019*, 1341370. https://doi.org/10.1155/2019/1341370

Zhang, Q., Wu, C., Wang, T., Sun, Y., Li, T., & Fan, G. (2019b). Improvement of biological activity of Morchella esculenta protein hydrolysate by microwave-assisted Selenization. *Journal of Food Science, 84*, 73–79. https://doi.org/10.1111/1750-3841.14411

Zhang, Y., Liu, F., & Ng, T. B. (2021a). Interrelationship among paraptosis, apoptosis and autophagy in lung cancer A549 cells induced by BEAP, an antitumor protein isolated from the edible porcini mushroom Boletus edulis. *International Journal of Biological Macromolecules, 188*, 313–322. https://doi.org/10.1016/j.ijbiomac.2021.07.169

Zhang, Y., Zhou, R., Liu, F., & Ng, T. B. (2021b). Purification and characterization of a novel protein with activity against non-small-cell lung cancer in vitro and in vivo from the edible mushroom Boletus edulis. *International Journal of Biological Macromolecules, 174*, 77–88. https://doi.org/10.1016/j.ijbiomac.2021.01.149

Zhang, H., Jiang, F., Tang, C., Liu, Y., & Zhang, J. (2025). Prospects and applications of efficient physical field processing technologies for polysaccharide extraction and quality improvement in edible mushrooms: A systematic review. *International Journal of Biological Macromolecules, 301*, 140412. https://doi.org/10.1016/j.ijbiomac.2025.140412

Zhao, J., & Kwan, H. S. (1999). Characterization, molecular cloning, and differential expression analysis of laccase genes from the edible mushroom Lentinula edodes. *Applied and Environmental Microbiology, 65*, 4908–4913. https://doi.org/10.1128/aem.65.11.4908-4913.1999

Zhao, S., Rong, C. B., Kong, C., Liu, Y., Xu, F., Miao, Q. J., Wang, S. X., Wang, H. X., & Zhang, G. Q. (2014). A novel laccase with potent antiproliferative and HIV-1 reverse transcriptase inhibitory activities from mycelia of mushroom Coprinus comatus. *BioMed Research International, 2014*, 417461. https://doi.org/10.1155/2014/417461

Zhao, S., Gao, Y., Wang, H., Fan, Y., Wang, P., Zhao, W., Wong, J. H., Wang, D., Zhao, X., & Ng, T. B. (2023). A novel mushroom (Auricularia polytricha) glycoprotein protects against lead-induced hepatoxicity, promotes lead adsorption, inhibits organ accumulation of lead, upregulates detoxifying proteins, and enhances immunoregulation in rats. *Frontiers in Nutrition, 10*, 1144346. https://doi.org/10.3389/fnut.2023.1144346

Zhao, Z. C., Gu, F. T., Li, J. H., Zhu, Y. Y., Huang, L. X., & Wu, J. Y. (2024). Fractionation, characterization, and assessment of nutritional and immunostimulatory protein-rich polysaccharide-protein complexes isolated from Lentinula edodes mushroom. *International Journal of Biological Macromolecules, 280*, 136082. https://doi.org/10.1016/j.ijbiomac.2024.136082

Zheng, C., Li, J., Liu, H., & Wang, Y. (2023). Review of postharvest processing of edible wild-grown mushrooms. *Food Research International, 173*, 113223. https://doi.org/10.1016/j.foodres.2023.113223

Zhou, F., Hansen, M., Hobley, T. J., & Jensen, P. R. (2022). Valorization of green biomass: Alfalfa pulp as a substrate for oyster mushroom cultivation. *Foods (Basel, Switzerland), 11*, 2519. https://doi.org/10.3390/foods11162519

Zhou, Y., Zhu, Y., Jin, X., Zhang, Y., Song, J., Wu, Z., Li, Y., Yi, J., Wang, D., & Hu, M. (2023). Chroogomphus rutilus regulates bone metabolism to prevent periodontal bone loss during orthodontic tooth movement in osteoporotic rats. *Nutrients, 15*. https://doi.org/10.3390/nu15234906

Zhou, X., Wang, J., Shao, G., Chang, X., Liu, Y., Xiang, T., Zhu, Q., Ren, A., Jiang, A., & He, Q. (2024). Bidirectional solid-state fermentation of highland barley by edible fungi to improve its functional components, antioxidant activity and texture characteristics. *Plant Foods for Human Nutrition, 79*, 308–315. https://doi.org/10.1007/s11130-024-01166-x

Zoglowek, M., Hansen, G. H., Lübeck, P. S., & Lübeck, M. (2015). Fungal consortia for conversion of lignocellulose into bioproducts. *Mycology: Current and Future Developments*, 329–365. https://doi.org/10.2174/9781681080741115010015

Zuzek, M. C., Macek, P., Sepcić, K., Cestnik, V., & Frangez, R. (2006). Toxic and lethal effects of ostreolysin, a cytolytic protein from edible oyster mushroom (Pleurotus ostreatus), in rodents. *Toxicon, 48*, 264–271. https://doi.org/10.1016/j.toxicon.2006.05.011

An Overview About Pharmaceutical Grade Fungal Protein and Peptides

Amro Abd Al Fattah Amara and Arti Gupta

Abstract The fungal pharmaceutical- and nutraceutical-grade products enrich the markets and the pharmacies. There are many proteins and peptides of fungal origin. The new technology enables recombinant proteins and peptides to be produced from fungi and from other creatures. Insulin, the human polypeptide hormone, is produced in yeast. Enzymes, active proteins, peptides, protein with nutritional values, and other forms are produced by fungi. The production process was optimized using different strategies such as classical and modern mutagenesis, biochemical engineering, fermentation optimization, use of recombinant strains, heterologous expression, and protein and peptide engineering. Other strategies are the use of fermented foods that proved to have nutraceutical use such as mushrooms, kefir, fermented milk, and other fermented foods. Microbes such as some yeast species, edible mushrooms, and some fungi can be consumed directly as single-cell proteins or mycoproteins with much nutritional and medicinal value. They have many useful ingredients, are high in protein content, peptides, vitamins, etc. A large number of useful fungi as well as their products are described in this chapter. Respective numbers of the fungal protein-based products, the production process, the ideas, the old and new technologies, etc., were discussed. As eukaryotic organisms, fungi have unique properties to support production of native or recombinant pharmaceutical-grade proteins and peptides.

Keywords Eukaryotic · Fungi · Mushrooms · Mutagenesis · Pharmaceutical grade · Protein engineering · Peptides · Proteins · Strain engineering · Strain improvement · Yeast

A. A. A. F. Amara (✉)
Protein Research Department, Genetic Engineering and Biotechnology Research Institute [GEBRI], City of Scientific Research and Technological Applications [SRTA-City], Universities and Research Center District, Alex, Egypt
e-mail: aamara@srtacity.sci.eg

A. Gupta
Department of Zoology, Shri Awadh Raj Singh Degree College, Ram Manohar Lohia University, Ayodhya, India

A. Gupta (ed.), *Prospects of Fungal Biotechnologies for Livestock Volume 2*,
Fungal Biology, https://doi.org/10.1007/978-3-032-06478-3_12

Abbreviations

ACE	Angiotensin-converting enzyme
ER	Endoplasmic reticulum
GRAS	Generally recognized as safe
LAP	Lyophyllum antifungal protein
MRSA	Methicillin-resistant *Staphylococcus aureus* strains
PTM	Post-translational modification
RIP	Ribosome-inactivating protein

1 Introduction

Fungal products from native or recombinant proteins (Molfetta et al., 2022) and peptides (Boot et al., 2007) can be collectively categorized into two types of products based on their origin: the first is a native fungal product (of fungal origin) (Ahmed et al., 2024) and the second is a transformed or a heterologous expressed recombinant product (Porro et al., 2011; Thak et al., 2020), either from another fungal source other than the native wild type mother product or from out kingdom products (e.g., Prokaryotic or Eukaryotic) (e.g., human insulin) (Thim et al., 1986; Amara & El-Baky, 2023). The expression 'pharmaceutical grade' highlights that these products with high purity are to be used as a medicine or as nutraceutical pure compounds. In some cases, it is named micro-pharmaceutical to signal its microbial origin (Al-Obaidi et al., 2021; Jiang et al., 2014). Usually, the native strain produces small amounts from the product of interest, which include the protein and the peptides. Therefore, during long years of development scientists, researchers, companies, producers, etc., have collectively collaborated to increase the amount of the proteins produced. They have examined nearly all the possible strategies. Such strategies are the norm (Amara, 2013a). Some processes show success in obtaining a high amount of certain proteins or peptides (Amara, 2013a). The fungal cell machinery enables extracellular secretion, which in turn enables their purification and production on a commercial scale. Generally, the highly biocompatible and bioavailable pharmaceutical-grade proteins and peptides in their pure form become essential part of the new drugs, drug design (Kan, 2002; Sen et al., 2022), and modern medicine.

The production of active protein or peptide starts by searching for its source, which in this case is a fungus. After that, the produced amount, activity, specificity, amino-acid sequences, structure, etc., are identified. Through gene technology, their genes can be cloned and can be retransferred to the mother cell wild type or to another better strain from the same species, or by the use of other species. The production process, gene, protein, the producer strains, the production conditions, the nutrients used (substrate), etc., all can be modified and all can be subjected to adjustment, improvement, optimization, and in some cases to deletion or blocking for some properties, aiming to increase the amount of the product produced. Besides proteins, fungi can also be used to produce amino acids. Common amino acids produced by fungi include glutamic acid, aspartic acid, and lysine. *Aspergillus* species (e.g., *A. oryzae* and *A. niger*) are commonly used to produce amino acids such as glutamic acid and aspartic acid. *Candida* species can also be utilized for amino-acid production. Fungi such as *A. niger* are used in fermentation processes to produce L-glutamic acid through the fermentation of carbohydrates. This chapter addresses some of those strategies, the producer strains, some of the important fungal protein and peptides, ideas, discussion, and will highlight important issues that can be useful for anyone who is starting from scratch in isolating a useful wild type fungus that can stand alone for producing useful products, or who is going to use modern technologies such as molecular biology, protein engineering, random mutagenesis, etc. The general aim is to highlight the importance of the fungi as a source of pharmaceutical-grade proteins and peptides.

2 How to Produce Proteins and Peptides from Fungi

2.1 Isolating the Best Strain

To optimize the production step, one should start from the first point and optimize the overall steps of production. For example, if a certain species of the fungi produces a certain product one can start by isolating as many different strains as possible, investigate them all, and select the best producer (Amara, 2013a). This step might appear simple but in fact screening different strains may save a lot of time and years of strain optimization. Of course, the best way is to isolate such strains from a source where the proper substrate exists either naturally or in the form of waste(s) or contamination(s). Or one can design a minimal medium that contains the proper substrate and growing many isolated strains on it. Enriching the sample where an expected strain exists with the proper substrate will increase the amount of this strain, which will facilitate its isolation in the laboratory. For example, when eating fresh sugary fruits. Yeasts exist on the surface of the fruits. The sugar content will attract the sugar consuming strains in the gut microbiota as well as the yeast strains on the surface. Being beneficial microbes, so, one could feel comfort. Similar strategies can be followed in many domains concerning microbial strains and their substrates. Such old strategies are of great interest and a wonderful wild type strain(s) can be isolated (Amara, 2013a). For example, *Aspergillus vadensis* is a new black aspergilli species that does not acidify the growth medium and produces very low levels of extracellular protease activity (de Vries et al., 2004). Additionally, it has a very high transformation frequency and enables a large number of promotors to process heterologous expression (Culleton et al., 2014; de Vries et al., 2005). These features make *A. vadensis* a very promising candidate as a host for protein production. *A. vadensis* is a rising alternative to *A. niger*. Feruloyl esterase B from *A. niger* was synthesized in *A. vadensis* at a 12-fold higher level than the *A. niger* overproducing strain (Alberto et al., 2009). Of course, one should design a medium that does not contain an alternative or a simpler substrate where in such cases the microbe might ignore the targeted one. The selection of the most suitable strain out of many promising isolates is another art. For example, for some it is an interesting issue to isolate a mesophilic microbe that could produce a thermophilic enzyme, which can save a lot of money during the production process. Another example, as above, wild type strains with very low protease production can be used to produce different kinds of recombinant proteins (de Vries et al., 2004). Other issues concern product optimization by improving the media components, either by enhancing each single component to optimize the overall process or running randomization for more than one variable at once using the statistical experimental design of Blackett-Burman or Box Behnken (Amara, 2013a). Other simple strategies can be followed where instead of using one strain one could use more than one strain where they can collectively collaborate to degrade a complex waste that might not be suitable for the strain of interest, but only when other strains exist. That mimics exactly what happens in our digestive system, which receives different forms of foods, and our microbiota react collectively to digest and to refine all those forms of substrates so that they are suitable for our body. Focusing on a single type of food will enrich some microbes over others, which may change the balance in our microbiota. Therefore, some people can eat any kinds of food, whereas others need time to be able to eat a new food. Typically, microbes may need an adaptation period to start to react with an uncommon substrate (Bigey et al., 2024).

2.2 Strain Improvement

A strategy named "strain improvement" is concerned with improving the microbial strain to produce certain products (Chiang, 2004) including proteins and peptides. It has been used basically in the improvement of antibiotic producer strains. Antibiotics to be produced need multiple steps collectively known as polyketides (Luo et al., 2019), where many proteins (more than hundreds in certain cases) collectively collaborate to build the antibiotic structure. Therefore, one could not practically clone all the related genes. Strain improvement strategies are usually carried out by applying a random mutagenesis protocol (Do et al., 2024; Greener et al., 1997; Saqib & Hamid Rashid, 2023; Suryadi et al., 2022). Some mutagenic chemical compounds (e.g., nitrosoguanidine) or physical conditions (e.g., UV mutagenesis) can simply be used (Amara, 2013a; Chiang, 2004). It was named "classical" to distinguish it from modern random mutagenesis tools (Greener et al., 1997). It did not deliver where mutation(s) occurred. In this process some metabolic pathways are completely blocked (owing to damage in certain genes in its initial step) or it stopped at a particular point to deliver a certain flux that can supply the targeted protein or the targeted metabolic pathway with an additional chemical structure (e.g., in peptide production). The mutation(s) can occur in any gene (Ermakova-Gerdes et al., 1996), so one should screen to select the best mutated strain. This method allows isolation of a positive strain (product increased) and a negative strain (product impaired) or a leaky strain (product decreased). All types of mutants are useful in particular molecular experiments. For example, negative mutant proteases and restriction enzymes are useful in cloning and protein production. Selection in the human histone hairpin-binding protein was performed using the yeast three-hybrid system (Martin et al., 2000). Classical mutagenesis could accelerate the process of obtaining strains with specific characters and improve the amount/quality of the active protein produced. After identifying the mutant, it is usually subjected to further cycles of the above kinds of optimization/mutagenesis. Classical mutagenesis is still popular and used by many. One could combine modern mutagenesis and classical mutagenesis. For example, *Kluyveromyces lactis* strain was recently constructed using CRISPR/Cas9 and UV mutagenesis and a mutant with multi-copy integration with enhanced bovine chymosin production was obtained (Song et al., 2024).

2.3 The Improvement on the Gene Level and Recombinant Strain Engineering

Fungal protein production today is mainly based on engineered strains (Askenazi et al., 2003) and improvements in fermentation technology for efficient production. Achieving high yields of the protein of interest often requires extensive optimization of fermentation conditions and genetic modifications. This development may involve design and alteration of the promoters. One or more than one nucleotide change in the promotor could increase the promotor efficacy. For example, virus promotors are used for protein overexpression. The maturation of the nascent polypeptide in the endoplasmic reticulum (ER), the secretory process, flux pathways, hypersecretion stress, unfolded protein response, protein modification in the Golgi, the secretion apparatus, the post-translation modification, the creation efficacy of the protein, and other cell activities related to protein production and secretion can be mutagenized or engineered to increase the protein yield (Kowalchuk et al., 2008). The nucleotides sequence could also be changed without changing the amino acids sequence or number to facilitate the transcription and the translation process. Genes with a high guanine-cytosine content can be decreased without changing the number and the order of the amino acids by changing the nucleotides sequence using alternative amino-acid sequence code. Fungal protein secretion is closely linked to the cell and mycelial morphology. Hyperbranching and tip density related to culture conditions can improve protein secretion (Bud, 1989). Modern genetic engineering also enables blocking of certain pathways by knocking out certain genes using transposons. The strategies are normal and one could use any or more than one strategies to achieve certain changes(s) to improve certain target(s). The respective amount of information facilitates selection of the best used protein or peptides. Mutants are designed and developed by genetic engineering (Altay et al., 2013). An example is the fungal defensin micasin from *Microsporum canis*, where a synthetically synthesized truncated version of the peptide binds the receptor binding domain of the COVID-19 spike protein six times better than the original mature peptide (Wang et al., 2012), potentially preventing or limiting the virus by penetrating into the cells.

2.4 Recombinant Strains

Heterologous expression means transferring genes from one strain to another to improve the production process or to enable the use of the new host genetic machinery. To enable that, the new strain, which hosts the transformed gene of a protein of interest, is in need of some adjustment. The restriction enzyme and proteases should be impaired. Proteolytic degradation by proteases could negatively affect any heterologous protein production by filamentous fungi (FAO, 2010). Protease-deficient strains of *Aspergillus*, *Trichoderma*, and *Myceliophthora* spp. have been described (FAO, 2010; Riviere, 1977; Chang et al., 1978; Rai, 2004). Classical mutagenesis is used to obtain specific mutants (Amara, 2013a; Song et al., 2024). Mutagenesis by direct gene knockout strategies is applied (Suwannarach et al., 2022; Kopke et al., 2010). The deletion of the secreted proteases and/or the transcription factors was performed. A transcription factor, *prtT*, was identified in *A. niger* and was found to have homologs in *A. oryzae* and *A. fumigatus* but was absent in *A. nidulans* (Chang et al., 1978; Rai, 2004). Deletion of *prtT* increased the production of a heterologous protein (Chang et al., 1978). Secretomic analysis is used to map the best design for a protein production cell factory (Fernández-Fueyo et al., 2016; Xu et al., 2023; Zhou et al., 2018). For example, comparative secretomics was used to identify and delete three target proteases in *T. reesei* (Boland et al., 2013). The overexpression of various ER proteins involved in protein production have been investigated, such as chaperones (Roongsawang et al., 2016), lectins, foldases, and nucleotide exchange factors (Song et al., 2017; Frey, 1930; Spencer, 1985). The membrane traffic, particularly the secretory and endocytic pathways, can be modified to improve the yield (Das et al., 2021). Novel gene-editing tools, including the CRISPR-based approaches, have been used (Nuhn et al., 2013; Hassan et al., 2020; Lee et al., 2018). Jarczynska et al., (2022) (Mihai et al., 2022) created a fungal gene-expression platform (DIVERSIFY) that makes it possible to simultaneously express a construct in different *Aspergillus* species to select the best producer. Other strategies include the use of isopropyl β-D-1-thiogalactopyranoside and similar analogues to keep the promoter in continuous work. Virus promoters (strong promoters) are used. Vectors that could not replicate in certain strains because of their replication origin were engineered. Besides the isolation of new wild types, genes are searched directly to isolate new genes from unknown microbes or uncultured microbes (e.g., from a soil sample) including fungi, a science named 'metagenomics' (Ceugniez et al., 2017).

2.5 Protein Engineering

Some proteins are so unique and it is hard to find their alternatives in nature. The production of some of them was developed through large amounts of investment, but a new feature is needed to be added to them or to be deleted. Protein engineering needs enough information and understanding of the protein structure function, or awareness about running a random mutagenesis protocol alongside a selection protocol to obtain the desired mutant. Many technical enzymes have been subjected to protein engineering (Amara, 2013a; Brannigan & Wilkinson, 2002; Evdokimov et al., 2008). Enzymes are any of several complex proteins that are produced by cells and act as catalysts in specific biochemical reactions (Amara, 2013a; Underkofler et al., 1958). Enzymes can accelerate a reaction by as much as 10^{17}-fold and can discriminate between closely related substrates (Hedstrom, 2001). Protein engineering is concerned with modifying the protein structure/function(s) or building protein from scratch. The engineered protein usually has new criteria and can be mediated at the level of the genes or proteins. There is a real need for the engineering of different types of active protein to improve their functions. Medically used proteins should be highly pure, active, correctly folded, and biocompatible (Tuncer Degim & Celebi, 2007). Properties such as stability, high activity, and the ability to fold and interact correctly with surfaces are necessary for a variety of active proteins (Goodenough, 1995; Arulmuthu et al., 2009; Crisman & Randolph, 2008). Protein engineers often found themselves having to choose between trying to make bigger libraries and focusing libraries on sequences most likely to be effective (Baker, 2011). Protein and metabolic engineering are both used to construct efficient microbial cell factories (Goodenough, 1995; Xu et al., 2020). A carotenoid cleavage dioxygenase engineered protein was used to optimize β-ionone biosynthesis in yeast cell factories. *S. cerevisiae* transporter Pdr5p was engineered to impair the *Fusarium* mycotoxin export and resistance to inhibition (Gunter et al., 2016). Protein engineering is used in the drug design of human immunodeficiency virus 1 (HIV-1) and human cytomegalovirus proteases (Kan, 2002).

3 Eukaryotic Recombinant Protein

Eukaryotic cells enable the processing of complex biological processes. They enable genetic issues related to higher eukaryotes to be investigated. For example, a novel light-induced promoter for recombinant protein production in *Pleurotus ostreatus* (oyster mushroom) was identified (Yin et al., 2020). Sensitive in situ recombinant bioluminescent *Yersinia enterocolitica* reporter mutant was monitored (in real time) on Camembert cheese (Maoz et al., 2002). Various recombinant-protein production systems for *Aspergillus* have been established (Fleißner & Dersch, 2010). Recombinant technical enzymes such as tannase is produced in *A. oryzae* using an improved glucoamylase gene promoter (Ichikawa et al., 2020).

Crude *Jatropha curcas* seed oil was converted into biodiesel using liquid recombinant *Candida rugosa* lipase isozymes (Kuo et al., 2015). Recombinant beta-glucosidase from *Volvariella volvacea* was studied (Li et al., 2005). The protein profile of heterologous enzymes was processed in recombinant overproducing *Aspergillus nidulans* strains (Zubieta et al., 2018). Recombinant prolyl aminopeptidase from *Aspergillus oryzae* has been produced (Matsushita-Morita et al., 2010). The bovine chymosin (CHY), human lysozyme (HLY), and recombinant antibodies (such as adalimumab) were also produced by *A. oryzae* to decrease their production costs (Daba et al., 2021). The production level of recombinant *Myceliophthora thermophila* trehalase and *Penicillium citrinum* nuclease P1 showed a great enzyme yield (Dong et al., 2020; Chen et al., 2019) in *A. niger*. The yield of human erythropoietin showed a 41-fold improvement over the original yield (Rojas-Sánchez et al., 2020).

3.1 Insulin as an Example

Insulin was the first protein to be sequenced, by Frederick Sanger, who won the Nobel Prize for this achievement in 1958. He succeeded in solving the insulin sequence by sequencing its content of amino acids. It was a complicated process, especially in the presence of disulfide bonds between A and B fragments. However, Sanger and Thompson have highlighted the importance of the insulin 3D structure (Sanger & Thompson, 1953a, b). This was a clear proof that eukaryotic proteins and peptides are different from those of prokaryotic cells. This success enabled later production of the recombinant insulin in prokaryotic and later in eukaryotic cells. The first approved recombinant protein manufactured using transformed *E. coli* was rDNA Lilly insulin (Humulin) from Eli Lilly &Co. in the 1970s (approved in the early 1980s) (Manning et al., 1989). This also explains how genes from any kingdoms could be translated to protein, which will perform the function if they are in suitable conditions. It happens to us upon our infection with certain viruses, microbes, or even when genes in the form of DNA or RNA enter to our bodies and reach our cells by one way or another. The eukaryotic cell is more complicated than the prokaryotic cell, because it is designed to do more complicated targeting. It is isolated from humans so it is logical to try to clone and produce it in a single-cell eukaryotic such as yeast. Lars Thim, a Novo protein chemist, and Niels P. Fiil, a microbiologist (University of Copenhagen), who had joined Novo in 1980, made contact in 1982 with the Zymogenetics company founded in 1981. Zymogenetics was acquired by Novo in 1988. In collaboration with Zymogenetics scientists, Thim established a method for the secretion of dibasic insulin precursors in *S. cerevisiae* (Thim et al., 1986). The construct used for yeast expression consisted of a double-stranded human preproinsulin cDNA, fused to MFα1, and inserted within the triose phosphate isomerase gene. These fragments provide presequences that direct the insulin precursors into the yeast secretory pathway and pass to the growth medium

(Thim et al., 1986). The entire expression unit was inserted into the plasmid CPOT (Thim et al., 1986). The purified product was converted to insulin by trypsin and carboxypeptidase B (Thim et al., 1986). Novo and Nordisk merged in 1989 to form Novo Nordisk, and the Novo Nordisk yeast DNA recombinant insulin was approved in 1991 under the name Novolin (De Meyts, 2017). The complication of the insulin is that it is produced naturally first in the form of proinsulin. It contains three fragments (A followed by C and then followed by B) in a U shape where C is at the bottom and A plus B are attached through two sulfide bonds (four cysteines, each two exist in one fragment). If one clones A alone or B alone, the A fragment during its folding will attach to each other through the sulfide in the two existing cysteine. The same will happen with the B fragments. But in the presence of the C fragment in the middle, the correct A–B two-cysteine to two-cysteine arrangement will happen through two sulfide bonds. Then one can remove the C fragment. Therefore, one should mimic the natural production steps where proinsulin is produced first, then a correct folding process occurs, and after that the C fragment is removed. Additionally, it is recommended to apply the protein engineering and random mutagenesis in a narrow range of applications. Preferably, in technical enzymes or in proteins that will be injected in vivo. For more information, refer to Amara (2013) (Amara, 2013b).

3.2 *Fungal Metabolism and Protein Secretion*

Production and secretion of proteins in fungi are complex and regulated processes (Kowalchuk et al., 2008; Song et al., 2017). The genes of extracellular proteins harbor a signal peptide (15–36 amino acids long) in the N-terminal end. In cytosol, the transcribed mRNA is translated to a polypeptide at the ribosomes, which attach to the membrane of the ER by the help from a signal-recognition particle. The produced polypeptide is translocated into the lumen of the ER and enters into the ER. The signal-peptide moiety is cleaved from the polypeptide by a signal peptidase. The resulting polypeptide is then folded. The polypeptide undergoes certain post-translational modifications (PTMs) such as glycosylation followed by transport to the Golgi bodies. In the Golgi bodies, further modification (O-mannosylation or hyperglycosylation) and folding occur. The almost final proteins are escorted by cytoplasmic vesicles, transported to the outer membrane, and then secreted out through fusion to the cytoplasm membrane. Highly secreted proteins contain efficient signal peptides. Differences between signal peptides can affect protein secretion (Kowalchuk et al., 2008; Song et al., 2017; Hildebrand et al., 2020). In the ER the quality of the proteins is coordinated (Wiebe, 2002; Koivurinta et al., 1979). The folding of the polypeptide is only facilitated by certain specific ER-resident proteins, e.g., foldases, chaperones, and protein disulfide isomerases, and correctly folded proteins pass through ER exit sites (Kowalchuk et al., 2008; Hyde et al., 2019). Protein glycosylation has two different variants: *N*- and *O*-glycosylation.

The *N*-glycosylation of fungal proteins occurs first in the lumen of the ER whereas *O*-glycosylation occurs later in the Golgi bodies (Frey, 1930). In fungi, protein secretion is believed to occur primarily at the hyphal tip (Meyer et al., 2008). Some fungi can be triggered to perform hyperbranching of hyphae and form larger numbers of hyphal tips to increase protein secretion (Bud, 1989; Ukaegbu-Obi, 2016).

4 Production of Active Peptides

Fungi are the source of many active peptides that have different applications (Umemura, 2020; Yimam et al., 2024; Zhou et al., 2023). Fungal peptides exhibit antiviral, antibacterial, or antifungal functions (Nout, 2000; Spinnler & Gripon, 2004). Antimicrobial peptides such as lipopeptides are produced by fungi such as *Aspergillus* species. Antioxidant (Zhu et al., 2022) and immunomodulatory peptides are produced by some fungi such as *Ganoderma lucidum* (Reishi mushroom) (Chen et al., 2018). Neuroactive peptides can also be produced by some fungi such as *Psilocybe* species (magic mushrooms). Some fungal peptides were found to induce apoptosis (programmed cell death) in cancer cells or inhibit tumor growth. Other forms contribute to the flavor and texture of fermented foods, such as soy sauce and cheese. Peptides could show cytotoxicity against nematodes (Tayyrov et al., 2018), insects, or human cells such as cancer cells (Nout, 2007; Mogensen et al., 2009). The peptides are encoded in the genomes, and many of them are not translated by the ribosomal system but are produced nonribosomally (Spinnler & Gripon, 2004). The peptides can either be constitutively expressed or expressed by induction (García Fontán et al., 2006). The peptides contain a signal peptide like other secreted proteins and translocate to the ER for transport to the exterior (Ghane et al., 2020). Peptides are either linear or cyclic, forming α-helices or β-sheets or combinations of the two (Spinnler & Gripon, 2004; Mogensen et al., 2009). Peptides of a cationic nature target different ion channels or cell-membrane components (Mogensen et al., 2009; García Fontán et al., 2006). Defensins are small, cationic, antimicrobial peptides 18–45 amino acids in length and most often contain three to four intramolecular disulfide bonds (Menezes et al., 2019). Eurocin, micasin, and plectasin are fungal defensin-like peptides that are being considered for potential pharmaceutical applications (Nout, 2007; Raveschot et al., 2020; Romero-Luna et al., 2020; Vasquez et al., 2020; Wang et al., 2019). The production of ETD151 in an *A. oryzae* host improved after altering the fungal surface structure (Wyder et al., 1999) by deleting a potential binding domain on the surface (Umemura, 2020; García Fontán et al., 2006; Youssef et al., 2019; Li et al., 2021; Kombrink et al., 2019; Seyfi et al., 2019). A peptide from *Penicillium* antifungal was involved in asexual development in *Penicillium chrysogenum* (Hegedüs et al., 2011) (Fig. 1).

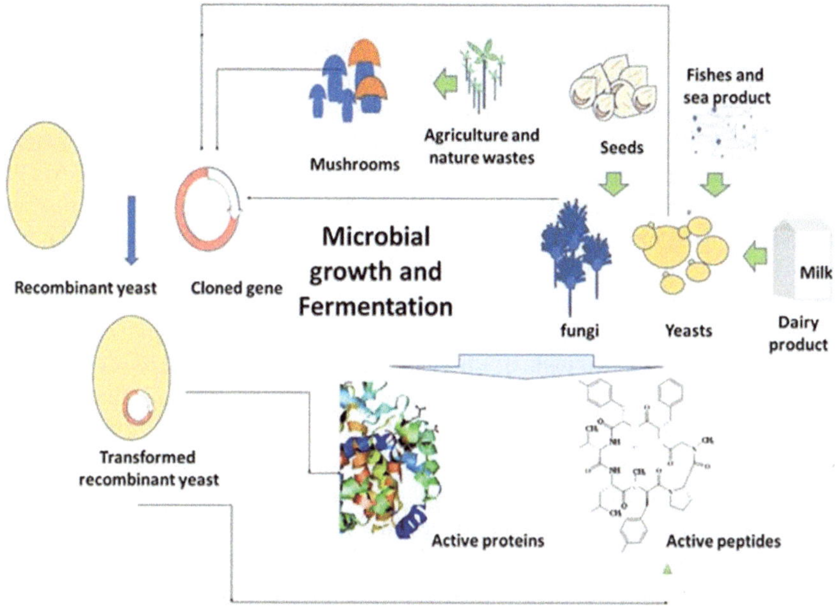

Fig. 1 Natural components that can be used as fungal substrates such as agriculture crops, fishes, seeds, dairy products, etc., are subjected to different types of fermentation by yeasts, fungi, and mushrooms to produce active proteins and peptides. Important genes from any of the microbes used can be cloned and transferred to recombinant cell factories such as yeast to produce the recombinant protein(s) and peptide(s)

4.1 Peptides Derived from Mushrooms

Mushrooms are known to be a source of various interesting proteins and peptides. The antihypertensive activity of mushroom-derived proteins and peptides via metabolomic and proteomic approaches has been reported (Li et al., 2024a). Edible mushrooms have antimicrobial activities against Gram-positive and Gram-negative bacteria (Ahmad et al., 2014), such as edible pink oyster mushrooms, *Pleurotus eous* (Dash et al., 2024), tropical milky white mushrooms, *Calocybe indica* (Agaricomycetes), an effective antimicrobial agent working in synergism with standard antibiotics (Datta et al., 2020). Antimicrobial potential, identification, and phylogenetic affiliation of wild mushrooms from two sub-tropical semi-evergreen Indian forest ecosystems have also been reported (et al., 2016). In addition, the nutritional composition, antioxidant compounds, and antimicrobial activity of wild culinary-medicinal mushrooms *Boletus edulis* and *Lactarius deliciosus* (Agaricomycetes) from Brazil have been investigated (Rosa et al., 2020). Phytochemical profile and antimicrobial activities of edible mushroom *Termitomyces striatus* have previously been described (Sitati et al., 2021; Branco et al., 2014). Proteins with suppressive effects on fungal growth are produced by mushrooms and other fungi. It was demonstrated that both angiosperm ribosome-inactivating protein (RIP) and antifungal proteins exert antifungal activity. Antifungal proteins and RIP alike can inhibit enzymes crucial to the HIV life cycle (Ng et al., 2002). The same occurs in the mushroom *L. shimeji* (Lam & Ng, 2001).

Lyophyllum antifungal protein (LAP) was isolated from fruiting bodies of *L. shimeji*. LAP suppresses cell-free translation with a low potency but inhibits HIV-1 reverse transcriptase with a high potency (Lam & Ng, 2001). Novel GAPDH-derived antimicrobial peptides (Branco et al., 2014) secreted by *S. cerevisiae* and involved in wine microbial interactions have been identified. Thaumatin-like proteins from the mushrooms *Lentinus edodes* and *Irpex lacteus* and the fungus *Rhizoctonia solani* inhibit the growth of *S. cerevisiae*, and are capable of hydrolyzing polymeric carboxymethylated pachyman in an in-gel β-1,3-glucanase assay (Grenier et al., 2000). An endo-1,3-β-glucanase from *Agaricus bisporus* was characterized with regard to substrate specificity, optimal temperature, and optimal pH. Antifungal activity was not investigated, however (Galan et al., 1999). Erygin (10 kDa) is an antifungal peptide from the mushroom *Pleurotus eryngii* with inhibitory activity toward *Fusarium oxysporum* and *Mycosphaerella arachidis* (Wang & Ng, 2004). A recombinant antitumor lectin from the *Agrocybe aegerita* (edible mushroom) is produced (Yang et al., 2005). A ribonuclease with antimicrobial, antimitogenic, and antiproliferative activities comes from the edible mushroom *Pleurotus sajor-caju* (Ngai & Ng, 2004). Novel angiotensin I-converting enzyme inhibitory peptides have been derived from an edible mushroom, *Pleurotus cystidiosus* O.K. Miller (Lau et al., 2013). The structural characterization and angiotensin-converting enzyme (ACE) inhibitory mechanism of *Stropharia rugosoannulata* mushroom peptides have been described (Li et al., 2022a). Revealing the ACE receptor binding properties and interaction mechanisms of salty oligopeptides from *Stropharia rugosoannulata* mushroom have also been studied (Li et al., 2024b). Extracellular polysaccharopeptides have been identified from fermented turkey tail medicinal mushroom (Lo et al., 2020). Flammin and velin are new ribosome-inactivating polypeptides from the mushroom *Flammulina velutipes* (Ng & Wang, 2004). *Lentinus squarrosulus* induces apoptosis in human lung-cancer cells (Prateep et al., 2017). Secretion of *Mucor* rennin, a fungal aspartic protease of *Mucor pusillus*, was achieved by recombinant yeast cells (Yamashita et al., 1987). There are proteins and peptides of mushroom origin and they find their way into various medicinal applications, as a source of antimicrobial, antifungal, anticancer, anti-inflammatory, etc., activities (Zhou et al., 2019).

4.2 Peptides Derived from Yeast

The production of recombinant proteins and metabolites in yeasts is well established (Porro et al., 2011). Recombinant Fps1 is produced in yeast with implications for the translational regulation of an aquaporin (Cartwright et al., 2017). The suppressors of amyloid-β toxicity improve recombinant-protein production in yeast by reducing oxidative stress and tuning cellular metabolism (Chen et al., 2022; Mehrab

Mohseni et al., 2019). Yeast is a preferable cell factory for the production of secretory recombinant proteins (Thak et al., 2020). Traditionally, *Saccharomyces bayanus* (for making wine) and *S. boulardii* (as a probiotic after antibiotic treatment) have been used as a cell factory (Jin et al., 2021; Rengel Dos Passos et al., 2021; van der Aa Kühle et al., 2005). *S. boulardii* is marketed in a lyophilized form to treat diarrhea. Most reports show a clinical benefit of *S. boulardii* in decreasing the duration of diarrhea (Vieira et al., 2012; Johnston et al., 2011; Dinleyici et al., 2012; Shan et al., 2013; Szajewska & Mrukowicz, 2005). Patients with irritable bowel syndrome are administered *S. boulardii* and have shown positive effects (Choi et al., 2011). It can prevent and treat relapses of inflammatory bowel disease and moderate symptoms of ulcerative colitis (Guslandi et al., 2000). Recurrent pseudomembranous colitis infection caused by *Clostridium difficile* can be reduced by daily doses of *S. boulardii* (with standard antibiotics). No malformations were reported during pregnancy (Dugoua et al., 2009). *Saccharomyces* yeast also creates symbiotic matrixes with bacteria to form kefir (Witthuhn et al., 2005) and is a component of kombucha (Marsh et al., 2014). *S. cerevisiae* is used for producing various other recombinant proteins. Different modifications in the recombinant strains, the vectors, transferring proteins, etc., occur to improve the protein production process in *S. cerevisiae.* The number of the recombinant proteins produced in *S. cerevisiae* has increased. For example, recombinant cyprosin B is produced from *S. cerevisiae* (W303-1A) strain (Sampaio et al., 2008). *S. cerevisiae* attracts the attention of different kinds of research. For example, comparative transcriptome analysis was performed between original and evolved recombinant lactose-consuming *S. cerevisiae* strains (Guimarães et al., 2008). To enhance its secretion a novel fusion partner of recombinant proteins is used in *S. cerevisiae* (Bae et al., 2016). *S. cerevisiae* strain is subjected to modification to be suitable for efficient conversion of lactose in salted and unsalted cheese whey into ethanol (Tahoun et al., 2002). Different expression systems were investigated aiming to study its ability to produce recombinant proteins (Liu et al., 2012). *S. cerevisiae* CCMI 885 secretes peptides that inhibit the growth of other types of yeasts during wine production (Alcaide-Hidalgo et al., 2007). Released *S. cerevisiae* bioactive peptides were studied under accelerated autolysis in a wine (Branco et al., 2014; Alcaide-Hidalgo et al., 2007; De Iseppi et al., 2024). *S. cerevisiae* accumulates GAPDH-derived peptides on its cell surface that induce death of non-*Saccharomyces* yeasts by cell-to-cell contact (Branco et al., 2017). *S. cerevisiae* antimicrobial peptides induce alterations in the intracellular pH, membrane permeability, and culture ability of *Hanseniaspora guilliermondii* cells (Branco et al., 2015). Increasing reports prove the symbiotic relationship between *Lactobacillus* and *S. cerevisiae* and other types of yeasts. The ACE inhibitory and anti-diabetic peptides from buffalo and camel milk are fermented by both lactobacilli and yeast (Khakhariya et al., 2023). Camel milk contains important proteins such as lactoferrin. There has been increasing interest in lactoferrin peptide, particularly for its antimicrobial and antiviral activity (Jean et al., 2016; Kuwata et al., 1998; Zarzosa-Moreno et al., 2020). Yeast and fungal proteolytic enzyme activity in fermented milk could lead to releasing various peptides, including lactoferrin peptide (Chaves-López et al., 2014; Cheng et al., 2025). In particular, milk fermented with fungi has for years been the subject of an increasing amount of research into producing active peptides (Chaves-López et al., 2014; Cheng et al., 2025; López-Expósito et al., 2008). Multiple-derived peptides produced by *S. cerevisiae* that are involved in malolactic fermentation inhibition have been identified (Rizk et al., 2018). Lentil protein has been isolated as a fermenting substrate for the production of bioactive peptides by neglected yeast species (Tonini et al., 2024). *S. cerevisiae* biomass was recognized as a safe source of next-generation food preservatives (Pereira et al., 2021). Recently, *Pichia pastoris* showed an increasing usage rate in recombinant protein production (Sørenson, 2010). A recombinant antimicrobial peptide hispidalin (Meng et al., 2019), and PaDef (Meng et al., 2017), were produced in *P. pastoris*. Different other recombinant proteins are included, such as *Oryctolagus cuniculus* chymosin, in *P. pastoris* (Ali Hanoğlu et al., 2021). The combined effect of the gene copy number and chaperone overexpression on the recombinant bovine chymosin production in *P. pastoris* with the mutant ADH2 promoter was investigated (Ersöz & İnan, 2025); a recombinant antimicrobial peptide (extracellular expression) and efficient purification of a functional recombinant *Volvariella volvacea* immunomodulatory protein (FIP-vvo) were performed using the *P. pastoris* system (Sun et al., 2014). The impact of ERAD on recombinant-protein secretion in *P. pastoris* (syn. *Komagataella* spp.) was studied (Zahrl et al., 2018). Recombinant expression of bovine chymosin in *P. pastoris* was produced (Zhang et al., 2009a) as well as a high-temperature cultivation of recombinant *P. pastoris,* which increased ER stress and decreased production of human interleukin-10 (Zhong et al., 2014). An engineered *P. pastoris* platform was established for the biosynthesis of silk-based nanomaterials with therapeutic potential (Tian et al., 2024).

4.3 Peptides Derived from Fungi

Filamentous fungi are known for their ability to produce enzymes that can find their way into industry. Industrial enzymes include cellulases, proteases, and amylases, which are used in various applications, including biofuels, food processing, and textiles. Fungi are important candidates in producing therapeutic proteins, including antibodies, hormones (such as insulin), and vaccines. For example, *A. niger* and *Trichoderma reesei* have been used to produce recombinant proteins for medical applications. Filamentous fungi can produce bioactive peptides with potential health benefits, including antimicrobial, antioxidant, and immunomodulatory properties. These peptides can be derived from the fungal metabolism or through the expression of imported genes. Correct protein folding and activity are important, especially for complex proteins. Filamentous fungi account for approximately 50% of all industrial enzymes (Song et al., 2017). Their ability to perform protein modification, maturation, and secretion is attracting attention to establishing them as protein cell factories (Hildebrand et al., 2020). PTM by glycosylation plays a great role in protein folding, localization, stability, and in quality control before final protein maturation and secretion. Even other eukaryotes such as yeast can perform PTM but the glycosylation of filamentous fungi was believed to be more efficient (FAO, 2010; Wunsch, 2022; Devlin, 2022). *A. niger*, *A. oryzae*, *T. reesei*, and *Neurospora crassa* are used widely as recombinant strains for protein production (Gao et al., 2011; Jaganmohan et al., 2013; Turnbull et al., 1992; Thrane, 2007). Their whole genomes have been sequenced, their transcriptomics are identified, and the other related metabolic and genetic issues are well known (Boland et al., 2013; Rodger, 2001; Miller & Dwyer, 2001; Williamson et al., 2006). Industrial filamentous fungi, e.g., *A. niger*, *A. oryzae*, and *T. reesei*, were accepted as generally recognized as safe (GRAS) strains (GMI Inc., 2022). Usually, homologous proteins are much more effectively expressed (Kowalchuk et al., 2008; Song et al., 2017). The amount of enzymes produced can reach tens of grams per liter (Anupma & Tamang, 2016). *T. reesei* cellulase are extra-optimized (Hashem et al., 2022). For facilitating high transcription and improving mRNA stability of heterologous genes, it should be integrated at transcriptionally active regions using the fungus genome (Kowalchuk et al., 2008; Song et al., 2017). Signal peptides are used to obtain high yields of heterologous proteins. A fungal host can secrete any protein (homologous or heterologous) provided that the protein has a recognized signal peptide (by the host secretion pathway) (Song et al., 2017). Homologous signal peptides are best recognized by the host (Kowalchuk et al., 2008; FAO, 2010). Using signal peptides from the highly secreted proteins glucoamylase in *A. niger*, and the *T. reesei* cellobiohydrolase fused to heterologous proteins led to much better secretion. The amino-acid codons can be changed without changing the main amino acids (degenerate), some codon usage is better than others, particularly affecting mRNA stability and translation efficiency (FAO, 2010). *Aspergillus* is a phenotypically polythetic genus and is widely distributed in the environment (Tsang et al., 2018), with some of its species having the ability to sporulate (Adekoya et al., 2007). Data for some useful fungal genera include *Aspergillus* (*A. acidus*, *A. oryzae*, *A. sojae*, *A. niger*, *A. sydowii*, *A. flavus*, and *A. versicolor*) are used in producing miso and soya sauce fermentations, sake and awamori liquors, and in fermenting Pu-erh tea respectively (Nout, 2000; Nout, 2007; Mogensen et al., 2009; Tamang & Samuel, 2010). *A. wentii* is used to process soybeans. *A. oryzae* and *Rhizopus oryzae* are broadly recognized as successful strains in the industrial sector. They have been used in a large number of pharmaceutical manufacturing and food-processing applications (Jo et al., 2023; Ding & Ye, 2023). *Aspergillus* species have biomedicine, bioenergy, health, and biotechnology applications. This is a large and diverse genus comprising ~400 recognized species (Romero et al., 2021). *Aspergillus* species are suitable fungal cell factories because of their ability to produce and secrete native and heterologous proteins, organic acids, and secondary metabolites (Ntana et al., 2020). *A. niger* and other aspergilli have been harnessed for biosynthesis of diverse enzyme cocktails (Cairns et al., 2018). As an example, phytases, which are used to improve the nutritional content of animal feed, were first marketed in 1991 (Jatuwong et al., 2020). *A. niger* was greatly involved in global markets of phytase production (Haefner et al., 2005). The *Myceliophthora thermophila* recombinant trehalase and *Penicillium citrinum* nuclease P1 achieved high production levels in *A. niger* (Dong et al., 2020).

The yield of human erythropoietin was a 41-fold improvement over the original yield (Rojas-Sánchez et al., 2020). Gigantin, a ribonuclease (Salvarelli et al., 1994; Wirth et al., 1997) is purified from the culture medium of *Aspergillus giganteus* IFO 5818 (Ng et al., 2003; Watanabe et al., 1995). Clavin, (17 kDa) also known as C-sarcin (Huang et al., 1997), is a type 1 RIP from *Aspergillus clavatus* (Ciechanover, 1994). Restrictocin and mitogillin are other well-known RIPs, also referred to as ribotoxins. Structurally related ribotoxins were demonstrated in other species (Martínez-Ruiz et al., 1999). Clavin (Polonsky et al., 1977; Nielsen et al., 2014; Fernández-Bodega et al., 2017) was constructed. Mushroom (Lam & Ng, 2001) antifungal protein was elaborated by soil-borne *Trichoderma viride*. It acts on the ribosomal RNA of reticulocytes to produce an α-sarcin RNA fragment (Lin et al., 1991). *A. oryzae* is considered safe as it has been used in fermented cuisine (miso, soy sauce, or douchi) for more than 10 centuries (Jin et al., 2021). In China and Japan, *A. oryzae* was extensively employed in the food industry. It is certified as GRAS by the US Food and Drug Administration (Frisvad et al., 2018; EFSA Panel on Food Contact Materials, Enzymes and Processing Aids (CEP), 2022). *A. oryzae* is used widely in protein secretion and PTM, enzyme production including glucoamylase, cellulase, and proteases (Saqib & Hamid Rashid, 2023; Chen et al., 2023), and in the fields of medicine, environment, and food production (Qiu et al., 2022; Nomura et al., 2021; Li et al., 2022b) in organic compound degradation. In 2005, Galagan et al. (Galagan et al., 2005) completed the whole-genome sequencing of *A. oryzae*. *A. oryzae* are used in producing multiple hydrolases, include α-amylases, endo-proteinases, and exo-peptidases. *A. oryzae* harbors at least 500 genes with potential practical value in industrial applications (Galagan et al., 2005). One of the advantages of *A. oryzae* is that it secretes large amounts of enzymes but produces very few secondary metabolites under typical cultural conditions (Meng et al., 2022).

Moreover, the bovine CHY, HLY, and recombinant antibodies (e.g., adalimumab) were produced by *A. oryzae* to decrease their production costs (Daba et al., 2021). Even koji and other fungi included in fermented food production served as a natural wild type strain over the years, but trials to increase their production were conducted using random mutagenesis, genome editing, gene deletion, etc. (Suryadi et al., 2022; Dong et al., 2020; Rojas-Sánchez et al., 2020; Katayama, 2022; Li et al., 2023; Nødvig et al., 2015; Schuster & Kahmann, 2019; Yang et al., 2020). *A. unguis* was demonstrated to be a host for heterologous protein production through the expression of the pharmaceutical human interferon gene (Madhavan et al., 2017). A gene coding for an a-galactosidase enzyme from *A. fumigatus* was expressed in *A. sojae*, leading to approximately three-fold higher production (Gürkök et al., 2010). The extracellular endo-ß-1,4-mannanase gene from *A. fumigatus* was successfully expressed in *A. sojae* with a 12-fold increase (Duruksu et al., 2009). *A. nidulans* is used for producing heterologous enzymes (Zubieta et al., 2018; Kumar, 2020). Endoxylanase- and arabinofuranosidase-encoding genes from *Penicillium funiculosum* and *A. niger* respectively were introduced into *A. nidulans* (Gonçalves et al., 2012). The genus *Fusarium* contains almost 200 accepted species, with the most common in the food industry (cheese fermentations) being *Fusarium domesticum*, *F. solani*, and *F. venenatum*. Species from the *Fusarium* genus are recognized as promising sources of several enzymes for industrial applications, "helpers" that promote the production of aromas and pigments to the final products. Some of the secondary metabolites produced by *Fusarium* spp. strains are mycotoxins that affect human and animal health, but studies have shown that producing mycotoxins stops under industrial conditions (Thrane, 2007; Anupma & Tamang, 2016). Strains of the genus *Rhizopus* (*R. oligosporus*, *R. microsporus*, *R. stolonifer*, *R. arrhizus*, *R. delemar*, *R. oryzae*, *R. delemar*) were detected in yaoqu from China and bánh men from Vietnam, which are strong amylase producers (Tamang & Samuel, 2010; Gryganskyi et al., 2018; Thanh et al., 2008). Also used in production of tempeh, a fermented soybean food, *R. oryzae* is considered a GRAS filamentous fungus, which is commonly used for producing some Asian fermented foods (Anupma & Tamang, 2016; Tamang & Samuel, 2010; Tamang et al., 2015; Ly et al., 2018; Lv et al., 2012; Bora et al., 2016). Species of the genus *Penicillium* (*P. camemberti*, *P. biforme*, *P. fuscoglaucum*, *P. palitans*, *P. commune* (as the wild type "ancestor" of *P. camemberti*), *P. biforme*, *P. fuscoglaucum*, *P. palitans*, *P. caseifulvum*, *P. nalgiovense*, *P. chrysogenum*, *P. commune*, *P. solitum*, *P. nordicum*, *P. expansum*, and *P. roqueforti*) are the most common species involved in the microbiota of several types of cheeses, including Camembert, Brie, and "blue cheeses," as well as fermented mild-ripened sausages (such as salami) and raw dry- or smoke-cured meat products (such as ham) (Pitt et al., 1986; Polonelli et al., 1987; Giraud et al., 2010) . All blue-mold cheeses are always made with *P. roqueforti*, which is inoculated and grows inside the product. *P. camemberti* is used in mold-ripened soft cheeses including Camembert and Brie. It is inoculated together with a lactic starter culture or, alternatively, sprayed on the product and grows on the surface of the cheese piece (Ropars et al., 2017). *P. roqueforti* can be produced secondary metabolite components that fall within the category of mycotoxins (Vallone et al., 2018). Today, starter cultures are available that do not have the ability to produce most of these mycotoxins, and the presence of roquefortine C and isofumigaclavines in the amounts usually found on cheeses does not seem to represent a risk to human health (Hymery et al., 2014). Alternatively, they are unstable in cheese and are converted to PR-imine (Siemens & Zawistowski, 1993). As an answer to the main concern regarding fungi as ripening agents and the possibility of mycotoxin production, starter cultures do not have the ability to produce most of the above-mentioned mycotoxins (Hymery et al., 2014; Engel & Prokopek, 1980; Engel & Prokopek, 1979).

4.4 Peptides Derived from Marine Fungi

Marine fungi are a source of bioactive peptides (Youssef et al., 2019). Asperterrestide A (cyclic tetrapeptide) derived from the marine *A. terreus* SCSGAF0162 has antiviral action on H1N1 and H3N2 (influenza A virus) (He et al., 2013; Abdelmohsen et al., 2017). Four antiviral peptides (a cyclic pentapeptide and three linear peptides), named aspergillipeptide D–G were extracted from a culture broth of

Aspergillus sp. SCSIO 41501 (aquatic) and presented antiviral action against HSV-1 (Ma et al., 2017). Simplicilliumtide J (a novel antiviral cyclic hexapeptide) and analogs verlamelins A and B (cyclic lipodepsipeptides) were obtained from *Simplicillium obclavatum* EIODSF 020 (a fungus found in the deep sea) and performed powerful anti-HSV-1 efficacy (Liang et al., 2017). Peptides isolated from *Acremonium* include cordyheptapeptide C, E (Chen et al., 2012), efrapeptin Eα, F, and G (Boot et al., 2007; Boot et al., 2006) anticancer peptides; efrapeptin G (Boot et al., 2007; Boot et al., 2006) antimicrobial peptide and RHM1 antibacterial peptide against *Staphylococcus epidermidis* (Boot et al., 2007; Boot et al., 2006). Peptides isolated from *Aspergillus* include aspergillicins A–E cytotoxic peptides (Capon et al., 2003), cyclo-(L-Trp-L-Tyr) cytotoxic peptide against HT-29 cancer cells (Zhang et al., 2009b); sclerotides A and B with antifungal peptide against *Candida albicans* (Zheng et al., 2009), similanamide weak cytotoxic peptide against MCF-7, A373, and NCI-H460 cancer cells (Prompanya et al., 2015); terrelumamide A and B peptides that can improve insulin sensitivity (Luo et al., 2019; He et al., 2013; You et al., 2015) and that are cytotoxic against U937 and MOLT4 human cancer cell lines and antiviral against influenza virus of H1N1 and H3N2 strains (Luo et al., 2019; He et al., 2013; You et al., 2015); psychrophilin E cytotoxic peptide against HCT116 (Ebada et al., 2014); psychrophilin G peptide with lipid-lowering activity (Peng et al., 2014); aspersymmetide A peptide weak cytotoxic activity against NCI-H292 as well as A431 cell lines; cotteslosin A has weak cytotoxic activity against human melanoma (Fremlin et al., 2009); diketopiperazine dimer an anti-inflammatory peptide with activity against IL-10 expression of the LPS-induced THP-1 cells (Liu et al., 2018); cyclic tetrapeptide with anti-inflammatory activity against IL-10 expression of the LPS-induced THP-1 cells (Liu et al., 2018), aspergillipeptide D with antiviral activity versus herpes simplex virus type 1 (HSV-1) and acyclovir-resistant clinical isolates; aspergillipeptide E with antiviral activity versus HSV-1 (Ma et al., 2017); 14-hydroxy-cyclopeptine with inhibition of nitric-oxide production (Zhou et al., 2015). Peptides isolated from *Asteromyces* include lajollamide A with weak antimicrobial activity against *Bacillus subtilis* and *Staphylococcus epidermidis* (Gulder et al., 2012). Peptides isolated from *Ceratodictyon* include dictyonamide A with inhibition on cyclin-dependent kinase 4 (Komatsu & Shigemori, 2001); clonostachysin A and B with inhibition of dinoflagellate *Prorocentrum micans* (Adachi et al., 2005). Peptides isolated from *Emericella* include unguisin A with antimicrobial activity versus methicillin-resistant *Staphylococcus aureus* strains (MRSA) (Oh et al., 2007); emericellamide B antimicrobial activity against MRSA strains (Oh et al., 2007). Peptides isolated from *Exserohilum* include rostratins A–D, which have cytotoxic activity versus HCT-116, the human colon carcinoma (Tan et al., 2004). Peptides isolated from *Microsporum* include microsporin A inhibition of histone deacetylase and with cytotoxic activity versus HCT-116 and the 60-cancer-cell panel of the National Cancer Institute (Gu et al., 2007); microsporin B with inhibition of histone deacetylase and cytotoxic activity against HCT-116 (Gu et al., 2007). Peptides isolated from *Penicillium* include penicimutide with cytotoxic activity versus HeLa cells (Wang et al., 2016); gliocladine C with cytotoxic activity versus HepG2 cells (Hong

& Yang, 2011); cis-Cyclo (leucyl-tyrosyl) with inhibition to biofilm formation (Scopel et al., 2013). Peptides isolated from *Scytalidium* such as halovir A with antiviral activity versus herpes simplex viruses of both type 1 and type 2 as well as HSV-1 (Rowley et al., 2004a, b); halovirs B–E with antiviral activity versus herpes simplex viruses of both type 1 and type 2 (Rowley et al., 2004a, b). Peptides isolated from *Simplicillium* such as simplicilliumtide A with weak cytotoxic activity versus human leukemia HL-60 and K562 cell lines; simplicilliumtide D with antifouling effect versus the larvae of *Bugula neritina*; simplicilliumtides E, G, and H with a weak cytotoxic effect versus human leukemia HL-60 and K562 cell lines (Liang et al., 2017; Liang et al., 2016); simplicilliumtide J with antifungal activity versus *Curvularia australiensis* and *A. versicolor* and antiviral activity versus HSV-1 (Liang et al., 2017; Liang et al., 2016). Peptides isolated from *Stachylidium* such as endolide A showed binding activity to the vasopressin receptor 1A (Almeida et al., 2016); endolide B with binding to serotonin receptor 5HT2b (Almeida et al., 2016). Peptides isolated from *Talaromyces* such as talaropeptide A with antibacterial activity versus *B. subtilis* (Dewapriya et al., 2018); talaropeptide B with antibacterial activity against *B. subtilis* (Dewapriya et al., 2018). Peptides isolated from *Zygosporium* include zygosporamide with cytotoxic activity against NCI's 60 cell-line panel, central nervous system cancer-cell line SF-268, and renal cancer-cell line RXF 393 (Torres-García et al., 2014; Oh et al., 2006).

5 Future Prospective

Fungi are one of many organisms that serve humans. Originally, they did a critical job by degrading waste in the nature. For a long time humans have observed positive and negative fungal activity. Human success benefited from both. Fermented food included both bacterial and fungal activity. Historically, different fermented foods have been produced in nearly all over the Old World, particularly in Asia and the Middle East. The art had wonderfully improved using simple strategies, such as preparing the preprepared koji. A huge number of fermented foods are produced globally. Currently, several companies have installed lines for producing some famous and popular fermented products. Kefir, which was a secret, is turn to a famous and popular food and sealed now in the shops, in USA, Canada, etc. Many Asian, old European, Indian, Egyptian, etc., fermented foods were transferred here and there and enrich the variety of food and dishes served worldwide. Fermented food is a real treasure for anyone to find one or more products that can establish a business. It is a gift for everyone. The probiotic, which as a term was known in a limited way, has now become the first choice for lay people. Moreover, after the success of insulin production in *E. coli*, eukaryotes have opened up a new horizon, whereas they have similar cell systems. The scientific focus changed from solving some technical points like how to open the inclusion bodies (in bacteria) and fold the protein correctly to how to add a signal peptide to the gene sequence of certain proteins to be transferred through the ER and to be secreted extracellularly. The successful knowledge accumulation and the perfect scientific system in western countries have enabled great progress in the field. Attracting brilliant brains from overall the world with their background in their local fermented food products and their positive effect encourage new research in this area and flourish the pharmacies with many nutraceuticals and medicinal peptides, active proteins, probiotics, mixed formulas, etc. The role of the companies and their products, research, technology improvement etc., flourishing this sector is clear. The collaboration between investors, politicians, planners, scientists, etc., have collectively opened the most scattered and richest treasure: fermented foods. The future of fermented foods, proteins, peptides, and other related products is clear. But all are invited to keep the origin of this treasure running beside or side by side with our modern production style, because for ages, fermented foods had been developed and, in the future, they will continue to develop, if kept running correctly.

6 Conclusion

Fungal species, including filamentous fungi, yeasts, and mushrooms, are involved in many fermentation processes. Fermented food can be consumed directly by humans. Only nonharmful species were suitable for certain biotechnological processes. That was the past. Nowadays, molecular-biology tools enable any unpleasant co-existing product to be bypassed, by transferring the gene of interest to another safer strain, or by impairing the genes of any harmful products to obtain the benefit of other useful genes in the targeted microbial strain. Human, animal, plant, invertebrate, any kind of gene, can nowadays be cloned, transferred, and expressed in a recombinant strain under a suitable promotor and nutrient condition to start its production. Various strategies could enrich the production process. Some are addressed in this chapter. This chapter describes some of the many active proteins and peptides that are either produced naturally by one of the fungal strains or transferred from one strain to others, as well as being transferred from other biological forms, or produced during fermentation of some types of food. The active proteins and active peptides addressed in this chapter are variant and diverse, either concerning their origin or concerning the method and the strain that were used during their production. Various pharmaceutical and nutraceutical forms are described.

References

Abdelmohsen, U. R., et al. (2017). Potential of marine natural products against drug-resistant fungal, viral, and parasitic infections. *The Lancet Infectious Diseases, 17*(2), e30–e41.

Adachi, K., et al. (2005). Clonostachysins A and B, new Antidinoflagellate cyclic peptides from a marine-derived fungus. *ChemInform, 36*(27).

Adekoya, I., et al. (2007). Contamination of fermented foods in Nigeria with fungi. *LWT, 86*, 76–84.

Ahmad, N., et al. (2014). Antioxidant activity via DPPH, gram-positive and gram-negative antimicrobial potential in edible mushrooms. *Toxicology and Industrial Health, 30*(9), 826–834.

Ahmed, T., et al. (2024). Research trends in production, separation, and identification of bioactive peptides from fungi – A critical review. *Journal of Functional Foods, 119*, 106343.

Alberto, F., et al. (2009). Technical advance in fungal biotechnology: Development of a miniaturized culture method and an automated high-throughput screening. *Letters in Applied Microbiology, 49*(2), 278–282.

Alcaide-Hidalgo, J. M., et al. (2007). Bioactive peptides released from *Saccharomyces cerevisiae* under accelerated autolysis in a wine model system. *Journal of Food Science, 72*(7), M276–M279.

Ali Hanoğlu, S., Ektiren, D., & Karaaslan, M. (2021). Recombinant expression and characterization of Oryctolagus cuniculus chymosin in Komagataella phaffii (Pichia pastoris). *Protein Expression and Purification, 183*, 105874.

Almeida, C., et al. (2016). Endolides A and B, vasopressin and serotonin-receptor interacting *N*-methylated peptides from the sponge-derived fungus *Stachylidium* sp. *Organic Letters, 18*(3), 528–531.

Al-Obaidi, J. R., Jambari, N. N., & Ahmad-Kamil, E. I. (2021). Mycopharmaceuticals and nutraceuticals: Promising agents to improve human well-being and life quality. *Journal of Fungi (Basel), 7*(7).

Altay, F., et al. (2013). A review on traditional Turkish fermented non-alcoholic beverages: Microbiota, fermentation process and quality characteristics. *International Journal of Food Microbiology, 167*(1), 44–56.

Amara, A. A. (2013a). Wild and engineered microbes and proteins in technical enzymes: Food for thought. In *Applications of microbial genes in enzyme technology* (pp. 327–352). Nova Science Publishers.

Amara, A. A. A. F. (2013b). Pharmaceutical and industrial protein engineering: Where we are? *Pakistan Journal of Pharmaceutical Sciences, 26*(1), 217–232.

Amara, A. A., & El-Baky, N. A. (2023). Fungi as a source of edible proteins and animal feed. *Journal of Fungi (Basel), 9*(1).

Anupma, A., & Tamang, J. P. (2016). Diversity of filamentous fungi isolated from some amylase and alcohol-producing starters of India. *Frontiers in Microbiology, 11*.

Arulmuthu, E., Williams, D., & Versteeg, H. (2009). The arrival of genetic engineering. *IEEE Engineering in Medicine and Biology Magazine, 28*(1), 40–54.

Askenazi, M., et al. (2003). Integrating transcriptional and metabolite profiles to direct the engineering of lovastatin-producing fungal strains. *Nature Biotechnology, 21*(2), 150–156.

Bae, J. H., et al. (2016). A novel fusion partner for enhanced secretion of recombinant proteins in *Saccharomyces cerevisiae*. *Applied Microbiology and Biotechnology, 100*(24), 10453–10461.

Baker, M. (2011). Protein engineering: Navigating between chance and reason. *Nature Methods, 8*(8), 623–626.

Bigey, F., et al. (2024). Differential adaptation of the yeast Candida anglica to fermented food. *Food Microbiology, 123*, 104584.

Boland, M. J., et al. (2013). The future supply of animal-derived protein for human consumption. *Trends in Food Science & Technology, 29*(1), 62–73.

Boot, C. M., et al. (2006). Highly N-methylated linear peptides produced by an atypical sponge-derived Acremonium sp. *Journal of Natural Products, 69*(1), 83–92.

Boot, C. M., et al. (2007). Four classes of structurally unusual peptides from two marine-derived fungi: Structures and bioactivities. *Tetrahedron, 63*(39), 9903–9914.

Bora, S. S., et al. (2016). Metagenomics analysis of microbial communities associated with a traditional rice wine starter culture (Xaj-pitha) of Assam, India. *3 Biotech, 6*(2).

Branco, P., et al. (2014). Identification of novel GAPDH-derived antimicrobial peptides secreted by *Saccharomyces cerevisiae* and involved in wine microbial interactions. *Applied Microbiology and Biotechnology, 98*(2), 843–853.

Branco, P., et al. (2015). Antimicrobial peptides (AMPs) produced by *Saccharomyces cerevisiae* induce alterations in the intracellular pH, membrane permeability and culturability of Hanseniaspora guilliermondii cells. *International Journal of Food Microbiology, 205*, 112–118.

Branco, P., et al. (2017). *Saccharomyces cerevisiae* accumulates GAPDH-derived peptides on its cell surface that induce death of non-Saccharomyces yeasts by cell-to-cell contact. *FEMS Microbiology Ecology, 93*(5).

Brannigan, J. A., & Wilkinson, A. J. (2002). Protein engineering 20 years on. *Nature Reviews Molecular Cell Biology, 3*(12), 964–970.

Bud, R. (1989). History of 'biotechnology'. *Nature, 337*(6202), 10–10.

Cairns, T. C., Nai, C., & Meyer, V. (2018). How a fungus shapes biotechnology: 100 years of Aspergillus Niger research. *Fungal Biology and Biotechnology, 5*, 13–13.

Capon, R. J., et al. (2003). Aspergillicins A–E: five novel depsipeptides from the marine-derived fungus *Aspergillus carneus*. *Organic & Biomolecular Chemistry, 1*(11), 1856–1862.

Cartwright, S. P., et al. (2017). Constitutively-stressed yeast strains are high-yielding for recombinant Fps1: Implications for the translational regulation of an aquaporin. *Microbial Cell Factories, 16*(1), 41.

Ceugniez, A., et al. (2017). Fungal diversity of "Tomme d'Orchies" cheese during the ripening process as revealed by a metagenomic study. *International Journal of Food Microbiology, 258*, 89–93.

Chang, S. T., Hayes, W. A., & Preface. (1978). *Biology and cultivation edible mushrooms* (pp. xix–xxi). Academic Press.

Chaves-López, C., et al. (2014). Impact of microbial cultures on proteolysis and release of bioactive peptides in fermented milk. *Food Microbiology, 42*, 117–121.

Chen, Z., et al. (2012). Cyclic heptapeptides, cordyheptapeptides C–E, from the marine-derived fungus *Acremonium persicinum* SCSIO 115 and their cytotoxic activities. *Journal of Natural Products, 75*(6), 1215–1219.

Chen, T. Q., et al. (2018). Antioxidant and Hepatoprotective activities of crude polysaccharide extracts from Lingzhi or Reishi medicinal mushroom, Ganoderma lucidum (Agaricomycetes), by ultrasonic-circulating extraction. *International Journal of Medicinal Mushrooms, 20*(6), 581–593.

Chen, S., et al. (2019). Innovative antigen carrier system for the development of tuberculosis vaccines. *The FASEB Journal, 33*(6), 7505–7518.

Chen, X., et al. (2022). Suppressors of amyloid-β toxicity improve recombinant protein production in yeast by reducing oxidative stress and tuning cellular metabolism. *Metabolic Engineering, 72*, 311–324.

Chen, C., et al. (2023). Improving protein utilization and fermentation quality of soy sauce by adding protease. *Journal of Food Composition and Analysis, 121*, 105399.

Cheng, L., et al. (2025). Potential prebiotic properties and proliferation mechanism of fermented milk-derived polypeptides. *Food Chemistry, 463*(Pt 3), 141335.

Chiang, S.-J. (2004). Strain improvement for fermentation and biocatalysis processes by genetic engineering technology. *Journal of Industrial Microbiology & Biotechnology, 31*, 99–108.

Choi, C. H., et al. (2011). A randomized, double-blind, placebo-controlled multicenter trial of *Saccharomyces boulardii* in irritable bowel syndrome. *Journal of Clinical Gastroenterology, 45*(8), 679–683.

Ciechanover, A. (1994). The ubiquitin-proteasome proteolytic pathway. *Cell, 79*(1), 13–21.

Crisman, R. L., & Randolph, T. W. (2008). Refolding of proteins from inclusion bodies is favored by a diminished hydrophobic effect at elevated pressures. *Biotechnology and Bioengineering, 102*(2), 483–492.

Culleton, H., et al. (2014). New promoters to improve heterologous protein production in *Aspergillus vadensis*. *Current Biotechnology, 3*(3), 244–251.

Daba, G. M., Mostafa, F. A., & Elkhateeb, W. A. (2021). The ancient koji mold (*Aspergillus oryzae*) as a modern biotechnological tool. *Bioresources and Bioprocessing, 8*(1), 52–52.

Das, A. K., et al. (2021). Edible mushrooms as functional ingredients for development of healthier and more sustainable muscle foods: A flexitarian approach. *Molecules (Basel, Switzerland), 26*(9), 2463.
Dash, P., et al. (2024). Antimicrobial properties of the edible pink oyster mushroom, *Pleurotus eous*: In-vivo and in-vitro studies. *Microbial Pathogenesis, 196*, 106915.
Datta, S., et al. (2020). Tropical milky white mushroom, *Calocybe indica* (Agaricomycetes): An effective antimicrobial agent working in synergism with standard antibiotics. *International Journal of Medicinal Mushrooms, 22*(4), 335–346.
De Iseppi, A., et al. (2024). Characterization and identification of yeast peptides released during model wine fermentation and lees contact. *Journal of Agricultural and Food Chemistry, 72*(44), 24749–24761.
De Meyts, P. (2017). Early recombinant protein therapeutics. In T. Vaughan, J. Osbourn, & B. Jallal (Eds.), *Protein therapeutics* (1st ed.). Wiley-VCH Verlag GmbH & KGaA.
de Vries, R. P., et al. (2004). A new black aspergillus species, A. vadensis, is a promising host for homologous and heterologous protein production. *Applied and Environmental Microbiology, 70*(7), 3954–3959.
de Vries, R. P., et al. (2005). *Aspergillus vadensis*, a new species of the group of black Aspergilli. *Antonie Van Leeuwenhoek, 87*(3), 195–203.
Devlin, E. (2022). *Quorn profits shrink as rivals eat in market leader's share of meat-free category. 2022.* Available online: https://www.thegrocer.co.uk/results/quorn-profits-shrink-as-rivals-eat-in-market-leaders-share-of-meat-free-category/671919.article. Accessed 16 Dec 2022.
Dewapriya, P., et al. (2018). Talaropeptides A-D: Structure and biosynthesis of extensively N-methylated linear peptides from an Australian marine tunicate-derived *Talaromyces* sp. *Frontiers in Chemistry, 6*, 6.
Ding, Q., & Ye, C. (2023). Microbial cell factories based on filamentous bacteria, yeasts, and fungi. *Microbial Cell Factories, 22*(1), 20–20.
Dinleyici, E. C., et al. (2012). Effectiveness and safety of saccharomyces boulardii for acute infectious diarrhea. *Expert Opinion on Biological Therapy, 12*(4), 395–410.
Do, S. H., Lee, T. G., & Kim, S. K. (2024). Enhancing protein content in wild-type *Saccharomyces cerevisiae* via random mutagenesis and optimized fermentation conditions. *Journal of Microbiology and Biotechnology, 34*(9), 1912–1918.
Dong, L., et al. (2020). High-level expression of highly active and thermostable trehalase from *Myceliophthora thermophila* in Aspergillus Niger by using the CRISPR/Cas9 tool and its application in ethanol fermentation. *Journal of Industrial Microbiology and Biotechnology, 47*(1), 133–144.
Dugoua, J.-J., et al. (2009). Probiotic safety in pregnancy: A systematic review and meta-analysis of randomized controlled trials of Lactobacillus, Bifidobacterium, and Saccharomyces spp. *Journal of Obstetrics and Gynaecology Canada, 31*(6), 542–552.
Duruksu, G., et al. (2009). Cloning, expression and characterization of endo-β-1,4-mannanase from *Aspergillus fumigatus* in *Aspergillus sojae* and *Pichia pastoris*. *Biotechnology Progress, 25*(1), 271–276.
Ebada, S. S., et al. (2014). Psychrophilin E, a new cyclotripeptide, from co-fermentation of two marine alga-derived fungi of the genus Aspergillus. *Natural Product Research, 28*(11), 776–781.
EFSA Panel on Food Contact Materials, Enzymes and Processing Aids (CEP), et al. (2022). Safety evaluation of the food enzyme phospholipase A1 from the genetically modified *Aspergillus oryzae* strain NZYM-LJ. *EFSA Journal, 20*(7), e07381–e07381.
Engel, G., & Prokopek, D. (1979). Kein Nachweis von Penicillium roqueforti-toxin in Kase. *Milschwissenschaft, 34*, 272–274.
Engel, G., & Prokopek, D. (1980). KeinNachweis von Patulin and Penicillinsaure in mit Patulin- und Penicillinsaure-bildenden Penicillium roqueforti-Stammenhergestell-ten Kasen. *Milchwissenschaft, 35*, 218–220.

Ermakova-Gerdes, S., S. Shestakov, and W. Vermaas, Random chemical mutagenesis of a specific psbDI region coding for a lumenal loop of the D2 protein of photosystem II in *Synechocystis* sp. PCC 6803. Plant Molecular Biology, 1996. 30(2): p. 243–254.

Ersöz, F., & İnan, M. (2025). The combined effect of the gene copy number and chaperone overexpression on the recombinant bovine chymosin production in Pichia pastoris, with mutant ADH2 promoter. *Protein Expression and Purification, 227*, 106636.

Evdokimov, A. G., et al. (2008). Rational protein engineering in action: The first crystal structure of a phenylalanine tRNA synthetase from *Staphylococcus haemolyticus*. *Journal of Structural Biology, 162*(1), 152–169.

FAO. (2010). *Climate-smart' agriculture: Policies, practices and financing for food security, adaptation and mitigation*. Food and Agriculture Organization.

Fernández-Bodega, Á., et al. (2017). Silencing of a second dimethylallyltryptophan synthase of Penicillium roqueforti reveals a novel clavine alkaloid gene cluster. *Applied Microbiology and Biotechnology, 101*(15), 6111–6121.

Fernández-Fueyo, E., et al. (2016). A secretomic view of woody and nonwoody lignocellulose degradation by *Pleurotus ostreatus*. *Biotechnology for Biofuels, 9*, 49.

Fleißner, A., & Dersch, P. (2010). Expression and export: Recombinant protein production systems for Aspergillus. *Applied Microbiology and Biotechnology, 87*(4), 1255–1270.

Fremlin, L. J., et al. (2009). Cottoquinazoline A and cotteslosins A and B, metabolites from an Australian marine-derived strain of Aspergillus versicolor. *Journal of Natural Products, 72*(4), 666–670.

Frey, C. N. (1930). History and development of the Mordern yeast industry. *Industrial & Engineering Chemistry, 22*(11), 1154–1162.

Frisvad, J. C., et al. (2018). Safety of the fungal workhorses of industrial biotechnology: Update on the mycotoxin and secondary metabolite potential of Aspergillus Niger, *Aspergillus oryzae*, and Trichoderma reesei. *Applied Microbiology and Biotechnology, 102*(22), 9481–9515.

Galagan, J. E., et al. (2005). Sequencing of Aspergillus nidulans and comparative analysis with *A. fumigatus* and *A. oryzae*. *Nature, 438*(7071), 1105–1115.

Galan, B., et al. (1999). News and notes: Production, purification and properties of an endo-1,3-beta glucanase from the basidiomycete *Agaricus bisporus*. *Current Microbiology, 38*, 190–193.

Gao, Y., Li, D., & Liu, Y. (2011). Production of single cell protein from soy molasses using Candida tropicalis. *Annals of Microbiology, 62*(3), 1165–1172.

García Fontán, M. C., et al. (2006). Microbiological and chemical changes during the manufacture of kefir made from cows' milk, using a commercial starter culture. *International Dairy Journal, 16*(7), 762–767.

Ghane, M., Babaeekhou, L., & Ketabi, S. S. (2020). Antibiofilm activity of kefir probiotic lactobacilli against Uropathogenic Escherichia coli (UPEC). *Avicenna Journal Of Medical Biotechnology, 12*, 221–229.

Giraud, F., et al. (2010). Microsatellite loci to recognize species for the cheese starter and contaminating strains associated with cheese manufacturing. *International Journal of Food Microbiology, 137*(2–3), 204–213.

GMI Inc. (2022). Single cell protein market share industry statistics—2030. Available online: https://www.gminsights.com/industry-analysis/single-cell-proteinmarket. Accessed 12 Dec 2022.

Gonçalves, T. A., et al. (2012). Functional characterization and synergic action of fungal xylanase and arabinofuranosidase for production of xylooligosaccharides. *Bioresource Technology, 119*, 293–299.

Goodenough, P. W. (1995). A review of protein engineering for the food industry. *Molecular Biotechnology, 4*(2), 151–166.

Greener, A., Callahan, M., & Jerpseth, B. (1997). An efficient random mutagenesis technique using an E. coli mutator strain. *Molecular Biotechnology, 7*(2), 189–195.

Grenier, J., Potvin, C., & Asselin, A. (2000). Some fungi express β-1,3-glucanases similar to thaumatin-like proteins. *Mycologia, 92*(5), 841–848.

Gryganskyi, A. P., et al. (2018). Phylogenetic and Phylogenomic definition of Rhizopus species. *G3 Genes|Genomes|Genetics, 8*(6), 2007–2018.
Gu, W., et al. (2007). Microsporins A and B: New histone deacetylase inhibitors from the marine-derived fungus Microsporum cf. gypseum and the solid-phase synthesis of microsporin A. *Tetrahedron, 63*(28), 6535–6541.
Guimarães, P. M., et al. (2008). Comparative transcriptome analysis between original and evolved recombinant lactose-consuming *Saccharomyces cerevisiae* strains. *Biotechnology Journal, 3*(12), 1591–1597.
Gulder, T. A. M., et al. (2012). Isolation, structure elucidation and total synthesis of lajollamide A from the marine fungus *Asteromyces cruciatus*. *Marine Drugs, 10*(12), 2912–2935.
Gunter, A. B., et al. (2016). Protein engineering of *Saccharomyces cerevisiae* transporter Pdr5p identifies key residues that impact Fusarium mycotoxin export and resistance to inhibition. *Microbiology, 5*(6), 979–991.
Gürkök, S., et al. (2010). Cloning and heterologous expression of the extracellular alpha-galactosidase from Aspergillus fumigatus in *Aspergillus sojae* under the control of gpdA promoter. *Journal of Molecular Catalysis B: Enzymatic, 64*(3–4), 146–149.
Guslandi, M., et al. (2000). Saccharomyces boulardii in maintenance treatment of Crohn's disease. *Digestive Diseases and Sciences, 45*(7), 1462–1464.
Haefner, S., et al. (2005). Biotechnological production and applications of phytases. *Applied Microbiology and Biotechnology, 68*(5), 588–597.
Hashem, M., et al. (2022). Valorizing food wastes: Assessment of novel yeast strains for enhanced production of single-cell protein from wasted date molasses. *Biomass Conversion and Biorefinery, 12*(10), 4491–4502.
Hassan, R. A., et al. (2020). Influence of oyster mushroom waste on growth performance, immunity and intestinal morphology compared with antibiotics in broiler chickens. *Frontiers in Veterinary Science, 7*, 333–333.
He, F., et al. (2013). Asperterrestide a, a cytotoxic cyclic Tetrapeptide from the marine-derived fungus *Aspergillus terreus* SCSGAF0162. *Journal of Natural Products, 76*(6), 1182–1186.
Hedstrom, L. (2001). Enzyme specificity and selectivity. In *Encyclopedia of life sciences*. Wiley.
Hegedüs, N., et al. (2011). The paf gene product modulates asexual development in *Penicillium chrysogenum*. *Journal of Basic Microbiology, 51*, 253–262.
Hildebrand, G., et al. (2020). Ultrasound-assisted processing of *Chlorella vulgaris* for enhanced protein extraction. *Journal of Applied Phycology, 32*(3), 1709–1718.
Hong, R., & Yang, Y. (2011). *Antitumor metabolites from marine sediment derived Penicillium sp. WF-06* (p. 2). World Notes Antibiots.
Huang, K. C., et al. (1997). Characterization of a new ribotoxin gene (c-sar) from *Aspergillus clavatus*. *Toxicon, 35*(3), 383–392.
Hyde, K. D., et al. (2019). The amazing potential of fungi: 50 ways we can exploit fungi industrially. *Fungal Diversity, 97*(1), 1–136.
Hymery, N., et al. (2014). Filamentous fungi and mycotoxins in cheese: A review. *Comprehensive Reviews in Food Science and Food Safety, 13*(4), 437–456.
Ichikawa, K., et al. (2020). Efficient production of recombinant tannase in *Aspergillus oryzae* using an improved glucoamylase gene promoter. *Journal of Bioscience and Bioengineering, 129*(2), 150–154.
Jaganmohan, P., Purushottam, D., & Prasad, S. V. (2013). Production of single cell protein (SCP) with aspergillus terreus using solid state fermentation. *European Journal of Biological Sciences, 5*, 38–43.
Jarczynska, Z. D., Garcia Vanegas, K., Deichmann, M., Nørskov Jensen, C., Scheeper, M. J., Futyma, M. E., Strucko, T., Jares Contesini, F., Sparholt Jørgensen, T., Blæsbjerg Hoof, J., Hasbro Mortensen, U. (2022). Versatile in Vivo DNA Assembly Toolbox for Fungal Strain Engineering. *ACS Synthetic Biology, 11*(10), 3251–3263. https://doi.org/10.1021/acssynbio.2c00159. Epub 2022 Sep 20. PMID: 36126183; PMCID: PMC9594312.

Jatuwong, K., et al. (2020). Bioprocess for production, characteristics, and biotechnological applications of fungal phytases. *Frontiers in Microbiology, 11*, 188–188.
Jean, C., et al. (2016). Antimicrobial activity of buttermilk and lactoferrin peptide extracts on poultry pathogens. *The Journal of Dairy Research, 83*(4), 497–504.
Jiang, S., et al. (2014). Medicinal properties of Hericium erinaceus and its potential to formulate novel mushroom-based pharmaceuticals. *Applied Microbiology and Biotechnology, 98*(18), 7661–7670.
Jin, Y., et al. (2021). Production of neoagarooligosaccharides by probiotic yeast *Saccharomyces cerevisiae* var. boulardii engineered as a microbial cell factory. *Microbial Cell Factories, 20*(1), 160.
Jo, C., et al. (2023). Unlocking the magic in mycelium: Using synthetic biology to optimize filamentous fungi for biomanufacturing and sustainability. *Materials Today Bio, 19*, 100560–100560.
Johnston, B. C., et al. (2011). Probiotics for the prevention of pediatric antibiotic-associated diarrhea. In *Cochrane database of systematic reviews*. Wiley.
Kan, C.-C. (2002). Impact of recombinant DNA technology and protein engineering on structure-based drug design: Case studies of HIV-1 and HCMV proteases. *Current Topics in Medicinal Chemistry, 2*(3), 247–269.
Katayama, T. and J.-i. Maruyama, CRISPR/Cpf1-mediated mutagenesis and gene deletion in industrial filamentous fungi *Aspergillus oryzae* and *Aspergillus sojae*. Journal of Bioscience and Bioengineering, 2022. 133(4): p. 353–361.
Khakhariya, R., et al. (2023). Production and characterization of ACE inhibitory and anti-diabetic peptides from buffalo and camel milk fermented with Lactobacillus and yeast: A comparative analysis with in vitro, in silico, and molecular interaction study. *Food, 12*(10).
Koivurinta, J., Kurkela, R., & Koivistoinen, P. (1979). Uses of Pekilo, a microfungus biomass from Paecilomyces varioti in sausage and meat balls. *International Journal of Food Science & Technology, 14*(6), 561–570.
Komatsu, K., Shigemori, H., & Kobayashi, J. (2001). Dictyonamides A and B, new peptides from marine-derived fungus. *The Journal of Organic Chemistry, 66*(18), 6189–6192.
Kombrink, A., et al. (2019). Induction of antibacterial proteins and peptides in the coprophilous mushroom Coprinopsis cinerea in response to bacteria. *The ISME Journal, 13*(3), 588–602.
Kopke, K., Hoff, B., & Kück, U. (2010). Application of the *Saccharomyces cerevisiae* FLP/FRT recombination system in filamentous fungi for marker recycling and construction of knockout strains devoid of heterologous genes. *Applied and Environmental Microbiology, 76*(14), 4664–4674.
Kowalchuk, G. A., Jones, S. E., & Blackall, L. L. (2008). Microbes orchestrate life on Earth. *The ISME Journal, 2*(8), 795–796.
Kumar, A. (2020). Aspergillus nidulans: A potential resource of the production of the native and heterologous enzymes for industrial applications. *International Journal of Microbiology, 2020*, 8894215–8894215.
Kuo, T. C., Shaw, J. F., & Lee, G. C. (2015). Conversion of crude Jatropha curcas seed oil into biodiesel using liquid recombinant Candida rugosa lipase isozymes. *Bioresource Technology, 192*, 54–59.
Kuwata, H., et al. (1998). Direct evidence of the generation in human stomach of an antimicrobial peptide domain (lactoferricin) from ingested lactoferrin. *Biochimica et Biophysica Acta, 1429*(1), 129–141.
Lam, S. K., & Ng, T. B. (2001). First simultaneous isolation of a ribosome inactivating protein and an antifungal protein from a mushroom (*Lyophyllum shimeji*) together with evidence for synergism of their antifungal effects. *Archives of Biochemistry and Biophysics, 393*(2), 271–280.
Lau, C. C., Abdullah, N., & Shuib, A. S. (2013). Novel angiotensin I-converting enzyme inhibitory peptides derived from an edible mushroom, *Pleurotus cystidiosus* O.K. Miller identified by LC-MS/MS. *BMC Complementary and Alternative Medicine, 13*, 313.
Lee, T.-K., et al. (2018). The use of fermented buckwheat to produce L-carnitine enriched oyster mushroom. *AMB Express, 8*(1), 138–138.

Li, X., et al. (2005). Expression, purification and characterization of a recombinant beta-glucosidase from Volvariella Volvacea. *Biotechnology Letters, 27*(18), 1369–1373.

Li, T., et al. (2021). Activity and mechanism of action of antifungal peptides from microorganisms: A review. *Molecules (Basel, Switzerland), 26*(11), 3438.

Li, W., et al. (2022a). Structural characterization and angiotensin-converting enzyme (ACE) inhibitory mechanism of Stropharia rugosoannulata mushroom peptides prepared by ultrasound. *Ultrasonics Sonochemistry, 88*, 106074.

Li, Q., et al. (2022b). Recent advances in the development of Aspergillus for protein production. *Bioresource Technology, 348*, 126768.

Li, Q., et al. (2023). CRISPR/Cas9-mediated multiplexed genome editing in *Aspergillus oryzae. Journal of Fungi (Basel, Switzerland), 9*(1), 109.

Li, W., et al. (2024a). Mechanistic insights into antihypertensive activity of mushroom-derived protein-peptides via metabolomic and proteomic approaches. *International Journal of Biological Macromolecules, 283*(Pt 1), 137189.

Li, W., et al. (2024b). Revealing the ACE receptor binding properties and interaction mechanisms of salty oligopeptides from Stropharia rugosoannulata mushroom by molecular simulation and antihypertensive evaluation. *Food & Function, 15*(10), 5527–5538.

Liang, X., et al. (2016). Eight linear peptides from the deep-sea-derived fungus *Simplicillium obclavatum* EIODSF 020. *Tetrahedron, 72*(22), 3092–3097.

Liang, X., et al. (2017). Antifungal and antiviral cyclic peptides from the deep-sea-derived fungus Simplicillium obclavatum EIODSF 020. *Journal of Agricultural and Food Chemistry, 65*(25), 5114–5121.

Lin, A., Chen, C. K., & Chen, Y. J. (1991). Molecular action of tricholin, a ribosome-inactivating protein isolated from Trichoderma viride. *Molecular Microbiology, 5*(12), 3007–3013.

Liu, Z., et al. (2012). Different expression systems for production of recombinant proteins in *Saccharomyces cerevisiae. Biotechnology and Bioengineering, 109*(5), 1259–1268.

Liu, J., et al. (2018). New anti-inflammatory cyclopeptides from a sponge-derived fungus Aspergillus violaceofuscus. *Frontiers in Chemistry, 6*, 226–226.

Lo, H. C., Hsu, T. H., & Lee, C. H. (2020). Extracellular Polysaccharopeptides from fermented Turkey tail medicinal mushroom, Trametes versicolor (Agaricomycetes), mitigate oxidative stress, hyperglycemia, and hyperlipidemia in rats with type 2 diabetes mellitus. *International Journal of Medicinal Mushrooms, 22*(5), 417–429.

López-Expósito, I., et al. (2008). Synergistic effect between different milk-derived peptides and proteins. *Journal of Dairy Science, 91*(6), 2184–2189.

Luo, X.-W., et al. (2019). Peptides and polyketides isolated from the marine sponge-derived fungus Aspergillus terreus SCSIO 41008. *Chinese Journal of Natural Medicines, 17*(2), 149–154.

Lv, X.-C., et al. (2012). Identification and characterization of filamentous fungi isolated from fermentation starters for Hong Qu glutinous rice wine brewing. *The Journal of General and Applied Microbiology, 58*(1), 33–42.

Ly, S., et al. (2018). Impact of microbial composition of Cambodian traditional dried starters (Dombea) on flavor compounds of rice wine: Combining amplicon sequencing with HP-SPME-GCMS. *Frontiers in Microbiology, 9*.

Ma, X., et al., Antiviral peptides from marine gorgonian-derived fungus Aspergillus sp. SCSIO 41501. Tetrahedron Letters, 2017. 58(12): p. 1151–1155.

Madhavan, A., Pandey, A., & Sukumaran, R. K. (2017). RETRACTED: Expression system for heterologous protein expression in the filamentous fungus *Aspergillus unguis. Bioresource Technology, 245*, 1334–1342.

Manning, M. C., Patel, K., & Borchardt, R. T. (1989). Stability of protein pharmaceuticals. *Pharmaceutical Research, 06*(11), 903–918.

Maoz, A., et al. (2002). Sensitive in situ monitoring of a recombinant bioluminescent Yersinia enterocolitica reporter mutant in real time on Camembert cheese. *Applied and Environmental Microbiology, 68*(11), 5737–5740.

Marsh, A. J., et al. (2014). Sequence-based analysis of the bacterial and fungal compositions of multiple kombucha (tea fungus) samples. *Food Microbiology, 38*, 171–178.

Martin, F., et al. (2000). Positive and negative mutant selection in the human histone hairpin-binding protein using the yeast three-hybrid system. *Nucleic Acids Research, 28*(7), 1594–1603.

Martínez-Ruiz, A., et al. (1999). Ribotoxins are a more widespread group of proteins within the filamentous fungi than previously believed. *Toxicon, 37*(11), 1549–1563.

Matsushita-Morita, M., et al. (2010). Characterization of recombinant prolyl aminopeptidase from *Aspergillus oryzae*. *Journal of Applied Microbiology, 109*(1), 156–165.

Mehrab Mohseni, M., et al. (2019). Immunogenicity evaluation of recombinant edible vaccine candidate containing HER2-MUC1 against breast cancer. *Iranian Journal of Allergy, Asthma, and Immunology, 18*(5), 511–522.

Menezes, A. G. T., et al. (2019). Probiotic potential, antioxidant activity, and phytase production of indigenous yeasts isolated from indigenous fermented foods. *Probiotics and Antimicrobial Proteins, 12*(1), 280–288.

Meng, D. M., et al. (2017). Recombinant expression, purification and antimicrobial activity of a novel antimicrobial peptide PaDef in Pichia pastoris. *Protein Expression and Purification, 130*, 90–99.

Meng, D. M., et al. (2019). Expression, purification and characterization of a recombinant antimicrobial peptide Hispidalin in Pichia pastoris. *Protein Expression and Purification, 160*, 19–27.

Meng, X., et al. (2022). Developing fungal heterologous expression platforms to explore and improve the production of natural products from fungal biodiversity. *Biotechnology Advances, 54*, 107866.

Meyer, V., et al. (2008). The polarisome component SpaA localises to hyphal tips of Aspergillus Niger and is important for polar growth. *Fungal Genetics and Biology, 45*(2), 152–164.

Mihai, R. A., et al. (2022). The edible gray oyster fungi *Pleurotus ostreatus* (Jacq. ex Fr.) P. Kumm a potent waste consumer, a biofriendly species with antioxidant activity depending on the growth substrate. *Journal of fungi (Basel, Switzerland), 8*(3), 274.

Miller, S. A., & Dwyer, J. T. (2001). Evaluating the safety and nutritional value of mycoprotein. *Food Technology, 55*, 42–47.

Mogensen, J. M., et al. (2009). Aspergillus acidus from Puerh tea and black tea does not produce ochratoxin A and fumonisin B2. *International Journal of Food Microbiology, 132*(2–3), 141–144.

Molfetta, M., et al. (2022). Protein sources alternative to meat: State of the art and involvement of fermentation. *Food, 11*(14).

Ng, T. B., & Wang, H. X. (2004). Flammin and velin: New ribosome inactivating polypeptides from the mushroom *Flammulina velutipes*. *Peptides, 25*(6), 929–933.

Ng, T. B., et al. (2002). Inhibitory effects of antifungal proteins on human immunodeficiency virus type 1 reverse transcriptase, protease and integrase. *Life Sciences, 70*(8), 927–935.

Ng, T. B., Lam, Y. W., & Wang, H. (2003). Calcaelin, a new protein with translation-inhibiting, Antiproliferative and Antimitogenic activities from the mosaic puffball mushroom *Calvatia caelata*. *Planta Medica, 69*(3), 212–217.

Ngai, P. H., & Ng, T. B. (2004). A ribonuclease with antimicrobial, antimitogenic and antiproliferative activities from the edible mushroom *Pleurotus sajor-caju*. *Peptides, 25*(1), 11–17.

Nielsen, C. A., et al. (2014). The important ergot alkaloid intermediate chanoclavine-I produced in the yeast *Saccharomyces cerevisiae* by the combined action of EasC and EasE from Aspergillus japonicus. *Microbial Cell Factories, 13*, 95.

Nødvig, C. S., et al. (2015). A CRISPR-Cas9 system for genetic engineering of filamentous fungi. *PLoS One, 10*(7), e0133085–e0133085.

Nomura, R., et al. (2021). Administration of *Aspergillus oryzae* suppresses DSS-induced colitis. *Food Chemistry Molecular Sciences, 4*, 100063–100063.

Nout, R. (2000). Useful role of fungi in food processing. In R. A. Samson, E. S. Hoekstra, J. C. Frisvad, & O. Filtenborg (Eds.), *Introduction to food- and airborne fungi* (6th ed.). Centraal Bureau Voor Schimmel Cultures.

Nout, R. (2007). The colonizing fungus as a food provider. In J. Dijksterhuis & R. A. Samson (Eds.), *Food mycology. A multifaceted approach to fungi and food* (Mycology) (pp. 335–352). CRC Press.

Ntana, F., et al. (2020). Aspergillus: A powerful protein production platform. *Catalysts, 10*, 1064.

Nuhn, M. E., et al. (2013). Phylogenetic overview of the Boletineae. *Fungal Biology, 117*(7–8), 479–511.

Oh, D.-C., Jensen, P. R., & Fenical, W. (2006). Zygosporamide, a cytotoxic cyclic depsipeptide from the marine-derived fungus *Zygosporium masonii*. *Tetrahedron Letters, 47*, 8625–8628.

Oh, D.-C., et al. (2007). Induced production of Emericellamides A and B from the marine-derived fungus *Emericella* sp. in competing co-culture. *Journal of Natural Products, 70*(4), 515–520.

Peng, J., et al. (2014). Psychrophilins E–H and versicotide C, cyclic peptides from the marine-derived fungus *Aspergillus versicolor* ZLN-60. *Journal of Natural Products, 77*(10), 2218–2223.

Pereira, P. R., Freitas, C. S., & Paschoalin, V. M. F. (2021). *Saccharomyces cerevisiae* biomass as a source of next-generation food preservatives: Evaluating potential proteins as a source of antimicrobial peptides. *Comprehensive Reviews in Food Science and Food Safety, 20*(5), 4450–4479.

Pitt, J. I., Cruickshank, R. H., & Leistner, L. (1986). Penicillium commune, P. camembertii, the origin of white cheese moulds, and the production of cyclopiazonic acid. *Food Microbiology, 3*(4), 363–371.

Polonelli, L., et al. (1987). Antigenic characterization of Penicillium camemberti and related common cheese contaminants. *Applied and Environmental Microbiology, 53*(4), 872–878.

Polonsky, J., Merrien, M. A., & Scott, P. M. (1977). Roquefortine and isofumigaclavine A, alkaloids from Penicillium roqueforti. *Annales de la Nutrition et de l'Alimentation, 31*(4–6), 963–968.

Porro, D., et al. (2011). Production of recombinant proteins and metabolites in yeasts: When are these systems better than bacterial production systems? *Applied Microbiology and Biotechnology, 89*(4), 939–948.

Prateep, A., et al. (2017). Peptides extracted from edible mushroom: Lentinus squarrosulus induces apoptosis in human lung cancer cells. *Pharmaceutical Biology, 55*(1), 1792–1799.

Prompanya, C., et al. (2015). A new cyclic hexapeptide and a new isocoumarin derivative from the marine sponge-associated fungus Aspergillus similanensis KUFA 0013. *Marine Drugs, 13*(3), 1432–1450.

Qiu, S., et al. (2022). *Aspergillus oryzae* accelerates the conversion of ergosterol to ergosterol peroxide by efficiently utilizing cholesterol. *Frontiers in Genetics, 13*, 984343–984343.

Rai, R. D. (2004). Chapter 21: Production of edible fungi. In D. K. Arora (Ed.), *Fungi biotechnology in agricultural, food, and environmental applications* (pp. 233–246). Marcel Dekker.

Raveschot, C., et al. (2020). Probiotic Lactobacillus strains from Mongolia improve calcium transport and uptake by intestinal cells in vitro. *Food Research International, 133*, 109201.

Rengel Dos Passos, F., et al. (2021). Production of a synbiotic composed of galacto-oligosaccharides and saccharomyces boulardii using enzymatic-fermentative method. *Food Chemistry, 353*, 129486.

Riviere, J. (1977). Microbial proteins. In M. O. Moss & J. E. Smith (Eds.), *Industrial applications of microbiology*. Surrey University Press.

Rizk, Z., et al. (2018). Identification of multiple-derived peptides produced by *Saccharomyces cerevisiae* involved in malolactic fermentation inhibition. *FEMS Yeast Research, 18*(7).

Rodger, G. (2001). Production and properties of mycoprotein as a meat alternative. *Food Technology, 55*, 36–41.

Rojas-Sánchez, U., et al. (2020). Enhancing the yield of human erythropoietin in Aspergillus Niger by introns and CRISPR-Cas9. *Protein Expression and Purification, 168*, 105570.

Romero, S. M., Giudicessi, S. L., & Vitale, R. G. (2021). Is the fungus Aspergillus a threat to cultural heritage? *Journal of Cultural Heritage, 51*, 107–124.

Romero-Luna, H. E., et al. (2020). Probiotic potential of lactobacillus paracasei CT12 isolated from water kefir grains (Tibicos). *Current Microbiology, 77*(10), 2584–2592.

Roongsawang, N., et al. (2016). A novel potential signal peptide sequence and overexpression of ER-resident chaperones enhance heterologous protein secretion in thermotolerant methylotrophic yeast Ogataea thermomethanolica. *Applied Biochemistry and Biotechnology, 178*(4), 710–724.

Ropars, J., López-Villavicencio, M., Snirc, A., Lacoste, S., & Giraud, T. (2017). Blue cheese-making has shaped the population genetic structure of the mould *Penicillium roqueforti. PLoS One, 12*(3), e0171387.

Rosa, G. B., et al. (2020). Investigation of nutritional composition, antioxidant compounds, and antimicrobial activity of wild culinary-medicinal mushrooms Boletus edulis and *Lactarius deliciosus* (Agaricomycetes) from Brazil. *International Journal of Medicinal Mushrooms, 22*(10), 931–942.

Rowley, D. C., et al. (2004a). Synthesis and structure–activity relationships of the halovirs, antiviral natural products from a marine-derived fungus. *Bioorganic & Medicinal Chemistry, 12*(18), 4929–4936.

Rowley, D. C., et al. (2004b). Halovirs A-E, new antiviral agents from a marine-derived fungus of the genus scytalidium. *ChemInform, 35*(6).

Salvarelli, S., Munoz, S., & Conde, F. P. (1994). Purification and characterization of a ribosome inactivating protein from Aspergillus giganteus IF 5818, the gigantin. Immunological and enzymic comparison with α-sarcin. *European Journal of Biochemistry, 225*, 243–251.

Sampaio, P. N., et al. (2008). Production and characterization of recombinant cyprosin B in *Saccharomyces cerevisiae* (W303-1A) strain. *Journal of Bioscience and Bioengineering, 105*(4), 305–312.

Sanger, F., & Thompson, E. O. P. (1953a). The amino-acid sequence in the glycyl chain of insulin. I. The identification of lower peptides from partial hydrolysates. *The Biochemical Journal, 53*(3), 353–366.

Sanger, F., & Thompson, E. O. P. (1953b). The amino-acid sequence in the glycyl chain of insulin. II. The investigation of peptides from enzymic hydrolysates. *The Biochemical Journal, 53*(3), 366–374.

Saqib, A., & Hamid Rashid, M. (2023). Random mutagenesis of koji mold (*Aspergillus oryzae*) to enhance the catalytic efficiency and thermostability of glucoamylase. *Pakistan Journal of Botany, 55*(5).

Schuster, M., & Kahmann, R. (2019). CRISPR-Cas9 genome editing approaches in filamentous fungi and oomycetes. *Fungal Genetics and Biology, 130*, 43–53.

Scopel, M., et al. (2013). Dipeptide cis-cyclo(Leucyl-Tyrosyl) produced by sponge associated Penicillium sp. F37 inhibits biofilm formation of the pathogenic *Staphylococcus epidermidis. Bioorganic & Medicinal Chemistry Letters, 23*(3), 624–626.

Sen, D., et al. (2022). Identification of potential edible mushroom as SARS-CoV-2 main protease inhibitor using rational drug designing approach. *Scientific Reports, 12*(1), 1503.

Seyfi, R., et al. (2019). Antimicrobial peptides (AMPs): Roles, functions and mechanism of action. *International Journal of Peptide Research and Therapeutics, 26*(3), 1451–1463.

Shan, L. S., et al. (2013). Prevention and treatment of diarrhoea with Saccharomyces boulardii in children with acute lower respiratory tract infections. *Beneficial Microbes, 4*(4), 329–334.

Siemens, K., & Zawistowski, J. (1993). Occurrence of PR imine, a metabolite of penicillium roqueforti, in blue cheese. *Journal of Food Protection, 56*(4), 317–319.

Sitati, C. N. W., et al. (2021). Phytochemical profile and antimicrobial activities of edible mushroom *Termitomyces striatus. Evidence-based Complementary and Alternative Medicine, 2021*, 3025848.

Song, H.-T., et al. (2017). Construction of recombinant Yarrowia lipolytica and its application in bio-transformation of lignocellulose. *Bioengineered, 8*(5), 624–629.

Song, Y., Zhou, J., & Ni, Y. (2024). Construction of a Kluyveromyces lactis strain with multi-copy integration for enhanced bovine chymosin production by CRISPR/Cas9 and UV mutagenesis. *Sheng Wu Gong Cheng Xue Bao, 40*(9), 2983–2997.

Sørenson, H. P. (2010). Towards universal system for recombinant gene expression. *Microbial Cell Factories, 9*, 27.

Spencer, D. M. (1985). The mushroom–its history and importance. In P. B. Flegg, D. M. Spencer, & D. A. Wood (Eds.), *The biology and technology of the cultivated mushroom*. Wiley.

Spinnler, H. E., & Gripon, J. C. (2004). Surface mould-ripened cheeses. In *Major cheese groups* (pp. 157–174). Elsevier.

Sun, X., et al. (2014). Extracellular expression and efficient purification of a functional recombinant Volvariella volvacea immunomodulatory protein (FIP-vvo) using Pichia pastoris system. *Protein Expression and Purification, 94*, 95–100.

Suryadi, H., Irianti, M. I., & Septiarini, T. H. (2022). Methods of random mutagenesis of Aspergillus strain for increasing kojic acid production. *Current Pharmaceutical Biotechnology, 23*(4), 486–494.

Suwannarach, N., et al. (2022). Impact of cultivation substrate and microbial community on improving mushroom productivity: A review. *Biology, 11*(4), 569.

Szajewska, H., & Mrukowicz, J. (2005). Meta-analysis: Non-pathogenic yeast Saccharomyces boulardii in the prevention of antibiotic-associated diarrhoea. *Alimentary Pharmacology and Therapeutics, 22*(5), 365–372.

Tahoun, M. K., el-Nemr, T. M., & Shata, O. H. (2002). A recombinant *Saccharomyces cerevisiae* strain for efficient conversion of lactose in salted and unsalted cheese whey into ethanol. *Nahrung, 46*(5), 321–326.

Tamang, J., & Samuel, D. (2010). Dietary cultures and antiquity of fermented foods and beverages. In *Fermented foods and beverages of the world* (pp. 1–40). CRC Press.

Tamang, J., et al. (2015). Microorganisms in fermented foods and beverages. In *Health benefits of fermented foods and beverages* (pp. 1–110). CRC Press.

Tan, R. X., et al. (2004). Isolation and structure assignments of Rostratins A–D, cytotoxic disulfides produced by the marine-derived fungus *Exserohilum ostratum. Journal of Natural Products, 67*(8), 1374–1382.

Tayyrov, A., et al. (2018). Toxicity of potential fungal defense proteins towards the fungivorous nematodes Aphelenchus avenae and Bursaphelenchus okinawaensis. *Applied and Environmental Microbiology, 84*(23).

Thak, E. J., et al. (2020). Yeast synthetic biology for designed cell factories producing secretory recombinant proteins. *FEMS Yeast Research, 20*(2).

Thanh, V. N., Mai, L. T., & Tuan, D. A. (2008). Microbial diversity of traditional Vietnamese alcohol fermentation starters (banh men) as determined by PCR-mediated DGGE. *International Journal of Food Microbiology, 128*(2), 268–273.

Thim, L., et al. (1986). Secretion and processing of insulin precursors in yeast. *Proceedings. National Academy of Sciences. United States of America, 83*, 6766–6770.

Thrane, U. (2007). Fungal protein for food. In J. Dijksterhuis & R. A. Samson (Eds.), *Food mycology. A multifaceted approach to fungi and food* (pp. 353–360). CRC Press.

Tian, E., et al. (2024). An engineered Pichia pastoris platform for the biosynthesis of silk-based nanomaterials with therapeutic potential. *International Journal of Biological Macromolecules, 269*(Pt 2), 131954.

Tonini, S., et al. (2024). Lentils protein isolate as a fermenting substrate for the production of bioactive peptides by lactic acid bacteria and neglected yeast species. *Microbial Biotechnology, 17*(1), e14387.

Torres-García, C., et al. (2014). Triazene as a powerful tool for solid-phase derivatization of phenylalanine containing peptides: *Zygosporamide analogues* as a proof of concept. *The Journal of Organic Chemistry, 79*, 11409–11415.

Tsang, C.-C., et al. (2018). Taxonomy and evolution of Aspergillus, Penicillium and Talaromyces in the omics era - Past, present and future. *Computational and Structural Biotechnology Journal, 16*, 197–210.

Tuncer Degim, I., & Celebi, N. (2007). Controlled delivery of peptides and proteins. *Current Pharmaceutical Design, 13*(1), 99–117.

Turnbull, W. H., Leeds, A. R., & Edwards, D. G. (1992). Mycoprotein reduces blood lipids in free-living subjects. *The American Journal of Clinical Nutrition, 55*(2), 415–419.

Ukaegbu-Obi, K. M. (2016). Single cell protein: A resort to global protein challenge and waste management. *Journal of Microbiology & Microbial Technology, 1*(1).

Umemura, M. (2020). Peptides derived from Kex2-processed repeat proteins are widely distributed and highly diverse in the Fungi kingdom. *Fungal Biology and Biotechnology, 7*, 11–11.

Underkofler, L. A., Barton, R. R., & Rennert, S. S. (1958). Production of microbial enzymes and their applications. *Applied Microbiology, 6*(3), 212–221.

Vallone, L., Giardini, A., & Soncini, G. (2018). Secondary metabolites from *Penicillium roqueforti*, a starter for the production of Gorgonzola cheese. *Italian Journal of Food Safety, 3*(3).

van der Aa Kühle, A., Skovgaard, K., & Jespersen, L. (2005). In vitro screening of probiotic properties of *Saccharomyces cerevisiae* var. boulardii and food-borne *Saccharomyces cerevisiae* strains. *International Journal of Food Microbiology, 101*(1), 29–39.

Vasquez, E. C., et al. (2020). New insights on the beneficial effects of the probiotic kefir on vascular dysfunction in cardiovascular and neurodegenerative diseases. *Current Pharmaceutical Design, 26*(30), 3700–3710.

Vieira, A.l. T., Teixeira, M. M., & Martins, F. S. (2012). The role of probiotics and prebiotics in inducing gut immunity. *Frontiers in Immunology*, 4.

Wang, H. X., & Ng, T. B. (2004). Eryngin, a novel antifungal peptide from fruiting bodies of the mushroom *Pleurotus eryngii*. *Peptides, 25*, 1–5.

Wang, S.-Y., et al. (2012). Investigation of microorganisms involved in biosynthesis of the kefir grain. *Food Microbiology, 32*(2), 274–285.

Wang, N., Cui, C.-B., & Li, C.-W. (2016). A new cyclic dipeptide penicimutide: The activated production of cyclic dipeptides by introduction of neomycin-resistance in the marine-derived fungus Penicillium purpurogenum G59. *Archives of Pharmacal Research, 39*(6), 762–770.

Wang, M.-C., et al. (2019). Putative probiotic strains isolated from kefir improve gastrointestinal health parameters in adults: A randomized, single-blind, placebo-controlled study. *Probiotics and Antimicrobial Proteins, 12*(3), 840–850.

Watanabe, H., et al. (1995). Primary structure of RNase from Irpex lacteus. *Bioscience, Biotechnology, and Biochemistry, 59*, 2092–2103.

Wiebe, M. G. (2002). Siderophore production by *Fusarium venenatum* A3/5. *Biochemical Society Transactions, 30*(4), 696–698.

Williamson, D. A., et al. (2006). Effects of consuming mycoprotein, tofu or chicken upon subsequent eating behaviour, hunger and safety. *Appetite, 46*(1), 41–48.

Wirth, J., et al. (1997). Sequence determination and molecular characterization of gigantin, a cytotoxic protein produced by the mould Aspergillus giganteus IFO5818. *Archives of Biochemistry and Biophysics, 343*, 188–193.

Witthuhn, R. C., Schoeman, T., & Britz, T. J. (2005). Characterisation of the microbial population at different stages of kefir production and kefir grain mass cultivation. *International Dairy Journal, 15*(4), 383–389.

Wunsch, N.-G. (2022). *Market share of meat substitute brands in Western Europe in 2020. 2021.* Available online: https://www.statista.com/statistics/1278538/market-share-of-the-leading-meat-substitute-brands-inwestern-europe/. Accessed 16 Dec 2022

Wyder, M.-T., Meile, L., & Teuber, M. (1999). Description of Saccharomyces turicensis sp. nov., a new species from kefyr. *Systematic and Applied Microbiology, 22*(3), 420–425.

Xu, N., et al. (2020). Combining protein and metabolic engineering to construct efficient microbial cell factories. *Current Opinion in Biotechnology, 66*, 27–35.

Xu, L., et al. (2023). Complete genome sequences and comparative secretomic analysis for the industrially cultivated edible mushroom Lyophyllum decastes reveals insights on evolution and lignocellulose degradation potential. *Frontiers in Microbiology, 14*, 1137162.

Yamashita, T., et al. (1987). Secretion of Mucor rennin, a fungal aspartic protease of Mucor pusillus, by recombinant yeast cells. *Molecular & General Genetics, 210*(3), 462–467.

Yang, N., et al. (2005). Molecular character of the recombinant antitumor lectin from the edible mushroom *Agrocybe aegerita*. *Journal of Biochemistry, 138*(2), 145–150.

Yang, L., et al. (2020). Metabolic engineering of Aspergillus Niger via ribonucleoprotein-based CRISPR-Cas9 system for succinic acid production from renewable biomass. *Biotechnology for Biofuels, 13*(1), 206–206.

Yimam, M. A., et al. (2024). The role of algae, fungi, and insect-derived proteins and bioactive peptides in preventive and clinical nutrition. *Frontiers in Nutrition, 11*, 1461621.

Yin, C., et al. (2020). Identification and characterization of a novel light-induced promoter for recombinant protein production in *Pleurotus ostreatus*. *Journal of Microbiology, 58*(1), 39–45.

You, M., et al. (2015). Lumazine peptides from the marine-derived fungus Aspergillus terreus. *Marine Drugs, 13*(3), 1290–1303.

Youssef, F. S., et al. (2019). A comprehensive review of bioactive peptides from marine fungi and their biological significance. *Marine Drugs, 17*(10), 559.

Zahrl, R. J., Mattanovich, D., & Gasser, B. (2018). The impact of ERAD on recombinant protein secretion in Pichia pastoris (syn Komagataella spp.). *Microbiology (Reading), 164*(4), 453–463.

Zarzosa-Moreno, D., et al. (2020). Lactoferrin and its derived peptides: An alternative for combating virulence mechanisms developed by pathogens. *Molecules, 25*(24).

Zhang, L., et al. (2009a). Recombinant expression of bovine chymosin in Pichia pastoris. *Sheng Wu Gong Cheng Xue Bao, 25*(8), 1160–1165.

Zhang, Y., et al. (2009b). Phenethyl- α -pyrone derivatives and cyclodipeptides from a marine algous endophytic fungus Aspergillus Niger EN–13. *Natural Product Research, 24*(11), 1036–1043.

Zheng, J., et al. (2009). Novel cyclic hexapeptides from marine-derived fungus, *Aspergillus sclerotiorum* PT06-1. *Organic Letters, 11*(22), 5262–5265.

Zhong, Y., et al. (2014). High-temperature cultivation of recombinant Pichia pastoris increases endoplasmic reticulum stress and decreases production of human interleukin-10. *Microbial Cell Factories, 13*, 163.

Zhou, X., et al. (2015). A novel cyclic dipeptide from deep marine-derived fungus *Aspergillus* sp. SCSIOW2. *Natural Product Research, 30*(1), 52–57.

Zhou, S., et al. (2018). Investigation of lignocellulolytic enzymes during different growth phases of Ganoderma lucidum strain G0119 using genomic, transcriptomic and secretomic analyses. *PLoS One, 13*(5), e0198404.

Zhou, R., et al. (2019). Research progress of bioactive proteins from the edible and medicinal mushrooms. *Current Protein & Peptide Science, 20*(3), 196–219.

Zhou, T., et al. (2023). A review on edible fungi-derived bioactive peptides: Preparation, purification and bioactivities. *International Journal of Medicinal Mushrooms, 25*(7), 1–11.

Zhu, L., et al. (2022). Process optimization, amino acid composition, and antioxidant activities of protein and polypeptide extracted from waste beer yeast. *Molecules, 27*(20), 6825.

Zubieta, M. P., et al. (2018). Protein profile in *Aspergillus nidulans* recombinant strains overproducing heterologous enzymes. *Microbial Biotechnology, 11*(2), 346–358.

Lallawmsanga, et al. (2016). Antimicrobial potential, identification and phylogenetic affiliation of wild mushrooms from two sub-tropical semi-evergreen Indian forest ecosystems. *PLoS One, 11*(11), e0166368.

Glossary

Adaptive Laboratory Evolution (ALE) A strategy in which fungal strains are cultured under defined selective pressures over multiple generations, allowing natural selection to yield strains with improved robustness and productivity.

Agaricus bisporus A common white-rot fungus used in bioremediation and biosorption of heavy metals.

Agricultural Waste By-products from agricultural processes such as stalks, straw, bark, sawdust, bran, molasses, often lignocellulosic and used in mushroom cultivation.

AI-Driven Bioprocess Modeling The application of artificial intelligence and machine learning to predict optimal fermentation conditions and identify metabolic bottlenecks, thereby enabling real-time process optimization.

Airlift Bioreactor A bioreactor type using gas sparging for mixing, suitable for shear-sensitive cultures.

Basidiomycete Fungi A group of fungi including mushrooms, known for enzymes and metabolites.

Bioactive Peptides Short chains of amino acids that exhibit specific biological activities (e.g., antioxidant, anti-inflammatory, or antihypertensive effects). The chapter discusses how engineered fungi can produce these health-promoting peptides.

Bioadsorption The process of toxic substances adhering to biomass surfaces.

Biocatalyst An enzyme or microorganism that speeds up a chemical reaction.

Bio-composites Sustainable materials made by combining natural fibers (e.g., plant or fungal-based) with polymer matrices, used as eco-friendly alternatives to synthetic composites.

Bioconversion Using biological agents to convert organic matter into other forms.

Bioconverters Organisms that transform organic waste into useful products such as biofuels, compost, enzymes, etc.

Biodegradable Packaging Packaging materials made from fungal mycelium that break down naturally, reducing environmental impact.

A. Gupta (ed.), *Prospects of Fungal Biotechnologies for Livestock Volume 2*, Fungal Biology, https://doi.org/10.1007/978-3-032-06478-3

Biodegradation The breakdown of organic substances by biological means, particularly by fungal enzymes in bioremediation.

Biodiesel A renewable fuel made from biological materials, such as lipid-containing agricultural wastes or fungal substrates.

Bioeconomy An economic framework that promotes the sustainable use of biomass, including agricultural residues and organic waste, to produce value-added products like food, fuels, and materials

Bioethanol A type of biofuel derived from fermenting sugars obtained from biomass like spent mushroom substrates.

Biofilm Reactor A bioreactor where microbes grow on a solid support.

Biogas A renewable energy source produced from the anaerobic digestion of organic matter, including waste mushroom compost.

Bioprocess Engineering Applying engineering principles to optimize biological production processes.

Bioreactor A vessel or system used to cultivate microorganisms under controlled conditions. Key design considerations include mixing efficiency, oxygen transfer, and temperature control, all of which are critical during scale-up.

Bioremediation The use of living organisms, especially fungi and bacteria, to degrade or neutralize, to detoxify and restore polluted environments such as soil and water.

Biosorption A process where fungi absorb and concentrate pollutants, such as heavy metals, from their environment, aiding in detoxification.

Biosynthetic Gene Cluster (BGC) A cluster of genes encoding enzymes for a specific metabolite pathway.

Biotransformation Fungi chemically modify pollutants into less harmful substances during bioremediation.

Brown-Rot Fungi Fungi that primarily degrade cellulose and hemicellulose in wood, using mechanisms like the Fenton reaction.

Carbon Catabolite Repression (CCR) Repression of gene expression by preferred carbon sources.

Cell Factory A cell engineered for producing specific compounds.

Chemical Epigenetics Using chemicals to induce expression of silent gene clusters.

Chitin A polysaccharide found in fungal cell walls, capable of binding metal ions.

Chromatin Remodeling Dynamic changes in chromatin structure affecting gene expression.

Circular Bio-economy An economic model using renewable biological resources and minimizing waste.

Circular Economy An economic model focused on minimizing waste and maximizing resource efficiency through recycling, reuse, and sustainable design.

Co-expression Strategies Techniques that involve the simultaneous expression of multiple genes—such as chaperones or foldases—to enhance the correct folding, processing, and secretion of the target protein.

Composting A biological process that decomposes organic waste into nutrient-rich soil amendments.

CRISPRa/CRISRi CRISPR-based systems for activating (a) or repressing (i) gene expression.

CRISPR-Cas9 A precise genome-editing tool that allows targeted modifications in fungal genomes. It is used to knock out or enhance specific genes in metabolic pathways, thereby increasing the yield of desired proteins and enzymes.

Cryptic Gene Cluster A gene cluster not expressed under standard conditions.

DBTL Cycle Design-Build-Test-Learn cycle for iterative strain development.

dCas9 Deactivated Cas9, used for gene regulation without cutting DNA.

Ecovative Design A company and technology platform that uses fungal mycelium to create sustainable materials.

Endophytic Fungi Fungi living within plant tissues.

Engineered Fungal Cell Factories Genetically modified fungal systems designed to produce proteins, peptides, enzymes, and other high-value bioproducts efficiently. These platforms leverage fungi's natural capabilities—such as rapid growth and robust secretion pathways—to enable sustainable production.

Enzyme Secretion Process of releasing enzymes outside the cell.

Epigenetic Modifications Heritable gene expression changes not involving DNA sequence alteration.

Fenton Reaction A chemical process involving iron ions and hydrogen peroxide producing reactive radicals that degrade pollutants, utilized by some fungi.

Ferroptosis Form of regulated cell death characterized by iron-dependent lipid peroxidation leading to lethal accumulation of reactive oxygen species (ROS) on cellular membranes.

Filamentous Fungi Filamentous fungi are generally saprophytic microorganisms which has the capability to synthesize versatile biological compounds of importance and growing as thread-like hyphae.

Flux Analysis Study of metabolic reaction rates.

Fouling Accumulation of material on bioreactor surfaces.

Fungal Bioremediation The use of fungi to degrade or remove environmental pollutants, such as heavy metals, hydrocarbons, or pesticides, from soil or water.

Fungal Biotechnology The application of fungi and their biological processes for industrial, environmental, and medical purposes, including waste recycling and production of bio-based materials.

Fungal Mycelium The vegetative network of fungal filaments used in biotextiles, packaging, construction materials, and sustainable products.

Genome Reduction Non-essential genes are frequently lost during evolution in obligate parasites such as microsporidia.

Genome-Scale Metabolic Model (GEM) A computational model of an organism's metabolism.

Genomics The study of the complete set of DNA (genome) in an organism, including gene structure, function, mapping, and evolution.

Golden Gate Cloning A modular DNA assembly method.

Heterologous Expression Expressing a gene from one organism in another.

Heterologous Protein Production The production of proteins in a host organism that does not naturally produce them. Fungal systems like *Pichia pasto-*

ris are engineered to produce complex proteins with specific post-translational modifications.

High-Throughput Screening Rapid testing of many samples.

Histone Acetylation Epigenetic modification associated with gene activation.

Histone Methylation Epigenetic modification affecting gene expression.

Holomycota A fungal lineage that includes fungi, microsporidia, and related taxa

Homologous Recombination (HR) A DNA repair mechanism used for targeted gene integration.

Hyphae Thread-like structures forming fungal mycelium.

Immunosuppressant A substance that reduces or inhibits the activity of the immune system, commonly used in organ transplantation and autoimmune disorders.

Kinetoplastida A group of protists with flagella that includes parasitic organisms like Leishmania and Trypanosoma

Laccases Enzymes produced by white-rot fungi that break down phenolic pollutants and complex aromatic compounds.

Lignocellulosic Biomass Plant material as a potential feedstock.

Lignocellulosic Waste Plant biomass composed of cellulose, hemicellulose, and lignin, used as a substrate in mushroom cultivation and biofuel production.

Mass Transfer Limitations Restrictions in movement of substances within a culture.

Metabolic Burden Negative impact of engineered processes on cell growth.

Metabolic Engineering The strategic modification of cellular metabolic pathways to optimize the allocation of resources toward the production of desired compounds while minimizing byproducts.

Metabolic Flux Rate of flow through a metabolic pathway.

Metabolic Modeling Using computational models to simulate metabolism.

Metabolomics The comprehensive analysis of small-molecule metabolites within cells, tissues, or organisms to understand metabolic pathways and responses.

Microsporidia Spore-forming, obligatory intracellular eukaryotic parasites that are related to fungi.

Mitosome Reduced mitochondrial organelle found in microsporidia, lacking its own genome

Morphology Physical form of an organism (e.g., dispersed hyphae vs. pellets).

Multiplexing Simultaneously targeting multiple genes.

Mycelium The network of fungal filaments that can degrade pollutants, aid in soil detoxification, and form the basis for sustainable materials.

Mycelium-Based Composites Materials grown from fungal mycelium and agricultural waste and then used as sustainable alternatives like insulation, packaging, and construction..

Myco-Biotechnology The application of fungi in industrial, agricultural, or medical biotechnology, including enzyme production, drug development, and biocontrol.

Mycoprotein Protein-rich fungal biomass used as a sustainable alternative to animal-derived proteins. Its production typically requires less land and water and results in a lower carbon footprint compared to traditional protein sources.

Mycoremediation A subset of bioremediation that specifically uses fungi and their enzymes to decompose and detoxify pollutants.

MycoTEX A biodegradable textile material produced from fungal mycelium developed by the Dutch company NEFFA.

MycoWorks A biotechnology company producing sustainable leather-like materials from fungal mycelium and agricultural by-products.

NHEJ (Non-Homologous End Joining) A DNA repair pathway.

Omics Technologies An integrated suite of approaches—including genomics, transcriptomics, proteomics, and metabolomics—that provides comprehensive insights into the biological systems of fungal cell factories, guiding pathway optimization and strain improvement.

Omics Large-scale biological data sets (genomics, transcriptomics, etc.).

Orthogonal CRISPR Activation Specific CRISPR-based gene activation.

p53 a tumor suppressor protein often called the "guardian of the genome." It plays a crucial role in regulating the cell cycle and preventing cancer formation.

Pathway Redesign Modifying or creating metabolic pathways.

Pellets Spherical aggregates of fungal hyphae in liquid culture.

Peroxidases Enzymes that degrade lignin and other complex pollutants, including lignin peroxidase (LiP), manganese peroxidase (MnP), and versatile peroxidase (VP).

Phylogenomic Analysis Study of evolutionary history using genomic data

Phytopathogenicity The ability of a microorganism (especially fungi or bacteria) to cause disease in plants, often studied in plant–pathogen interactions.

Plastic Biodegradation The breakdown of synthetic polymers by fungi through enzymatic processes.

Polar tube A specific infection tool that microsporidia utilize to inject host cells with their sporoplasm.

Post-translational Modifications (PTMs) Chemical modifications made to proteins after synthesis (including phosphorylation, glycosylation, ubiquitination, methylation, and acetylation) that influence their activity, stability, and cellular localization.

Post-translational Modifications The process causing changes in the amino acid sequence after the synthesis of protein.

Precision Fermentation Using microbes to produce specific molecules with high purity.

Proteomics The large-scale study of proteins, including their structures, functions, and interactions, typically using mass spectrometry-based techniques.

Rate-Limiting Enzyme Enzyme catalyzing the slowest step in a pathway.

Recombinant Proteins Proteins produced through recombinant DNA technology, where the gene encoding the protein is inserted into a host (such as yeast or filamentous fungi) for large-scale production.

Regenerative resource utilization Approach of restoring or renewing resources through sustainable practices using biological systems.

Regulatory Circuit Network controlling gene expression and cell processes.

Resource Optimization The efficient and effective use of available resources to achieve the desired outcome with minimal waste.

Rheology Study of flow and deformation of matter.

RNA Interference (RNAi) A gene-silencing mechanism employing small RNA molecules (siRNAs or shRNAs) to downregulate the expression of targeted genes. This technique helps redirect cellular resources toward the production of the desired bioproducts.

Secondary Metabolite Compound not essential for primary growth but with other roles.

Shear Stress Force exerted on a fluid or surface due to flow.

Signal Intestinal Gluconeogenesis A metabolic process where glucose is synthesized in the intestine, which can send signals affecting systemic energy balance and insulin sensitivity.

Smart Fermentation Technologies Advanced fermentation systems that incorporate real-time sensors, automation, and IoT connectivity to continuously monitor and adjust cultivation parameters, ensuring optimal production conditions in liquid medium.

Solid-State Fermentation (SSF) A method where fungi are grown on solid substrates (often agro-industrial residues), offering advantages like reduced water usage and lower contamination risk while sometimes boosting the production of secondary metabolites or fermentation on a moist solid substrate.

Spent Mushroom Substrate (SMS) The substrate left after mushroom cultivation, which can be reused as fertilizer, biofuel feedstock, etc.

Spliceosome A ribonucleoprotein complex, primarily found in the nucleus of eukaryotic cells. It's responsible for removing introns, non-coding sequences, from pre-messenger RNA (pre-mRNA). This process is essential for producing functional messenger RNA (mRNA).

Sporoplasm The infectious material that enters the host cell from a microsporidian spore

Sporulation Production of fungal spores.

Strain Engineering Modifying a microbial strain's genetics.

Submerged Fermentation (SmF) A fermentation strategy in which fungi are cultivated in liquid media, allowing precise control of parameters (such as pH, temperature, and oxygen levels) to maximize enzyme and protein yields.

Substrate Range Types of nutrients an organism can use.

Sustainability in Textiles The use of fungal materials and biotextiles as eco-friendly alternatives to traditional textiles to reduce environmental impact.

Sustainability Meeting present needs without compromising the ability of future generations to meet theirs, often by reducing waste and environmental impact.

Sustainable Development Goals (SDGs) A set of global goals set by the United Nations to achieve a better and more sustainable future

Sustainable Protein Production The generation of proteins using methods that minimize resource use and environmental impact. Fungal systems, for example, require far less land and water than conventional animal farming while yielding high-quality proteins.

Synthetic Biology in Fungi A field combining biology and engineering to redesign fungal metabolic pathways for producing bio-based materials and chemicals.

Synthetic Biology The design and construction of new biological components or systems, and the re-engineering of existing ones. In fungal biotechnology, synthetic biology enables the creation of modular gene circuits and the tailoring of strains for enhanced production.

Synthetic Gene Cluster Artificially constructed gene cluster.

Systems Biology Study of biological systems as integrated networks.

Targeted Gene Amplification Increasing the copy number of specific genes.

Transcription Factor Protein controlling gene transcription.

Transcriptomics Study of all RNA molecules.

Transesterification Chemical reaction used in biodiesel production.

Velvet Complex Regulatory complex in filamentous fungi.

Wall Growth Biomass accumulation on bioreactor surfaces.

Waste Mushroom Compost Residual compost after mushroom harvesting, usable as biogas feedstock or biofuel substrate.

Waste Upcycling Reuse of wastes to form new products with higher environmental or economic value than the original one.

Waste Valorization Process of converting waste materials into more valuable products, through sustainable technologies.

White-Rot Fungi Fungi that degrade lignin and a variety of pollutants using phenoloxidase enzymes.

Yeast Unicellular fungi.

Index

A. Gupta (ed.), *Prospects of Fungal Biotechnologies for Livestock Volume 2*, Fungal Biology, https://doi.org/10.1007/978-3-032-06478-3

MIX
Papier aus verantwortungsvollen Quellen
Paper from responsible sources
FSC® C105338

If you have any concerns about our products,
you can contact us on
ProductSafety@springernature.com

In case Publisher is established outside the EU,
the EU authorized representative is:
Springer Nature Customer Service Center GmbH
Europaplatz 3, 69115 Heidelberg, Germany

Printed by Libri Plureos GmbH
in Hamburg, Germany